LIST OF SAMPLE DOCUMENTS AND FORMS

TECHNICAL 6th *CANADIAN EDITION*

JOHN M. LANNON
UNIVERSITY OF MASSACHUSETTS, DARTMOUTH

DON KLEPP
MYKON COMMUNICATIONS

LAURA J. GURAK
UNIVERSITY OF MINNESOTA

COMMUNICATION

PEARSON

Toronto

Vice-President, Editorial Director: Gary Bennett
Editor-in-Chief: Michelle Sartor
Acquisitions Editor: David Le Gallais
Marketing Manager: Loula March
Program Manager: John Polanszky
Developmental Editor: Patti Altridge
Media Editor: Simon Bailey
Media Producer: Simon Bailey
Project Manager: Kimberley Blakey
Full-Service Project Management: Electronic
 Publishing Services Inc., NYC

Copy Editor: Rodney Rawlings
Proofreader: Rachel Redford
Compositor: Aptara®, Inc.
Permissions Project Manager: Daniela Glass
Photo Researcher: PreMediaGlobal
Permissions Researcher: Electronic Publishing
 Services Inc., NYC
Art Director: Zena Denchik
Cover Designer: Anthony Leung
Interior Designer: Anthony Leung
Cover Image: Bocos Benedict/Getty

10 9 8 7 6 5 4 3 2 1 CKV

Library and Archives Canada Cataloguing in Publication

Lannon, John M., author
 Technical communication / John M. Lannon, University of Massachusetts, Dartmouth, Don Klepp, Mykon Communications. — 6th Canadian edition.

Includes bibliographical references and index.
ISBN 978-0-205-92584-1 (pbk.)

 1. Technical writing—Textbooks. 2. Communication of technical information—Textbooks. I. Klepp, Don, 1944-, author II. Title.

T11.L36 2014   808.06'66   C2013-907541-0

ISBN 978-0-205-92584-1

BRIEF CONTENTS

CONTENTS

PREFACE

Thank you for using this text. Pearson Canada and your authors have collaborated on a text that you can use as a workplace tool for many years in your career. Its advice on communications analysis, writing processes, document structures, and effective language will remain relevant, because that advice is based on practical experience and on solid rhetorical theory.

Communications theory is rarely discussed directly in this text because we are primarily interested in practical applications, not in theory for its own sake. So, for example, *Technical Communication* does not discuss genre analysis; instead, Chapters 21 and 22 show how the *action structure* can be profitably adapted to a wide variety of business and technical documents.

WHAT'S NEW?

One of our main challenges in creating a new edition of this text is to keep abreast of rapid changes in communications technology. Those changes have altered the way that technical communicators complete their tasks. In response, this sixth edition contains new material or more emphasis on the following topics:

- Chapter 1 now discusses the expanded role of the technical communicator, which is necessitated by the trend toward Big Data analysis, by complex, expanded requirements for delivering technical information, and by the use of structured documents that use hidden commands to translate content to multiple delivery media.
- Chapter 1 also introduces material regarding the communicator's contributions to product and service development. That discussion inevitably leads to increased coverage of project management.
- Growing out of Chapter 1's discussion of virtual teams, a new Chapter 4 provides guidelines for managing academic and workplace collaborative projects. In addition to material that had been placed in the opening chapter of previous editions, this new chapter advises how to conduct meetings and collaborate in global teams. Chapter 4 complements Chapter 1's introduction to virtual team meetings.
- Chapter 9 now features sample sources cited or references sections of four documentation formats: MLA, APA, CSE, and IEEE.
- Chapters 12, 13, and 15 feature new examples of graphics, webpage design, and marketing literature.
- A new Chapter 25 discusses design features and phrasing requirements for electronic media. The chapter covers webpages, blogs, podcasts, wikis, and social media.
- In order to compensate for the extra material introduced in Chapters 1, 4, and 25, the chapter on editing for readable technical style has been removed and

placed in the companion workbook, *Grammar and Style at Work*. Full-colour, high-quality textbooks such as *Technical Communication* are expensive to produce. So, to help control the textbook's cost, we have moved this lengthy chapter to the workbook where it fits in nicely with that resource book's emphasis on writing readable, error-free prose.

To complement new material on communications technology and new processes for sharing information, this edition of *Technical Communication* includes 19 new quotes from technical communication practitioners in the On the Job boxed feature of this text.

SUPPLEMENTS
For Students

Grammar and Style at Work. This supplement will help you communicate clearly and effectively in business and technical settings, whether you're working part-time, volunteering, or applying for jobs. It keeps its focus on the essential English skills that you need to succeed in today's workplace. See the inside back cover of this text for details.

CourseSmart. CourseSmart goes beyond traditional expectations—providing instant, online access to the textbooks and course materials you need at an average savings of 60 percent. With instant access from any computer and the ability to search your text, you'll find the content you need quickly, no matter where you are. And with online tools like highlighting and note-taking, you can save time and study efficiently. See all the benefits at **www.coursesmart.com/students**.

For Instructors
MyWritingLab

Instructors who package **MyWritingLab: Technical Communication** provide their students with a comprehensive resource that offers the very best multimedia support for technical writing in one integrated, easy-to-use site. Features include interactive model documents, case studies, multimedia resources, and more. For more information, please visit **www.mywritinglab.com**.

MyTest. MyTest from Pearson Education Canada is a powerful assessment generation program that helps instructors easily create and print quizzes, tests, and exams, as well as homework or practice handouts. Questions and tests can be authored entirely online, allowing instructors ultimate flexibility and the ability to efficiently manage assessments at any time, from anywhere. MyTest for *Technical Communication* includes multiple-choice and short-answer questions. To access MyTest, please go to **www.pearsonmytest.com**.

Instructors' Resources. Instructors can download the following resources from a password-protected location on Pearson Canada's online catalogue (**www.pearsoncanada.ca/highered**). Search for this text, then click on "Instructors" under "Resources" in the left-hand menu. Contact your local Pearson Canada sales representative for more information.

Instructor's Manual. The Instructor's Manual includes teaching notes and resources and answers to end-of-chapter exercises.

Test Item File. The Test Item File includes all the questions from the MyTest in Microsoft Word format.

CourseSmart for Instructors. CourseSmart goes beyond traditional expectations—providing instant, online access to the textbooks and course materials you need at a lower cost for students. And even as students save money, you can save time and hassle with a digital eTextbook that allows you to search for the most relevant content at the very moment you need it. Whether it's evaluating textbooks or creating lecture notes to help students with difficult concepts, CourseSmart can make life a little easier. See how when you visit **www.coursesmart.com/instructors**.

Technology Specialists. Pearson's Technology Specialists work with faculty and campus course designers to ensure that Pearson technology products, assessment tools, and online course materials are tailored to meet your specific needs. This highly qualified team is dedicated to helping schools take full advantage of a wide range of educational resources by assisting in the integration of a variety of instructional materials and media formats. Your local Pearson Canada sales representative can provide you with more details on this service program.

For More Resources

You will find extensive resources in **MyWritingLab: Technical Communication**, but if you have comments or requests for material not covered by MyWritingLab: Technical Communication, email me at **dklepp@shaw.ca**. I will respond.

ACKNOWLEDGMENTS

This text results from a collaboration of many people. Pearson Canada editors David LeGallais, Patti Altridge, and Kimberly Blakey, have all made valuable contributions to this edition. Copy editor Rodney Rawlings has clarified phrasing, updated material, and identified technical errors.

I acknowledge the contributors who provided quotes for the text's On the Job boxes—you'll see their names in those boxes. I also thank Ken Langedyk, Roger Webber, Flightcraft, JLG Liftpod's Sarah Riley, and Norco's Peter Stace-Smith for graciously providing print and website material that appears either in the text or in MyWritingLab: Technical Communication.

I appreciate the valuable suggestions and challenges posed by reviewers from universities and colleges across Canada:

Sue Ackerman, *Nova Scotia Community College*
Kimberly Alexander, *St. Clair College*
Denise Blay, *Fanshawe College*

Sara Frampton, *Cambrian College*
Chris Legebow, *St. Clair College*
Michael Lutz, *McMaster University*
Richard McMaster, *Ryerson University*
Robert Riches, *Fanshawe College*
Lara Sauer, *George Brown College*
Darlene Webb, *British Columbia Institute of Technology*

My wife, Betty Chan Klepp, has supported this project from the very start, and I thank her for her encouragement.

1

Introduction to Technical Communication

LEARNING OBJECTIVES

After reading this chapter, you should be able to

- appreciate the nature, purpose, and range of technical writing
- understand the challenges faced by technical writers
- appreciate the value of collaborating with others in producing technical documents
- understand how advances in technology and changing customer expectations require technical writers to develop an expanded tool box

As a technical communicator, you interpret and communicate specialized information for your readers and listeners, who may need your information to perform a task, answer a question, solve a problem, or make a decision. Your email, letter, report, manual, or online post must advance the goals of your audience and of the company or organization you represent. You often collaborate with others to plan, prepare, and present important documents and oral presentations.

Successful communicators know that the communication process is complex and subject to misunderstanding, so they choose their communication media carefully. They also choose words, accompanying non-verbal cues, and images that will help readers and listeners grasp the intended meaning. Each chapter of this text is designed to help you make those choices.

> Communication is complex and subject to misinterpretation

As a first step, this chapter introduces technical communication environments and communication media, with special emphasis on technical writing. The chapter also examines the challenges faced by workplace communicators, as technological change has increased its pace. To further place this discussion in context, Chapter 1 reviews three careers that require strong communication skills: video game development, independent consulting, and project management.

TECHNICAL WRITING SERVES PRACTICAL NEEDS

Unlike poetry and fiction, which appeal mainly to our *imagination*, technical documents appeal to our *understanding*. Technical writing rarely seeks to entertain, create suspense, or invite differing interpretations. If you have written a lab or research report, you know that technical writing must be clear, audience-oriented, and efficient.

> Technical documents are quite different from essays

When novice technical writers first encounter the task of producing scientific, technical, or business documents, they often rely on the strategies they've used to create essays or personal documents. Some of those strategies can be readily transferred to technical and business documents. For example, arguments and opinions usually require detailed support in the form of evidence, examples, statistics, or expert testimony.

However, essays and personal correspondence use structures and writing styles that can differ quite markedly from technical documents, which emphasize clearly defined structures and concise, readable prose. Those emphases result from meeting the reader's needs and priorities, not the writer's needs and goals.

Technical Documents Meet the Audience's Needs

Instead of focusing on the *writer's desire* for self-expression, a technical document addresses the *audience's desire* for information. This requirement should not make your writing sound like something produced by a robot, without any personality (or *voice*) at all. Your document may in fact reveal a lot about you (your competence, knowledge, integrity, and so on), but it rarely focuses on you personally. Readers are interested in who you are only to the extent that they want to know *what you have done*, *what you recommend*, or *how you speak for your company*. A personal essay, then, is not technical writing. Consider this essay fragment:

Focuses on the writer's feelings

Computers are not a particularly forgiving breed. The wrong key struck or the wrong command entered is almost sure to avenge itself on the inattentive user by banishing the document to some electronic trash can.

This personal view communicates a good deal about the writer's resentment and anxiety but very little about computers themselves. The following example can be called technical writing because it focuses on the subject (see italics), on what the writer has done, and on what readers should do:

Focuses on the subject

On MK 950 terminals, the BREAK key is adjacent to keys used for text editing and special functions. Too often, users inadvertently strike the BREAK key, causing the program to quit prematurely. To prevent the problem, we have modified all database management terminals: to quit a program, you must now strike BREAK twice successively.

This next example also can be called technical writing because it focuses on what the writer recommends:

Focuses on the recommendation

We should develop Help 2.0 support sites to (a) establish online communities of customers and support staff, (b) create a hub for collaborating on solutions for issues that arise, and (c) provide our customers with an easy way to send feedback about our products and services.

As the above examples illustrate, while a technical document never makes the writer "disappear," it does focus on what is most important to readers.

Technical Documents Strive for Efficiency

Educators read to *test* our knowledge, but colleagues, customers, and supervisors read to *use* our knowledge. Workplace readers hate wasting time and demand efficiency; instead of reading a document from beginning to end, they are more likely to use it for reference and want only as much as they need: "When it comes to memos, letters, proposals, and reports, there's no extra credit for extra words. And no praise for elegant prose. Bosses want employees to get to the point—quickly, clearly, and concisely" (Spruell 32). Efficient documents save time, energy, and money in the workplace.

No reader should have to spend 10 minutes deciphering a message worth only five minutes. Consider, for example, this wordy message:

An inefficient message

> At this point in time, we are presently awaiting an on-site inspection by vendor representatives relative to electrical utilization adaptations necessary for the new computer installation. Meanwhile, all staff members are asked to respect the off-limits designation of said location, as requested, due to liability insurance provisions requiring the online status of the computer.

Inefficient documents drain a reader's energy—they are too easily misinterpreted and they waste time and money. Notice how hard you had to work with the previous message to extract information that could be expressed this efficiently:

A more efficient message

> Hardware consultants soon will inspect our new computer room to recommend appropriate wiring. Because our insurance covers only an operational computer, this room must remain off limits until the computer is fully installed.

When readers sense they are working too hard, they tune out the message—or they stop reading altogether.

Inefficient documents have varied origins. Even when the information is accurate, errors like the following make readers work too hard:

Causes of inefficient documents

- more (or less) information than readers need
- irrelevant or uninterpreted information
- confusing organization
- jargon or vague technical expressions readers cannot understand
- more words than readers need
- uninviting appearance or confusing layout
- no visual aids when readers need or expect them

An efficient document sorts, organizes, and interprets information to suit readers' needs, abilities, and interests. Instead of merely happening, an efficient document is carefully designed to include these elements:

Elements of efficient documents

- *content* that makes the document worth reading
- *organization* that guides readers and emphasizes important material

ON THE JOB...

The Importance of Written Communication

"As an environmental consultant, I spend 75 percent of my workweek writing and editing correspondence, proposals, and reports. I'd like to spend more time in the field, but it's critical to communicate what you've learned and what you think it means. That's what the client is paying for! And in providing that information and analysis, we're dealing with some serious health and safety issues, so we have to be very careful in interpreting and reporting the data we collect. . . ."

—Dave Ayriss, Associate, Senior Occupational Hygienist, Golder Associates Ltd., Calgary

- *style* that is economical and easy to read
- *visuals* (graphs, diagrams, pictures) that clarify concepts and relationships, and that substitute for words whenever possible
- *format* (layout, typeface) that is accessible and appealing
- *supplements* (abstracts, appendices) that enable readers with different needs to read only those sections required for their work

Efficiency and audience orientation are not abstract rules: writers are accountable for their documents. In questions of liability, faulty writing is no different from any other faulty product. If your inaccurate, unclear, or incomplete information leads to injury, damage, or loss, you can be held legally responsible.

WRITING IS PART OF MOST CAREERS

Although you might not anticipate a career as a "writer," your writing skills will be tested routinely in situations like these following:

Ways in which your career may test your writing skills

- proposing various projects to management or to clients
- contributing articles to employee newsletters
- describing a product to employees or customers
- writing procedures and instructions for employees or customers
- justifying to management a request for funding or personnel
- editing and reviewing documents written by colleagues
- designing material that will be read on a computer screen or transformed into sound and pictures

ON THE JOB...

Writing on the Job

"When I was a co-op student, I worked in Kelowna at Industry Canada, which places a high value on interpersonal and writing skills. I was quite surprised at how much writing I did. About 20 to 25 percent of my time was spent writing reports, manuals, letters, and memos.

For example, I wrote letters and memos upgrading our database of who has transmitters and receivers and where they're located. I reported on repairs and equipment modifications. I reported LAN maintenance. . . ."

—**Dave Parsons, software consultant and trainer**

So, whatever your career plans, you can expect to write as part of your job.

Your value to any organization will depend on how clearly and persuasively you communicate. Many working professionals spend at least 40 percent of their time writing or dealing with someone else's writing. The higher their position, the more they write (Barnum and Fisher).

Good writing gives you and your ideas *visibility* and *authority* within your organization. Bad writing, on the other hand, is not only useless to readers and potentially damaging to the writer, but also expensive: written communication in North American business and industry costs billions of dollars every year. More than 60 percent of the writing is inefficient; it is unclear, misleading, irrelevant, deceptive, or otherwise wasteful of time and money (Max).

In your career, you'll have to write well. Employers first judge your writing by your application letter and résumé. In a large organization, your future may be decided by executives you've never met. One concrete measure of your job performance will be your correspondence and reports. As you advance, communication skills become more important than technical background. The higher your goals, the better you need to communicate.

Even in high-tech jobs, where once it was sufficient to have strong technical skills, workers now need "soft skills." Jonathan Wray, a communication specialist with IBM Canada, says that IT workers need to be able to write clear documents and speak well. Technical skills get people hired, says Wray, but soft skills help advance careers. He also points out that workers who publish papers, speak at conferences, and work well with clients are the ones who get ahead (qtd. in Kavur).

ELECTRONIC TECHNOLOGIES HAVE TRANSFORMED INFORMATION SHARING

Information has become our prized commodity and, to a large extent, that information is stored and transmitted electronically. Electronic technologies, collectively known as *information technology* (IT),[1] enhance the speed, volume, and varieties of creating and transmitting messages:

- *Integrated software* facilitates the inclusion of verbal, graphic, and video elements from word processors, internet sources, digital recording and storage media, electronic spreadsheets and databases, and oral presentation software.
- *Basic email* helps people exchange ideas and information, while email attachments allow for sophisticated formatting of documents.
- *Wireless laptop computers* and *tablets* allow users to receive and send email from locations (hotel conference rooms, airport lounges, classrooms, etc.) that support wireless transmission.
- *Smartphones*, which are now full-fledged computers, combine all the advantages of wireless communication (phone, email, text messaging, instant messaging, internet access, entertainment) with organizer applications and more. In March 2010, Canalys, an online analyst of high-tech products, predicted sales of 65.1 million smartphones for 2010. By 2012, annual smartphone sales worldwide had risen to over 400 million.

  The only real drawback to smartphones is their size, so developers have created tablets. Over 100 tablet models are available, attesting to the growing demand for devices that are capable of performing all the functions of portable computers, and more. Tablets are now used for educational settings, business and industrial training, sales information and quotes, multi-party conferencing, social networking, Web browsing, video recording and streaming, phone calls, and myriad other applications. One manufacturer in particular, Cisco, has developed its Cius tablet specifically for business users.
- *Virtual private networks (VPNs)* connect home-based workers to their office intranet, without allowing their messages to be monitored by anyone outside that intranet.
- *Video conferencing* and web conferencing provide live online meetings in which participants at different sites can present their ideas and comment on the ideas of others while observing one another's non-verbal messages.
- *Voice over Internet Protocol (VoIP)* technology is in the process of replacing traditional phone systems, partly because of massive long-distance phone call savings, and partly because it enables a host of communication convergence

---

1. Howard Solomon reports that companies, such as Hewlett-Packard, are trying to replace the term *information technology* with the term *business technology* to reflect the integration of software, hardware, and related services in meeting a company's internal and external business communication needs. Others argue that the term *information and communications technology* (ICT) better suits the wide range of hardware and software applications that are now available.

ON THE JOB...

Video versus Audio Conferencing

"Easy-to-use video conferencing has been a real game changer. Audio conferencing can lead to inattentiveness and other activities, but when you introduce video close-ups and support that with good team norms and behavioural patterns, it's easier to keep everybody focused and productive. Old-style video conferencing, with the camera way back, allows participants to check their BlackBerrys but when webcams zero in on participants, it's like we're all around the same table."

—Craig Gorsline, President and Chief Operating Officer, ThoughtWorks, a custom software developer

features (Lima "Calling All Workstations" 22). *Converged messaging* (also known as *unified messaging*) allows different types of messages, such as voice mail and email, to be automatically forwarded to and accessible on a smartphone or on a tablet.

Google Voice and other phone management systems take VoIP one step further. Such systems allow subscribers to link their land line and cellphones to a single number at a web browser that manages all aspects of phone usage. Google Voice forwards and records calls, provides text alerts, and transcribes voice mails. This service could be a real boon for entrepreneurs, consultants, and service providers (Bertolucci).

◆ *Electronic document sharing* uses file transfer software for sharing and editing drafts. The best software allows participants to comment on one another's work online so that each participant's comments are distinct from everyone else's and from the original text.

◆ *Teleconferencing,* using speakerphones, is good for small rooms with 15 or fewer people seated around a table. Many job interviews are now conducted via teleconference.

◆ *Social networking sites,* such as Facebook, Twitter, and LinkedIn, offer unprecedented opportunities for making and managing personal, professional, and business contacts.

These technologies have had a profound effect on office communication:

IT has changed the office environment

◆ Small and medium-sized companies, which make up 95 percent of all Canadian businesses, are slower than large companies to adopt emerging technologies, such as VoIP, wireless networking, web conferencing, and video conferencing. However, almost all have desktop computers and email access, and many have laptops, cellphones, and websites (Lima, "Small Firms" E2). Also, in what Steve Jobs called a "post-PC age," sales of tablets are starting to outstrip sales of portable computers.

The virtual office may take several forms

◆ Instead of being housed in one location, the virtual company may have branches in widespread locations, or just one central office, to which employees "commute" electronically. Such arrangements require workers to be skillful communicators who must master the latest technologies, use proper e-communication etiquette, and know how to compensate for the lack of face-to-face contact (Marron, "Close Encounters" C1). A study by mobile technology company Citrix Systems surveyed companies in Canada and 10 other countries. Citrix reports that 93 percent of those companies "expect to have policies to let staff work remotely" by the end of 2013, which is up from 37 percent in early 2012 ("Working remotely gains popularity").

◆ A special class of worker has emerged—the "virtual assistant," who works on contract, at home, for one or more firms. A virtual assistant might answer phones, direct emails, produce and edit documents, update websites, or maintain databases and inventories. He or she

ON THE JOB...

The One-Man Virtual Office

"My one-man operation can't afford a 'normal' office with someone to answer the phone and send and receive concept drawings and photographed samples. So, when I'm away from the shop, I rely on my smartphone to receive calls and emails. I can open email attachments in the phone, so I don't have to wait until I get back to my office. Later, at my workplace, I can send graphic representations and other information to my clients and suppliers."

—Blair Peden, owner of Awards and Trophies HQ

can even go into a client firm's computer (via an internet connection) to work on material stored on that computer's hard drive (Buckler).

♦ Instead of relying on secretaries, most managers compose and send their own messages via email and messaging. Many also compose, design, and deliver their own reports and proposals.

♦ On desktop publishing networks, the composition, layout, graphic design, and printing of external documents and webpages are done in-house.

Today's technology produces documents that may have a short life cycle

♦ Paper documents (such as résumés) can be optically scanned and stored electronically in file formats, such as PDF. However, electronically stored documents may have a short life cycle if new software does not properly reproduce documents stored in earlier formats. Many experts are deeply concerned that a whole generation's store of technical and historical data could become lost because today's storage media will eventually become obsolete (Reagan).

♦ Computer-supported cooperative work systems, instant messaging, and video conferencing enable employees worldwide to converse in real time. Email listservs announce daily developments in prices, policies, and procedures. In particular, video conferencing has benefited from recent improvements. Users now talk about "telepresence," which provides life-sized, sharp video images and "near-direct eye contact between participants" (Bradbury par. 9).

ON THE JOB...

The Value of Video Conferencing

"Hardly any of us can work alone in an increasingly complex world that requires specialists to join forces. Further, we need to supplement written online collaboration with face-to-face conversation, Skype calls, and video conferencing. I'm involved in video conferencing nearly every day. An amazing amount of information is transmitted by one's voice and facial expressions. Personal contact is not only more fun; it's more instructive. Let's not forget that business decisions are made partly on an emotional level, and emotion is conveyed non-verbally."

—**Wayne Pagens, CMC, senior IT consultant**

Globally outsourced offices face intercultural challenges

♦ On another front, "enterprise rights management (ERM) technology allows a company to restrict the access to and use of all documents and e-mails throughout the organization" (Gooderham, "It Was" E1). Restricted access is especially important when part of a firm is halfway around the world, in a location where "ethics [may be] questionable and corruption is rampant" (E1). Overseas outsourcing has become possible for small companies as well as large ones, according to business reporter Mary Gooderham. Canadian firms might outsource administrative functions, call centres, design work, or even product design to workers in India, Finland, or Singapore. One main challenge in managing the outsourced office is fitting those workers into the company culture while respecting the workers' own cultural values and practices.

Today's workers depend heavily on computer technology

However, there's a price to be paid for the efficient, convenient electronic office. For example, knowledge workers depend on IT professionals (network technicians and others), who are the *wunderkinds* of the new office environment. Their expertise is essential to maintaining not only the computers, but also the hardware networks and associated software. In a short time, today's workers have become totally dependent on computer technology.

Electronic multitasking can exact a heavy personal price

We pay another price for full-time connectivity and electronic multitasking— we have scant time for quiet reflection. Dr. Edward Hallowell, author of *Crazy Busy: Overstretched, Overbooked, and About to Snap*, sees symptoms of attention-deficit disorder in adult executives who are "caught up in a dust storm of information competing for our attention" (qtd. in Immen, "The Next Great Curse" C1).

Researchers have also found evidence that "when people keep their brains busy with digital input, they are forfeiting downtime that could allow them to better learn or remember information or to come up with new ideas" (Richtel).

In reaction, legions of "plugged-in" people are taking "unplugged days" when they turn off their cellphones and computers for a full day or more (Serjeant). Companies, such as Loblaws and Intel, have banned smartphone and computer use for one day a week to help employees have more time for deep thinking and long-term planning (Dube).

These companies recognize that despite the opportunities for multitasking brought by electronic devices, those tools facilitate interruptions. The Basex research firm found that, on average, interruptions consume 28 percent of a knowledge worker's day. Further, a study of Microsoft employees found that they took an average of 15 minutes to return to "serious mental tasks" after responding to emails or text messages (Galt, "Drive-by Interruptions").

There is no substitute for human critical thinking

And, despite the tremendous advantages brought by IT, information does not write itself. Information technology provides tools; it does not substitute for human interaction. Only humans can answer the following questions:

- Which information is most relevant?
- What does this information mean?
- Can I verify the accuracy of this information?
- How will others interpret it?
- With whom should I share it?
- What action does it suggest?

A relatively new form of high-tech communication, video gaming presents special challenges for those who propose, develop, and document video gaming hardware and software.

ON THE JOB...

The Exciting World of Game Development

"I tried game development at Disney and Electronic Arts (EA) in my University of Victoria co-op terms and was immediately hooked. EA hired me when I graduated and I've been there 8½ years. I've worked on various games; classified as a generalist, I can program various areas of the game, but my main interest and expertise is in AI [artificial intelligence]. I also work on the physics of games. For example, for the game Need for Speed Undercover, *I worked on all aspects of the car driving—acceleration rates, suspension sway, and tire grip on the corners.*

I've worked on AI for how pedestrians respond when faced with various choices in Skate 3. There's quite a knack to making AI that's not too intelligent, because you don't want to produce a game that gamers can't win. You have to make the AI in characters somewhat fallible, so that they make some dumb choices, just like people in real life.

I think there'll continue to be a strong video game industry in Canada, with good employment opportunities, but I think companies need to respond quickly when new technologies come along. We've seen this recently with games for the new platforms offered by Facebook and the iPhone."

— **John Wheeler, Vancouver-based game developer**

IN BRIEF

The Video Gaming Industry

Video game development and production has become an important Canadian industry. Some 247 Canadian firms employed over 14 000 people as of March 2009, according to a study by the Electronic Software Association of Canada (ESAC). Most positions are "high-paying software development jobs," says ESAC executive director Danielle Parr. Moreover, "Canadian video game developers hope to add as many as 29% more employees by 2011" (Pilieci).

The main development centres are Ubisoft's Montreal campus and the Electronic Arts (EA) campus in Burnaby, British Columbia. Education and training are available at institutes such as Concordia University (Montreal), The Art Institute of Vancouver, triOS College (Brampton), and the Centre for Arts and Technology (Kelowna). Also, computer science programs at many Canadian universities prepare students for game development careers.

On a more advanced level, the IT University of Copenhagen offers a master's degree in Media Technology and Games and a Ph.D. program that encourages research into such topics as "Serious Games in a Global Marketplace."

The impetus for all this activity comes from the penetration of video gaming in the entertainment industry. Recently, strong demand has also come from educators and corporate trainers.

The NPD Group's "Kids and Games 2009" market research report indicates that 82 percent of Americans aged 2 to 17—over 55 million kids—were gamers as of March 2009 (Gaudiosi). Other surveys have reported even higher participation: a 2008 report released by the MacArthur Foundation reported that 97 percent of respondents aged 12 to 17 were gamers (FOX News).

The wave of computer and console gaming has prompted educators to use video games to help students learn various subjects, including math, chemistry, molecular biology, and history.

Also, "organizations are finding that games can provide a painless way to train their staff" (Flood). The e-Learning Centre, an international hub for learning technology and game development businesses in Sheffield, England, has links to over 30 articles about games and training (**http://archive. e-learningcentre.co.uk/eclipse/Resources/ games.htm**).

The game developers who will need to meet all this demand require strong communication skills. Many of today's popular video games are built by large teams that include computer programmers, audio and game technicians, visual artists, and script writers. The contributions of each team member affect the scope and nature of the contributions of other team members, so all must communicate clearly and efficiently. And, of course, each game requires detailed documentation.

EA game developer John Wheeler adds that "you have to be a good oral communicator and function well in groups. Also, we write proposals, time estimates, and risk assessments. Often, we write white papers and knowledge articles that describe what we've learned from developing a particular game."

Some development teams work from *design documents*, which are similar in some respects to film scripts. Others employ a software program known as a *game engine* to provide a game's basic platform. For example, Swedish studio DICE has developed the Frostbite engine that its parent company, EA, has used as the basis for games in its *Battlefield* series. Wheeler says that "when you use an engine like Frostbite, the game building process is a matter of generating the game's content—the characters, the action, the story line, the artificial intelligence."

TECHNICAL WRITERS FACE INTERRELATED CHALLENGES

No matter how sophisticated our communication technology, computers cannot *think* for us. More specifically, computers cannot solve all the challenges faced by people who write in the technical professions. These challenges include the following:

◆ *The information challenge.* Different people in different situations have different information needs.

◆ *The persuasion challenge.* People often disagree about what the information means and about what should be done.
◆ *The ethics challenge.* The interests of your company may conflict with the interests of your audience.
◆ *The global context challenge.* Diverse people work together on information for a diverse audience.

Information has to have meaning for its audience, but people differ in their interpretations of facts, and so they may need persuading that one viewpoint is preferable to another. Persuasion, however, can be powerful and unethical. Even the most useful and efficient document could deceive or harm. Therefore, solving the persuasion challenge doesn't mean manipulating your audience by using "whatever works"; rather, it means building a case from honest and reasonable interpretation of the facts. Figure 1.1 offers one way of visualizing how these challenges relate.

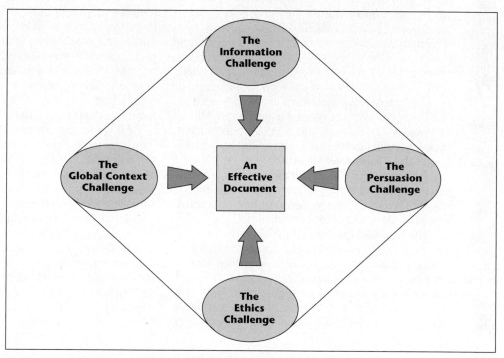

Figure 1.1 Writers Face Four Related Challenges

The scenarios that follow illustrate the challenges faced by a professional in her own day-to-day communication on the job.

The Information Challenge

What do the data really mean?

Erika Song, who has a background in biology and environmental studies, works for Enviro Associates, an Edmonton-based environmental assessment firm. On contract to the Alberta Environment ministry, Enviro monitors water flow and water quality in the Athabasca River near the Fort McMurray oil sands projects.

Erika's task is to compile and analyze Enviro's 2010 monitoring data and compare that data with both Enviro's 2009 data and with 2010 data made public by the Wood Buffalo Environmental Association and the Regional Environmental Monitoring Association (REMA). Alberta Environment wants Enviro's data and conclusions, *and* it wants to know whether there is any redundancy in the three collections of data.

After analyzing the reams of data, Erika's next challenge is to organize and present the data for technical and non-technical audiences. She has to answer such questions as *How much explaining do I need to do? Do I need visuals? How much of the hundreds of pages of raw data should I include in the appendices? What conclusions can I draw?*

To some extent, Erika also faces a persuasion challenge. She sees some striking discrepancies between the Enviro data and data released by REMA, an initiative funded by oil sands producers. Further examination reveals that some discrepancies have resulted from the contrasting locations of monitoring stations. (Most of REMA's stations are downstream from the oil sands developments, while half of Enviro's stations are upstream and half are downstream.)

The Persuasion Challenge

Sources of collaborative conflict

Before writing her report, Erika must persuade her boss, Dr. Russ Klingbeil, to include the data discrepancies, only some of which resulted from the location of monitoring stations. For 16 days in July 2010, Enviro's main downstream station recorded wild fluctuations in water flows because of an electrical short that was eventually diagnosed and repaired. Klingbeil is reluctant to admit this equipment failure. He is about to propose a contract extension to Alberta Environment.

Erika's associates have asserted that Enviro's monitoring methods are superior to REMA's methods, but Erika sees little evidence in the 2009 and 2010 data to support this belief. Enviro's stations might be better placed to record sediment deposits, but not to measure stream flow or water levels. Also, it seems that REMA and Enviro use similar methods to determine benthic invertebrate communities (a biological indicator that is an important component of fish habitats).

Finally, Erika concludes that measuring sediment quality is not very useful and should be discontinued—varying flows affect which parts of the riverbanks get eroded; shifting sand bars affect sedimentation and obscure the effect of releasing water from the oil sands production facilities; and deposited hydrocarbon sediments, which occur naturally from the hydrocarbons found in the riverbanks, have historically varied very little.

If Erika persuades Alberta Environment to discontinue the sediment measurements, the contract will be less profitable for her firm, so she will face pressure from her boss to "re-examine" the data and her conclusions.

The Ethics Challenge

The ethics of omitting information

Erika can choose what data to include and what to omit. Does she report the 16-day equipment malfunction and the faulty data, or simply omit the information that reflects badly on Enviro Associates? Admitting the malfunction may jeopardize her firm's continuing contract.

What is more
important—personal
integrity or business
success? (See Case
Study Exercise 1 on
page 95.)

By choosing certain data and omitting other facts, Erika could show that Enviro's methods are superior to the methods employed by REMA, thus strengthening her firm's case for continuing to monitor the Athabasca River, and she could show that multiple sets of monitoring data are not redundant. Further, recommending that sediment measurements be discontinued may affect her firm's bottom line and perhaps her own employment.

Situations that jeopardize truth and fairness present the hardest choices of all. Some aspects of Erika's ethical challenges are as follows: *How much am I obligated to report? What do I feel is fair? What would be the consequences of omitting some of the data and interpreting the data to favour Enviro Associates?*

In addition to meeting these various challenges, Erika collaborates with others to produce the final report. Technologist Avre Ostif, who collected and downloaded most of the data, will review Erika's data presentation; Klingbeil will proofread and edit the report; and a graphic artist will help to create professional, bound copies.

Finally, Erika's audience will extend beyond her own group of technical experts. Government bureaucrats will use the report to aid their decision making. Erika's audience will even extend beyond her own culture.

The Global Context Challenge

Content and phrasing
choices need to reflect
the needs of multiple
audiences

Eventually, Alberta Environment will place the report on its website, where it will be available to a worldwide audience of interested oil-industry business people and environmentalists. Also, Enviro Associates is negotiating with Alice Earth, an Australian-based company that wants to buy Enviro and incorporate it into its global environmental practice. If the acquisition occurs, all of Enviro's reports will be placed in Alice Earth's database for associates in Australia, the United States, China, Singapore, and Malaysia to read. Thus, Erika and her colleagues have been urged to write reports that will be understood by an international audience.

Possibly, Erika will soon start to develop working relationships with people she has never met and whose cultural expectations differ from hers.

IN BRIEF

Writing Reaches a Global Audience

Our linked global community shares social, political, and financial interests that demand cooperation as well as competition. Multinational corporations often use parts that are manufactured in one country and shipped to another for assembly into a product that will be marketed elsewhere. Medical, environmental, and other research crosses national boundaries, and professionals in all fields deal with colleagues from other cultures.

Here is a sample of documents that might address global audiences (Weymouth 143):

♦ studies of global pollution and industrial emissions
♦ specifications for hydroelectric dams and other engineering projects
♦ operating instructions for appliances and electronic equipment

(continued)

◆ catalogues, promotional literature, and repair manuals
◆ contracts and business agreements

To communicate effectively across cultural and national boundaries, any document must respect not only language differences but also cultural differences. One writer offers this helpful definition of culture:

> Our cultures, our accumulated knowledge and experiences, beliefs and values, attitudes and roles, shape us as individuals and differentiate us as a people. Inbred through family life, religious training, and educational and work experiences . . . cultures manifest themselves . . . in our thought and feelings, our actions and reactions, and our views of the world . . . [and] in our information needs and our styles of communication. . . . Our cultures define our expectations as to how information should be organized, what should be included in its content, and how it should be expressed. (Hein 125)

Cultures define appropriate social behaviours, business relationships, contract negotiation, and communication practices. A communication style considered perfectly acceptable in one culture may be offensive elsewhere.

Effective communicators recognize these differences but withhold judgment or evaluation, focusing instead on similarities. For example, needs assessment and needs satisfaction know no international boundaries; technical solutions are technical solutions without regard to nationality, creed, or language; courtesy and goodwill are universal values. In a diverse global context, the writer must establish trust and enhance human relationships.

Outsourcing of products and services brings special challenges. North American culture tends to focus on tasks and deadlines for task completion. Meanwhile, Asian cultures are more concerned with group cohesion and solidarity. The specifics of the product or service to be delivered may be interpreted differently by the vendor and client, purely because of language differences and inaccurate translations. Also, rules and regulations about product quality, labour standards, and documentation vary from culture to culture.

Because of the potential pitfalls inherent in contracting services and product components from other parts of the globe, many North American companies hire outsourcing consultants who understand the vendor's cultural values and practices.

TECHNICAL COMMUNICATION IS CHANGING

New technologies have created new customer expectations. For example, many customers who use manuals prefer online documentation to printed manuals. Users might also prefer 3D graphics, video demonstrations, or interactive high-reality simulations. Regardless of the technology, technical communicators need to process and understand the technical information. Their role further requires them to structure and "write" the content of reports, instructions, proposals, and other varieties of "technical" content.

ON THE JOB...

The New Challenge

*"We are in the middle of a content development revolution. To attract and engage the next generation of tech-savvy customers, we must do more than just write content—we must deliver user-optimized content when the customer wants it, and in the **format** the customer wants. That takes an effective content strategy—and really, really smart use of available resources."*

—Jack Molisani, Executive Director, LavaCon Conference on Digital Media and Content Strategies

According to Adobe Systems trainer Matt Sullivan, "the term Technical Writer is becoming increasingly inaccurate. Writers are producing the print, the online- and application-based help, and the self-paced training books, along with instructor-led materials. In some cases, the writer is even the instructor as well!"

For example, the aviation manual writing process used by technical writer Roger Webber for clients, such as Bombardier and Flightcraft, requires collaboration at nearly every one of the many stages outlined in Figure 1.2 on page 15.

Successful collaboration combines the best that each team member has to offer. It enhances critical thinking by providing feedback, new perspectives, group support, and the chance to test ideas in group discussion.

Not all members of a collaborative team do the actual writing—some might research, edit, proofread, or test the document's *usability*. The more important the document, the more it will be reviewed. Notice the engineering approval, technical review, verification, and quality control checks in Figure 1.2.

As technologies advance, writers have to become proficient in many areas that were once the realm of graphic artists and computer specialists. Here are some of the major trends that are now affecting the role of the technical communicator.

The Advent of Big Data

> ### ON THE JOB...
>
> **The Challenge of Big Data**
>
> *"For the first time in human history, we have the ability to collect information, process it, visualize it, and respond to it while it's still happening. . . . We've reached a tipping point in history: today more data is being generated by machines—servers, cell phones, GPS-enabled cars—than by people. I think Big Data is going to have a bigger effect on humanity than even the internet."*
>
> —Rick Smolan, Creator of the Day in the Life series, as he welcomed participants on October 2, 2012 to the conference The Human Face of Big Data (http://the_human_face_of_bigdata.com)

Decision makers are challenged by the sheer volume of available data and the speed at which data is produced. In addition to the standard data sources (news sources, government, academic, and business reports, and in-house research), internet sources and social media data streams add to the avalanche of daily information. This increasing pool of data and data sources is labelled *Big Data*.

A sizeable chunk of Big Data is analyzed, structured, and stored; but these processes require skilled workers who tackle the time-consuming task of finding patterns and drawing useful inferences in the reams of data they mine. In other words, big businesses can devote considerable resources to making sense of Big Data. Those who can afford to pay for data processing services profit from the expenditure, in terms of increased workforce productivity, product development, and overall profitability (Gravelle 1).

Government and business mining of Big Data has created technological and sociological challenges

However, small businesses and individuals lack the resources to mine and make sense of data-intensive scientific domains. The other side of the coin deals with the constant streams of user-generated data on social media, such as Facebook, Google+, Flickr, and LinkedIn. Various manufacturers, employers, retail chains, entertainment purveyors, governments, and the social media companies themselves are able to collect and analyze useful data about individuals. Thus, there are privacy implications as big business and government have more tools to conduct surveillance.

The explosion of data requires new methods of analyzing and using information

Solution providers, such as IBM, are developing and grouping platforms that analysts and managers minimize risk and improve decision making. IBM, for example, is

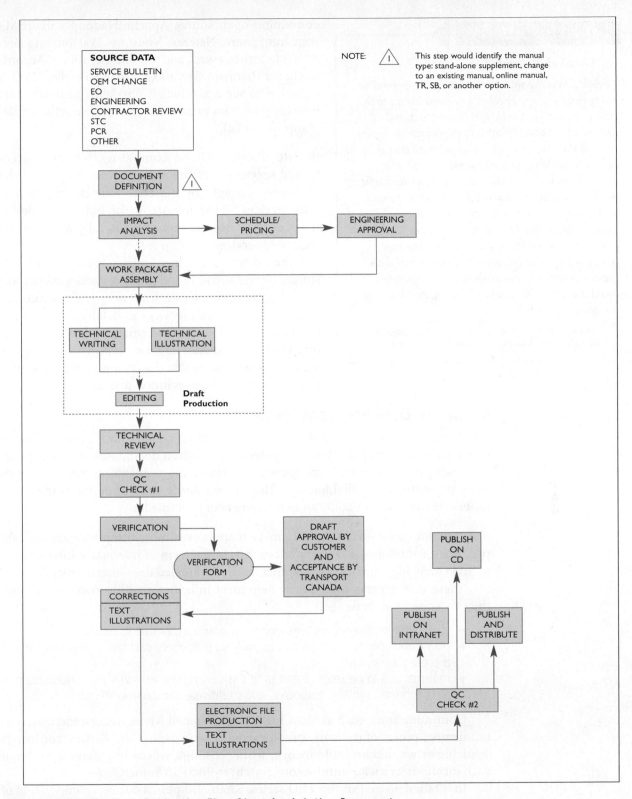

Figure 1.2 Publications Production Flow Chart for Aviation Documents

Source: Courtesy of Roger Webber, Torbay Technical Services.

Big Data Left Behind

"Think about the social DNA you leave behind every time you use Facebook or even log on to the internet. You leave little bits of unstructured data. Hundreds of millions of people are doing that all the time. The challenge with all that data is how to gather and make sense of it all. For example, how does an airline track Twitter traffic in order to determine attitudes and trends that will affect the airline industry? In traditional data warehouses, everything is organized and clearly related to other data as it is entered. The new world of Big Data includes that structured data, but it's dwarfed by an avalanche of disparate, unrelated data that could be very useful if it were analyzed properly."

—Craig Gorsline, President and Chief Operating Officer, ThoughtWorks, a custom software development company

combining open source Apache Hadoop with IBM's own InfoSphere, Netezza, Vivisimo, Watson, and Big-Sheets to gather, assess, and present useful data. According to the Branham Group's Christian Gravelle, "2012 is expected to see a significant uptake in solutions that help organizations extract insights hidden in their data" ("Top 10 . . ." 12).

IN THE CLOUD.   Cloud computing, which delivers hosted services over the internet, may help meet the challenges brought by Big Data. The cloud provider sells services by the minute or the hour, provides as much as the customer requests, and fully manages the requested service.

The customer might purchase server usage and storage or rent software and product development tools hosted on the provider's infrastructure. The customer might outsource such services as database processing, inventory control, web-based email, or third-party billing. The cloud provider might even provide a complete suite of software tools that the customer uses to author content and distribute technical drafts.

Complex Delivery Requirements

Traditionally, manuals, reports, proposals, and other content have been printed or made available to be placed on computers. Now a given document—whether product descriptions, software help, operating instructions, or analytical reports—needs to be presented in parallel formats. The content needs to go to online readers and mobile device users, in addition to consumers of print media.

MULTIMEDIA COMMUNICATION.   To make matters even more complex, audiences are pushing for multimedia delivery of content, in the form of interactive instructions, online video, high quality 3-D graphics, and/or animated demonstrations.

These delivery methods clearly help meet information consumers' needs, but they also pose real challenges:

- video and animation are time-consuming and expensive to produce;
- non-print formats must be searchable, so that users can find particular parts of a presentation; and
- material has to be reformatted to fit the varying screen sizes and orientations of smartphones, tablets, netbooks, and traditional computers.

Communicators, such as Hans Rosling and Salman Khan, understand the communicative power of dynamic graphics, audio, and video. The Khan Academy is available at **www.khanacademy.org**, while Rosling's videos populate YouTube at such locations as **www.youtube.com/watch?v=jbkSRLYSojo**.

In a talk delivered in the TED series, Khan also points out two components of the "humanizing power" of technology: (1) viewers can use the Academy's tutorial videos on their own time and at their pace; (2) teachers assign his lectures as homework, and then class time is used to work on assignments and problems while students and teacher interact (**www.youtube.com/watch?v=gM95HHI4gLk**).

E-LEARNING. Such considerations are especially important for e-learning, whether that learning takes place in an academic setting or in business/industrial settings. E-learning is one of the most rapidly growing industries as more and more people turn to portable devices. A recent study by Ambient Insight Research shows that "the worldwide market for e-learning products was $32.1 billion (£20.5 billion) in 2010 and is expected to rise to $49.9 billion by 2015" (Roundpeg).

Managers love e-learning because it's cost-effective—workers don't have to leave the office to take a training course. Such courses are increasingly necessary as new technology begets new kinds of work. In this ever-changing environment, workers frequently change careers or take on new responsibilities.

Structured Documents

The need to adapt documents for different media or for delivery to different cultures and varying levels of expertise has led to the use of structured documents that "use some method of embedded coding or markup to provide structural meaning to an agreed upon organizational structure or schema" (Gravelle "Top 10 . . ." 1). Adobe FrameMaker, Ideapi, and Autotag are tools that ease and speed up the processes of reusing content or publishing for multi-channels. For example, information about a new tablet designed for the construction industry might find its way into news releases, brochures, printed manuals, online help, website product descriptions, and video or animated tutorials.

Single sourcing allows content to be used in different documents and converted quickly from one format to another. However, not all single-sourcing uses "embedded coding or markup" to help convert the content.

The TechComm Generalist

Subject-matter experts will continue to write the content of technical documents, but increasingly technical communicators will need to go beyond research and writing. This trend will be especially noticed in small organizations where the technical writer will need to be proficient in text, layout, structured documentation, video, and animation. Large companies will be able to afford specialists for those roles, but even in large companies technical communicators will be involved in most components of product and service communications.

ON THE JOB...

A Sample Technical Writer Job Description

"**Responsibilities**:

◆ Use your advanced writing skills to design, schedule, write, and edit guides, online help, and release notes, following company standards and house style.
◆ Use source material and interviews with subject matter experts to research documentation.
◆ Manage your projects independently. Closely track product-development progress and news.
◆ Attend meetings with other technical writers and product development teams."

—Part of a WowJobs Calgary Posting, September 18, 2012

PRODUCT DEVELOPMENT.   One area where technical communicators are assuming increased responsibility is *product development*. In the **initial planning stages** a writer can help to shape and write the project plan. Later, in the **design and development phase**, the technical communicator might produce documents for building a prototype, purchasing components, and integrating systems into a product. Then, the **test phase** usually requires a written test plan.

During the new product's **sell phase**, writers launch material, press releases, advertising copy, catalogue copy, and product documents. Writers continue to contribute to the new product development process during the **service phase**, which requires documents, such as an installation guide, a theory of operation, adjustment and replacement procedures, training manuals, and diagnostic guides.

As the project draws to a close, a technical writer might even compile a project history file of lessons learned, consumer surveys, and recommendations for future projects (Shenouda).

COMMUNICATION SKILLS REQUIRED BY CONSULTANTS.   The communication skills discussed in this chapter certainly apply to *consultants*—people who are paid to provide assessments, advice, technical solutions, and business solutions for external clients. Although many prefer the security of working for consulting firms, others choose an *entrepreneurial* path. These risk takers are willing to work long and hard to identify and meet clients' needs.

Many types of clients require consultant services

People provide consulting services in almost any field—engineering, environmental assessment and improvement, manufacturing, project management, health care, criminal justice, municipal services, business management, education, and computer technology. The list could go on and on.

Consultants provide assessment and other analytical services; they determine causes of problems and recommend solutions; they plan and lead projects; they provide skills assessment and training; they guide their clients through procedures that the consultants have recommended; and they write reports for private and government organizations.

Whether they work for a firm or are self-employed, consultants rely on several critical communication skills:

1. *They must be active listeners.* In conversation or in writing, they help clients identify needs and priorities. All successful consulting begins with an understanding of the pressing issues and problems. Often, the consultant supplements this knowledge with secondary research into the history and/or technical context of the situation.
2. *Consultants must be good analysts and problem solvers.* Often, they will need creative, innovative strategies to deal with unique situations. There are very few "cookie cutter" solutions in the world of consulting. So consultants need to be open to new ways of doing things.

ON THE JOB...

Successful IT Consulting

"In many ways, a good consultant operates like a reporter, gathering information that answers the standard journalistic questions: what, when, where, who, why, how, and how much? Beginning consultants listen to clients and then go away to do what the client has requested. Intermediate consultants are not satisfied merely with the 'what.' They also find out why the client requests certain outcomes. Senior consultants go further; they consider how the requested actions will affect the client's business and what challenges the client will face in implementing proposed solutions. Successful consultants also understand that in order to solve a business problem or effect a substantial improvement, there will usually have to be changes in people, technology, and processes. By itself, technological change is not sufficient."

—**Wayne Pagens, senior IT consultant, CMC**

Consultants must be skilled communicators

3. *Consultants must handle the pressure of looming deadlines and client expectations.* Even the least assertive client expects high-quality work in return for what he or she may perceive as high fees.
4. *Consultants must project a professional image*, through their written correspondence, personal appearance, and timely communication. Phone calls, text messages, and emails should be acknowledged and returned promptly. Consultants must fully prepare for client meetings.
5. *Consultants must have excellent interpersonal skills* to work with all levels of client understanding and a variety of personality types.
6. *Consultants must possess team-building and leadership skills.* Almost every project, at some stage or other, requires collaborative effort. Often, the consultant will lead or supervise that team effort.
7. *Above all, consultants must display strong oral and written skills for networking, marketing, reporting, and proposal writing.* The independent consultant will learn about opportunities through his or her contact network or through advertised requests for proposals. The next step will be to prepare and present a proposal that likely will compete with rival proposals.

PROJECT MANAGEMENT COMMUNICATIONS. Specialized consultants who plan and direct the collaborative efforts of people in a defined project are known as *project managers* (or, colloquially, *PMs*). They manage processes, resources, and people in projects, such as a study, the construction of a building, or the selection and installation of IT equipment and software.

These consultants might be either freelancers or company employees. Most project managers have learned their professional skills on the job, but increasingly they're gaining the required knowledge and skills from colleges and training institutes, such as the Canadian Management Centre and Procept Associates.

On the job and in the classroom, project managers learn the critical importance of communication. According to Calgary-based project manager Rob Corbett, "you need the right types and amount of resources, but the other main factor in successful project management is good communication, which develops understanding and trust. Having that trust with the client and the members of your project team is essential."

Effective project communication starts in the planning stages. The project manager must solicit and listen carefully to the opinions and information provided by clients, team members, and subcontractors: "measuring time and resources is always a challenge, so when planning a project it's important to get input from everyone who has pertinent information and requirements" (Corbett).

As a project progresses, successful project managers keep clients and team members

Consultants win most of their contracts through proposals

Good communication develops trust

Project managers report progress to clients and team members

ON THE JOB...

Communicating with the Team

"When everyone gets involved early, they're more likely to get engaged and stay engaged. The net result is that the work gets done on time, and we don't have breaks in the process. To reinforce that level of commitment and inter-dependence, I keep everyone up to date with status reports, spreadsheets, and Gantt charts that show each participant's contributions."

—**Rob Corbett, professional project manager**

informed through progress reports and project planning tools, such as Gantt charts. These charts can also be used to manage budgets. Gantt-chart proponents, such as Jon Peltier, know how to maximize Microsoft Office to produce such charts.

Some find Gantt charts too limiting, so they turn to more comprehensive project-management tools, such as PERT charts, Microsoft Project, GanttProject, Lighthouse, or Intervals, to track and manage all aspects of a project. Such tools facilitate flexible planning and help managers react promptly to unexpected problems and delays.

NOTE   **MyWritingLab: Technical Communication** *has additional information and advice about project-management processes and techniques.*

EXERCISES

1. Locate a brief example of a technical document (or a section of one). Make a photocopy, bring it to class, and explain why your selection can be called technical writing.
2. Research the kinds of writing you will do in your future career. You might interview a member of your chosen profession. Why will you write on the job? For whom will you write? Explain in a memo to your instructor.
3. In a memo to your instructor, describe the skills you seek to develop in your technical writing course. How exactly will you apply these skills to your career?

COLLABORATIVE PROJECT

Divide the class into three groups. Conduct an informal survey of communications technologies used by group members' acquaintances who are employed in a business setting.

The groups will be responsible for learning more about usage of the following communications tools:

♦ Group 1—smartphones
♦ Group 2—tablets
♦ Group 3—small laptops/netbooks

As a group, compile the information you have gathered and appoint one member to present your findings to the class.

Ask such questions as:

♦ Approximately what percentage of your written business messages (email, text message, social media post) is sent via this device? (Group 1 members will ask about smartphone usage; Group 2 members will ask about tablets; Group 3 members will ask about laptops).
♦ What are the main advantages and limitations of this device for gathering business-related information?
♦ What are the main advantages and limitations of this device for displaying detailed information, such as tables, reports, and spreadsheets?
♦ What are the main advantages and limitations of this device for viewing graphics and video?
♦ What are the main advantages and limitations of this device for keeping supervisors, colleagues, and clients informed about your progress on a project?
♦ What are the main advantages and limitations of this device for maintaining your social networks?

MyWritingLab

MyWritingLab: Technical Communication offers the best multimedia resources for technical communication in one, easy-to-use place. You can find a variety of interactive model documents and case studies. There are also extensive guidelines, tutorials, and exercises for Document Design, Writing and the Writing Process, and Research, and a large bank of diagnostics and practice for grammar review.

2

Preparing to Write: Audience/Purpose Analysis

Explore

Click here in your eText to access a variety of student resources related to this chapter.

LEARNING OBJECTIVES

After reading this chapter, you should be able to

- identify the factors to consider in analyzing the communication in a given situation
- understand why communication can easily break down
- choose the appropriate level of technicality for a given message and audience
- develop an audience/purpose profile to help choose the content, structure, design, and tone for a document

As a technical writer, you'll need to consider your audience's needs. Perhaps you want your audience to support a proposal; perhaps you are responding to your supervisor's request for information about your progress on a project. Whatever you write, you must think very carefully about your audience and your purpose for writing in *any* kind of technical or business communication, written or spoken.

For example, when preparing for a job interview, consider more than the points you hope to make about your qualifications and skills; also consider what the interviewer hopes to gain. Analyzing the interviewer's role—to find an ideal candidate for the job—will help you anticipate the questions you will be asked.

Can you predict how a listener will react to statements in a job interview, or how a reader will interpret a report's facts and conclusions? The answer is that you can't know *for certain*. However, you *can* make some insightful guesses. The following communication model can help you anticipate receiver reactions so that you can adapt the content and presentation of your messages.

USE A COMMUNICATION MODEL

Whether you have a quiet talk with a friend or write a major report for an important client, certain key factors affect the nature and outcome of that communication. The model discussed in this section presents those factors to show how they interact and contribute to the success of that exchange.

Certain key factors affect communication

CASE	**Using a Communication Model to Understand a Conversation**

Let's analyze a conversation between a consulting mechanical engineer, Daphne McCrae, and her client, Max Lauder, the maintenance supervisor for the Trendmark Ski Resort. Max has heard about a neighbouring ski hill's problem with the massive cast-metal gripping mechanisms that clamp the chairs of a high-speed chairlift onto the cable of that lift. Max wants Daphne's advice about whether to replace the grips on Trendmark's high-speed quad lift. Thus, Daphne has tested several grips and is giving a preliminary oral report of her findings.

Identify the primary message

First, in the communication model, let's identify the conversation's primary *message* as McCrae's semi-technical answer to Lauder's question. McCrae is the *sender* of that message; Lauder is the *receiver*. Their conversation, which occurs in the ski resort's maintenance building, uses an oral verbal *channel* and several non-verbal *channels*.

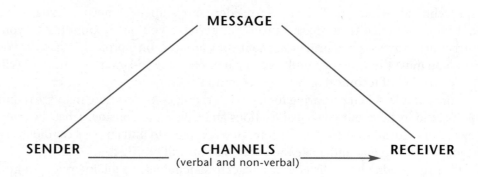

"Content" verbal messages are mediated by "relational" non-verbal messages

The conversation between McCrae and Lauder seems to depend on the verbal channel of speech. After all, McCrae uses words to convey her knowledge about the factors that might contribute to cast-alloy failure. Such *content* messages are often assumed to constitute the entire message.

However, several non-verbal channels, such as facial expression, vocal inflections, posture, and gestures, send accompanying messages about the sender's confidence in his or her own knowledge, degree of certainty, and level of concern for the client's current problem. How Lauder receives these non-verbal *relational* messages affects how he judges the accuracy and value of McCrae's words.

Has McCrae chosen an appropriate channel to convey her initial reaction to her client's question? The answer is yes, especially if she wants to alleviate her receiver's immediate concerns. However, if McCrae were to communicate a full technical analysis of the grips she has tested, she should use a formal written report to present her findings. McCrae's choice of communication channels illustrates a basic communication principle: senders need to choose their primary communication channels carefully in order to reach their receivers.

Here are some other questions to help assess which channels should be used in a given situation:

1. *Has the sender chosen a channel used by the receiver?* For example, McCrae won't reach Lauder via email on days when Lauder is working on the lifts, away from his office computer. Lauder doesn't have a smartphone.
2. *Has the sender taken advantage of the chosen channel's particular strengths?* The main advantage of presenting McCrae's initial analysis *orally* is that it allows her to accompany her semi-technical comments with reassuring vocal tones and other non-verbal messages.
3. *Have both the sender and receiver blocked out external "noise"?* The maintenance building could be very noisy, as workers repair and maintain equipment, so perhaps the conversation should be held in Lauder's office. However, in this case, the two participants need to examine the equipment as they talk, so they must meet in the maintenance shed, where they need to block out the noise in order to focus on each other's messages.
4. *Has the sender considered both the verbal and non-verbal aspects of the transmission?* McCrae's subsequent written analysis will need to use a formal report format, to correspond to the subject's serious nature and to signal the writer's credibility.

Choosing the right channel is only the beginning. The sender also has to encode the message so that the receiver has at least a hope of decoding it as the sender intended. Proper encoding means choosing the words, sentence patterns, and message structures to suit the sender's purpose and the receiver's interests and needs. This collection of choices requires careful thought. (That's why this book emphasizes audience/purpose analysis for all writing and speaking assignments.)

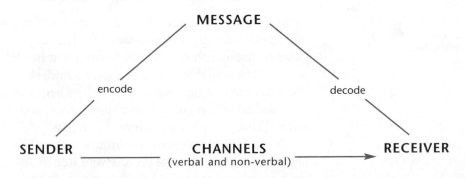

Decode messages
carefully

Does Max Lauder understand Daphne McCrae's explanations? Perhaps. If McCrae uses a great deal of technical jargon, Lauder may be perplexed. And if McCrae oversimplifies her message, Lauder won't get the full picture and thus will not understand McCrae's advice about the grips.

Indeed, decoding messages is often harder than encoding them. To start with, many receivers do not know how to listen or read effectively, so they miss much of the intended message. Even the alert, skilled receiver may have difficulty following the sender's train of thought if the encoded message has been poorly organized, if the sender has chosen an inappropriate level of phrasing or detail, or if the sender's verbal and non-verbal messages contradict each other.

Even when the receiver is confident that she or he has understood the message, communication can break down. That's because senders and receivers often have different meanings for the same words. (Your English professor, for example, might intend the word *decode* to interpret written and spoken words, but you might think of deciphering Morse code and other coded messages if you have a military background or a passion for spy novels.)

Now let's introduce more psychology to our communication model:

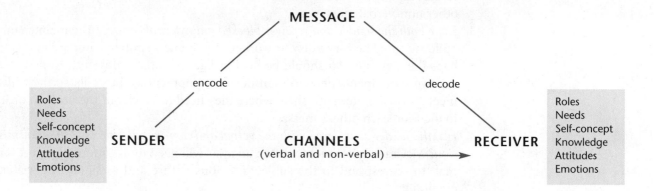

The terms beside *sender* and *receiver* in the model refer to factors that may affect each person at the time the communication occurs. These terms will now be discussed. The discussion will analyze the McCrae–Lauder conversation, but you might also like to stop at various points to think of a recent conversation you've had or a letter you've sent in order to see how each of the following factors may have affected the communication's outcome in each of those cases.

Consider the factors that affect the encoding and decoding processes

1. *Roles.* Daphne McCrae's role is the expert analyst, which compels her to think and speak carefully and rationally. In that role, she must not jump to conclusions or make rash statements. Why? First, her job role requires her to make sense of the available technical data. Second, she has to maintain credibility in order to convince her listener to consider her expert advice.

   Lauder's role requires him to understand and absorb McCrae's analysis and advice. Thus, Lauder listens very attentively.

   Now, think of how your role influences your behaviour at this moment. Probably, you're reading this page because your student role requires you to do so, making your receiving behaviour role-dominated. But what if you find yourself getting really interested in this subject? What if you are now starting to think of how *role* affects your communication with your friends and your colleagues at work or school? If so, your *personal needs* are starting to influence your receiving behaviour.

2. *Needs.* It's obvious that we all have strong reasons for communicating with others: *practical needs* associated with making a living, basic *physical needs*, and *social needs* for acceptance, affection, and control. Also, *identity needs* seem always to influence our behaviour, even when we're taking care of our basic physical needs or our practical business needs. Much of our communication has us trying to determine who we are and then trying to assert that identity to others. If we allow our identity needs to dominate, however, problems may result. For example, if

McCrae were to have a strong identity need to appear forceful and infallible, she might be driven to make definite conclusions even if there's insufficient data to support such conclusions. If her identity need overruns her role requirement, she might provide disastrous advice to her client.

3. *Self-concept.* Our self-beliefs affect our sending and receiving behaviours. If, for example, McCrae sees herself as analytical and intelligent, her word choices and speaking pace will reflect those self-perceptions. And if Lauder sees himself as very practical but lacking education, he may defer to some of McCrae's conclusions even if his instincts tell him that she is wrong.

4. *Knowledge.* A sender like McCrae has to know her subject well before she can successfully explain, for example, resistance to stress in cast-metal chairlift grips. Similarly, Lauder has to have a rudimentary knowledge in order to understand any of the complexities in McCrae's analysis. From a practical viewpoint, Lauder probably understands these stresses well: he has observed the chairlift in action. However, he probably does not understand how to measure shear forces or how metal breaks down.

5. *Attitudes.* In the conversation we've been analyzing, both participants have a serious attitude toward the topic being discussed. They concentrate on the technical and practical aspects of a potential problem. Another factor lies in their degree of respect for each other. As it turns out, both are in their late thirties with over 15 years' experience in their fields. Each is aware of the other's expertise, so each chooses words carefully.

   Actually, our attitudes are usually shown non-verbally. In the McCrae–Lauder conversation, the most likely channels revealing their attitudes toward the situation and each other would be posture, facial expressions, and tone of voice.

6. *Emotions.* You have experienced many situations where your emotions have affected how you spoke or how you listened. Indeed, one's emotions can totally block effective communication. In business, we should not allow that to happen, and it's not likely to happen in the conversation between Lauder and McCrae.

Reasons for communication breakdown

As this chapter has hinted, communication can break down for many reasons:

◆ poor choice of channels
◆ receiver inattentiveness
◆ poor sender encoding
◆ lack of knowledge in sender or receiver
◆ conflicting roles or contrasting personal needs

Feedback can help communication

However, most misunderstandings can be prevented by timely and useful *feedback*, which is very important for two-way channels like face-to-face conversations, telephone calls, and emails. Let's look at that feedback loop next (see page 26).

Feedback could be particularly useful in the McCrae–Lauder meeting—Lauder could immediately signal that he didn't understand some aspect of the message he heard, and McCrae could use different terminology or a different order of explanation to convey her message.

While two-way channels provide timely, direct feedback, written channels (letters, emails, reports) do not. That's why you should consider the receiver's role, needs, and knowledge level when you choose the content, structure, and style of your written messages.

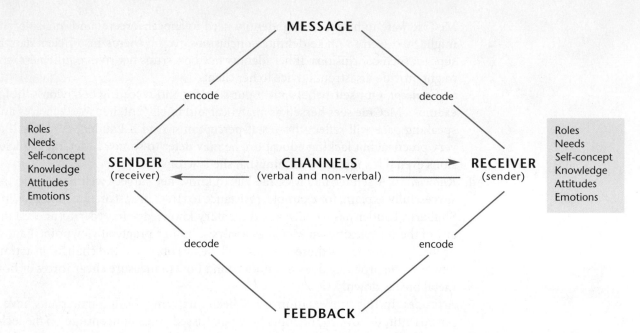

Environmental noise can be distracting

The above model does not include the impact of the *environment* in which the communication occurs. The *physical environment* can introduce audio or visual "noise." Some people can receive and send messages in noisy environments; others cannot. Employers are now recognizing that the noise generated by phone rings, fax machines, printers, speakerphones, and email prompts, creates undue stress for many employees. Consequently, a number of employers are taking steps to reduce and/or mask noise in the workplace (Johne). Visual stimuli, such as loud wallpaper, bright and shiny walls, too many pictures, a constant flow of people passing by, bright lighting, or superfluous computer monitors are sources of additional noise.

Social conventions affect sending behaviours

The *social environment* further affects how we act and how we interpret messages. For example, we use a lot more slang in conversations with friends than we do in business meetings.

Senders and receivers should consider each other's cultural filters

Another important environmental consideration is the *cultural context*. Friends in social settings usually share culturally biased perceptual patterns. But the workplace, especially in Canada's cultural mosaic, can present a bewildering array of belief systems, non-verbal communication behaviours, and ways of interpreting messages. Many business encounters misfire when sender and receiver interpret their exchange through different cultural filters. (The guidelines on pages 36–37 provide advice on communicating with people from different cultural backgrounds.)

Small group communication is more complex than dyadic communication

It should be noted that *small group communication* (3 to 15 people) is quite different from *dyadic communication* (one on one). Groups provide significantly more opportunities for interaction than two-person dyads do. For example, a three-person group has potential for four sets of interaction as against just one potential interaction in a dyad. A four-person group produces 10 potential interaction sets. Increasing group complexity enhances the potential not only for information sharing but also for disagreement and conflict (Barker et al.).

Small groups always contain an audience or observer, unlike one-on-one dyads, and speakers often modify their comments when they're aware of being observed. When they do consider the others in the group, speakers reveal a persuasive motive—in the small group, majority coalitions can form, and alliances may therefore be more important than evidence or other forms of logical persuasion.

Now, let's look at assessing an audience's information needs.

ASSESS THE AUDIENCE'S INFORMATION NEEDS

Good writing connects with an audience by recognizing its different backgrounds, needs, and preferences. A single message may appear in several versions for several audiences. For instance, an article describing a new cancer treatment might appear in a medical journal read by doctors and nurses. A less technical version might appear in a medical textbook read by medical and nursing students. An even simpler version might appear in your local newspaper. All three versions treat the same topic, but each meets the needs of a different audience.

Technical writing is intended to be used. You become the teacher, and the reader becomes the student. Because your audience knows less than you, it will have questions, including the following.

TYPICAL AUDIENCE QUESTIONS ABOUT A DOCUMENT

- Why should I read this document?
- What is the purpose of this document?
- What information can I expect to find here?
- What happened, and why?
- What action should be taken, and why?

- What are the risks?
- When will the action happen?
- How much will it cost?
- Do I need to respond to this document? If so, how?

Always write to enable a specific audience to grasp the information and follow the discussion. To be useful, the writing must connect with the audience's level of understanding.

IDENTIFY LEVELS OF TECHNICALITY

When you write for a close acquaintance (friend, computer crony, co-worker, instructor, or supervisor), you know a good deal about that person's background. You deliberately adapt your document to his or her knowledge, interests, and needs. But sometimes you write for less-defined audiences, particularly when the audience is large (e.g., when you are writing a journal article, a computer manual, a set of first-aid procedures, or a report of an accident). Even though you have only a general notion about your audience's background, you must decide whether your document should be *highly technical*, *semi-technical*, or *non-technical*, as depicted in Figure 2.1.

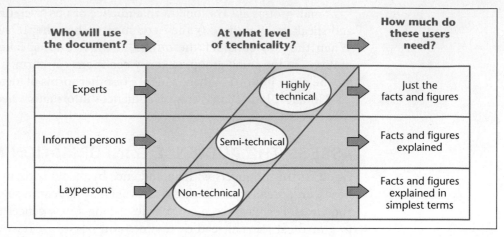

Figure 2.1 Deciding on a Document's Level of Technicality

The Highly Technical Document

Readers at a specialized level expect the technical facts and figures they need, without long explanations. The following report of treatment given to a heart attack victim is highly technical. The writer, an emergency room physician, is reporting to the patient's doctor. This reader needs an exact record of the patient's symptoms and treatment.

CASE	**A Highly Technical ECG Document**

Expert readers need only the facts and figures, which they can interpret for themselves

The patient was brought to the emergency room by ambulance at 0100 hours, September 27, 2013. The patient complained of severe chest pains, dyspnea, and vertigo. Auscultation and ECG revealed a massive cardiac infarction and pulmonary edema marked by pronounced cyanosis. Vital signs: blood pressure, 80/40; pulse, 140/min; respiration, 35/min. Lab: WBC, 20 000; elevated serum transaminase; urea nitrogen, 60 mg%. Urinalysis showed 4+ protein and 4+ granular casts/field, indicating acute renal failure secondary to the hypotension.

The patient received 10 mg of morphine stat, subcutaneously, followed by nasal oxygen and D5W intravenously. At 0125 the cardiac monitor recorded an irregular sinus rhythm, indicating left ventricular fibrillation. The patient was defibrillated stat and given a 50 mg bolus of Xylocaine intravenously. A Xylocaine drip was started and sodium bicarbonate administered until a normal heartbeat was established. By 0300, the oscilloscope was recording a normal sinus rhythm.

As the heartbeat stabilized and cyanosis diminished, the patient received 5 cc of heparin intravenously, to be repeated every six hours. By 0500 the BUN had fallen to 20 mg% and vital signs had stabilized: blood pressure, 110/60; pulse, 105/min; respiration, 22/min. The patient was then conscious and responsive.

This highly technical report is clear only to the medical expert. Because her reader has extensive background, this writer defines no technical terms (*pulmonary edema, sinus rhythm*). Nor does she interpret lab findings (*4+ protein, elevated serum transaminase*). She uses abbreviations her reader understands (*WBC, BUN, D5W*).

Because her reader knows the reasons for specific treatments and medications (*defibrillation*, *Xylocaine drip*), she includes no theoretical background. Her report answers concisely the main questions she can anticipate from her reader: *What happened? What treatment was given? What were the results?*

The Semi-technical Document

One broad class of readers may have some technical background but less than the experts. For instance, first-year medical students have specialized knowledge, but not as much as second-, third-, and fourth-year students. Yet students in all four groups could be considered semi-technical readers. When you write for a semi-technical audience, identify the *lowest* level of understanding in the group and write to that level. Too much explanation is better than too little.

Here is a partial version of the earlier medical report. Written at a semi-technical level, it might appear in a textbook for first-year medical or nursing students, in a report for a medical social worker, in a patient's history for the medical technology department, or in a monthly report for the hospital administration.

| CASE | **A Semi-technical Document** |

Informed but non-expert readers need enough explanation to understand what the facts mean

Examination by stethoscope and electrocardiogram revealed a massive failure of the heart muscle along with fluid buildup in the lungs, which produced a cyanotic **discolouration of the lips and fingertips from lack of oxygen**.

The patient's blood pressure at 80 mm Hg (systolic)/40 mm Hg (diastolic) was **dangerously below its normal measure of 130/70**. A pulse rate of 140/minute was **almost twice the normal rate of 60–80**. Respiration at 35/minute was more than **twice the normal rate of 12–16**.

Laboratory blood tests yielded a white blood cell count of 20 000/cu mm (normal value: 5000–10 000), **indicating a severe inflammatory response by the heart muscle**. The elevated serum transaminase enzymes **(produced in quantity only when the heart muscle fails)** confirmed the earlier diagnosis. A blood urea nitrogen level of 60 mg% (normal value: 12–16 mg%) indicated **that the kidneys had ceased to filter out metabolic waste products**. The 4+ protein and casts reported from the urinalysis (normal value: 0) **revealed that the kidney tubules were degenerating as a result of the lowered blood pressure**.

The patient immediately received morphine **to ease the chest pain**, followed by oxygen **to relieve strain on the cardiopulmonary system**, and an intravenous solution of dextrose and water **to prevent shock**.

This semi-technical version explains (in boldface) the raw data. Exact dosages are not mentioned because the readers are not treating the patient. Normal values of lab tests and vital signs, however, make interpretation easier. (Expert readers would know these values.) Knowing what medications the patient received would be especially important to the lab technician because some medications affect test results. For a non-technical audience, however, the message needs further translation.

The Non-technical Document

Readers with no specialized training expect technical data to be translated into terms they understand. Non-technical readers are impatient with abstract theories but want enough background to help them make the right decision or take the right action. They are bored by long explanations but frustrated by bare facts not explained or interpreted. They expect a report that is clear on first reading, not one that requires review or study.

The following is a non-technical version of our medical report. The physician might write this version for the patient's spouse who is overseas on business, or as part of a script for a documentary film about emergency room treatment.

CASE	**A Non-technical Document**

General readers need everything translated into terms they understand

Heart sounds and electrical impulses both were abnormal, **indicating a massive heart attack caused by failure of a large part of the heart muscle**. The lungs were swollen with fluid and the lips and fingertips showed a bluish discolouration from **lack of oxygen**.

Blood pressure was dangerously low, **creating the risk of shock**. Pulse and respiration were **almost twice the normal rate, indicating that the heart and lungs were being overworked** in keeping oxygenated blood circulating freely.

Blood tests confirmed the heart attack diagnosis and **indicated that waste products usually filtered out by the kidneys were building up in the bloodstream. Urine tests showed that the kidneys were failing as a result of the lowered blood pressure**.

The patient was given medication to ease the chest pain, oxygen to ease the strain on the heart and lungs, and intravenous solution to prevent the blood vessels from collapsing and causing irreversible shock.

This non-technical version explains (in boldface) the situation using everyday language. It omits any mention of medications, lab tests, or normal values because these have no meaning to readers. The writer merely summarizes events and explains the causes of the crisis and the reasons for the particular treatment.

In some other situation, however (say in a jury trial for malpractice), the non-technical audience might need information about specific medication and treatment. Such a report would, of course, be much longer—a short course in emergency coronary treatment.

Each version of the medical report is useful *only* to readers at a specific level. Doctors and nurses have no need for the explanations in the two latter versions, but they do need the specialized data in the first. Beginning medical students and paramedics might be confused by the first version and bored by the third. Non-technical readers would find both the first and second versions meaningless.

Primary and Secondary Audiences

Whenever you prepare a single document for multiple audiences, classify each audience as *primary* or *secondary*. Usually, the primary audience consists of those who requested the document and who will use it as a basis for decisions or actions. The secondary audience consists of those who will carry out the project, who will advise the primary audience, or who will somehow be affected by the decisions of the primary audience. The secondary audience will read your document (or perhaps only part of it) for information that will help it get the job done, for educated advice, or to keep up with new developments.

Often these two audiences differ in technical background. The primary audience may require highly technical messages, and the secondary audience may require semi-technical or non-technical messages—or vice versa. When you must write for audiences at different levels, follow these guidelines:

How to tailor a single document to multiple audiences

◆ If the document is short (a letter, memo, or anything under two pages), rewrite it at various levels.

◆ If the document exceeds two pages, address the primary audience. Then provide appendices for the secondary audience (technical appendices when the secondary audience is technical, or non-technical versions when it is not). Letters of transmittal, informative abstracts, and glossaries are other supplements that help a non-specialized audience understand a highly technical report.

The document in the next scenario must be tailored to both primary and secondary audiences.

CASE	**Tailoring a Document to Different Audiences**

When Daphne McCrae writes the results of her tests on potentially damaged cast-metal chairlift grips, her primary audience will be the client, Trendmark Ski Resort. That audience will include Max Lauder, the maintenance supervisor who has some technical knowledge of metal and metal fatigue. It will also include Trendmark's management board and its lawyer, all of whom have little or no technical knowledge.

Daphne's report may eventually have legal implications, so it must be presented in meticulous detail. But her non-specialist readers will need explanations of her testing methods. These readers will also need photos of the test equipment to fully understand the processes involved. The report will have to define specialized terms, such as *fractographs* (microscopic photographs of fractured surfaces), and *HSLA* (high-strength, low-alloy) steel, such as ASTM grade A 242 (which has 0.4% copper alloyed to the steel to provide greater weathering resistance).

The report's secondary audience will include Daphne's supervisor and outside consulting engineers who may evaluate Daphne's test procedures and assess the validity of her findings. Consultants will focus on various parts of the report to verify that Daphne's procedure has been exact and faultless. For this audience, she will have to include appendices spelling out the technical details of her analysis: *how* light-microscopic fractographs revealed the presence and direction of fractures, and *how* the pattern of these fractures indicates a casting flaw in the original grip, not torsional fatigue. Finally, Daphne must present the technical details of her finding that only one grip has the casting flaw and that the other grips are safe for operation.

In McCrae's situation, the primary audience needs to know *what her findings mean*, whereas the secondary audience needs to know *how she arrived at her conclusions*. If she serves each group's needs, her information will be worthwhile.

DEVELOP AN AUDIENCE/PURPOSE PROFILE

When you write for a particular reader or defined group of readers, you can focus sharply on your audience by using an analytical tool, like the Audience/Purpose Profile Sheet in Figure 2.2. This form can also be used to identify and analyze a variety of *secondary* audiences, whose varied roles, self-perceptions, attitudes toward you and your role, attitudes toward the document's subject matter, and emotional states could produce diverse reactions to your document.

Audience Characteristics

Identify people in the primary audience by name, job title, and specialty (e.g., Martha Jones, Director of Quality Control, B.S. and M.S. in mechanical engineering). Are they superiors, colleagues, or subordinates? Are they inside or outside your organization? What is their likely attitude toward this topic? Are they apt to accept or reject your conclusions and recommendations? Will you present good or bad news? How might their cultural backgrounds affect their expectations and interpretations?

Also identify people in the secondary audience who are interested in or affected by your document or who might affect the primary audience's perceptions or use of your document.

Purpose of the Document

Learn why people want the document and how they will use it. Do they merely want a record of activities or progress? Do they expect only raw data, or conclusions and recommendations as well? Will they act immediately on the information? Do they need step-by-step instructions? Will the document be read and discarded, filed, published, or distributed electronically? In your audience's view, *what* is most important? What purpose should this document achieve?

> ## ON THE JOB...
>
> **Varying Technical Knowledge**
>
> *"One of the cases I was involved in had seven engineers and seven insurance companies with their lawyers and adjusters. Our moderator was an engineer. We engineers understood each other, but the other people in the room were very frustrated. We could have simplified for their benefit, but some of the analysis had to be explained through engineering math. The nuances of the failure analysis would have been lost if expressed in lay terms. Also, we had to go into detail when challenged by another engineer. . . ."*
>
> —**Tom Guenther, consulting structural engineer**

Audience's Technical Background

Colleagues who speak your technical language will understand raw data. Supervisors responsible for several technical areas may want interpretations and recommendations. Managers who have limited technical knowledge expect definitions and explanations. Clients with no technical background expect versions that spell out what the facts mean to *them* (to their health, pocketbook, or business prospects).

Audience's Knowledge of the Subject

Do not waste time rehashing information your audience already has. Readers expect something *new* and *significant*. Writing has informative value[1] when it (1) conveys knowledge that will be new *and* worthwhile to the intended audience; (2) reminds the audience about something they know but ignore; or (3) offers fresh insight about something familiar.

---

1. Adapted from Kinneavy.

AUDIENCE/PURPOSE PROFILE

Audience Identity and Needs

Primary audience: _____ *(name, title)*

Secondary audience: _____

Relationship: _____ *(client, employer, other)*

Intended use of document: _____ *(perform a task, solve a problem, other)*

Prior knowledge about this topic: _____ *(knows nothing, a few details, other)*

Additional information needed: _____ *(background, only bare facts, other)*

Probable questions: _____

_____

_____

_____

_____

_____

Audience's Probable Attitude and Personality

Attitude toward topic: _____ *(indifferent, skeptical, other)*

Probable objections: _____ *(cost, time, none, other)*

Probable attitude toward this writer: _____ *(intimidated, hostile, receptive, other)*

Persons most affected by this document: _____

Temperament: _____ *(cautious, impatient, other)*

Probable reaction to document: _____ *(resistance, approval, anger, guilt, other)*

Risk of alienating anyone: _____

_____

Audience Expectations about the Document

Reason document originated: _____ *(audience request, my idea, other)*

Scope: _____ *(comprehensive, concise, other)*

Material important to this audience: _____ *(interpretations, costs, conclusions, other)*

_____

Most useful arrangement: _____ *(problem-causes-solutions, other)*

Tone: _____ *(businesslike, apologetic, enthusiastic, other)*

Intended effect on this audience: _____ *(win support, change behaviour, other)*

Due date: _____

Figure 2.2 Audience/Purpose Profile Sheet

The informative value of any document is measured by its relevancy to the writer's purpose and the audience's needs. As a member of this text's audience, for instance, you expect to learn about technical writing, and our purpose is to help you do so. In this situation, which of these statements would you find useful?

1. Technical writing is hard work.
2. Technical writing responds to a specific situation. In the writing process, you carefully choose and present information, analysis, and supporting materials that meet your audience's needs *and* serve your own purpose.

Statement 1 offers no news to anyone who has ever picked up a pencil, and so it has no informative value for you. But statement 2 offers a new perspective on something familiar. No matter how much you might have struggled through decisions about punctuation, organization, and grammar, chances are you haven't viewed writing as entailing the critical thinking discussed above. Because statement 2 provides new insight, you could say it has informative value.

The more non-essential information an audience receives, the more it is likely to overlook or misinterpret the important material. Take the time to determine what your audience needs, and try to give it just that.

Appropriate Details and Format

The amount of detail in your document (*How much is enough?*) will depend on what you have learned about your audience's needs. Were you asked to "keep it short" or to "be comprehensive"? Can you summarize, or does everything need spelling out? What length will your audience tolerate? Is the primary audience most interested in conclusions and recommendations or all of the details? What has been requested? A letter, a memo, a short report, or a long, formal report with supplements (title page, table of contents, appendices, and so on)? What kinds of visuals (charts, graphs, drawings, photographs) make this material more accessible? What level of technicality will connect with the primary audience?

For example, which level of technicality in the following example means more to you? Which level would be more appropriate for an automotive sales brochure?

High Technicality	The diesel engine generates 10 BTUs per gallon of fuel as opposed to the conventional gas engine's 8 BTUs.
Low Technicality	The diesel engine yields 25 percent better fuel mileage than its gas-burning counterpart.

Due Date

Does your document have a deadline? Workplace documents almost always do. Allow plenty of time to collect data, to write, and to revise. Whenever possible, ask the primary audience to review an early draft and to suggest improvements.

Audience's Cultural Background

Some information needs are culturally determined. For example, some cultures might value thoroughness and complexity above all: lists of data, with every relevant detail included and explained. Other cultures might prefer an overview of material, with multiple perspectives and the liberal use of graphics (Hein 125–26).

Canadian business culture generally values plain talk that spells out the meaning directly, but some cultures prefer indirect and somewhat ambiguous messages, which leave explanations and interpretation for readers to decipher (Leki 151; Martin and Chaney 276–77). To avoid seeming impolite, some readers might hesitate to request clarification or additional information. Even disagreement or refusal might be expressed as "We will do our best" or "This is very difficult," instead of "No"—to avoid offending and to preserve harmony (Rowland 47).

Correspondence practices vary from culture to culture. In British business letters, for example, the salutation is followed by a comma (Dear Ms. Morrison,); in Canada and The United States, it is followed by a colon (Dear Ms. Morrison:). Also, European data formats vary from Canadian and American practices, as Table 2.1 illustrates.

Table 2.1 Differing Data Formats

	Canada/United States	United Kingdom	Germany
Long Distance Calls	1-250-555-8780	Intl. code + country + city or region (011) 44 1344 771615 (UK)	Intl. code + country + city or region + local number (011) 49 1244 40 35 16 15 (Ger)
Address and Postal Code	493 Marmot Court Vernon BC V1B 3R8 73 Melville Lane Ramona CA 92065	726 Abingdon Road Sandhurst, Berkshire, UK GU47 9RP	Stoekhartstrasse 33 20146 Hamburg Deutschland
Time	The 12-hour system is preferred (9:15 p.m.), except for the military, health services, and emergency services (2115).	The 12-hour system (9:15 p.m.) is preferred, except for documents going to other European nations (21:15).	The 24-hour system is preferred in most of Europe: 21.15 Uhr or Uhr 21.15.
Date	June 15, 2011, or 15 June 2011 15/6/11 (Can.) 6/15/11 (U.S.)	15th June 2011 15/6/11	15 Juni 2011 15.06.11
Currency	$225.95 CDN$225.95 US$225.95	The decimalized pounds sterling is used in England, Scotland, Wales, and Northern Ireland. £225.95 = 225 pounds, 95 pence	The euro is used in Germany and 15 other European Union nations. € 225,95 = 225 euros, 95 cents)
Large Numbers	2,373,866.99 OR 2 373 866.99	2,373,866.99	2 373 866.99

Source: Partly based on European Commission Directorate-General for Translation, *English Style Guide*, 5th ed., rev. April 2010. Web. 20 May 2010. **http://ec.europa.eu/translation/writing/style_guides/english/style_guide_en.pdf**.

GUIDELINES FOR INTERCULTURAL COMMUNICATION

The following advice may help with intercultural communication. Although it seems aimed at our encounters with people outside Canada, our own culture is not homogeneous, so the advice also applies to inter-Canadian communication.

1. Where feasible, hire a translator to convert your proposal or report to the *buyer's* language. Where that's not feasible, or where your reader prefers to receive your document in English, keep your sentences and paragraphs short. Use direct, simple, precise words. Use relative pronouns such as *that*, *which*, and *who* to introduce clauses. Avoid technical jargon and idioms (e.g., "hit the ceiling," "A-OK," "par for the course").

2. Pay special attention to the openings of letters and memos and to the introductions of reports—Many Canadians like to get straight to the point; other cultures may prefer to build relationships first.

3. When writing recommendations, consider whether your client's culture favours careful, deliberate team consultation or quick, individual action.

4. Look at the physical format of reports produced in the reader's culture, and incorporate some aspects of that format in your report.

5. In oral communication, use perception-checking to learn whether your message has been interpreted the way you hoped: (1) pay attention to the receiver's reaction as you see and hear it (e.g., is there silence and little facial expression?); (2) ask yourself what this tells you (e.g., has the receiver understood and assented?); and (3) confirm your perception (it may well be inaccurate if your receiver comes from a culture that values silence or is reluctant to admit a lack of understanding).

Certain factors affect each method of communication

6. Remember that much of what we communicate orally is conveyed non-verbally, and interpretations of non-verbal cues vary widely from culture to culture, even within the same nation. For example, many Canadians may see eye contact as an indicator of openness and friendliness, but some Canadians do not like to make eye contact with authority figures, such as teachers, supervisors, or coaches.

Consider the meaning of non-verbal cues

7. In both writing and speaking, keep in mind Edward T. Hall's differentiation of "low-context cultures" and "high-context cultures"—American, German, and Scandinavian cultures are at the low-context end, and Arab, Chinese, Japanese, and Mexican cultures are at the high-context end. British and French cultures are placed in the middle of the continuum. People from low-context cultures value individualism and direct, explicit communication, while high-context cultures prefer collectivist approaches and indirect, implicit communication that relies on shared understandings and non-verbal cues.

   Individualists tend to prefer linear, analytical thinking that shows how various parts relate to each other. However, collectivists from high-context cultures like to use synthetic thinking, which favours holistic approaches—elements are seen as part of the larger whole (E. Hall; Gudykunst).

(continued)

8. In dealing with people from other cultures, avoid *ethnocentrism*—the tendency to believe that the cultural practices, standards, and values of one culture are superior to those of other cultures. Ethnocentrism often leads to misunderstandings and conflict, especially when all communicators believe their culture is superior.

EXERCISES

1. Using internet or print sources, choose a piece written at the highest level of technicality you understand and then translate the piece for a layperson, as in the example on page 30. Exchange translations with a colleague from a different major. Read your colleague's translation and write a paragraph evaluating its level of technicality. Submit to your instructor a copy of the original, your translated version, and your evaluation of your colleague's translation.

2. Assume that a new employee is taking over your job because you have been promoted. Identify a specific problem in the job that could cause difficulty for the new employee. For the employee, write instructions for avoiding or dealing with the problem. Before writing, create an audience/purpose profile by answering (on paper) the questions on page 33. Then brainstorm for details. Submit to your instructor your audience/purpose analysis, brainstorming list, and instructions.

COLLABORATIVE PROJECT

Write a survival guide aimed at students coming to your college or university from a specific country. Discuss housing, food, campus student resources, how to get the most from one's professors, local entertainment, aspects of campus life, and healthy living choices, among other topics. Note that the internet has plenty of sites giving advice that seems aimed at Canadians. You will need to adapt this advice to consider your audience's knowledge of your city, knowledge of Canadian customs, food preferences, attitudes toward study and the professor–student role set, and familiarity with Canadian idioms and the English language itself.

MyWritingLab

MyWritingLab: Technical Communication offers the best multimedia resources for technical communication in one, easy-to-use place. You can find a variety of interactive model documents and case studies. There are also extensive guidelines, tutorials, and exercises for Document Design, Writing and the Writing Process, and Research, and a large bank of diagnostics and practice for grammar review.

3

Writing Efficiently

Explore

Click here in your eText to access a variety of student resources related to this chapter.

LEARNING OBJECTIVES

After reading this chapter, you should be able to

- understand that an organized writing process leads to quickly produced, effective documents
- identify and use the stages of an efficient writing process
- write first drafts that require minimal editing
- apply these efficiencies to collaborative writing projects

At work, writers need to produce effective documents quickly. Most employers will not tolerate inefficient work habits, including writing habits. Here are three scenarios illustrating that it's just as important to write efficiently as it is to create effective documents.

Bill, a Halifax civil engineering technologist, returns to his office from a site inspection. As Bill sits down at his desk, his office manager tells him that he must write a proposal that afternoon for a soil-testing contract. Bill checks his watch; he has two hours to gather relevant data from company files and produce a two-page proposal before he has to catch a plane for a company meeting in Toronto.

Bill spends 20 minutes gathering, choosing, and arranging the data and supporting arguments. Then, working from a standard proposal structure for soil-testing contracts, he composes a 550-word proposal on his personal computer in 35 minutes. He spends another 15 minutes polishing the document and sends it to two colleagues for proofreading. They find four mechanical errors and two minor errors in logic. Bill corrects the errors and prints three copies of the proposal; he takes two copies to the office manager. In total, Bill produces the proposal in 90 minutes.

George works for a national research organization in Saskatoon. He has a master's degree in chemistry and a doctorate in biology. His company bills his services for $85 an hour, a rate that his company's clients are glad to pay because George's research methods, data, and analyses are thorough and accurate. His reports feature clear structures and phrasing.

However, George is required to take a three-day technical writing course because typically it takes him a full workweek (plus his own time in the evenings) to produce a project-completion report that other researchers could produce in half the time.

Rita has a background in electronics and computers; she has a 10-month contract to write software documentation manuals for an Ottawa-based software development company. The company is very pleased with the quality of Rita's work, but its director of development, Martin Lefebvre, has told Rita that her contract will be renewed only if she can decrease the time it takes her to produce a manual. Martin suggests a minimum improvement of 25 percent in writing efficiency.

Bill is an efficient writer. George and Rita are not, partly because both have perfectionist tendencies, but more because each has learned bad writing and time-wasting habits. Rita, for example, writes through a discovery method that requires several complete rewrites of each document. George's main problem is that he frequently reorganizes his reports as he composes drafts and thus wastes time rewriting whole sections in order to make the material read smoothly.

Bill writes more efficiently than Rita and George because Bill has learned to

- identify several related but separate writing tasks
- focus on one task at a time and perform each task well
- identify the best sequence for completing the various writing tasks
- reduce writing time by starting quickly and by writing a first draft that requires relatively little revision

Efficient writers such as Bill have learned, primarily through trial and error, to use a writing process that is broken down into the stages shown in Figure 3.1.

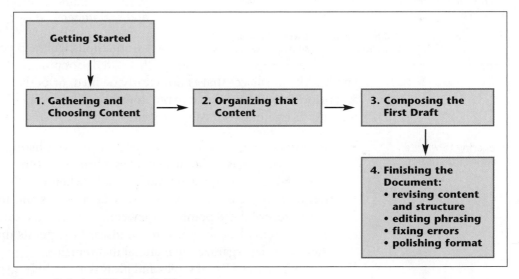

Figure 3.1 An Efficient Writing Process

GETTING STARTED

Figure 3.1 shows the first stage, Getting Started, as preliminary to the other four stages. Technical and business writers benefit from immediately asking, *What does my audience need and expect?* and *What purpose am I trying to fulfill?* (The audience/ purpose profile described in Chapter 2 provides an excellent starting point for people who write business correspondence and technical documents.)

Technical writing differs from essay writing. Essayists often have to discover their subject as they progress. This process of writing until the writer discovers what she or he really wants to say is called *free-writing*; it may be necessary in some cases, but it results in many crumpled pages and in lengthy writing sessions.

Rewriting draft after draft is simply not necessary for most business and technical writing projects. Nor do you have to wait for that first "golden phrase" to start the river of words flowing. As a technical/business writer, you can be productive within 30 seconds of sitting down to write. Here's how:

> ## ON THE JOB...
>
> ### A Productive Writing Process
>
> *"These days, I do more editing than writing. But when I did write a lot, I started off with a very detailed outline, which I had the client approve. That way, if I misunderstood a concept, it was straightened out before the actual writing. I always researched the topic quite thoroughly, so I understood it clearly. I also wrote (even long pieces) by hand. My first cuts happened as I typed the first draft. This process seemed to work: rewrites were usually minimal. All the upfront work paid off!"*
>
> **—Judith Whitehead, writer and editor,**
> **Vankellers Editorial and Writing Services**

- *Use an audience/purpose profile* to determine the types of information and analysis to include. List the types of questions that your audience would ask (or that your audience has already asked);
- *Choose an appropriate, proven structure* and then "fill in the blanks." Several of this text's chapters suggest structures for frequently written documents; progress reports, proposals, feasibility reports, and application letters are among those described. Use elements of an audience/purpose profile to modify the suggested structure to meet your audience's needs and preferences.

These two methods will help you start writing almost any job-related document. However, occasionally you'll tackle a subject that is not clearly defined, or perhaps your audience's needs and priorities will be difficult to pin down. For "open-ended" situations, here are two methods for getting your mind in gear:

Preparing to write a product description

1. *Brainstorm a list of ideas and topics.* A random listing of possible topics and ideas works precisely because it takes advantage of the natural chaos that exists in our minds. Often, we are most creative when we allow free thought association to generate a series of loosely related points and topics. It's important to simply record these points as they come and not to edit them. Later, when the creative frenzy has abated, you can discard the points that don't seem relevant. Then, you can organize the material that remains.

   Let's look at a list of topics and ideas that might be generated by the writer of an in-house product description of a new smartphone scanner. (The writer is part of a design team at GlobeTech that has recently redesigned a smartphone in response to a request from GlobeTech's general manager to produce a new consumer product.)

   The writer, Thanh Pham, hasn't previously written such a product description. Also, the product is quite different from the standard phones manufactured by his company, so he doesn't have a model to emulate. Here's Pham's list of ideas and topics for his description of the new "Smart Scan":

   - limitation of previous scanners: inadequate for scanning images in the field
   - solution: appearance and size; weight; features and controls
   - how the new device works (an overview): (1) the digital process and (2) image storage—emailed to office computer; on-board storage

◆ downloading images to a PC; software needed to link phone to PC
◆ performance specs: resolution, size of scanned images, storage capacity

Notice that Pham's thoughts contain some order and "connections," even though he was just letting the ideas "flow." The first items in the list reflect a problem–solution sequence, and the last three items a chronological pattern.

2. *Brainstorm ideas in a cluster diagram.* Cluster diagramming suits people who think visually. It also suits those who are used to following hyperlinks through the internet. Here's how Pham could use clustering to generate ideas for his product description. He could

◆ circle the main topic ("Smart Scan") in the centre of a clean page
◆ record any ideas that pop into his mind and circle those ideas
◆ avoid censoring ideas, but simply record them
◆ join related ideas with lines, but do not focus on these connections—instead, keep on recording ideas until the flow stops

The resulting diagram might look like Figure 3.2:

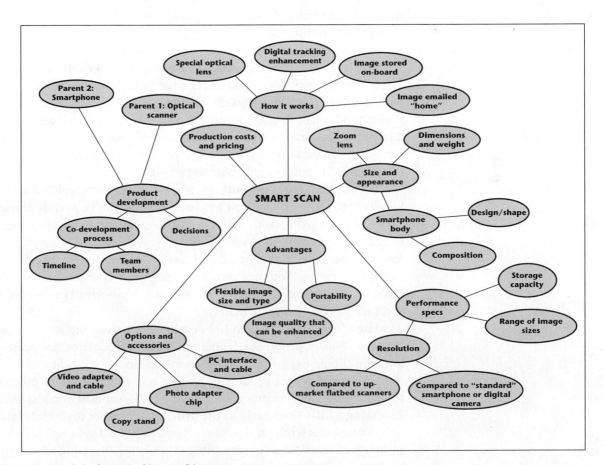

Figure 3.2 Sample Cluster Diagram

HOW TO SAVE WRITING TIME

Whether you have generated ideas by brainstorming or by one of the more structured approaches described earlier, you are now well underway and can complete these four writing tasks in turn:

1. *Choose the content*, based on

    ◆ technical or research notes
    ◆ personal observations
    ◆ arguments and evidence
    ◆ deductions and conclusions
    ◆ available illustrations

    Then, check to see if anything has been omitted, or if any material should be deleted. **Revising content at this point takes less time than revising material later in the process.**

2. *Organize the blocks of material.* If you're writing a letter, decide what goes in each paragraph. If you're writing a longer document, start with larger blocks of material and work your way to the paragraph level. Next, organize the ideas and information within each block.

    Again, determine whether any material should be added, rearranged, or omitted; **such changes take much more time after a draft's been phrased**.

3. *Write the first draft.* Since you know what to include and where to place it, you will be able to concentrate on the best way to phrase each sentence, and you will understand where to inject transitional statements. If you have done steps 1 and 2 properly, this draft will be close to a finished product. The very act of phrasing sentences can sometimes change your perception of your message, so you might have to modify your writing plan as you go.

    By working from an outline, you will save composition time. After all, you have to think only of how to phrase things; you've already chosen the content and arranged it. Your mind is free to concentrate on phrasing; you're not burdened by three writing tasks at once.

4. *Evaluate, revise, edit, proofread, and correct.* Wait as long as possible before polishing the writing. You may gain a new perspective on the best way to structure and express parts of the document. Also, you'll proofread more effectively if you distance yourself from the material.

    Use objective indexes to help evaluate your writing and to make it more readable. Finally, proofread the material at least three times to locate errors. Ideally, you should use two proofreaders in addition to yourself.

    Compared with a process of writing and rewriting (and more rewriting), this four-stage writing process will save you a great deal of revision and editing. **Investing a little time early in the process pays large dividends later on.**

    Writing efficiently in the manner just described also will help you produce more *effective* writing; concentrating on one task at a time allows you to better perform each of those tasks.

COMPOSING WITH A WORD PROCESSOR

If you type as fast as you handwrite and if you're accustomed to reading a computer screen, consider using a word processor to plan and compose the first draft of a document. Composing with a computer has several advantages:

◆ *Brainstorming* lists of ideas suits word processing, especially if you type quickly.

◆ Choosing content is easy to do—you can add, delete, or move points with little effort.

◆ Today's sophisticated word-processing software makes it easy to arrange the *chosen content into an outline*. Most software includes an outliner function that helps you divide topics into main headings and levels of subheadings. The computer tracks levels in the outline so that you can easily add, subtract, or rearrange parts of the outline. The outliner function is especially useful for long, complicated reports.

◆ *Key phrases* can be placed within the outline to represent paragraphs. Later, when you compose each paragraph, each key phrase gets expanded into a topic sentence for its corresponding paragraph. Essentially, then, the computer eliminates the drudgery of retyping headings and key phrases. You simply fill in the paragraphs under each heading or key phrase. (Some writers type several key notes for each intended paragraph and then expand the notes into a series of closely linked sentences. They find that a premeditated list of points for each paragraph helps them write tight, clear paragraphs that require little subsequent revision.)

◆ Word processors help *identify and correct errors in spelling, punctuation, and grammar*. However, automated checkers have many limitations: a synonym offered by an electronic thesaurus may not accurately convey your intended meaning; the spell-checker cannot differentiate between incorrectly used words such as "they're," "their," and "there," or "it's" and "its." And although spelling and grammar checkers help, they cannot evaluate those subtle choices of phrasing that can be very important—no automated checker will tell you whether "you can reach me at . . ." or "call me at . . ." is more appropriate in a given situation.

THE PROCESS IN ACTION

The following situation illustrates how a busy person, who must balance the technical and management components of the job, uses time efficiently to write a proposal to his or her supervisor.

CASE	**Process to Write a Proposal**

The company is MMT Consulting, an engineering firm with its headquarters in Calgary and with branch offices in Sudbury, Winnipeg, Edmonton, Kelowna, and Prince George. MMT specializes in feasibility studies, design projects, and construction management for the mining and petroleum sectors.

Art Basran manages MMT's Kelowna branch. He is responsible for MMT's contract to help the Jackson Mining Company choose a method of hauling coal from Jackson Mining's projected new mine site in the mountains north of Grand Forks, British Columbia. As Art and his staff start to investigate the project, they learn that they need to use specialized accounting methods to evaluate four alternatives for hauling the coal to the railhead.

Art's team, at Jackson Mining's request, examines the proposed route and calculates grade resistances and energy requirements along the route. The team then researches capital costs for constructing road beds and a railroad. Capital costs are also calculated for purchasing diesel trucks, a diesel train, an electric train, and a fleet of electric trains. Finally, annual operating costs are computed for the four options.

After gathering all of the data, Art realizes that he does not have a uniform method of applying the three main evaluation criteria (capital costs, annual operating costs, and potential for expanding the delivery system) to all four transportation options. He consults Brendan Winters, a chartered accountant who works with mining companies. Brendan recommends an accounting vehicle called Equivalent Uniform Annual Costs (EUAC), which combines all the variables and thus allows a uniform comparison of the four haul methods.

Now, Art faces the task of convincing his regional manager, Brenda Backstrom, to accept the EUAC method. Brenda, a conservative thinker, tends to resist new ways of doing things. However, she has accepted her company's newly developed business strategy—MMT presents itself as an innovative, advanced engineering firm.

Art gets started by

- checking notes of his meeting with Brendan Winters
- jotting questions that Brenda Backstrom will have
- brainstorming a list of points to make in explaining EUAC

Time to get started: 10 minutes

Art's next step is to **choose an approach**. He decides to write a direct proposal because he knows that his reader prefers direct communication, not a lengthy analysis that eventually leads to a recommendation. Art also knows that his reader prefers short messages, so Art chooses a memo that will not exceed two pages.

With the format decision made, Art turns to the **choice of content**. His word-processed list includes

- situation more complex than usual—financial considerations
- *solution*: EUAC (Brendan's idea)

- ◆ process of preparing EUAC (see my notes)
- ◆ *note*: we already do 5 of the 6 steps
- ◆ advantages of EUAC
  (1) relatively easy—Brendan's software
  (2) qualitative and quantitative
- ◆ our qualifications: Peter, John, Brendan
- ◆ why our "normal" method won't work: too complex, not reliable, too diffuse
- ◆ request authorization

Before thinking about the order of his material, Art **edits the content** he's listed. He decides that discussing the inadequacies of the "normal" method could be too negative and that it would take the memo beyond two pages, so he deletes that part. He also thinks of a third EUAC advantage (using the EUAC method makes MMT look good), and he adds that point, so his list now includes

- ◆ situation more complex than usual—financial considerations
- ◆ *solution*: EUAC (Brendan's idea)
- ◆ process of preparing EUAC (see my notes)
- ◆ *note*: we already do 5 of the 6 steps
- ◆ advantages of EUAC
  (1) relatively easy—Brendan's software
  (2) qualitative and quantitative
  (3) using this innovative method gives us a competitive advantage
- ◆ our qualifications: Peter, John, Brendan
- ◆ request authorization

Time to choose and edit content: 8 minutes

At this point, Art's writing is interrupted by a phone call, which leads to other conversations between Art and two technologists in the office. After 40 minutes, Art returns to organize the content of his proposal. As he looks at his edited contents list, Art realizes that he has jotted down his points in the order he has successfully used for proposals in the past:

1. reader connection and statement of problem
2. proposed solution
3. description of proposal
4. supporting arguments
5. request for authorization

However, Art also realizes that it would be better to place his team's qualifications *before* his supporting arguments, to establish that his Kelowna team can handle the proposed approach. Using the cut-and-paste function on his word processor, he places his list of points as follows:

- ◆ situation more complex than usual—financial considerations
- ◆ *solution*: EUAC (Brendan's idea)

- ◆ process of preparing EUAC (see my notes)
- ◆ *note*: we already do 5 of the 6 steps
- ◆ our qualifications: Peter, John, Brendan
- ◆ advantages of EUAC
  - (1) relatively easy—Brendan's software
  - (2) merges qualitative and quantitative methods
  - (3) using this innovative method gives us a competitive advantage
- ◆ request authorization

Before using this outline to compose his memo, Art **reviews its structure**. He realizes that he should start his supporting arguments with his strongest point, so he rearranges the EUAC advantages as follows: (1) merges qualitative and quantitative methods; (2) gives us a competitive advantage; (3) Brendan's software—easy to use and inexpensive.

Time to organize content: 3 minutes

Feeling in control of the writing process, Art **composes the first draft**, working quickly from his outline and detailed notes. At that point, Art has to leave for a meeting, which takes the rest of the afternoon. Art's draft appears in Figure 3.3, pages 47–48.

Time to compose the first draft: 33 minutes

His evaluation leaves him satisfied with the content of the proposal, but he sees wordy phrases and some phrases whose tone needs changing. Some of these are shown in the table that appears on page 51.

Art's paring of unnecessary words reduces the word count from 753 to 661 and makes several sentences easier to read. Then, confident that he has established the right tone in the memo, Art puts the document through his software's spell-checker, which detects several errors ("preform," "enginerring," "inovative," "caclations"). But, not trusting either his own proofreading or the spell-checker, Art gives the document to a colleague, who finds a comma splice in paragraph 3 and a pronoun agreement error (". . . advantage of using EUAC comparisons is that they provide . . .") in paragraph 6.

Art's revision, editing, and proofreading time: 18 minutes

Finally, Art takes four minutes to apply some final touch-ups and print the proposal.

Art's finished document is shown in Figure 3.4, on pages 49–50. **His time to produce the memo, including discussions with colleagues, totals 84 minutes.** Every minute of that time has been productive.

Re: Recommending a Coal-Haul Method to Jackson Mining

As you requested, I have begun to prepare a recommendation for Jackson Mining's consideration. Its situation, however, is considerably much more complex than other transportation issues we have analyzed in the past. In this case, we must consider the factor of fluctuating interest rates as well as varying amortization periods for carrying the capital debt. This problem is additionally compounded by not knowing how long Jackson Mining expects to operate the proposed Othello mine and haul coal from it.

After consulting with Brendan Winters, a Penticton C.A. who has a special interest in mining projects, I propose we should use Equivalent Uniform Annual Costs (EUAC) as the main method of comparing Jackson Mining's four alternative transportation methods. EUAC provides a comprehensive, as well as a clear, way of comparing the four alternatives.

The Process

Preparing a comprehensive set of EUAC requires a six-step process. You will note from the following list of steps that we would normally preform the first five steps in this kind of analysis, the sixth step, the actual EUAC calculations, uses data generated during the first five steps of the process. These, then, are the steps:

1. Gather field data about the proposed road and railroad routes. This step will require doing our own surveys as well as gathering data from existing maps and surveys.
2. Research specifications of the diesel trucks and train options that are suitable for the type of terrain found on the proposed routes. We will need physical and mechanical specifications as well as the capital costs for this equipment.
3. Determine the fuel consumption figures for each of the four options. This set of calculations will be quite extensive, particularly in this case, because of the length of the haul road and because of the grade variances on that road.
4. Combine the fuel costs with projected labour and maintenance costs to determine the annual operating costs for each alternative.
5. Determine the capital costs for the four alternatives.
6. Combine the annual operating costs and the capital costs with varying interest rates to produce an EUAC for each of the study periods of one to 20 years. Please see the attached sample table and sample figure, which show the results for three of the 20 years that could be considered. These samples are based on roughly estimated data.

Figure 3.3 Draft Document

(continued)

Qualifications

I believe our team is well equipped to handle this innovative method of assessing Jackson Mining's transportation needs. Peter Bondra, the engineer who will lead the research team, has degrees in both civil and mechanical engineering, and useful experience in railbed construction. His main assistant will be John Housley who graduated two years ago from the Civil Engineering Technology program here in Kelowna. John has 15 years' experience in road construction and knows the area north of Grand Forks because of his frequent hunting trips in that area. In addition, John is well versed in a variety of surveying techniques. He'll be able to deal with the area's rough terrain.

After the data have been collected, Brendan Winters has indicated his willingness to be available to direct the cost analysis. He has particular expertise in calculating and presenting EUAC data.

Advantages of Using the EUAC Comparisons

The main advantage of using EUAC comparisons is that it provides qualitative comparisons as well as quantitative data. In effect, an EUAC creatively merges enginerring analysis with the maximal kind of accounting techniques to help our client see what the data really mean.

Providing this kind of inovative analysis will position our firm as a creative, advanced engineering firm. Brendan Winters and I have recently surveyed a variety of resource-based companies, and not one of them had used the EUAC method. I believe its use will give us a competitive advantage in this bid and in others.

Furthermore, the complicated EUAC caclations will only take an extra day of work on this project because Brendan Winters has modified an accounting software program to help do the computations. His contribution will cost our firm an additional $1200.00, a relatively small part of our budget for this project.

Conclusion

I need to talk to you about using Equivalent Uniform Annual Costs in the Jackson Mining project before the end of this week. If I'm out of the office when you call, call me at my new cellular phone number, (250-863-2999. If you would like a more comprehensive view of how EUAC can be used, I can prepare such a document in three hours and email it to you immediately.

Art Basran

Figure 3.3 Draft Document

MMT Consulting   Inter-office MEMORANDUM

To: Brenda Backstrom   **Date:** December 3, 2014
           Western Regional Manager

From: Art Basran
           Kelowna Branch Manager

Re: Method of Comparing Transportation Alternatives for Jackson Mining's
           Proposed Othello Mine

As you requested, I'm preparing a proposal for Jackson Mining's consideration. Its situation, however, is more complex than other transport issues we have analyzed. In this case, we must consider fluctuating interest rates and varying amortization periods for carrying the capital debt. Also, we don't know how long Jackson Mining expects to operate the proposed Othello mine.

After consulting with Brendan Winters, a Penticton C.A. who has a special interest in mining projects, I propose that we use Equivalent Uniform Annual Costs (EUAC) as the main method of comparing Jackson Mining's four alternative transportation methods. EUAC provides a complete, clear comparison of the four options.

The Process
Preparing a comprehensive set of EUAC requires a six-step process, the first five of which we would normally perform in this kind of analysis; the sixth step, the actual EUAC calculations, uses data generated during the first five steps of the process. Here are the steps:

1. Gather field data about the proposed road and railroad routes. This step will require doing our own surveys as well as gathering data from existing maps and surveys.
2. Research specifications of the diesel trucks and train options that are suitable for the type of terrain found on the proposed routes. We will need physical and mechanical specifications as well as the capital costs of this equipment.
3. Determine the fuel consumption figures for each of the four options. This set of calculations will be quite extensive for the Othello Mine road, which is about 80 kilometres and which has many grade variances.
4. Combine the fuel costs with projected labour and maintenance costs to determine the annual operating costs for each alternative.
5. Determine the capital costs for the four alternatives.
6. Combine the annual operating costs and the capital costs with varying interest rates to produce an EUAC for each of the study periods of one to 20 years. Please see the attached sample table and figure that show the results for three of the 20 years that could be considered. These samples are based on estimated data.

Figure 3.4 An Efficiently Produced Document   *(continued)*

Brenda Backstrom
Date: December 3, 2014   Page 2

Qualifications

Our team is well equipped to handle this innovative method of assessing Jackson Mining's transportation needs. Peter Bondra, the engineer who will lead the research team, has degrees in both civil and mechanical engineering and useful experience in railbed construction. His main assistant will be John Housley, who graduated two years ago from the Civil Engineering Technology program here in Kelowna. John has 15 years' experience in road construction and knows the area north of Grand Forks because of his frequent hunting trips in that area. He'll be able to deal with the area's rough terrain. In addition, John is well versed in a variety of surveying techniques.

After the data have been collected, Brendan Winters will be available to direct the cost analysis. He's an expert in calculating and presenting EUAC data.

Advantages of Using the EUAC Comparisons

The main advantage of using EUAC is that it provides qualitative comparisons as well as quantitative data. EUAC combines engineering and accounting ideas to help our client see what the data really mean. Our competitors have not yet realized the value of EUAC analysis, so I believe its use will give us a competitive advantage in this and future bids.

This kind of innovative analysis helps position MMT as a creative, advanced engineering firm, a strategy that matches the positioning objective established at our recent management summit.

Furthermore, the complicated EUAC calculations will take only one day of work on this project because Brendan Winters has modified some accounting software to do the computations. His work will cost our firm $1200.00, which is just 2.5 percent of our project budget.

Conclusion
May we discuss using EUAC for Jackson Mining by this Friday? If I'm out of the office when you call, please call my new cellular phone, (250) 863-2999. If you would like a fuller explanation of how EUAC can be used, I could email you.

Art Basran

Attachments: Sample Calculations

Figure 3.4 An Efficiently Produced Document

Paragraph	Phrasing	Concern/problem	Improvements
1	Its situation, however, is *considerably much more* complex than other transportation issues we have analyzed in the past. . . . This problem is *additionally compounded by not knowing* . . . operate the proposed Othello mine *and haul coal from it.*	wordy, pompous wordy, pompous	Its situation, however, is more complex than other transport issues we have analyzed. Also, we don't know . . . operate the proposed Othello mine.
2	EUAC provides a comprehensive, *as well as* a clear, way of comparing the four alternatives.	wordy	EUAC provides a complete, clear comparison of the four options.
5	. . . Brendan Winters has indicated his willingness to be available . . .	wordy	. . . Brendan Winters will be available . . .
7	. . . surveyed a variety of resource-based companies . . .	inexact and therefore not persuasive	. . . surveyed 12 mining and forestry companies . . .
9	I need to talk to you about using Equivalent Uniform Annual Costs . . . before the end of this week.	the tone is too aggressive for addressing one's supervisor	May we discuss using EUAC . . . by this Friday?

EXERCISES

1. Use the efficient writing process described on page 39 to complete your next writing assignments. Adapt that process to suit your skills and preferences. For example, you may prefer to complete all stages of the process on a computer. Or, if you type very slowly, you may prefer to handwrite the content list, the outline, and the first draft. Then, you could edit and correct that draft as you key it into the computer.

2. In a one-page memo or email to your writing instructor, describe your writing habits and outline your plans for becoming a more efficient writer. Use your own version of the writing process described on page 39 to create this memo. You should produce a printed memo within 60 minutes!

3. Use the form in Figure 3.5 to audit your writing habits as you complete your next few writing assignments. The time log should give you an idea of how efficiently you write and where you could begin to save time in the writing process.

COLLABORATIVE PROJECT

See this book's accompanying website for a collaborative writing exercise. Go to **MyWritingLab: Technical Communication**, "Writing Exercises," and then "Two Collaborative Writing Projects."

Stage of Writing Process	Time Spent	Methods and Results
Getting started		
1. Gathering/choosing content: information, ideas, and analysis		
2. Organizing content		
3. Phrasing first draft		
4. Revising content and structure Editing style and paragraphs Finding and correcting errors		

Figure 3.5 Time Audit

MyWritingLab: Technical Communication offers the best multimedia resources for technical communication in one, easy-to-use place. You can find a variety of interactive model documents and case studies. There are also extensive guidelines, tutorials, and exercises for Document Design, Writing and the Writing Process, and Research, and a large bank of diagnostics and practice for grammar review.

Collaborating at Work

LEARNING OBJECTIVES

After reading this chapter, you should be able to

- understand the factors that contribute to successful group work
- appreciate the advantages and challenges faced by virtual teams
- identify the skills needed to collaborate with others in producing technical documents
- follow guidelines for conducting successful meetings
- understand the special aspects of conducting virtual meetings

Complex documents (especially long reports, proposals, and manuals) are rarely produced by one person working alone. For instance, an instruction manual for a medical device (such as a cardiac pacemaker) is typically produced by a team of writers, engineers, graphic artists, editors, reviewers, marketing personnel, and lawyers. Other team members might research, edit, and proofread.

Teams also collaborate to create and deliver oral presentations, such as reports, proposals, and training sessions.

Traditionally composed of people from one location, teams are increasingly distributed across different job sites, time zones, and countries. The internet, via email, video conferencing, instant messaging, blogs, and other communication tools, offers the primary means for distributed (or virtual) teams to interact. In addition, tools, such as computer-supported cooperative work (CSCW) software, and project management software, such as Microsoft *Project*, allow distributed teams to collaborate. But whether the team is on-site or distributed, members have to find ways of expressing their views persuasively, of accepting constructive criticism, and of getting along and reaching agreement with others who hold different views.

Teamwork is successful only when there is strong cooperation, a recognized team structure, and clear communication. To manage a team project you need to (a) spell out the project goal, (b) break the entire task down into manageable steps, (c) create a climate in which people work well together, and (d) keep each phase of the project under control.

Factors in successful teamwork

Starting on page 59, this chapter presents guidelines for effective collaborative writing. First, though, let's examine the forces and roles that govern groups and their work.

OPERATING IN GROUPS

Many different interactions and lines of communication can occur within a group. Interactions depend on such factors as group size, group member similarity, and the structure that has been established by leadership and organizational culture. Experience shows that smaller groups can more easily be cohesive, agree on goals, and work toward goals than larger groups. Groups also have a better chance of being successful if group members come from similar backgrounds and share values.

Another key factor in group communications is leadership. Leaders must foster collaboration within their organizations. However, a 2011 study by ESI International found that "while the majority of organizations value high-impact team collaboration, less than one out of three organizations actually provides the proper framework for it" (Marketwire).

ESI International's study concluded that

- "the majority of organizations, in fact, do not work collaboratively, despite the value that they realize would come from better teamwork;
- rigid work structures . . . within companies . . . keep people from working together; and
- organizations are not investing in the right mix of skills training needed to improve collaboration on projects and initiatives" (Marketwire).

Effective Roles in Groups[1]

Most groups, including collaborative writing teams, function best when certain key roles are established and followed. Two main types of roles have been identified: *task roles* and *maintenance roles*. Often, such roles are established informally.

TASK ROLES.   In order to accomplish its set task, a group has to have its members successfully complete various task roles. Some people play one or two of these roles exclusively, but most people slide easily in and out of most of the roles.

- *Initiators* propose and define tasks; they also suggest solutions to problems or ways of solving problems.
- *Information seekers* notice where facts are needed, and they push the group to find those facts. *Information givers* have the information at hand, or they know how to find the needed information.
- *Opinion seekers* actively canvass group members for their ideas and opinions concerning a problem. *Opinion givers* volunteer input, respond readily when asked, and help set the criteria for solving problems.
- *Summarizers* draw together various ideas, facts, and opinions into a coherent whole; they review solutions, decisions, or problem-solving criteria. Often, the group's elected chair or appointed manager assumes this role.

CAUTION   Although group consensus is a worthwhile goal and although summarizers can play a very useful role in achieving consensus, groups must be careful to not slide into **groupthink**, a state in which group pressure prevents individuals from questioning or criticizing the group's conclusions and decisions.

---

1. Adapted from Barker et al. 1987.

GROUP MAINTENANCE ROLES.　All successful groups need a supportive group climate in order for members to give their best efforts. A positive group climate does not usually happen by accident; members have to consciously perform some of the following roles in order to develop and sustain a good working relationship.

- *Encouragers* help other group members feel accepted and valued. In particular, they go out of their way to reward those group members who are shy about contributing to the group's tasks.
- *Feeling expressers* encourage group members to state their feelings about the group and those members' roles in the group.
- *Harmonizers* help deal with unproductive conflict by removing the personal, emotional aspects of disputes and concentrating on the objective issues. They recognize the good points made by the respective disputants, and they help the "warring members" recognize the merits of each other's position.
- *Gatekeepers* try to draw quiet group members into the discussion so that the louder, more aggressive group members do not dominate. When everyone contributes, two main advantages result: (1) the group has a better chance of getting the ideas and information it needs, and (2) group morale improves.

Overcoming Differences by Active Listening

One of the best ways to foster positive group climates is to listen actively to one's colleagues. Listening is a critical skill in getting along with co-workers, building relationships, and learning. Many of us seem more inclined to speak, to say what's on our minds, than to listen. We often hope someone else will do the listening. Effective listening requires active involvement—instead of merely passive reception, as explained in the guidelines below.

GUIDELINES　　**FOR ACTIVE LISTENING**

1. *Don't dictate.* If you are leading the group, don't express your view until everyone else has had a chance.
2. *Be receptive.* Take in different views and evaluate them later.
3. *Keep an open mind.* Judgment stops thought (Hayakawa 42). Reserve judgment until everyone has had his or her say.
4. *Be courteous.* Don't smirk, roll your eyes, whisper, or wisecrack.
5. *Show genuine interest.* Non-verbals are vital (eye contact, nodding, smiling, leaning toward the speaker). Make it a point to remember everyone's name.
6. *Hear the speaker out.* Instead of "tuning out" a message you find disagreeable, allow the speaker to continue without interruption (except to ask for clarification). Delay your own questions, comments, and rebuttals until the speaker has finished.

(continued)

7. *Focus on the message.* Instead of thinking about what you want to say next, try to get a clear understanding of the speaker's position.

8. *Ask for clarification.* If anything is unclear, say so: "Can you repeat that?" To ensure accuracy, paraphrase the message: "So you're saying that …"

9. *Be agreeable.* Don't turn the conversation into a contest, and don't insist on having the last word.

Source: Adapted from Armstrong 24; Bashein and Markus 37; Cooper 78–84; Dumont and Lannon 648–51; Pearce, Johnson, and Barker 28–32.

Conflict within Groups

Interpersonal sources of collaborative conflict

In any group effort, conflict can arise. Members might fail to get along because of differences in personality, working style, commitment, standards, or ability to take criticism. Some might disagree about exactly what or how much the group should accomplish, who should do what, or who should have the final say. Some might feel intimidated or hesitant to speak out. These interpersonal problems actually can worsen when the group communicates exclusively via email.

NOTE   *As students, we often cannot choose our project partners, just as we often have little say in choosing work colleagues. Indeed, it's rare to find a co-worker with whom you feel totally compatible.*

Generation gaps produce "natural" conflict

Members of *collaborative writing groups* are susceptible to the normal sources of human discord—defensiveness, rivalry, mistrust, clashing roles, and contrasting personal needs. In addition, today's younger workers are more likely than previous generations to question authority; these workers also want their opinions to be heard and their contributions to be valued. They present a special challenge for "old-school" managers (Moses, "The Challenge").

Misused power corrodes group effectiveness

A special form of unnecessary conflict ensues when persons in positions of power misuse their organizational authority, their prestige, or their personal charisma to get their way. Mistrust and sagging morale inevitably follow, often with negative effects on employee productivity and job satisfaction.

Any collaborative effort stands the best chance of succeeding when each group member feels included. A big mistake is to ignore personal differences, to assume everyone shares one viewpoint, one communication style, and one approach to problem solving. Gender and cultural differences in collaborative groups can create perceptions of inequality and can lead to serious misunderstandings. Some sources of gender and cultural misunderstanding are presented in the following In Brief box, which is then followed by guidelines for communicating in a global team.

Managing and Resolving Group Conflict

Constructive conflict has many benefits

Well-managed conflict can lead to many positive outcomes. To start with, facing a conflict situation shows that you care about the work you do and the people with

Gender and Cultural Differences in Collaborative Groups

Gender Differences

Research on ways men and women communicate in meetings indicates a definite gender gap. Communication specialist Kathleen Kelley-Reardon offers this assessment of gender differences in workplace communication:

> Women and men operate according to communication rules for their gender, what experts call "gender codes." They learn, for example, to show gratitude, ask for help, take control, and express emotion, deference, and commitment in different ways. (88–89)

Professor Kelley-Reardon describes specific elements of a female gender code: women are more likely than men to take as much time as needed to explore an issue, build consensus and relationships among members, use tact in expressing views, use care in choosing their words, consider the listener's feelings, speak softly, allow interruptions, make requests instead of giving commands ("Could I have the report by Friday?" versus "Have this ready by and Friday"), and preface assertions in ways that avoid offending ("I don't want to seem disagreeable here, but . . .").

One study of mixed-gender interaction among peers indicates that women tend to be more agreeable, solicit and admit the merits of other opinions, ask questions, and express uncertainty (e.g., with qualifiers, such as *maybe*, *probably*, *it seems as if*) more often than men (Wojahn 747).

None of these traits, of course, is gender-specific. Some people—regardless of gender—are more soft-spoken, contemplative, and reflective. But such traits most often are attributed to the "feminine" stereotype. Moreover, any woman who breaches the gender code, for instance, by being assertive, may be seen by peers as "too controlling" (Kelley-Reardon 6). In fact, studies suggest that women have less freedom than their male peers to

alter their communication strategies: less assertive males often are still considered persuasive, whereas more assertive females often are not (Perloff 273).

Cultural Differences

International business expert David A. Victor describes cultural codes that influence interaction in collaborative groups: some cultures value silence more than speech, intuition and ambiguity more than hard evidence or data, politeness and personal relationships more than business relationships.

Cultures differ in their attitudes toward time. Some want to get the job done immediately; others take as long as needed to weigh all the issues, engage in small talk and digressions, and inquire about family, health, and other personal matters.

Cultures may differ in their willingness to express disagreement, question or be questioned, leave things unstated, touch, shake hands, kiss, hug, or backslap.

Direct eye contact is not always a good indicator of listening. In some cultures, it is considered offensive. Other eye movements, such as squinting, closing the eyes, staring away, or staring at legs or other body parts are acceptable in some cultures but insulting in others.

Differences Remain in Cyberspace Meetings

When collaborators discuss issues and contribute to documents online, "status cues" such as age, gender, appearance, or ethnicity virtually disappear (Wojahn 747–48). However, those cues are quite apparent when audio and video is added to the meeting. Advances in hardware and converged collaborative software facilitate discussions that closely mimic in-person meetings. New web-conferencing and video conferencing technologies allow participants to see and hear each other very clearly, so gender and cultural differences can be just as apparent as in face-to-face meetings.

whom you work. Conflict that's confronted within a supportive climate often leads to creation of new ideas, new ways of solving problems. In the course of dealing with conflict, participants gain new perspectives, and they often learn more about themselves and more about their colleagues.

However, when faced with conflict, many groups seek short-term solutions instead of dealing with the underlying causes of the disagreement. Often, then, the unresolved conflict poisons subsequent working relationships, igniting severe conflicts that ought to have been easily handled. In some cases, unresolved conflict is allowed to fester because the protagonists aren't sufficiently committed to dealing with the situation. In other words, they don't care enough to confront the situation.

AVOIDING CONFLICT.   In other cases, people are afraid of conflict, so they avoid confronting the situation that makes them uncomfortable. Unfortunately, the avoidance response fails to recognize that conflict is natural and inevitable, even in the most compatible work groups.

People who avoid conflict tend to discount the benefits that successfully managed conflict can bring, especially the increased understanding that accrues from carefully examining the nature and causes of a dispute, a process that usually produces better decisions (Graves).

Avoidance is a common response to conflict

USING POWER TO RESOLVE CONFLICT.   Power is used to resolve conflict most commonly through an appeal to authority and the formation of voting blocs. In the first approach, a problem might be referred up the chain of command, or a disputant might fall back on regulations of the law. Lawsuits represent an extreme version of this approach, and they often leave deep wounds that inflame subsequent conflicts. The second approach, which is found in more democratic organizations, usually involves lobbying behind the scenes to help a "combatant" form a successful voting bloc. This approach also erodes the quality of working relationships.

Both types of power plays can result in "win–lose" outcomes, although if both sides of the dispute are able to muster enough support, the battle may turn into a dispute that neither can win—a "lose–lose" outcome.

Power plays sow the seeds of future conflicts

RESOLVING CONFLICT COLLABORATIVELY.   "Win–win" outcomes rarely result from exercising power and influence. The parties in a dispute are more likely to be satisfied by collaborative approaches. The stage can be set for resolving potential conflicts by maintaining a supportive communication climate in the workgroup—in such a climate, workers feel welcome to express their concerns and proposals, and they know that they will be rewarded, not censured, for confronting issues.

In collaborative approaches, participants emphasize solving problems, not fighting battles. They use their common goals as a starting point, and they get good information as the basis for a productive discussion. They show support for others in the discussion, and, more to the point, they discuss issues; they don't debate with the intention of winning a battle. Participants explain their reactions and the source of their disagreement or discomfort. Sometimes, a mediator may be employed to help group members through this process.

Collaborative conflict resolution shares many features with case study problem-solving methods taught at Harvard University (hence the name, Harvard Case Study Model) and at Canadian business schools. Sample cases may be found at **MyWritingLab: Technical Communication.**

Advice on negotiation, including a five-stage structured approach, is available at the federal government's Canada Business website. Although it is aimed at entrepreneurs looking to succeed in business negotiations, the advice applies to any negotiation (see **www.canadabusiness.ca/eng/page/2656**).

"Win–win" strategies focus on solving problems, not on "winning battles"

Ethical Issues in Workplace Collaboration

Teamwork versus "survival of the fittest"

Our lean, downsized corporate world spells competition among workers:

> Many companies send mixed signals . . . saying they value teamwork while still rewarding individual stars, so that nobody has any real incentive to share the glory. (Fisher, "My Team Leader" 291)

The resulting mistrust interferes with fair and open teamwork and promotes unethical behaviour, which may be especially tempting to ambitious employees eager to climb the organizational ladder.

INTIMIDATING CO-WORKERS.   A dominant personality may intimidate peers into silence or agreement, or the group leader may allow no other viewpoints (Matson, "The Seven Sins" 30). Intimidated employees resort to "mimicking"—merely repeating what the boss says (Haskin 55).

How to ensure that the deserving team members get credit

CLAIMING CREDIT FOR OTHERS' WORK.   Workplace plagiarism occurs when the team or project leader claims all the credit. Even with good intentions, "the person who speaks for a team often gets the credit, not the people who had the ideas or did the work" (Nakache 287–88). Team expert James Stern describes one strategy for avoiding plagiarism among co-workers:

> Some companies list "core" and "contributing" team members, to distinguish those who did the heavy lifting from those [who] were less involved. (Qtd. in Fisher 291)

Stern advises groups to decide beforehand—and in writing—what credit will be given for which contributions.

Information that people need to do their jobs

HOARDING INFORMATION.   Surveys reveal that the biggest obstacle to workplace collaboration is people's "tendency to hoard their own know-how" (Cole-Gomoloski 6) when confronted with questions like these:

- ◆ Whom do we contact for what?
- ◆ Where do we get the best price, the quickest repair, the most dependable service?
- ◆ What's the best way to do X?

Despite all the technology available for information sharing, fewer than 10 percent of companies succeed in persuading employees to share ideas on a routine basis (Koudsi 233). People hoard information that they think gives them power or self-importance, or when having exclusive knowledge might provide job security (Devlin 179). They might even withhold information in order to sabotage peers.

PRODUCTIVE COLLABORATIVE WRITING

The challenges of writing efficiently as a solitary writer are magnified when a group produces a document (or prepares a collaborative oral presentation). As this chapter shows, many sources of misunderstanding and disagreement can arise, which will interfere with the timely completion of a project. The pressure of deadlines can exacerbate friction caused by differing levels of commitment, varying degrees of liking and respect that group members have for one another, and differing work habits.

GUIDELINES   **FOR MANAGING A COLLABORATIVE PROJECT**

These guidelines focus on projects in which people meet face to face, but they can apply as well to electronically mediated collaboration.

1. *Appoint a project or group manager.* This person assigns tasks, enforces deadlines, conducts meetings, and consults with supervisors.

2. *Define a clear and definite goal.* Compose a purpose statement that spells out the project's goal and the group's plan for achieving it. Be sure each member understands the goal.

3. *Decide how the group will be organized.* Here are two possibilities:
   ◆ The group researches and plans together, but each person writes a different part of the document.
   ◆ Some members plan and research; one person writes a complete draft; others review, edit, revise, and produce the final version.

Whatever the arrangement, the final revision should display a consistent style throughout—as if written by one person only.

4. *Divide the task.* Who will be responsible for which parts of the document or which phases of the project? Should one person alone do the final revision? Which jobs are the most difficult? Who is best at doing what (writing, editing, layout, graphics, oral presentation)? Who will make final decisions?

5. *Establish a timetable.* Specific completion dates for each phase will keep everyone focused on what is due and when.

6. *Decide on a meeting schedule and format.* How often will the group meet, and for how long? In or out of the office? Who will take notes (or minutes)? Will the supervisor attend or participate?

Group Processes

7. *Establish a procedure for responding to the work of other members.* Will reviewing and editing be done in writing, face to face, as a group, one on one, or via computer? Will the project manager supervise this process?

8. *Establish procedures for dealing with group problems.* How will disagreements be aired (to the manager, the whole group, the offending individual)? How will disputes be resolved (e.g., by vote, the manager)? How will irrelevant discussion be avoided or curtailed?

9. *Decide how to evaluate each member's contribution.* Will the manager assess each member's performance and, in turn, be evaluated by each member? Will members evaluate each other? What are the criteria? **MyWritingLab: Technical Communication** presents two possible forms for a manager's evaluation of members. Equivalent criteria for evaluating the manager include open-mindedness, ability to organize the team, fairness in assigning tasks, ability to resolve conflicts, or ability to motivate.

10. *Prepare a project management plan.* Figure 4.1 presents a sample plan sheet for managing a collaborative project.

(continued)

11. *Submit progress reports regularly.* Progress reports enable everyone to track activities, problems, and rate of progress.

Beyond these guidelines, respect for other people's views and willingness to listen are essential ingredients for successful collaboration.

Source: Adapted from Debs 38–41; Hill-Duin 45–46; Hulbert, "Developing" 53–54; McGuire 467–68; Morgan 540–41.

Management Plan Sheet

Project title:
Audience:
Project manager:
Team members:
Purpose statement:

Specific Assignments   **Due Dates**
  Research: Research due:
  Planning: Plan and outline due:
  Drafting: First draft due:
  Revising: Reviews due:
  Preparing final documents: Revisions due:
  Presenting oral briefing: Final document due:
                                                  Progress report(s) due:

Work Schedule
  Group meetings: Date   Place   Time   Note-taker
    #1
    #2
    #3
    etc.
  Mtgs. w/manager:
    #1
    #2
    etc.

Miscellaneous
  How will disputes and grievances be resolved?
  How will performances be evaluated?
  Other matters (internet searches, email routing, video conferencing, etc.)?

Figure 4.1 Sample Plan Sheet for Managing a Collaborative Project

In any group, then, members have to find ways of persuasively expressing their views, of accepting criticism, and of working with others who hold different views. The Guidelines box on page 63 aims to provide a context for productive group work.

These guidelines must be established early in the collaborative process. The more complex the project, the more important such guidelines become. However, even the simplest group project benefits from clearly defined roles, project goals, and timelines. Also, a written record should be kept.

ON THE JOB...

Closely Collaborating to Produce Technical Documents

"We're working hard to refine a process that combines the messaging and technical information at every step of the product development and production. We reduce re-work and disconnect by closely collaborating throughout the process. First, our product researchers and our marketing researchers work together to determine the feasibility of a proposed product. Together, they produce a report at the end of that stage. In stage 2, the product development stage, product literature is drafted as software and hardware are developed. The marketing messages crafted in stage 1 will affect the look and feel of the product. Also, during hardware development and testing, we produce technical instructions for the people who will assemble the product. As the testing phase concludes and a ready-to-go prototype is made, the accompanying marketing literature helps prepare for a product launch. Finally, the first set of units is run through production, and we write a launch plan that coordinates market literature, training materials, technical data, and launch details, such as dates, locations, and distribution. Finally, as part of the launch process we integrate internal training for our sales force, our reps, our wholesalers, and everybody along the distribution line.

I want to stress that our descriptions and instructions must be really clear right from the beginning. That clarity is possible only if the designers, the marketers, and the writers collaborate from start to finish."

—**Don Gibbs, CPO, Tekmar Control Systems, which designs and manufactures controls for heating and cooling systems, solar thermal, and snow-and-ice melting systems**

Reviewing and Editing Others' Work

Documents produced collaboratively are reviewed and edited extensively. *Reviewing* means evaluating how well a document connects with its intended audience and meets its intended purpose with these qualities:

- accurate, appropriate, useful, and legal content
- material organized for the audience's understanding
- a clear, easy-to-read, and engaging style
- effective visuals and page design
- a document that is safe, dependable, and easy to use

In reviewing, you explain how you respond as a reader, which helps writers to think about ways of revising. Criteria for reviewing various documents appear in checklists throughout this text.

Ways in which editors improve writing

Editing makes copy more precise and readable. Editors typically suggest improvements like these:

- rephrasing or reorganizing sentences
- clarifying a topic sentence
- choosing a better word or phrase
- correcting spelling, usage, or punctuation, and so on

GUIDELINES   **FOR PEER REVIEWING AND EDITING**

1. *Read the entire piece at least twice before you comment.* Develop a clear sense of the document's purpose and its intended audience. Try to visualize the document as a whole before you evaluate specific parts or features.

2. *Remember that correctness does not guarantee effectiveness.* Poor usage, punctuation, or mechanics do distract readers and harm the writer's credibility. However, "correct" writing might still contain faulty rhetorical elements (inappropriate content, confusing organization, wordy style, etc.).

3. *Understand the acceptable limits of editing.* In the workplace, editing can range from cleaning up and fine-tuning to an in-depth rewrite (in which case editors are cited prominently as consulting editors or co-authors). In school, however, rewriting someone else's piece to the extent that it ceases to belong to that writer may constitute plagiarism.

4. *Be honest but diplomatic.* Most of us are sensitive to criticism—even when it is constructive—and we all respond more favourably to encouragement. Be supportive instead of judgmental.

5. *Explain why something doesn't work.* Instead of "this paragraph is confusing," say "this paragraph lacks a clear topic sentence, so I had trouble discovering the main idea." Help the writer identify the cause of the problem.

6. *Make specific recommendations for improvements.* Write out suggestions in enough detail for the writer to know what to do.

7. *Be aware that not all feedback has equal value.* Even professional reviewers and editors can disagree. Your job as a reviewer or editor is to help clarify and enhance a document—without altering its original meaning.

Single Sourcing

Single sourcing facilitates collaboration

The term "single sourcing" has been coined to refer to a variety of software and related processes that facilitate virtual collaboration. The fourth, fifth, and sixth editions of this text, for example, have been produced with WriteRAP, proprietary software that Pearson Education Canada has adapted. WriteRAP allows authors, reviewers, editors, and page designers to view and work on the text while it is displayed on a password-protected website.

Single-Sourcing Technology "Rewrites" the Writing Process

At one time, technical writers were responsible for researching material, writing it, formatting it, and printing it. And that was it. These days, the "writer" converts paper documents to electronic formats, stores data in multiple formats, creates and manipulates graphics, retrieves documents and lost hyperlinks, and updates data on various media. No wonder technical communicators have turned to *single sourcing* to help manage their work.

Single sourcing may be defined as using "one document source to produce multiple outputs" (Wardin). But what does that mean? Technical marketing writer Marc Arellano talks about "a well of data and images from which [he draws] appropriate material in appropriate formats for the application at hand—a proposal, a brochure, a website page, a magalogue" ("Single Sourcing"). Other technical writers think of creating a variety of documents (annual reports, user manuals, online tutorials, technical updates) all from one database that has been planned in such a way that "certain types of information can be presented or omitted without affecting the presentation of other types of information" (Davidovic).

Single sourcing can actually be quite simple. You might convert a Word file to a PDF document and then plug that PDF file into a website and onto a CD-ROM. Or you might create modules or "chunks" of information for your base résumé, and then use cut-and-paste to select and place those chunks that suit the résumé for a particular job application. Milan Davidovic, a member of the Toronto chapter of the Society for Technical Communication (STC), believes that planning one's résumé as a series of related chunks is a good way to learn the basic principles of single sourcing.

The trick is in the planning, says Davidovic, who lists three types of tools for single sourcing a résumé:

1. "create your source content as a word processor template and delete unneeded information from each document you create based on that template";
2. "put the content into a database, and then use the report function of the database to output your résumé"; and
3. "use help-authoring tools or … online résumé tools found at job search sites, such as Monster.ca."

So how do you know when to use single-sourcing methods? Speaking at the March 2003 meeting of the Toronto chapter of the STC, Rob Hanna suggested one should ask questions such as the following:

◆ Is this a valuable document, perhaps one tied to a product?
◆ Will this document require updating?
◆ Will this document have a long lifespan?
◆ Will parts of this document be used elsewhere?

Here are some other tools that facilitate single sourcing, at various levels of complexity:

1. Screen-capture tools, says Carla Wardin, "are specifically designed to save time due to their systematic and automatic features." Wardin points to other advantages of using screen-capture programs:

   ◆ They eliminate editing time.
   ◆ They allow the user to capture objects, such as toolbar buttons and icons.
   ◆ Programs, such as Snagit, set up a catalogue of items that's easy to access.
   ◆ Some screen-capture software programs have batch-conversion features.
   ◆ Most screen-capture programs have a multiple output feature to "simultaneously send captured material . . . to the printer, email, and the web."

2. Microsoft Word has features that make single sourcing in Word possible: main and subdocuments, mail merge, database links, search and replace, templates, and others.
3. More technically complex projects can, for example, "convert FrameMaker source files to HTML Help *.chm files," which can also be "the source of a printed user guide and the hyperlinked PDF of the user guide placed on the distribution CD" (Finger). TechWhirl offers a good introduction to FrameMaker's capabilities at **http://techwhirl.com/software/publishing/single-sourcing-with-framemaker/.** Mif2Go can be used to convert FrameMaker to HTML, XML, Word RTF, and Help formats.
4. Still more complex single-sourcing models employ various combinations of markup

(continued)

language (HTML, XML, XSL, XSLT, XSL-FO, and/or CSS) to build architectures that transform and format information for simultaneous applications—print, website, e-newsletter, online help, and so on.

5. "Dynamic customized content" sets up documentation to meet users' needs in three main modes:

   ◆ A user profile is assigned to each login name so that when the user logs in, he or she sees only that part of the information package that has been defined as relevant to that user.
   ◆ User selection modes allow users to select from a form.
   ◆ Pattern-recognition software "learns" the user's preferences and then provides that kind of data or provides selectable links (Rockley 191).

6. The highest level of single sourcing uses electronic performance system software (EPSS) to provide the right information (training, stock quotes, data updates, answers to user questions, reference material) when users need the information, "often before they know they need it" (Rockley 191).

Whatever the level of single-sourcing technology, its use will grow because of its advantages:

   ◆ Reusing content saves time and money— Sean Brierley describes how a software manufacturer reduced its production time for an online version of its printed manual from 80 hours to three hours (Brierley 16).
   ◆ Single sourcing reduces errors.
   ◆ Coded single sourcing simplifies the document-creation processes (once the code is set).
   ◆ Single sourcing suits the needs of e-commerce, especially when information can be customized for individual consumers and automatically sent to them as the information becomes available. (For example,

an airline can alert a customer about special flights and fares from that customer's "home" airport.)

   ◆ Single sourcing can automatically update documents as new information becomes available.
   ◆ Some single-sourcing tools can do more than reproduce content in different formats; they can be programmed to reduce the amount of detail and change the appearance or size of illustrations automatically for such applications as online quick-reference guides.

Looking at single sourcing from the point of view of a small firm that writes proposals and contracts for clients, communications consultant Stan Chung says, "an intranet can be used to make sure that a company's profile, projects, clients, and résumés are always current. For a large client, say a bridge builder or engineering firm, it is really important to have all the specs, photos, and reports in one place."

Ann Rockley, a Toronto-based expert in single-sourcing applications, predicts a paradigm shift in the way *writers* will function, as single sourcing becomes prevalent. More and more, writers will focus on creating the content (researching, organizing, composing), while *information designers* will set up the structures required for effective communication in various media. Meanwhile, *editors* will regain an important role as they ensure standards and consistency from platform to platform. And *information technologists* will take care of the technical aspects of delivering information to users, says Rockley (192–93). In this scenario, writers will go back to what they used to do—researching, thinking, and writing!

A variety of documents describing single sourcing can be accessed through the STC's Single-Sourcing Special Interest Group, at **http://singlesourcingsig.org.** Note, though, that the material is available only to members of the Society for Technical Communication.

WORKING IN VIRTUAL TEAMS

In order to understand far-flung markets and to tap into the global talent pool, companies are forming work teams that have workers in widely separated locations. Meeting primarily in cyberspace, such groups share their expertise and collaborate to design and develop products. They use technology to stay in touch.

Tripwire Magazine's Rohitha Perera lists six choices that he says "take collaboration to the next level": *e-tipi*, *Podio*, *Creately*, *Basecamp*, *TextFlow*, and *DimDim*. Many other collaborative tools are available, including Skype and a variety of online video conferencing tools. DimDim is particularly useful, says Perera, for "real-time collaboration and meetings" wherein users can "show presentation slides, share desktops, share, and even talk and broadcast via webcam."

Minigroup, a Calgary company, has developed a cloud-based tool to help workers manage and discuss projects online. The tool originally launched as a private social network, but became popular with businesses that employ workers in remote locations. Co-founder Chris Niecker says, "Good collaboration isn't just about doing tasks. It's really the conversations and the communication that lead to the . . . work that needs to get done" (Rockel).

That opinion is reiterated by Claire Sookman, of Toronto-based Virtual Team Builders. Her LinkedIn site advises how to lead virtual work teams:

1. Each employee communicates differently, so leaders need to observe team members' communication styles and adjust their own communication to meet the needs of team members.
2. Observing is not quite enough, so the next step is to talk to employees about their preferences and needs. Also, leaders need to express their own communication style preferences to employees.
3. Each employee needs to work in a compatible environment. Some prefer a structured, supportive environment; others like challenges and the freedom to be rewarded for taking initiative.

CASE	**Differing Experiences with the Virtual Office**

TELUS encourages the virtual office

When the organizers of the 2010 Winter Olympics asked the TELUS Corp. to help reduce traffic by having the TELUS Vancouver workforce work from home, TELUS was happy to comply. Spokesman Shawn Hall says the move was very popular with employees and "resulted in significant cost savings," including having to own or lease less real estate. The company now plans to have 70 percent of its employees telecommuting by 2015 (El Akkad and Bowness B6).

Yahoo Inc. has returned to a more traditional office model

Recently, Yahoo Inc. has taken the opposite stance, relocating its home-based employees back to company offices. Despite trying to encourage a creative, "fun" culture with a variety of morale-building initiatives, Yahoo believed that too many of its telecommuter workers were abusing their freedom. Yahoo's leaders wanted to encourage a hard-working, collaborative approach that may be fostered by having workers see each other daily. Yahoo came to believe that its productivity is improved by having people work in close proximity (Silcoff).

Crowdsourcing

Successful crowdsourcing depends on clear directions and expectations

One method of outsourcing work is to farm it out to a crowd of workers, usually online. Wikipedia is a very good example; instead of hiring writers and editors, it invites online contributions. Wikipedia is a non-profit enterprise, but those that operate for profit also use the technique, especially for very large projects or for areas or expertise that the business does not have. However, clear instructions and expectations are essential to get the desired products and product quality.

Face-to-Face versus Electronically Mediated Collaboration

Benefits of face-to-face meetings

Should groups meet in the same physical space or in virtual space? Face-to-face collaboration is usually preferable when people don't know one another, when the issue is sensitive or controversial, or when people need to interact on a personal level (Munter 81).

Advantages of electronic collaboration

Electronically mediated (virtual) collaboration might be better when people are in different locations or have different schedules. Virtual collaboration might also work when it is important to avoid personality clashes, to encourage shy participants, or to prevent intimidation by dominant participants (Munter 81, 83).

Chapter 1 lists several technologies that erase distance and enable people to work together in the same virtual space.

Focuses on the subject, actions taken, and actions required

Research indicates that electronic meetings are more productive than face-to-face meetings (Tullar, Kaiser, and Balthazard 54). Written ideas can be more carefully considered and expressed, and they provide a durable record for feedback and reference. On the negative side, equipment crashes are disruptive. And the lack of personal contact (such as a friendly grin or a handshake) makes it hard for trust to develop (T. Clark 49–50).

Why telecommuters need to meet face to face

Wayne Pagens, a senior consultant, agrees: "I'm a great believer in telecommuting to the virtual office. But you have to go to the real office from time to time. For a lengthy period, all 10 of us in our developmental team worked from our home offices. However, we learned that if you're not at the company office for people to see you and work with you, you can become irrelevant. So, our team met at the office from time to time, in order to maintain team morale and to remind the rest of the company that we still existed!"

NOTE   *Many cultures value the social (or relationship) function of communication as much as its informative function (Archee 41). Therefore, a recipient might consider one communication medium more appropriate than others, preferring a phone conversation to text messaging or email, for example.*

CONDUCTING MEETINGS

As the above comment by Wayne Pagens indicates, despite many digital tools for collaboration, face-to-face meetings are still a fact of life because they provide vital personal contact. Meetings are usually scheduled for two purposes: to convey or exchange information or to make decisions. Informational meetings tend to run smoothly because there is less cause for disagreement. But decision-making meetings often fail to reach clear resolution about various debatable issues.

Such meetings often end in frustration because the leader has never managed to take charge.

Taking charge doesn't mean imposing one's views or stifling opposing views, but it does mean moving the discussion along and keeping it centred on the issue, as explained in the following guidelines.

NOTE   *The following guidelines apply equally well to virtual meetings. However, old-style video conferencing that uses a single, wide-angle camera allows participants to tune out from time to time as they check their smartphones and so on. New-style virtual meetings that train a webcam on each participant encourage everyone to stay engaged!*

GUIDELINES   **FOR RUNNING A MEETING**

- Set an agenda. Distribute copies to members beforehand: "Our 10 a.m. Monday meeting will cover the following items: ..." Spell out each item, set a strict time limit for discussion, and stick to this plan.
- Ask each person to prepare as needed. A meeting works best when each member prepares a specific contribution.
- Appoint a different "observer" for each meeting. At Charles Schwab & Co., the designated observer keeps a list of what worked well during the meeting and what didn't. The list is added to that meeting's minutes (Matson, "The Seven Sins" 31).
- Begin by summarizing the minutes of the last meeting.
- Give all members a chance to speak. Don't allow anyone to monopolize.
- Stick to the issue. Curb irrelevant discussion. Politely nudge members back to the original topic.
- Keep things moving. Don't get hung up on a single issue; work toward a consensus by highlighting points of agreement; push for a resolution.
- Observe, guide, and listen. Don't lecture or dictate.
- Summarize major points before calling for a vote.
- End the meeting on schedule. This is not a hard-and-fast rule. If you feel the issue is about to be resolved, continue.
- Take minutes at each meeting.

for Collaborating with Others

Teamwork

◆ Have we appointed a team manager?
◆ Does the team agree on the type of document required?
◆ Do we have a plan for how to divide the tasks?
◆ Have we established a timetable and decided on a meeting schedule?
◆ Do we have an agenda for our first meeting?
◆ Do we have a clear understanding of how we will share drafts of the document and how we will name the files?
◆ Have we decided what technology to use (Track Changes, wiki, blog)?
◆ Are we using the Project Planning Form?

Running a Meeting

◆ Has the team manager created an agenda and circulated it in advance?
◆ Do members understand their individual roles on the team so they can be prepared for the meeting?
◆ Has someone been appointed to take meeting minutes?
◆ Are all members given the opportunity to speak?
◆ Does the team manager keep discussion focused on agenda items?
◆ Does the meeting end on schedule?

Active Listening

◆ Are team members receptive to each other's viewpoints?
◆ Does everyone communicate with courtesy and respect?
◆ In face-to-face settings, are all team members allowed to speak freely?

◆ Are people able to listen to all ideas with an open mind?
◆ Are interruptions discouraged?
◆ On email, do people take time to reflect on ideas before responding?
◆ Do people observe the 90/10 rule (listen 90 percent of the time; speak 10 percent of the time)?

Peer Review and Editing

◆ Have I read the entire document twice before I make comments?
◆ Have I focused on content, style, and logical flow of ideas before looking at grammar, spelling, and punctuation?
◆ Do I know what level of review and editing is expected of me (focus only on the content, or focus on style, layout, and other factors)?
◆ Am I being honest but polite and diplomatic in my response?
◆ Do I explain exactly why something doesn't work?
◆ Do I make specific recommendations for improvements?

Global Considerations

◆ Do I understand the communication customs of the international audience for my document?
◆ Is my document clear and direct, so that it is easy to translate?
◆ Have I avoided humour, idioms, and slang?
◆ Have I avoided stereotyping of different cultures and groups of people?
◆ Does my document avoid cultural references (such as TV shows and sports), which may not make sense to a global audience?

EXERCISE

Describe the role of collaboration in a company, organization, or campus group where you have worked or volunteered. Among the questions: What types of projects require collaboration? How are teams organized? Who manages the projects? How are meetings conducted? Who runs the meetings?

How is conflict managed? Summarize your findings in a one- or two-page memo.

Hint: If you have no direct experience, interview a group representative, say a school administrator or faculty member or editor of the campus newspaper. (See page 117–120 for interview guidelines.)

COLLABORATIVE EXERCISE

1. **Listening Competence:** Use the questions above and the Guidelines on pages 55 and 56 to

   ◆ assess the listening behaviours of one member in your group during collaborative work,
   ◆ have another member assess your behaviours, and
   ◆ assess your own listening behaviours.

   Record the findings and compare each self-assessment with the corresponding outside assessment.
   Discuss your group's findings with the class.

2. **Gender Differences:** Divide into small groups of mixed genders. Review page 57. Then test the hypothesis that women and men communicate differently in the workplace.

Each member prepares the following brief messages—without consulting other members:

◆ A thank-you note to a co-worker who has done you a favour.
◆ A note asking a co-worker for help with a problem or project.
◆ A note asking a team member to be more cooperative or stop interrupting or complaining.
◆ A note in which you attempt to take control, because a meeting has gotten out of hand.
◆ A note by which you attempt to deal with some group members are procrastinating on a project.

As a group, compare the information and the tone conveyed by the messages in each category. Then draw conclusions about the hypothesis that men and women communicate differently. Appoint one member to present your findings to the class.

COLLABORATIVE PROJECT

See this text's accompanying website for a collaborative writing exercise. Go to **MyWritingLab: Technical Communication**, "Writing Exercises," and then "Two Collaborative Writing Projects."

MyWritingLab

MyWritingLab: Technical Communication offers the best multimedia resources for technical communication in one, easy-to-use place. You can find a variety of interactive model documents and case studies. There are also extensive guidelines, tutorials, and exercises for Document Design, Writing and the Writing Process, and Research, and a large bank of diagnostics and practice for grammar review.

Writing Persuasively

> **LEARNING OBJECTIVES**
>
> **After reading this chapter, you should be able to**
>
> - understand that receivers may resist persuasive messages and how to overcome that resistance
> - practise techniques for presenting effective arguments and evidence
> - consider the cultural and "political" contexts in which messages will be received*

Writers can face persuasion challenges even when their primary goal is to inform. Here are some persuasion challenges:

- A reader might dispute the *facts*: that reader might not agree that all the factual statements are indeed correct or may believe that key facts have been omitted. Or, the reader might believe that your information sources are unreliable.
- You might have difficulty convincing the reader to accept your *interpretation and evaluation of the facts*.
- Your reader might not accept your *recommended actions*.

Whenever an audience disagrees about what things mean, what is better or worse, or what should be done, you have a persuasion challenge to overcome.

As you read this chapter, you will benefit by referring to the audience/purpose profile found in Chapter 2 on page 33. The profile's categories are explored in the following advice for persuading others.

ASSESS THE POLITICAL REALITIES

If you have worked with others, you already know something about office politics: how some people seek favour, influence, status, or power; how some resent, envy, or intimidate others; how some are easily threatened. Writing consultant Robert Hays sums up the writer's political situation this way: "A writer must labour under political pressures from boss, peers, and subordinates. Any conclusion affecting other people can arouse resistance" (19). Some readers might resist your suggestion for shortening lunch breaks, cutting expenses, or automating the assembly line. Or your document might be seen as an attempt to undermine your supervisor.

---

*This chapter uses the term *politics* to refer to office and organizational politics rather than the politics of government.

EXPECT AUDIENCE RESISTANCE

People who haven't made up their minds about what to do or think are more likely to be receptive to persuasive influence:

We are all consumers as well as providers of persuasion. Daily, we open OURSELVES to the persuasion of others. We need others' arguments and evidence. We're busy. We can't and don't want to discover and reason out everything for ourselves. We look for help, for short cuts, in making up our minds. (Gilsdorf, "Write Me" 12)

We rely on persuasion to help us make up our minds

In a world overwhelmed by information, persuasion can help people "process" the information and decide on its meaning.

People who have already decided what to do or think, however, don't like to change their minds without good reason. Sometimes, people refuse to budge. Whenever you question people's stand on an issue or try to change their behaviour, expect resistance. The bigger their stake in an issue, the more personal their involvement will be, and the more resistance you can expect. Research indicates that inducing permanent change in behaviour is especially difficult because people tend to revert to the familiar patterns and activities that are part of their lifestyle or work habits (Perloff 321).

Any document can evoke different reactions—depending on a reader's temperament, preferences, interests, fears, biases, preconceptions, misconceptions, ambitions, or general attitude. Whenever people feel their views are being challenged, they respond with questions such as the following:

> ### ON THE JOB...
>
> **The Key to Persuading Others**
>
> *"Everything is about the relationship! If you want to persuade others to accept your argument, you must have open and honest communication that they can empathize with. Once they truly understand your position, they can be persuaded to reconsider theirs. . . ."*
>
> —**Mike Lane, Controller, Flair Airlines**

TYPICAL AUDIENCE QUESTIONS ABOUT A DOCUMENT THAT ATTEMPTS TO PERSUADE

- ◆ Says who?
- ◆ So what?
- ◆ Why should I?
- ◆ What's in it for me?
- ◆ What's in it for you?
- ◆ What does this really mean?
- ◆ Will it mean more work for me?
- ◆ Will it make me look bad?

Some ways of yielding to persuasion are better than others

When people do yield to persuasion, they yield either grudgingly, willingly, or eagerly (e.g., as shown in Figure 5.1 on page 73). Researchers categorize these responses as *compliance*, *identification*, or *internalization* (Kelman 51–60):

- ◆ *Compliance*: "I'm yielding to your demand in order to get a reward or to avoid punishment. I really don't accept it, but I feel pressured and so I'll go along."
- ◆ *Identification*: "I'm yielding to your appeal because I like and believe you, I want you to like me, and I feel we have something in common."
- ◆ *Internalization*: "I'm yielding because what you're saying makes good sense and it fits my goals and values."

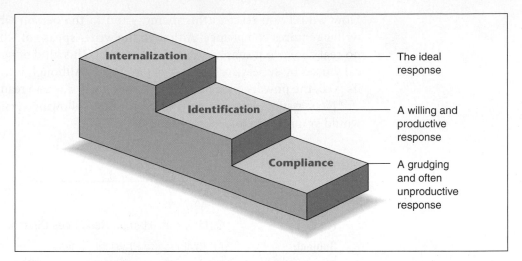

Figure 5.1 The Levels of Response to Persuasion

Although compliance is sometimes necessary (as in military orders or workplace safety regulations), nobody likes to be coerced. Effective persuasion relies on identification or internalization. If people merely comply because they feel they have no choice, then you probably have lost their loyalty and goodwill—and as soon as the threat or reward disappears, you will lose their compliance as well.

KNOW HOW TO CONNECT WITH THE AUDIENCE

Persuasive people know when to merely declare what they want, when to reach out and create a relationship, when to appeal to reason—or when to employ some combination of these strategies. Let's call these three connecting strategies the *power connection*, the *relationship connection*, and the *rational connection*.

To illustrate these different connections, picture the following situation:

CASE	**Selling a Fitness Program**

Your company has just developed a fitness program, based on findings that healthy employees work better, take fewer sick days, and cost less to insure. This program offers clinics for smoking, stress reduction, and weight loss, along with group exercise. In your second month on the job, you receive this notice through email:

Memo

To:　　　　All Employees　　　　　　　　　　　　　　　　5/20/14
From:　　　　G. Maximus, Human Resources Director
Subject:　　Physical Fitness

POWER CONNECTION: Orders readers to show up

On Monday, June 7, all employees will report to the company gymnasium at 8:00 A.M. for the purpose of choosing a walking or jogging group. Each group will meet 30 minutes three times weekly during lunchtime.

How would you react to this memo—and to the person who wrote it? Here, the writer requires compliance. Although the writer speaks of "choosing," you are given no real choice but simply ordered to show up. This kind of *power connection* is typically used by supervisors and others in power. Although it may or may not achieve its goal, the power connection almost surely will alienate readers.

Now assume instead that you received the following version of our memo. How would you react to this message and its writer?

Memo

To:	All Employees	5/20/14
From:	G. Maximus, Human Resources Director	
Subject:	An Invitation to Physical Fitness	

RELATIONSHIP CONNECTION: Invites readers to participate

Leaves the choice to readers

I realize most of you spend lunch hour enjoying a bit of well-earned relaxation in the middle of a hectic day. But I'd like to invite you to join our lunchtime walking/jogging club.

We're starting this club in hopes that it will be a great way for us all to feel better. Why not give it a try?

This second version evokes a sense of identification, of shared feelings and goals. Instead of being commanded, readers are invited—they are given a real choice. This *relationship connection* establishes goodwill.

Often, the biggest factor in persuasion is a reader's perception of the writer. Readers are more receptive to people they like, trust, and respect. Of course, you would be unethical in appealing to the relationship or in faking the relationship merely to hide the fact that you had no evidence to support your claim (R. Ross 28). Audiences still need to find the claim believable ("Exercise will help me feel better") and relevant ("I personally need this kind of exercise").

Here is a third version of our memo. As you read, think about the ways it makes a persuasive case.

Memo

To:	All Employees	5/20/14
From:	G. Maximus, Human Resources Director	
Subject:	Invitation to Join One of Our Jogging or Walking Groups	

RATIONAL CONNECTION: Presents authoritative evidence

Offers alternatives

Offers a compromise

A recent Health Canada study reports that adults who walk 3 kilometres a day could increase their life expectancy by three years. Other research shows that 30 minutes of moderate aerobic exercise at least three times weekly has a significant and long-term effect in reducing stress, lowering blood pressure, and improving job performance.

As a first step in our exercise program, our company is offering a variety of daily jogging groups: the 1.5-kilometre group, the 5-kilometre group, and the 8-kilometre group. All groups will meet at designated times on our brand-new, half-kilometre, rubberized clay track.

For beginners or skeptics, we're offering daily 3-kilometre walking groups. For the truly resistant, we offer the option of a Monday-Wednesday-Friday 3-kilometre walk.

Leaves the choice to readers

Offers incentives

> Coffee and lunch breaks can be rearranged to accommodate whichever group you select.
>
> Why not take advantage of our hot new track? We will reimburse anyone who signs up as much as $100 for running or walking shoes, and will even throw in an extra 15 minutes for lunch breaks. And with a consistent turnout of 90 percent or better, our company insurer may be able to eliminate everyone's $200 yearly deductible in medical costs.

Here, the writer shows willingness to compromise ("If you do this, I'll do that"). This *rational connection* communicates respect for the reader's intelligence *and* for the relationship by presenting good reasons, a variety of alternatives, and attractive incentives—all framed as an invitation. Whenever an audience is willing to listen to reason, the rational connection stands the best chance of succeeding.

ASK FOR A SPECIFIC DECISION

Unless you are giving an order, diplomacy is essential in persuasion. But don't be afraid to ask for the specific decision you want, preferably at the end of the message:

Let people know exactly what you want

> Studies show that the moment of decision is made easier for people when we show them what the desired action is, rather than leaving it up to them. . . . Without this directive, people may misunderstand or lose interest in the entire message. No one likes to make decisions: there is always a risk involved. But if the writer asks for the action, and makes it look easy and urgent, the decision itself looks less risky and the entire persuasive effort has a better chance of succeeding. (Cross 3)

Let people know what you want them to do or think.

NEVER ASK FOR TOO MUCH

No amount of persuasion will move people to accept something they consider unreasonable. And the definition of *reasonable* depends on the individual. In response to an attempt to persuade workers to exercise, employees will differ as to which walking/jogging option they might accept. To the jock writing the memo, a daily eight-kilometre jog might seem perfectly reasonable, but some employees would think it outrageous. The company program therefore has to offer something most of its audience will accept as reasonable (except for, say, couch potatoes and those in poor health). Any request that exceeds its audience's "latitude of acceptance" (Sherif et al. 39–59) is doomed, as Figure 5.2 on page 76 suggests.

RECOGNIZE ALL CONSTRAINTS

Persuasive communicators observe certain limits or restrictions imposed by their situation. These *constraints* govern what should or should not be said, who should say it and to whom, when and how it should be said, and through which medium (e.g., printed document, online, telephone, face to face, and so on).

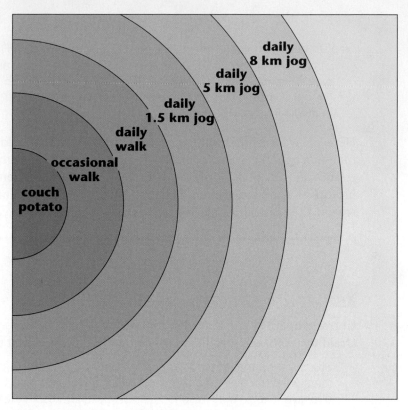

Figure 5.2 Options and Latitude of Acceptance

Organizational Constraints

Organizations often have their own official constraints, such as schedules, deadlines, budget limitations, writing style, the way a document is organized and formatted, and its chain and medium of distribution throughout the organization. But writers also face unofficial constraints:

Decide carefully when to say what to whom

> Most organizations have clear rules for interpreting and acting on (or responding to) statements made by colleagues. Even if the rules are unstated, we know who can initiate interaction, who can be approached, who can propose a delay, what topics can or cannot be discussed, who can interrupt or be interrupted, who can order or be ordered, who can terminate interaction, and how long interaction should last. (Littlejohn and Jabusch 143)

The exact rules vary among organizations, depending on whether communication channels are open and flexible or closed and rigid, on whether employee participation in decision making is encouraged or discouraged. Although the rules of the game are mostly unspoken, anyone who ignores them invites disaster.

Airing even the most legitimate gripe in the wrong way through the wrong medium to the wrong person can be fatal to your work relationships and your career. The following memo, for instance, is likely to be interpreted by the chief executive officer as petty and whining behaviour, and by the maintenance director as a public attack.

To: P. Susan Singh, Chief Executive Officer

Wrong way to the wrong person

Please ask the Maintenance Director to get his people to do their job for a change. I realize we're all short-staffed, but I've received 50 complaints this week about the filthy restrooms and overflowing wastebaskets in my department. If he wants us to empty our own wastebaskets, why doesn't he let us know?

cc: Robert Stuart, Maintenance Director

Instead, why not address the memo directly to the key person—or better yet, phone the person?

To: Robert Stuart, Maintenance Director

A better way to the right person

I wonder if we could meet to exchange some ideas about how our departments might be able to help one another during these staff shortages.

Legal Constraints

Sometimes, what you can say is limited by contract or by laws protecting confidentiality or customers' rights, or by laws affecting product liability. For example, in a collection letter for non-payment, you can threaten legal action but cannot threaten any kind of violence or threaten to publicize the refusal to pay, or pretend to be a lawyer (Varner and Varner). If someone requests information on one of your employees, you can "respond only to specific requests that have been approved by the employee. Further, your comments should relate only to job performance which is documented" (Harcourt 64). In sales literature or manuals, you and your company are liable for faulty information that leads to injury or damage.

Ethical Constraints

While legal constraints are defined by federal and provincial laws, ethical constraints are defined by good conscience, honesty, and fair play. For example, it may be perfectly legal to promote a new pesticide by emphasizing its effectiveness, while downplaying its carcinogenic effects; whether such action is *ethical*, however, is another issue entirely.

Persuasive skills carry tremendous potential for abuse. There is a difference between honestly presenting your best case and using deception to manipulate people.

Time Constraints

Persuasion often depends on good timing. Does your proposal match current needs and priorities? Is your reader preoccupied with other matters at the moment? If so, you should wait until that reader is ready for your message.

Also consider the time required to complete the proposed work. Project proposals are often rejected by decision makers who believe that too much of their time or your time will be required during the project. Perhaps there's not enough time to complete a project by a certain date—say, a research project by the end of a semester, or road repairs before the summer tourist season.

Social and Psychological Constraints

Too often, what we say can be misunderstood or misinterpreted. Here are just a few of the human constraints routinely encountered by communicators.

◆ *Relationship between communicator and audience.* Are you writing to a superior, a subordinate, or an equal? How well do you and your audience know each other? Can you joke around or should you be serious? Do you have a history of conflict? Do you trust and like one another? What you say and how you say it—and how it is interpreted—will be influenced by the relationship.

◆ *Audience's personality.* Researchers claim that "some people are easier to persuade than others, regardless of the topic or situation" (Littlejohn 136). A person's ability to be persuaded might depend on such personality traits as confidence, optimism, self-esteem, willingness to be different, desire to conform, degree of openness, or regard for power (Stonecipher 188–89).

◆ *Audience's sense of identity and affiliation as a group.* How close-knit is the group? Does it have a strong sense of identity (as in union members, conservationists, or engineering majors)? Will group loyalty or pressure to conform prevent certain appeals from working?

◆ *Perceived size and urgency of the problem or issue.* In the audience's view, how big is this issue or problem? Has it been understated or overstated? Big problems are more likely to cause exaggerated fears, anxieties, loyalties, and resistance to change—or desperate attempts at quick and easy solutions.

Writers who can assess a situation's constraints will avoid serious blunders and develop their message for greatest effectiveness.

SUPPORT YOUR CLAIMS CONVINCINGLY

The strength of your case depends on the *reasons* you offer to support your claims.

> *Persuasive claims are backed up by reasons that have meaning for the audience*

When we seek a project extension, argue for a raise, interview for a job, justify our actions, advise a friend, speak out on issues of the day . . . we are involved in acts that require good reasons. Good reasons allow our audience and ourselves to find a shared basis for cooperating. . . . In speaking and writing, you can use marvelous language, tell great stories, provide exciting metaphors, speak in enthralling tones, and even use your reputation to advantage, but what it comes down to is that you must speak to your audience with reasons they understand. (Hauser 71)

To sell a plan to supervisors and co-workers, you will need good reasons, in the form of *evidence* and *appeals to readers' needs and values* (Rottenberg 104–06).

Offer Convincing Evidence

Evidence is any factual support from an outside source. It is a powerful element in persuasion—as long as it measures up to readers' standards. Discerning readers evaluate evidence by using these criteria (Perloff 157–58):

◆ *The evidence has quality.* Instead of sheer quantity, people expect evidence that is strong, specific, new, or different.

◆ *The sources are credible.* People want to know where the evidence comes from, how it was collected, and who collected it.

◆ *The evidence is considered reasonable.* It falls within the audience's "latitude of acceptance" (Sherif et al. 39–59).

Evidence includes factual statements, statistics, examples, and expert testimony.

FACTUAL STATEMENTS.   *Facts* can be shown by observation, experience, research, or measurement.

Offer the facts

> Many of our competitors have already installed wireless networks.

When your space and your audience's tolerance are limited—as they usually are—be selective. Decide which facts best support your case.

STATISTICS.   Numbers can be highly convincing. Before considering other details of your argument, many workplace readers are interested in the "bottom line": costs, savings, and profits (Goodall and Waagen 57).

Give the numbers

> The proposed wireless network will cost 20 percent less than a wired network and will provide a 17 percent improvement in upload times

Numbers can also be highly misleading. Any statistics you present have to be accurate, trustworthy, and easy to understand and verify. Always cite your source.

EXAMPLES.   By showing specific instances of your point, examples help people visualize an idea or a concept. For example, the best way to explain what you mean by "flexibility of the wireless network" is to show a map of the locations where users' laptops would be "connected" to the network.

Show what you mean

Good examples have persuasive force; they give readers something solid, a way of understanding even the most surprising or unlikely claim.

EXPERT TESTIMONY.   Expert testimony lends authority and credibility to any claim.

Cite the experts

> Ron Catabia, a consultant who designed CanJet's systems, has studied our needs and strongly recommends we move ahead with a wireless network.

To be credible, however, an expert has to be unbiased and considered reliable by the audience.

Although solid evidence can be persuasive, evidence alone isn't always enough to influence a person. In many companies, the bottom line might be very persuasive for company executives but could mean little to some managers and employees, who will be asking, *Does this threaten my authority? Will I have to work harder? Will I fall behind? Is my job in danger?* These people will have to perceive some benefit beyond company profit.

Appeal to Common Goals and Values

Audiences expect a writer to share their goals and values. If you hope to create any kind of consensus, you have to identify at least one goal you and your audience have in common: "What do we all want most?"

Most people want job security, a sense of belonging, control over their jobs and destinies, and a growing and fulfilling career. Any persuasive recommendation will have to take these goals into account. For example,

Appeal to shared goals

Learning this new software will enhance our skill set and increase career mobility for all of us.

Our goals are shaped by our values (qualities we believe in, ideals we stand for): friendship, loyalty, ambition, honesty, self-discipline, equality, fairness, and achievement, among others (Rokeach 57–58). Beyond appealing to common goals, you can appeal to shared values.

You might appeal to a shared commitment to quality and achievement:

Appeal to shared values

None of us needs reminding of the fierce competition in the software industry. Improved communication among networking departments will result in better manuals, keeping us on the front line of quality and achievement.

Give your audience reasons that have real meaning for *them* personally. For example, in a recent study of teenage attitudes about the negative consequences of smoking, respondents listed these reasons for not smoking: bad breath, difficulty concentrating, loss of friends, and trouble with adults. None of them listed dying of cancer—presumably because this last reason carries little meaning for young people personally (Baumann et al.).

CONSIDER THE CULTURAL CONTEXT

Reaction to persuasive appeals can be influenced by a culture's customs and values.[1] Cultures might differ in their willingness to debate, criticize, or express disagreement or emotion. They might differ in their definitions of "convincing support," or they might observe special formalities in communicating. Expressions of feelings and concern for one's family might be valued more than logic, facts, statistics, research findings, or expert testimony. Some cultures consider the *source* of a message as important as its content. Establishing trust and building a relationship might weigh more heavily than proof and might be an essential prelude to getting down to business.

Cultures might differ in their attitudes toward the environment, big business, technology, or competition. They might value delayed gratification more than immediate reward, stability more than progress, time more than profit, politeness more than candour, or age more than youth.

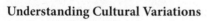

ON THE JOB...

Understanding Cultural Variations

"When I run international projects, my rule is to get to know the client before doing any work. In the process of 'connecting' with the client, one has to be aware of the cultural variations in meaning that get revealed through vocal tone and other signals. For example, I found that Thai business people often express three variations of 'yes':

1. *'Yes, I hear what you're requesting';*
2. *'Yes, I hear what you're requesting and I understand what's required, but I won't necessarily meet your request'; and*
3. *'Yes, I hear what you're requesting, I understand what needs to be done, and I'll do what you ask.'"*

—**Wayne Pagens, Business Analyst, Telus**

---

1. Adapted from Beamer 293–95; Gestelend 24; Hulbert, "Overcoming" 42; Jameson 9–11; Kohl et al. 65; Martin and Chaney 271–77; Nydell 61; Thrush 276–77; Victor 159–66.

One essential element of audience expectations in all cultures is the need for *face saving*:

Some cultures put greater emphasis than others on face saving

> Face saving [is] the act of preserving one's prestige or outward dignity. People of all cultures, to a greater or lesser degree, are concerned with face saving. Yet . . . [its] importance . . . varies significantly from culture to culture. . . . Indirectness in high face-saving cultures is viewed as consideration for another's sense of dignity; in low face-saving cultures, indirectness is seen as dishonesty. (Victor 159–61)

GUIDELINES   **FOR ANALYZING CULTURAL DIFFERENCES**

Take the time to get to know your audience's culture, to appreciate its frame of reference, and to establish common ground. The following questions can get you started.

What Are the Values and Attitudes?
- attitude toward the environment, big business, technology, competition, risk taking, status, youth versus age, rugged individualism versus group loyalty
- preference for immediate reward or delayed gratification, progress or stability
- importance of gender equality, interaction, and differences in the workplace
- importance of the personal relationship in a business transaction
- importance of time ("Time is money!" or "Never rush!")
- importance of feelings versus logic and facts, results versus relationships
- importance of candor versus saving face and sparing other people's feelings
- extent of belief in fate, luck, or destiny
- view of our culture (admiration, contempt, envy, fear)

How Do the Social and Legal Systems Work?
- social/political system inflexible or open
- class distinctions
- democratic, egalitarian ideals or rank-conscious, authoritarian system
- relative importance of the law versus interpersonal trust
- formality of the contract process: a handshake or extensive legal documents
- extent of lawyer involvement
- extent to which the legal system enforces contractual agreements
- role of gift giving (viewed as bribery or as a display of respect)

What Is Accepted Behaviour?
- formalities for making requests, expressing disagreement, criticism, or praise
- preferred form for greetings or introductions (first or family names, titles)
- willingness to criticize or request clarification

(continued)

- ◆ willingness to argue, debate, or express disagreement
- ◆ willingness to be contradicted
- ◆ willingness to express emotion (pleasure, gratitude, anger)
- ◆ importance of trust and relationship building
- ◆ importance of politeness, euphemism, and leaving certain things unsaid
- ◆ preference for casual or formal interaction
- ◆ preference for directness and plain talk or for indirectness and ambiguity
- ◆ preference for rapid decision making or for extensive analysis of a topic

Source: This list was largely adapted from Caswell-Coward 265; Weymouth 144; Beamer 293–95; Martin and Chaney 271–77; Victor 159–61.

A rich source of information about many of the world's cultures

To gain a preliminary understanding of how to apply the above guidelines to a given culture, visit Communicaid's introduction to national cultures, at **www.communicaid.com/access/pdf/index.php.** The website's PDF Library provides overviews of 80+ countries, including Canada, along with some description of each country's key cultural concepts, core values, and local business culture. Of course, this information just begins to delve into the intricacies of each national culture. Companies, such as Communicaid, provide in-depth workshops that help people work or do business in most of the world's major trading nations. Such workshops are very popular, because learning a different culture from trial and error while "on the ground" is very inefficient, often costly (in terms of lost goodwill and lost business), and sometimes even dangerous.

Training workshops are not restricted to international cultural studies. In Canada, for example, a variety of universities, colleges, and private trainers conduct training that helps Aboriginal people and other Canadians understand how to live and work with one another.

GUIDELINES FOR PERSUASION

Later chapters in this book offer specific guidelines for various persuasive documents. For all types of documents, documents that are meant to persuade, remember this principle:
No matter how brilliant, any argument rejected by its audience is a failed argument.

If readers find cause to dislike you or conclude that your argument has no meaning for them personally, they usually reject *anything* you say. Connecting with people means being able to see things from their perspective. The following guidelines can help you make that connection.

(continued)

1. *Assess the political climate.* Can you be outspoken? Who will be affected by your document? How will they react? How will your motives be interpreted? Will the document enhance your reputation or damage it? The better you assess readers' political feelings, the less likely your document will backfire. Do what you can to earn confidence and goodwill:

   ◆ Be diplomatic; try not to make anyone look bad or lose face.

   ◆ Be aware of your status in the organization; don't overstep.

   ◆ Don't expect anyone to be perfect—including yourself.

   ◆ Ask your intended readers to review early drafts.

   　　When reporting company negligence, dishonesty, stupidity, or incompetence, expect political fallout. Decide beforehand whether you want to keep your job or your integrity.

2. *Learn the unspoken rules.* Know the constraints on what you can say, to whom you can say it, and how and when you can say it.

3. *Be clear about what you want.* Diplomacy is important, but people won't like having to guess about your purpose.

4. *Never make a claim or ask for something you know people will reject outright.* Be sure your audience can live with whatever you're requesting or proposing. Offer a genuine choice.

5. *Anticipate your audience's reaction.* Will people be defensive, surprised, annoyed, angry, or pleased? Try to address their biggest objections beforehand. Express your judgments ("We could do better") without making people defensive ("It's your fault").

6. *Decide on a connection (or combination of connections).* Does the situation call for you to merely declare your position, appeal to the relationship, or appeal to common sense and reason?

7. *Avoid an extreme persona.* Persona is the image or impression of the writer's personality suggested by a document's tone. Resist the urge to "sound off," no matter how strongly you feel, because audiences tune out aggressive people, no matter how sensible the argument. Try to be likeable and reasonable. Admit the imperfections in your case—a little humility never hurts. Don't hesitate to offer praise when it's deserved.

8. *Find points of agreement with your audience.* Focusing on a shared value, goal, or concern can reduce conflict and help win agreement on later points.

9. *Never distort the opponent's position.* A sure way to alienate people is to cast the opponent as more of a villain or simpleton than the facts warrant.

10. *Try to concede something to the opponent.* Surely the opposing case is based on at least one good reason. Acknowledge the merits of that case before arguing for your own. Instead of seeming like a know-it-all, show some empathy and willingness to compromise.

(continued)

11. *Use only your best material.* Not all your reasons or appeals will have equal strength or significance. Decide which material—from your *audience's* view— best advances your case.

12. *Make no claim or assertion unless you can support it with good reasons.* "Just because" does not constitute adequate support!

13. *Use your skills responsibly.* Persuasive skills are easily abused. People who feel they have been bullied, manipulated, or deceived most likely will become your enemies. Know when to back off.

14. *Seek a second opinion of your document before you release it.* Ask someone you trust and who has no stake in the issue at hand.

15. *Decide on the appropriate medium.* Given the specific issue and audience, should you communicate in person, in print, by phone, email, newsletter, or bulletin board? Should all recipients receive your message via the same medium?

NOTE   *A social networking site such as Twitter can help inform customers about product developments, brief technical updates, and special offers. Clearly, marketing messages of 140 characters or less cannot present much technical detail, but they can help to create a "buzz" about your products and services.*

CASE   Applying Persuasive Guidelines in a Technical Sales Letter

Figure 5.3 on pages 85-86 illustrates how our guidelines are employed in an actual persuasive situation. This letter is from a company that distributes and installs a variety of home-heating systems. General manager Manson Harding writes a persuasive answer to a potential customer's question: "Why should I invest in a geothermal heat pump system you propose for my home?" As you read the letter, notice the kinds of evidence and appeals that support the opening claim. Notice also how the writer focuses on reasons important to the reader.

Harding Heartland HVAC

**Bay 17, 1699 Larose Avenue, St. Boniface, Manitoba
R4C 2Y1
(204) 948-6662**

May 12, 2014

Mr. Pascual di Santos
245 Carleton Way
Winnipeg, Manitoba R3T 6L7

Dear Mr. di Santos:

Bill Neville has asked me to suggest a heating/cooling system for the new house he has contracted to build for you. I understand you had planned to install a conventional natural gas furnace and standard air conditioner, but I recommend a geothermal heat pump system as the best long-term system for your needs.

Geothermal heat pumps use electricity to move heat; thus, the same pumps would take heat from the earth in winter and dissipate heat from the house to the earth in summer. Enclosure A describes a horizontal-loop geothermal system that would best suit your house design and your six-acre location.

A geothermal system would result in major savings for your proposed residence. We've calculated heat loss for your home design, and we've projected the costs for both a conventional system and a geothermal system. Over the next 10 years, your 4950-square-foot residence would average $3500 annually in heating costs and $2450 in annual air conditioning costs. By contrast, a geothermal system would cost $1500 annually. All calculations are listed in Enclosure B.

A geothermal system would cost $14 250 more to install than a conventional system. However, I understand that you have a contingency fund for "necessary modifications." More to the point, at an annual operating saving of $4450, the geothermal system's additional cost would be regained in less than four years.

The writer states his claim

Provides background, in case the reader has little previous knowledge

States major reason and gives supporting figures

Acknowledges potential reader resistance and refutes reason for reader concern

Figure 5.3 Supporting a Claim with Good Reasons   *(continued)*

Mr. di Santos   May 12, 2014   Page 2

States a second "reason"—appeals to emotion

Geothermal systems also offer increased comfort and a healthy environment. My clients comment about the air feeling cleaner, and they've observed that the temperature doesn't vary like it does with gas furnaces. Allergy sufferers tell me that they notice fewer symptoms in winter, now that they've switched to a geothermal system. If you wish, I can put you in contact with clients who've reported feeling better about the air they breathe with a geothermal system. Further, geothermal systems contribute to a cleaner external environment, because fossil fuels aren't burned and because geothermal systems use less electricity than conventional air conditioners.

Offers additional proof

Deals with another potential cause of reader resistance

You'll be pleased to know that geothermal systems require little maintenance. Thus, warranties are impressive—10 years for heat pumps and 50 years for the remainder of the system. A typical closed-loop system has a life expectancy of 100 years.

Appeals to reader's need for security

Gives reasons to trust the writer

To be totally effective and dependable, any HVAC system must be installed by skilled, knowledgeable people. We've been designing and installing heating systems for 14 years; over the past four years, we've designed and installed 17 geothermal systems. My file of letters from satisfied customers is available for your perusal. Bomber Drilling, which installs the underground piping for our projects, has had extensive geothermal loop field experience in Manitoba and in northern Minnesota.

Leads to suggested action

For more information about geothermal heating and cooling, and for several Canadian case studies, see **www.geothermop.ca**. Please call me at 948-6662 to discuss any aspect of the HVAC system for your new home.

Sincerely,

Manson Harding
General Manager

Enclosures:

A: Geothermal system description
B: Operating cost calculations

Figure 5.3 Supporting a Claim with Good Reasons

EXERCISES

1. Find an example of an effective persuasive letter, email, or sales brochure. In a memo to your instructor, explain why and how the message succeeds. Base your evaluation on the persuasion guidelines listed on pages 82–84. Attach a copy of the letter to your evaluation memo. Be prepared to discuss your evaluation in class.

   Now, evaluate a poorly written document, explaining how and why it fails.

2. Think about some change you would like to see on your campus. Perhaps you would like to make something happen, such as a campus-wide policy on plagiarism, changes in course offerings or requirements, more access to computers, a policy on sexist language, or a daycare centre. Or perhaps you would like to improve something, such as the grading system, campus lighting, or the system for student evaluation of instructors. Decide whom you want to persuade, and write an email to that audience. (Note: If the persuasive message is longer than about 250 words, format it as a letter or memo and attach it to a short email message that introduces the attached persuasive message.)

   Anticipate carefully your audience's implied questions, such as:

   ◆ Do we really have a problem or need?
   ◆ If so, should we care enough about it to do anything?
   ◆ Can the problem be solved?
   ◆ What are some possible solutions?
   ◆ What benefits can we anticipate? What liabilities?

   Can you envision additional audience questions? Complete an audience/purpose profile (see Chapter 2, page 33).

   Don't think of this message as the final word, but as a consciousness-raising introduction meant to the reader to acknowledge that the issue deserves attention.

At this early stage, highly specific recommendations would be premature and inappropriate.

COLLABORATIVE PROJECT

Assume that you work for an environmental consulting firm that is under contract with various countries for a range of projects, including these:

◆ a plan for rainforest regeneration in Latin America and sub-Saharan Africa
◆ a plan for "clean" industries in developing countries
◆ a plan for organic agricultural development in Africa and India

Each project will require environmental impact statements, feasibility studies, grant proposals, and a legion of other documents. These are often prepared in collaboration with members of the host country, and in some cases prepared by your company for audiences in the host country: from political, social, and industrial leaders to technical experts and so on.

For such projects to succeed, people from different cultures have to communicate effectively and sensitively, creating goodwill and cooperation. Before your company begins work in earnest with a particular country, your co-workers will need to develop cultural awareness. Your assignment is to select a country and to research that culture's behaviours, attitudes, values, and social system in terms of how these variables influence the culture's communication preferences and expectations. What should your colleagues know about this culture in order to communicate effectively and diplomatically? Do the necessary research based on the questions in the guidelines on pages 81–82.

MyWritingLab

6

Writing Ethically

✱ Explore

Click here in your eText to access a variety of student resources related to this chapter.

LEARNING OBJECTIVES

After reading this chapter, you should be able to

- recognize the potential for unethical communication in a variety of situations
- apply guidelines for ethical communication
- understand how to deal with group pressure and other sources of unethical communication
- know how and when to report unethical communication

Effective messages (that achieve their informative and persuasive purposes) are not necessarily *ethical*. For example, an advertisement that says "our potato chips contain no cholesterol" may be technically accurate but misleading, because potato chips contain saturated fats that produce cholesterol. Or a technical report may present accurate facts and conclusions about the causes of a mining accident but may not describe a history of neglect and poor safety practices at the mine.

When people "do the right thing," what causes them to do so? Which of the following bases of ethical behaviour prompt you to act honestly and fairly, in others' as well as your own interests?

"What makes me act ethically?"

- ◆ *The moral basis.* "Treat others as you would have them treat you" is a basic tenet of many belief systems' moral teaching.
- ◆ *A practical basis.* "What consequences will result from my actions?" "How will my behaviour affect others?"
- ◆ *The authoritative stance.* "Follow the listed rules of ethical behaviour, or face penalties!"
- ◆ *The institutional stance.* "This is the way we do things here."
- ◆ *The peer group pressure stance.* "Our group believes in fair, open, honest behaviour; we expect you to share those beliefs."
- ◆ *The familial/environmental approach.* "I was raised to be honest and to respect others; these values are part of who I am."

The above list assumes ethical thinking and behaviour, but can unethical persons learn to behave ethically? Socrates didn't think so, but he still spent his life trying to teach virtue. Perhaps a more practical approach is based on the possible consequences of ethical and unethical actions. Critical-thinking skills, then, help us to recognize unethical behaviour and find ways to act ethically in the workplace.

RECOGNIZE UNETHICAL COMMUNICATION

Thousands of people are injured or killed yearly in avoidable accidents, the result of faulty communication that prevented intelligent decision making. The following tragedies were caused ultimately by unethical communication.

CASE	**The *Columbia* Accident**

Unethical communication has consequences

On February 1, 2003, the space shuttle *Columbia*, with seven crew members, disintegrated as it re-entered Earth's atmosphere on its way home. A seven-month investigation by the independent *Columbia* Accident Investigation Board (CAIB) identified the primary physical cause as "a breach in the Thermal Protection System on the leading edge of the left wing, caused by a piece of insulating foam" during the launch on January 16, 2003 (CAIB 9).

The CAIB also concluded that "the organizational causes of this accident are rooted in the Space Shuttle Program's history and culture" (9). The CAIB report describes "compromises . . . resource constraints . . . schedule pressures. . . . and organizational barriers that prevented effective communication of critical safety information and stifled differences of opinion" (9).

Similar pressures had played a part in a previous shuttle disaster, the 1986 explosion of the space shuttle *Challenger*, 43 seconds into its launch. For a chronology of the event, see the webpage maintained by the Online Ethics Center for Engineering and Science, at **www.onlineethics.org/cms/7123.aspx**. Roger Boisjoly, who worked for the manufacturer of the solid rocket boosters for the space shuttle program, gives his account and analysis of the *Challenger* disaster.

CASE	**The Westray Coal Mining Accident**

On May 9, 1992, 26 men were killed in an explosion at the Westray coal mining operation in Plymouth, Nova Scotia. The mine, employing the latest mining technology, had been operating for about eight months when the explosion occurred. In a subsequent report of the Westray Mine Public Inquiry, Mr. Justice K. Peter Richard said that the Westray disaster was a "complex mosaic of actions, omissions, mistakes, incompetence, apathy, cynicism, stupidity, and neglect." As the evidence emerged during the inquiry, it became clear that many people and entities had not communicated appropriately, had not acted on clear directives, and had not applied or enforced coal mine and occupational health and safety regulations. While criminal negligence and manslaughter charges were withdrawn late in 1998, the families of the victims of the disaster commenced a negligence lawsuit in 1999 against the federal and Nova Scotia governments, former mine officials, and equipment manufacturers.

These catastrophes make for dramatic headlines, but more routine examples of deliberate miscommunication rarely are publicized. Outcomes like the following succeed because the people communicate unethically by saying whatever works:

- ◆ A person lands a great job by exaggerating credentials, experience, or expertise.
- ◆ A marketing specialist for a chemical company negotiates a huge bulk sale of the company's powerful new pesticide by downplaying its carcinogenic hazards.
- ◆ To meet the production deadline on a new auto model, the test engineer suppresses, in the final report, data indicating that the fuel tank could explode upon impact.
- ◆ A manager writes a strong recommendation to get a friend promoted, while overlooking someone more deserving.

Can you recall some instances of deliberate miscommunication from your personal experience or that you learned about in the news? What were some of the effects of this miscommunication? How might the problems have been avoided?

Ethical decisions are not always black and white

To save face, escape blame, or get ahead, anyone might be tempted to say what people want to hear or to suppress bad news or make it seem "rosier." Some of these decisions are not simply black and white. Here is one engineer's description of the grey area in which issues of product safety and quality often are decided:

The company must be able to produce its products at a cost low enough to be competitive. . . . To design a product that is of the highest quality and consequently has a high and uncompetitive price may mean that the company will not be able to remain profitable, and be forced out of business. (Burghardt 92)

Do you emphasize to a customer the need for extra-careful maintenance of a highly sensitive computer—and risk losing the sale? Or do you downplay maintenance requirements, focusing instead on the computer's positive features? Do you tell a white lie so as not to hurt a colleague's feelings, or do you "tell it like it is" because you're convinced that lying is wrong in any circumstance? The decisions we make in these situations are often influenced by the pressures we feel.

EXPECT SOCIAL PRESSURE TO PRODUCE UNETHICAL COMMUNICATION

Pressure to get the job done can cause normally honest people to break the rules. At some point in your career, you might have to choose between doing what your employer wants ("just follow orders" or "look the other way") and doing what you know is right. Maybe you will be pressured to ignore a safety hazard in order to meet a project deadline:

Pressure to "look the other way"

Just as your company is about to unveil its hot new pickup truck, your safety engineering team discovers that the reserve gas tanks (installed beneath the truck but outside the frame) can explode on impact in a side collision. The company has spent a small fortune developing this new model and doesn't want to hear about this problem.

Companies often face the contradictory goals of *production* (whose goal is *making* money on the product) and *safety* (which means *spending* money to avoid accidents that may or may not happen). When productivity receives exclusive priority, safety concerns may suffer (Wickens 434–36). Thus, it comes as no surprise that well over 50 percent of managers studied feel "pressure to compromise personal ethics for company goals" (Golen et al. 75). These pressures come in varied forms (Lewis and Reinsch 31):

◆ the drive for profit
◆ the need to beat the competition (other organizations or rival co-workers)
◆ the need to succeed at any cost, as when superiors demand more productivity or savings without questioning the methods
◆ an appeal to loyalty—to the organization and to its way of doing things

Figure 6.1 depicts how such pressures can add up.

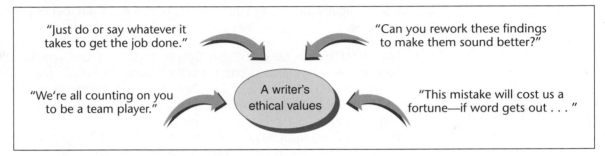

Figure 6.1 How Workplace Pressures Can Influence Ethical Values

Following is a tragic example of how organizational pressure and/or personal agendas can help create a climate where important information is not reported or not heeded by those who receive the reports.

CASE | The Walkerton Tragedy

On May 17, 2000, many of the residents of Walkerton, Ontario, began to display symptoms of bacterial poisoning—bloody diarrhea, vomiting, cramps, and fever. Eventually, more than 2300 of the town's fewer than 5000 citizens became seriously ill. Seven died. The tragic results were preventable, if various public officials had fulfilled their obligations.

The outbreak was primarily caused by *Escherichia coli O157:H7* (*E. coli*) and by *Campylobacter jejuni*, traced to bacteria from cattle manure that was washed into the No. 5 well, which supplied most of Walkerton's water supply during the time that supply became contaminated.

Sadly, many of Walkerton's residents might have escaped the illness if Stan Koebel, Walkerton's water manager, had notified the public, the Ministry of the Environment, or the public health office that tests of water sampled May 15 had revealed *E. coli* contamination. Koebel sat on these results from May 18 to 23. Indeed, when health officials called on May 19 and 20 to inquire about Walkerton's water supply, he lied; he said that the water was safe to drink. So the health officials looked for other possible causes of contamination.

On May 21, in response to more and more cases of *E. coli* poisoning, and with increasing likelihood of a water contamination, medical officer Dr. Murray McQuigge issued a "boil water" advisory. McQuigge's unit took samples on May 21 and on May 23, and learned on May 23 that Walkerton's water was indeed *E. coli*–contaminated. When confronted with the health unit's findings, Koebel admitted that he had known about the contamination since May 18 and also that another well's chlorination equipment had not been working for several days (Evenson).

Mr. Justice Dennis O'Connor was commissioned to head an inquiry into the disaster. His report concluded that "the outbreak would have been prevented by the use of continuous chlorine residual and turbidity monitors at Well 5. The failure to use continuous monitors at Well 5 resulted from shortcomings in the approvals and inspections programs of the Ministry of the Environment" (O'Connor, "Summary" 3). The Walkerton Inquiry, in May and June 2001, uncovered a number of possible contributing factors:

◆ From 1997 through 1999, a London, Ontario, lab found *E. coli* bacteria in Walkerton water on five separate occasions; the lab also noticed unusually low levels of chlorine disinfectant. Microbiologist Gary Palmateer faxed the results to Walkerton officials and the Ministry of the Environment, but no corrective actions were taken (Canadian Press, "Notifying Government").

◆ Despite knowing about water safety issues in Walkerton and elsewhere, the Ministry of the Environment did not "release [water safety] reports and test results documenting the severity of the situation" (Canadian Press, "Ontario Ministry").

◆ Dr. Richard Schabas was Ontario's chief medical officer of health in May 1997 when he tried to warn then-premier Mike Harris, among others (at a meeting of the provincial cabinet's policy and priorities committee), that eliminating provincial funding for regional health boards would allow small-town politicians to bully or circumvent local medical health officers. Ignoring such warnings, the government downloaded public health boards to municipalities.

◆ Also, at Schabas's request, health minister Jim Wilson sent, on August 20, 1997, a letter to Norm Sterling, minister of the environment. In that letter, Wilson urgently recommended that the *Ontario Water Resources Act* be amended to require local waterworks to inform medical health officers of tests showing water contamination. In the 2001 inquiry, Sterling said he didn't recall receiving the letter; at the time, he was "distracted by other issues" (Mittelstaedt, "Tory Didn't See"). But Sterling said he would accept some blame for the Walkerton tragedy if the inquiry established that his failure to react to the letter contributed to the *E. coli* outbreak.

◆ Previous to Norm Sterling's tenure at the Ministry of the Environment, Brenda Elliott was the minister responsible for chopping the provincial environmental-protection budget by nearly 50 percent from June 1995 to August 1996. Under Elliott's leadership, the ministry moved to privatize water-testing services in two months, not in the three years recommended in reports submitted by ministry staffers. At the Walkerton Inquiry, Elliott said that she had been "acting as part of a team" and that she should not be held personally responsible for the Walkerton *E. coli* outbreak (Mittelstaedt, "Tory 'Team' Made Cuts").

◆ For years, Stan Koebel had not reported any test results that would reflect badly on his municipal water operation. And, along with his brother Frank (waterworks maintenance foreman), Koebel for years had "routinely falsified water samples, test results, and chlorination records" (Canadian Press, "Doctor"). Indeed, in his annual reports, Stan Koebel had described a conscientious, competent staff and a safe, secure water system (Blatchford).

◆ At the apex of the pyramid of responsible officials, Premier Harris initially claimed at the Walkerton Inquiry that he didn't know of risks associated with eliminating government water-testing labs. Later at the inquiry, he admitted that he did know of such risks but considered them "manageable" and that therefore the public didn't need to be informed.

◆ Justice O'Connor's report concluded, however, that "there is no evidence that the specific risks, including the risks arising from the fact that the notification protocol was a guideline rather than a regulation, were properly assessed or addressed" (O'Connor, "Summary" 35). In effect, concluded O'Connor, the government's drive to reduce "red tape" led to the Walkerton tragedy because privatizing lab testing of water samples led directly to the events of May 2000, and reducing the Ministry of the Environment's approvals and inspections indirectly helped cause the outbreak by reducing the chance of reporting negative test results (O'Connor, "Full Report" 406–08).

◆ Dr. Murray McQuigge, who blew the whistle on Stan Koebel, faced intense criticism from Koebel's lawyer at the Walkerton Inquiry. Also, inquiry witnesses suggested that McQuigge's office knew about "problems with Walkerton's water system two years earlier but did nothing about it" (T. Blackwell), and therefore McQuigge was not above reproach himself. Such criticism illustrates one of the potential difficulties faced by whistle-blowers.

NEVER CONFUSE TEAMWORK WITH GROUPTHINK

Any successful organization relies on teamwork, everyone cooperating to get the job done. But teamwork is not the same as *groupthink* (Janis 9).

Groupthink occurs when group pressure prevents individuals from questioning, criticizing, or "making waves." Group members feel a greater need for acceptance and a sense of belonging than for critically examining the issues. In a conformist climate, critical thinking is impossible. Anyone who has lived through adolescent peer pressure has already experienced a version of groupthink.

Yielding to pressure can be especially tempting in a large company or on a complex project, where individual responsibility is easy to hide in the crowd:

How some corporations evade responsibility for their actions

Lack of accountability is deeply embedded in the concept of the corporation. Shareholders' liability is limited to the amount of money they invest. Managers' liability is limited to what they choose to know about the operation of the company. The corporation's liability is limited by governments . . . by insurance, and by laws allowing corporations to duck liability by altering their . . . structure. (Mokhiber 16)

The Ethics Resource Center (ERC), a non-profit group that promotes organizational ethics, conducted an extensive 2007 survey that found "companies with a weak ethical culture experienced more frequent workplace misconduct compared with companies with a strong ethical culture" (Karakowsky). The ERC study cited the Enron corporate corruption (and false reporting) as the result of a culture that had "many team players" and that "cultivated a culture that rewarded unbridled ambition. This kind of culture creates huge pressure to perform and typically leads to ethical compromises" (Karakowsky).

RELY ON CRITICAL THINKING FOR ETHICAL DECISIONS

Ethical decisions require critical thinking

Because of their impact on people and on your career, ethical decisions challenge your critical-thinking skills:

- How can I know the "right thing" in this situation?
- What are my obligations, and to whom, in this situation?
- What values or ideals do I want to stand for in this situation?
- What is likely to happen if I do X, or Y?

Can you rely on more than intuition or conscience in navigating the grey areas of ethical decisions? How will you make a convincing case against danger or folly to a roomful of people caught up in groupthink? Although ethical issues resist simple formulas, you can avoid two major fallacies that obstruct good judgment: the fallacy of "doing one's thing" (personal preference or *relativism*) and the fallacy of "one rule fits all" (regardless of the circumstances; *absolutism*).

Reasonable Criteria for Ethical Judgment

Somewhere between the extremes of relativism and absolutism are *reasonable criteria* (standards of measurement that most people would consider acceptable). These criteria for ethical judgment take the form of *obligations*, *ideals*, and *consequences* (Ruggiero 55–56; Christians et al. 17–18).

Obligations are the responsibilities we have to everyone involved:

Obligations help us to determine what's ethical

- *obligation to ourselves*, to act in our own self-interest and according to good conscience
- *obligation to clients and customers*, to stand by the people to whom we are bound by contract—and who pay the bills
- *obligation to our company*, to advance its goals, respect its policies, protect confidential information, and expose misconduct that would harm the organization
- *obligation to co-workers*, to promote their safety and well-being
- *obligation to the community*, to preserve the local economy, welfare, and quality of life
- *obligation to society*, to consider the national and global impact of our actions

When the interests of these parties conflict—as they often do—we have to decide very carefully where our primary obligations lie.

Ideals are "notions of excellence" (Ruggiero 55), the positive values that we believe in or stand for: loyalty, friendship, courage, compassion, dignity, and fairness.

The *consequences* of our actions may be beneficial or harmful, immediate or delayed, intentional or unintentional, obvious or subtle (Ruggiero 56). Some consequences are easy to predict; some are impossible to know ahead of time.

Figure 6.2 depicts the relationship among these three criteria.

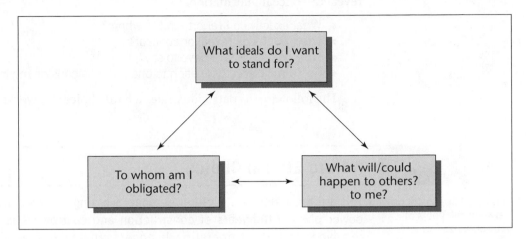

Figure 6.2 Reasonable Criteria for Ethical Judgment

The criteria help us understand why even good intentions can produce bad judgments, as in the following situation:

What seems like the right thing might be the wrong thing

Someone observes . . . that waste from the local mill is seeping into the water table and polluting the water supply. This serious situation requires a remedy. But before one can be found, extremists condemn the mill for lack of conscience and for exploiting the community. People get upset and clamour for the mill to be shut down and its management tried on criminal charges. The next thing you know, the plant does close, 500 workers are without jobs, and no solution has been found for the pollution problem. (Hauser 96)

Because of their zealous dedication to the *ideal* of a pollution-free environment, the extremist protestors failed to anticipate the *consequences* of their protest or to respect their *obligation* to the community's economic welfare.

Ethical Dilemmas

Ethics decisions are especially frustrating when no single answer seems acceptable:

Ethical questions often resist easy answers

[An ethical] dilemma exists whenever the conflicting obligations, ideals, and consequences are so very nearly equal in their importance that we feel we cannot choose among them, even though we must. (Ruggiero 91)

In private and in public, such dilemmas are inescapable. For example, politicians speak of drastic plans to eliminate the federal deficit in five years. One could argue that such a dedication to the *consequences* (or results) would violate our *obligations* (to the poor, the sick, etc.) and our *ideals* (of compassion, fairness, etc.). On the basis of our three criteria, how else might the deficit issue be considered?

ANTICIPATE SOME HARD CHOICES

Communicators' ethical choices basically are concerned with honesty in choosing to reveal or conceal information:

- ◆ What exactly do I report, and to whom?
- ◆ How much do I reveal or conceal?
- ◆ How do I say what I have to say?
- ◆ Could misplaced obligation to one party be causing me to deceive others?

The following scenario illustrates a hard choice at a workplace.

CASE	**A Hard Ethical Choice**

You may have to choose between the goals of your organization and what you know is right

You are an assistant structural engineer working on the construction of a nuclear power plant. After years of construction and cost overruns, the plant finally has received its limited operating licence from the Atomic Energy Control Board (AECB).

During your final inspection of the nuclear core containment unit on February 15, you discover a 3-metre-long hairline crack in a section of the reinforced concrete floor, within 6 metres of the area where the cooling pipes enter the containment unit. (The especially cold and snowless winter likely has caused a frost heave under a small part of the foundation.) The crack either has just appeared or was overlooked by AECB inspectors on February 10.

The crack could be perfectly harmless, caused by normal settling of the structure, and this is, after all, a "redundant" containment system (a shell within a shell). But then again, the crack could signal some kind of serious stress on the entire containment unit, which furthermore could damage the entry and exit cooling pipes or other vital structures.

You phone your supervisor, who is just about to leave on a ski vacation, and who tells you, "Forget it—no problem," and hangs up.

You know that if the crack is reported, the whole start-up process scheduled for February 16 will be delayed indefinitely. More money will be lost; excavation, reinforcement, and further testing will be required; and many people with a stake in this project (from company executives to construction officials to shareholders) will be furious—especially if your report turns out to be a false alarm. All segments of plant management are geared up for the final big moment. Media coverage will be widespread. As the bearer of bad news—and bad publicity—you suspect that, even if you turn out to be right, your own career could be hurt by what some people will see as your overreaction, which has made them look bad.

On the other hand, ignoring the crack could compromise the system's safety, with unforeseeable consequences. Of course, no one would ever be able to implicate you. The AECB has already inspected and approved the containment unit, leaving you, your supervisor, and your company in the clear. You have very little time to decide. Start-up is scheduled for tomorrow, at which time the containment system will become intensely radioactive.

Working professionals often face similar choices caused by conflicting goals and expectations and the pressure to meet deadlines and achieve results. What the employee feels is right may be countered by the need to be a loyal employee and team player and to consider the bottom line. Often, choices have to be made alone or on the spur of the moment, without the luxury of meditation or consultation. It's a good idea, then, to be prepared to make hard choices *before* sticky situations occur.

Many business schools and other post-secondary disciplines work hard to prepare graduates to recognize and navigate ethical dilemmas. Some, like Queen's University and the Richard Ivey School of Business, expose students to white-collar criminals and others who have experienced unethical behaviour first-hand (Ebden).

NEVER DEPEND ONLY ON LEGAL GUIDELINES

Can the law tell you how to communicate ethically? Sometimes. If you stay within the law, are you being ethical? Not always. Legal standards "sometimes do no more than delineate minimally acceptable behaviour." In contrast, ethical standards "often attempt to describe ideal behaviour, to define the best possible practices for corporations" (Porter 183). Consider this legal paradox: It is perfectly legal to advertise a cereal made with oat bran (which allegedly lowers cholesterol) without mentioning that another ingredient in the cereal is coconut oil (which raises cholesterol).

Lying is rarely illegal, except in cases of lying under oath or breaking a contractual promise (Wicclair and Farkas 16). But putting aside these and other illegal lies, such as defamation of character or lying about a product so as to cause injury, we see plenty of room for the kinds of legal lies depicted in Figure 6.3 other kinds of legal lying, such as page design that distorts the real emphasis or words that are deliberately unclear, misleading, or ambiguous.

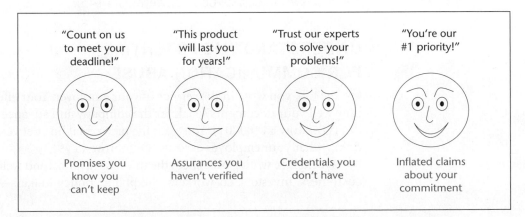

"Count on us to meet your deadline!"

Promises you know you can't keep

"This product will last you for years!"

Assurances you haven't verified

"Trust our experts to solve your problems!"

Credentials you don't have

"You're our #1 priority!"

Inflated claims about your commitment

Figure 6.3 Some Legal Lies in the Workplace

What, then, are a communicator's legal guidelines? Workplace writing is regulated by the types of laws described below.

- *Laws against libel* prohibit any false written statement that attacks or ridicules anyone. A statement is considered libellous when it damages someone's reputation, character, career, or livelihood or when it causes humiliation or mental suffering. Material that is damaging but *truthful* would not be considered libellous unless it was used intentionally to cause harm. In the event of a libel suit, a writer's ignorance is no defence; even when the damaging material has been obtained from a source presumed reliable, the writer and publisher are legally accountable.[1]

- *Copyright laws* protect the ownership rights of authors—or of their employers, in cases where the writing was done as part of an individual's employment. You will find the Copyright Act posted at **http://laws.justice.gc.ca/en/C-42/index.html**.

- *Laws protecting software* provide penalties for illegally duplicating copyrighted software. (Guidelines for ethical, legal use of software are provided through Access Copyright's website at **www.accesscopyright.ca**.) Also, in Canada and internationally, the Software and Information Industry Association fights software piracy through education and enforcement measures.

- *Laws against deceptive or fraudulent advertising* make it illegal, for example, to falsely claim or imply that a product or treatment will cure cancer, or to represent and sell a used product as new. Fraud can be defined as "lying that causes another person monetary damage" (Harcourt 64).

- *Liability laws* define the responsibilities of authors, editors, and publishers for damages resulting from the use of incomplete, unclear, misleading, or otherwise defective information. The misinformation might be about a product (e.g., failure to warn about the toxic fumes from a spray-on oven cleaner) or a procedure (misleading instructions in an owner's manual for using a tire jack). Even if misinformation is given out of ignorance, the writer is liable (Walter and Marsteller 164–65).

- *Legal standards that govern product literature* vary from country to country. A document must satisfy the legal standards for safety, health, accuracy, language, or other issues for any country in which it will be distributed. For example, instructions for any product requiring assembly or operation have to carry warnings as stipulated by the laws of the country in which the product will be sold. Inadequate documentation, as judged by that country's standards, can result in a lawsuit (Caswell-Coward; Weymouth 145).

UNDERSTAND THE POTENTIAL FOR COMMUNICATION ABUSE

On the job, you write in the service of your employer. Your effectiveness is judged by how well your documents speak for the company and advance its interests and agendas (Ornatowski 100–01). You walk the proverbial line between telling the truth and doing what your employer expects (Dombrowski 79).

Workplace writing influences the thinking, actions, and welfare of different people: customers, investors, co-workers, the public, policy makers—to name a few. These

---

1. Thanks to Peter Owens for the material on libel.

people are victims of communication abuse whenever we give them information that is less than the truth as we know it, as in the sections below.

Suppressing Knowledge the Public Deserves

Except for disasters that make big news (*Columbia*, Walkerton, Westray), people hear plenty about the *wonders* of technology: how fluoride eradicates tooth decay, how smart bombs and cruise missiles never miss their targets, how nuclear power will solve our energy problems— but they rarely hear about the failures or the dangers (Staudenmaier 67).

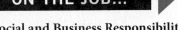

ON THE JOB...

Social and Business Responsibilities

"Although we feel we're ethical and we really do try to be socially responsible in our designs and our construction, we can fall into a trap. Clients make demands because they have their own pressures. Some of these demands may result in reduced quality if we follow the principle that 'the client gets what he wants'! I'm not always comfortable with that attitude, especially if we don't exercise due diligence in our pre-project research and in our project designs. . . ."

—Jan Bath, CADD Manager, MMM Group

In fact, the pressure to downplay failure sometimes results in censorship. For instance, some prestigious science journals have refused to publish studies linking chlorine and fluoride in drinking water with cancer risk, and fluorescent lights with childhood leukemia. The papers allegedly were rejected as part of widespread suppression of news about dangers of technological products (Begley 63).

In January 2008, the *New England Journal of Medicine* "published a study showing that a lot of negative research on antidepressant drugs never sees the light of day . . . 88 percent of clinical trials showing the drugs didn't work weren't published or were spun to make them look positive" (Taylor).

Another dramatic example occurred during the massive April 2010 British Petroleum oil spill in the Gulf of Mexico. BP misled the U.S. government and its own investors about the spill's severity, claiming that about 5000 barrels a day were spilling, even though BP's own engineers secretly estimated about three times that amount were escaping. Later, government and independent scientists estimated the spill at about 60 000 barrels per day. Kurt Mix, a BP engineer, was indicted for deleting a string of text messages that discussed the flow rate (Krauss and Schwartz).

Exaggerating Claims about Technology

Organizations that have a stake in a particular technology (say, bioengineered foods) may be especially tempted to exaggerate its benefits, potential, or safety, and to downplay the technology's risks.

Unwarranted claims help technology sell

An entrepreneur needs financiers. Scientists in a large corporation need advocates high enough in the hierarchy to allocate funds. And government-supported researchers at universities and national labs have an obvious incentive to overstate their progress and understate the problems that lie ahead: the better the chances for success, the more money an agency is willing to shell out. (Brody 40)

Stealing or Divulging Proprietary Information

Proprietary information is any document or idea that can be considered the exclusive property of the company in which it originated. Proprietary documents may include company records, test and experiment results, surveys paid for by clients, market research, minutes of meetings, plans, and specifications (Lavin 5). In theory, such information is legally protected, but it remains vulnerable to sabotage or theft. Rapid developments in technology create fierce competition among rival companies for the very latest intelligence, giving rise to measures like these:

Examples of corporate espionage

> Companies have been known to use business school students to garner information on competitors under the guise of conducting "research." Even more commonplace is interviewing employees for slots that don't exist and wringing them dry about their current employer. (Gilbert 24)

Moreover, employees within a company can leak confidential information to the press or to anyone else who has no legal right to know.

Hiding Conflicts of Interest

Can scientists and other experts who have a financial stake in a particular issue or experiment provide fair and impartial information about the topic?

- In one analysis of 800 scientific papers, Tufts University's Sheldon Krimsky found that 34 percent of authors had "research-related financial ties," but none had been disclosed (King B1).
- *Los Angeles Times* medical writer Terence Monmaney recently investigated 36 drug-review pieces in a prestigious medical journal and found "eight articles by researchers with undisclosed financial links to drug companies that market treatments evaluated in the articles" (qtd. in Rosman 100).

Falsifying or Fabricating Data

Research data might be manipulated or invented to support specific agendas (say by a scientist seeking grant money). Sometimes, timing is the issue: developments in fields, such as biotechnology, often occur too rapidly to allow for adequate peer review of articles before they are published (Turner, "Misconduct Scandal" 2).

Using Visual Images that Conceal the Truth

Pictures are generally more powerful than words and can easily distort the real meaning of a message. For example, U.S. law requires that TV commercials for prescription medications identify a drug's side effects, which can sometimes be serious. But drug commercials typically show images of smiling, healthy people at the same time as the side effects are listed. The happy images eclipse the sobering verbal message. (Direct-to-consumer advertising of prescription drugs is prohibited in Canada and in most other industrialized countries.)

Misusing Electronic Information

Much personal information is stored in databases (by schools, governments, banks, mail-order retailers, credit card companies, insurance companies), and, of course, all employers keep data about their employees. So how we combine, use, and share that information raises questions about privacy (Finkelstein 471). Also, a database is easy to alter; one simple command can change or wipe out facts. Private or inaccurate information can be sent from one database to many others, but "correcting information in one database does not guarantee that it will be corrected in others" (Turner, "Online Use" 5).

Here are some examples of the potential for communication abuse connected with web usage:

◆ plagiarizing or republishing electronic sources without getting permission or giving proper credit—although internet-based information may be free, the source of paraphrased or copied material must still be properly cited
◆ failing to safeguard the privacy of personal information about a website visitor's health, finances, buying habits, or affiliations
◆ publishing anonymous attacks or smear campaigns against people, products, or organizations
◆ offering inaccurate medical advice or information

Withholding Information People Need to Do Their Jobs

Nowhere is the adage that "information is power" more true than among co-workers. One sure way to sabotage a colleague is to withhold vital information about the task at hand.

Beyond the deliberate communication abuse of withholding information is this reality: *all* information is a matter of personal or social interpretation (Dombrowski 79); therefore, "objective reporting," practically speaking, is impossible. What we say on the job and how we say it are influenced by the expectations of our employer and by our own self-interest.

Exploiting Cultural Differences

Cross-cultural documents carry great potential for communication abuse. Based on its level of business experience, technological development, or financial need, a particular culture might be especially vulnerable to manipulation or deception. Some countries, for instance, are persuaded to purchase, from Canadian or American companies, pesticides and other chemicals whose use is banned in Canada or the U.S. Other countries witness depletion of their natural resources or exploitation of their labour force by more developed countries—all in the name of progress. All communication in all cultural contexts should embody universal standards of honesty and fairness.

APPLY GUIDELINES FOR ETHICAL COMMUNICATION

How do you balance self-interest with the interests of others—the organization, the public, your customers? How can you be practical and responsible at the same time? Here are two practical guidelines for ethical communication (G. Clark 194):

1. *Give the audience everything it needs to know.* To see things as clearly as you do, people need more than just a partial view. Don't bury readers in needless details, but do make sure they get all the key facts and get them straight.
2. *Give the audience a clear understanding of what the information means.* Even when all of the facts are known, they can be misinterpreted. Do all you can to ensure that your readers understand the facts as you do.

The Checklist for Ethical Communication on page 104 incorporates additional guidelines. Use the checklist for any document you prepare or for which you are responsible. Also, see science ethics resources at **www.files.chem.vt.edu/chem-ed/ethics**.

DECIDE WHERE AND HOW TO DRAW THE LINE

Suppose your employer asks you to do something unethical, such as alter data to cover up a violation of federal pollution standards. If you decide to resist, your choices seem limited: resign or go public (i.e., blow the whistle).

Walking away from a job isn't easy, however, and whistle-blowing can spell career disaster (Rubens 330). Many organizations refuse to hire anyone blacklisted as a whistle-blower (Wicclair and Farkas 19). Even if you aren't fired, expect your job to become hellish. Here is one communicator's gloomy assessment, based on personal experience:

> Most of the ethical infractions [you] witness will be so small that blowing the whistle will seem fruitless and self-destructive. And leaving one company for another may prove equally fruitless, given the pervasiveness of the problem. (Bryan 86)

Canadian law tends to protect employees who blow the whistle and are then dismissed. An employee has the right to bring a wrongful dismissal lawsuit and, in certain circumstances, to include human rights and/or labour-relations claims. The outcome of these claims varies.

Even if an employer doesn't act against a whistle-blowing employee, that employee might still feel compelled to leave the company, as in the case of Melvin Crothers, who revealed that his employer, WestJet, had spied electronically on its main competitor, Air Canada. In December 2003, Crothers accidentally discovered the spying. Unable to contact the WestJet CEO (who was on vacation) and morally shaken, he contacted a former WestJet manager who had moved to Zip Air, an Air Canada subsidiary. On April 6, 2004, Air Canada filed suit against WestJet. Four days later, Crothers felt that he had to quit working for WestJet, even though he loved working in the airline industry. He found it very difficult to find another airline job (Jang and Brethour).

Gordon McAdams, a Nelson, B.C.–based biologist, blew the whistle on a proposed road that would destroy the nesting sites of a population of endangered turtles. He was fired for revealing confidential provincial government documents. Eventually, he won the right to receive some of the salary and pension benefits that he lost when he was fired, and he was awarded the 2007 Whistleblower Award by Canada's Freedom of Information and Privacy Association. Also, the released documents helped convince the Supreme Court of British Columbia to halt government plans for the road. Still, Mr. McAdams found the whole experience very stressful (Hume).

ON THE JOB...

Ethics in Environmental Consulting

"We don't struggle with ethical reporting because of our professional approach to business. We carefully gather and interpret sufficient data to form reasonable, logical, and defensible conclusions. There's no advantage to our staff, our company, or our clients in cutting corners or steering data or conclusions towards a perceived desired outcome. In addition, most of our staff are professionals and as such are bound by a Code of Ethics that requires ethical conduct at all times."

—**Dr. Brian Guy, vice-president and general manager, Summit Environmental Consultants**

Employees covered by a contract or collective agreement can seek advice from their union or from a lawyer. Employees not covered by a contract or collective agreement often seek initial advice from local departments of labour or from their personal lawyer.

Anyone who reports employer violations to an agency such as the Canadian Centre for Occupational Health and Safety (CCOHS) and who is punished can ask the province's labour department to investigate. Employees whose claims are ruled valid can win reinstatement and reimbursement for back pay and legal expenses.

Even with such protections, an employee who takes on a company without the backing of a labour union or other powerful group can expect lengthy court battles, high legal fees (which may or may not be recouped), and disruption of life and career. Exactly where you draw the line (on having your integrity or health instead of your job) will be strictly your own decision.

If you do take a stand, be reasonable and cautious, and follow these suggestions (Unger 127–30):

- *Get your facts straight, and get them on paper.* Don't blow matters out of proportion, but do keep a paper trail in case of legal proceedings.
- *Appeal your case in terms of the company's interests.* Instead of being pious and judgmental ("This is a racist and sexist policy, and you'd better get your act together"), focus on what the company stands to gain or lose ("Promoting too few women and minorities makes us vulnerable to legal action").
- *Aim your appeal toward the right person.* If you have to go to the top, find someone who knows enough to appreciate the problem and who has enough clout to make something happen.
- *Get professional advice.* Contact a lawyer and your professional society for advice about your legal rights.

Before accepting a job offer, discreetly research the company's ethical reputation. (Of course, you can learn only so much about a company before actually working there.) Lynn Brewer, a former Enron executive who blew the whistle on Enron's unethical and illegal practices, describes early warning signals in her book *Confessions of an Enron Executive* and in her talks at university campuses.

Some companies have ombudspersons who help employees lodge complaints. Others offer hotlines for advice on ethics problems or for reporting violations. Also, realizing that good ethics is good business, companies increasingly are developing codes for personal and organizational behaviour. Without such supports, don't expect to last long as an ethical employee in an unethical organization.

Remember that very few employers tolerate any public statement, no matter how truthful, that makes the company look bad.

A final note: Sometimes the right choice is obvious, but often it is not so obvious. No one has any sure way of always knowing what to do. This chapter is only an introduction to the inevitable hard choices that, throughout your career, will be yours to make and to live with. For further guidance, see the website maintained by the Online Ethics Center for Engineering and Science, established by the Institute of Electrical and Electronics Engineers (IEEE), at **www.onlineethics.org**. This site covers a wide range of ethics issues for technical people trying to cope with difficult

This site is highly recommended!

situations. The site includes over 400 cases in addition to various codes of ethics, methods of evaluating ethical versus non-ethical behaviour, and projects that can help people develop ethical awareness and actions.

NOTE   *In July 2010, the WikiLeaks website became well known when it posted more than 90 000 internal U.S. military documents about the war in Afghanistan. WikiLeaks engages in whistle-blowing on a global scale, but it presages a time when internet sites will exist to facilitate whistle-blowing by insiders in any organization, big or small. WikiLeaks has its critics, who claim that publically releasing some information puts individuals and groups in harm's way.*

CHECKLIST

for Ethical Communication

Use this checklist [2] to help you write documents that reflect reasonable, ethical judgment.

◆ Am I reasonably sure this document will harm no innocent persons or damage their reputation?
◆ Am I respecting all legitimate rights to privacy and confidentiality?
◆ Do I inform people of the consequences or risks (as I am able to predict) of what I'm advocating?
◆ Do I state the case clearly, instead of hiding behind jargon and generalities?
◆ Do I give candid feedback or criticism, if it is warranted?
◆ Am I distributing copies of this document to every person who has the right to know about it?
◆ Do I credit all contributors and sources of ideas and information?
◆ Do I avoid exaggeration, understatement, sugarcoating, or any distortion or omission that leaves people at a disadvantage?

◆ Do I make a clear distinction between "certainty" and "probability"?
◆ Am I being honest and fair?
◆ Have I explored all sides of the issue and all possible alternatives?
◆ Are my information sources valid, reliable, and unbiased?
◆ Do I actually believe what I'm saying, instead of being a mouthpiece for groupthink or advancing some hidden agenda?
◆ Would I still advocate this position if I were held publicly accountable for it?
◆ Do I provide enough information and interpretation for people to understand the facts as I know them?

EXERCISES

In your workplace communications, you may end up facing hard choices concerning what to say, how much to say, how to say it, and to whom. Whatever your choice, it will have definite consequences. Discuss the following cases in terms of

◆ obligations faced by the person at the centre of the case (see page 94)
◆ appropriate moral values or ideals (see page 94)
◆ likely consequences of the action(s) you discuss (see page 94)

Case 1. In the scenario described in Chapter 1 on pages 10–12, Erika Song faces some difficult decisions. Should she omit or include data that reflects badly on the firm that employs her? Even though the data do not support conclusions that favour her company's bid for a contract extension and that show that her company's data gathering is not redundant, should she write those conclusions and improve her own chances of continuing employment? Should she "tell the truth," knowing that her boss will likely edit the report to enhance his firm's business potential?

Case 2. Review the scenario entitled, "CASE: A Hard Ethical Choice" (page 96). What would you do? Come to the class

---

2. Adapted from Brownell and Fitzgerald 18; Bryan 87; Johannesen 21–22; Larson 39; Unger 39–46; Yoos 50–55.

prepared to justify your decision on the basis of the obligations, ideals, and consequences involved.

Case 3. You are in the last semester of your college or university program. Your classmate, a hard worker who has earned good grades until recently, has been neglecting her classwork for the last month as she spends time with a younger sister who is dying of cancer. She approaches you to complete a report that she has researched and partly organized, but not written. The report, which is a major assignment in a technical writing course, is worth 40 percent of the final grade. It is due in three days. Your classmate did poorly on the last assignment and mid-term exam, and will likely fail the course if she doesn't get at least a C on this report. You have submitted your report, but you had planned to spend the next week studying hard for your final exams. She has offered to pay you $200 to write the report that she will submit under her name. What should you do?

Case 4. You are one of three employees being considered for a yearly production bonus, which will be awarded in six weeks. You've just accepted a better job, which you can start any time in the next two months. Should you wait until the bonus decision is made before announcing your plans to leave?

Case 5. While travelling on assignment that is being paid for by your employer, you visit an area in which you would really like to live and work, one where you have lots of contacts but never can find time to visit on your own. You have five days to complete your assignment, and then you must report on your activities. You complete the assignment in three days. Should you spend the remaining two days checking out other job possibilities, without reporting this activity?

Case 6. You have been authorized to hire a technical assistant, so you are about to prepare an advertisement. This is a time of threatened cutbacks for your company. People hired as temporary, however, have never seemed to work out well. Should your ad include the warning that this position could be only temporary?

Case 7. You are employed by a biotechnology company working on an AIDS vaccine. At a national conference, a researcher from a competing company secretly offers to sell your company crucial data that could speed discovery of an effective vaccine. Should you accept the offer?

COLLABORATIVE PROJECTS

1. In a group of three to five people, discuss the "right thing to do" in any one of the seven cases described above. In a memo to your supervisor, outline what action you would recommend and explain why. Be prepared to defend your group's ethical choice in class on the basis of the obligations, ideals, and consequences involved.

2. At **MyWritingLab: Technical Communication**, click on "Writing Exercises" and then on "Applying the Harvard Case Study Model to an Ethics Case." There, you'll find an example of how to adapt the Harvard model to help make hard ethical choices. Then apply the model to one or more of the above seven cases.

3. Discuss the ethical implications of taking money to "review" a product or service in a personal blog or discussion forum. Should the blogger disclose that he or she receives payment to discuss a product or service? Before accepting payment from an advertiser, should the blogger insist that the comments must state an honest opinion, not contrived praise for a product that the blogger has never tried or believes to be second-rate? Does "blogvertising" betray loyal readers of that blog? (To get an idea of the commercial aspect of this issue, visit **www.reviewme.com** or **http://johnchow.com**.)

MyWritingLab

MyWritingLab: Technical Communication offers the best multimedia resources for technical communication in one, easy-to-use place. You can find a variety of interactive model documents and case studies. There are also extensive guidelines, tutorials, and exercises for Document Design, Writing and the Writing Process, and Research, and a large bank of diagnostics and practice for grammar review.

7

Gathering Information

Explore

Click here in your eText to access a variety of student resources related to this chapter.

LEARNING OBJECTIVES

After reading this chapter, you should be able to

- adapt the standard stages of the research process to suit a particular project
- identify, select, and explore primary and secondary information sources
- plan and conduct surveys, interviews, and questionnaires

ON THE JOB...

Trends in Research

"The sheer amount of quality information in structured databases is increasing and readily available in electronic formats. It used to be mostly articles, but now reports are available, such as the Conference Board of Canada's e-library, which made its whole library available online and provided access to current information on industries, companies, and business trends in Canada. Statistics are becoming more and more available from a variety of government and private sources. So, the types of sources, the range of sources, and the sheer volume of sources are forcing students and other researchers to be very particular in the early stages of their research. . . ."

—Ross Tyner, director of library services, Okanagan College

Major workplace decisions are typically based on careful research, with the findings recorded in a report. The report's readers expect current information that helps answer their questions.

Research is classified as *primary* or *secondary*. Primary research involves an original, first-hand study of your topic or problem: observations, interviews, questionnaires, inquiry emails, personal experiments, analysis of samples, fieldwork, or company records. Secondary research includes materials published by other researchers: journal articles, books, handbooks, reports, online articles, electronic databases, government documents, internet sites, and material held by public agencies and special-interest groups.

Research strategies and resources differ widely among disciplines. This chapter focuses on research for preparing a formal technical report.

THINKING CRITICALLY ABOUT THE RESEARCH PROCESS

Research is a deliberate form of inquiry, a process of problem solving in which certain procedures follow a recognizable sequence, as shown in Figure 7.1 on page 107.

But research does not simply follow a numbered set of procedures ("First, do this; then, do that"). The procedural stages depend on the many decisions that accompany any legitimate inquiry (see Figure 7.2 on page 107).

ON THE JOB...

Advice for an In-Depth Library Search

"Step back, before typing words into a database or search engine or library catalogue. Think carefully about (a) the questions to be answered, (b) the types of sources that will have answers, and (c) where you can find those sources. Then plan a search strategy and write a list of key words. Approach the process methodically and systematically. The 'natural mistake' is to accept the first source that pops up. . . ."

—Ross Tyner, director of library services, Okanagan College

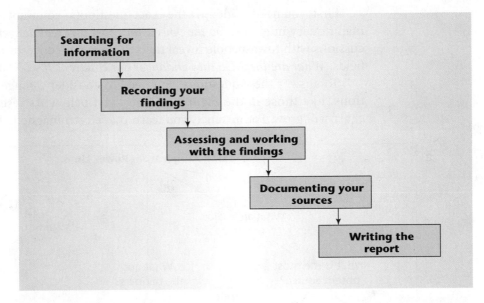

Figure 7.1
Procedural Stages of
the Research Process

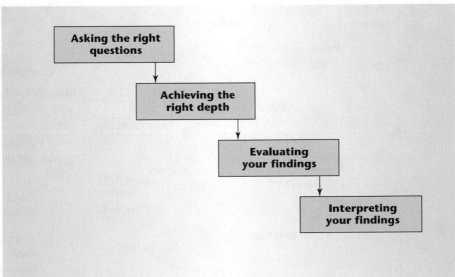

Figure 7.2
Inquiry Stages of the
Research Process

ASKING THE RIGHT QUESTIONS

The answers you uncover will depend on the questions you ask. Assume, for example, that you are faced with the following scenario.

| CASE | **Defining and Refining a Research Question** |

You are the public health manager for a small New Brunswick town in which high-tension power lines run within 30 metres of the elementary school. Parents are concerned about the danger from electromagnetic radiation (EMR) emitted by these power lines in energy waves known as electromagnetic fields (EMFs). Town officials ask you to research the issue and prepare a report.

First, you need to identify the exact question(s) you want answered. Initially, the major query might be, *Do the power lines pose any real danger to our children?* Discussions with townspeople reveal their three major concerns about electromagnetic fields: *What are they? Do they endanger our children? If so, what can be done?*

To answer these questions, you need to consider a range of subordinate questions, like those in the Figure 7.3 tree chart below. As research progresses, this chart will grow. For instance, you learn that electromagnetic fields radiate not only

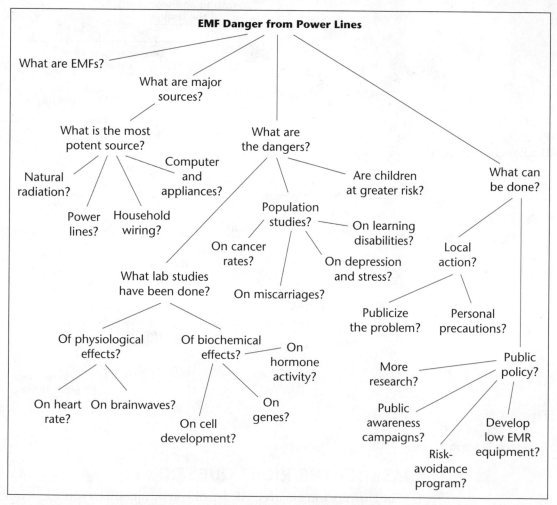

Figure 7.3 How the Right Questions Help Define a Research Problem

from power lines but also from all electrical equipment, and even from Earth itself. So you face an additional question: *Do power lines present the greatest hazard as a source of EMFs?* You now wonder whether the greater hazard comes from power lines or from other sources of EMF exposure. Critical thinking has enabled you to define and refine the essential questions.

Achieving Adequate Depth in Your Search

Balanced research examines a broad *range* of evidence; *thorough research*, however, examines that evidence at an appropriate *depth*. As depicted in Figure 7.4 below, different types of secondary information about any topic occupy different levels of detail and dependability.

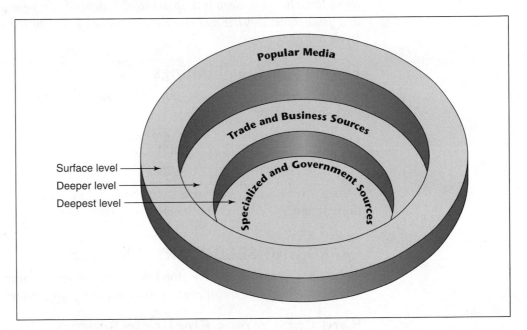

Figure 7.4 Effective Research Achieves Adequate Depth

1. At the surface level are items from the popular media (newspapers, radio, TV, general magazines). Designed for general consumption, this layer of information often contains more journalistic interpretation than factual detail.
2. At the next level are trade and business publications, such as *Construction Safety* magazine, published by the Construction Safety Association of Ontario, and *Dreams Alive*, an online home and garden design magazine. Designed for readers who range from moderately informed to highly specialized, this level focuses more on practice than on theory, on items newsworthy to group members, on issues affecting the field, on public relations, on viewpoints that tend to reflect the particular biases of that field.
3. At a deeper level is specialized literature (journals from professional associations: medical, legal, engineering, etc.). Designed for practising professionals, this level of information focuses on theory as well as practice, on descriptions of the latest studies—written by the researchers themselves and scrutinized by

others for accuracy and objectivity—on debates among scholars and researchers, and on reviews, critiques, and refutations of prior studies and publications.

Also at this deeper level are government sources and corporate documents. Designed for anyone willing to investigate its complex resources, this layer of information offers hard facts and highly detailed and (in many instances) *relatively* impartial views of virtually any issue or topic in any field. For example, the U.S. Environmental Protection Agency addresses the topic of EMF radiation at **www.epa.gov/radtown/power-lines.html.**

NOTE *Webpages, of course, offer links to increasingly specific levels of detail. But the actual "depth" and quality of a website's information depends on the sponsorship and reliability of that site. How deep is deep enough? It depends on your purpose, your audience, and your topic. But the real story and the hard facts more likely reside at the deeper levels of information.*

Evaluating Your Findings

As your research progresses, you should start to consider whether your gathered data are legitimate:

♦ Does the research answer the question you've posed on behalf of your reader?
♦ Are the data consistent? Reliable? Verifiable?
♦ Do you need more information?

Also, as you conduct your research, you should start to interpret the data to determine if findings conflict and whether you should reconsider your approach.

EXPLORING SECONDARY SOURCES

Although electronic searches for information have become the norm, a *thorough* search may require careful examination of hard-copy sources as well.

Hard Copy versus Electronic Sources

Every search medium has its advantages and drawbacks (as Table 7.1 shows), so there are good reasons for exploring both types.

Table 7.1 Hard Copy versus Electronic Sources: Benefits and Drawbacks

	Benefits	Drawbacks
Hard-Copy Sources	♦ discovered and organized by librarians ♦ easier to preserve and keep secure	♦ time-consuming and inefficient to search ♦ difficult to update
Electronic Sources	♦ more current, efficient, and accessible ♦ searches can be narrowed or broadened ♦ can offer material that has no hard-copy equivalent	♦ access to recent material only ♦ not always reliable: sources may be very biased ♦ user might get lost ♦ material may disappear

Types of Hard-Copy Sources

Many hard-copy sources are now published electronically

Where you begin your hard-copy search depends on whether you are searching for background and basic facts or the latest information. If you are an expert in the field, you might simply do a computerized database search or browse through specialized journals and listservs.

If you have limited knowledge or you need to focus your topic, you probably will want to begin with general reference sources, such as reference works (dictionaries, handbooks, almanacs, directories, indexes, abstracts, and bibliographies). Increasingly, however, such reference books are being published only in online versions, to which you will have access via your college or university library's subscription—ask the library's reference librarians for assistance in identifying the appropriate publications for your search.

> ### ON THE JOB...
>
> **Researching Electronic Sources**
>
> *"Research is far more convenient now. Most of our students do their research from home, or from wireless-enabled locations on campus. However, the electronic environment has made the process much more complex. In the old days, you could get, for example, Canadian population statistics from numbered publications on a bookshelf. But now that same data is buried in an online table somewhere, and it may even be a 'dynamic table' that requires the searcher to combine some variables. In such situations, research librarians can help students and faculty locate highly specific sorts of data. . . ."*
>
> —Ross Tyner, director of library services, Okanagan College

The Library Catalogue

Access points for an electronic card catalogue

All books, reference works, journal articles, and other materials held by a library are usually listed in a catalogue under author, title, and subject.

Nearly all libraries have automated their card catalogues, and many allow their users electronic access from outside locations, even from home. These electronic catalogues offer additional access points (beyond *author*, *title*, and *subject*):

> ### ON THE JOB...
>
> **The Role of the Library Catalogue**
>
> *"The library catalogue was once the first and most important tool for locating credible research sources, but now it is just one of many in the range of available tools. People are focusing more on material available in various online formats. The catalogue was designed as an access system for the book warehouse and the journal warehouse found in each library, and it is still important for research that requires book material."*
>
> —Ross Tyner, director of library services, Okanagan College

◆ *Descriptor*—for retrieving works on the basis of a keyword or phrase (e.g., *electromagnetic* or *power lines and health*) in the subject heading, in the work's title, or in the full text of its bibliographic record (its catalogue entry or abstract)
◆ *Document type*—for retrieving works in a specific format (e.g., video, audio, compact disk, or film)
◆ *Organizations and conferences*—for retrieving works produced under the name of an institution or professional association (e.g., Brookings Institution or Canadian Heart and Stroke Foundation)

- *Publisher*—for retrieving works produced by a particular publisher (e.g., Pearson Education Canada.)
- *Combination*—for retrieving works by combining any available access points (a book about a particular subject by a particular author or institution)

The library catalogue has an expanded role to play

New technology has expanded the role of the library catalogue, which now offers services that can be accessed at any time. A user, working at any location where internet access is available, can order interlibrary loans, renew books, or preview a book through Google Books.

ON THE JOB...

The Future of Academic Searches

"Academic research requires a linear approach that follows clear lines of thinking but many people these days are not linear thinkers. They prefer the mind map approach that reaches out in many directions. We're in a transitional stage and eventually academic database companies will find a way to coordinate their offerings so that free form hyperlinked searches through databases will be possible."

—**Margaret Hodgins, academic librarian, Okanagan College**

Two Modes of Academic Searches

Technological advances have allowed students and faculty to do more than gain remote access to their college or university library. Now, many educational institutions offer two parallel routes to find and explore sources: database searches and federated searches.

DATABASE SEARCHES. Many newspaper, magazine, and professional journal articles are stored as "full-text" electronic articles accessible through dozens of databases to which your university or college library subscribes. Database collections include JSTOR, Scopus, EBSCO, and the Applied Science and Technology Index.

"The main advantage of databases is that they're organized around subject areas, so they're relatively easy to search. Also, they've been screened for their quality, and while you're a student, you get them free through your institution's library" (Tyner).

Interlibrary loan

Most academic libraries in Canada subscribe to the same collections of journals and databases, "especially the ones that have proven to be the most relevant and useful. However, the big academic packages tend to have general application and lack articles with interest for specific regions." (Hodgins) Some libraries have special collections that your library does not have.

If your library does not hold the article you need, you can search an online database to identify a holding library, and then request the article through interlibrary loan. It may be sent as a photocopy, as a PDF attachment, or as an electronic full-text journal article.

Federated searches may not find all the resources held by the library

FEDERATED SEARCHES. Using an approach that mimics search engines, such as Google, academic researchers can now search multiple information resources using a single query and receiving a single list of resources available in different formats and different locations. This straightforward approach, which is known as *federated searching*, supplements the library catalogue. Although federated searches usually yield copious results, they do not necessarily search everything the library owns. Some databases can be accessed only by individual searches of those specialized databases.

Federated search engines at Canadian schools include the following examples:

- *McGill University*—McGill WorldCat Catalogue
- *University of Toronto*—Search All
- *Humber College*—Central Search
- *Ryerson University*—Search Everything
- *University of Saskatchewan*—USearch

- ◆ *Northern Alberta Institute of Technology*—Library Search Tool
- ◆ *Okanagan College*—OCtopus

Internet Sources

Today's internet connects about 2.5 billion computer users who have access to about 200 million *active* websites and addresses (uniform resource locators, or URLs). The number of sites increases by thousands daily across the globe.

ON THE JOB...

Google versus a Federated Search

"Google gives the impression that it provides access to everything, but it doesn't. Also, Google never says 'no' so you can get a lot of hits that have only peripheral interest or that are suspect sources. By contrast, a federated search produces sources that have been carefully reviewed by experts and selected by faculty in consultation with library staff. Finding the most relevant sources requires useful search terms and Boolean logic. That's why students should start their research in the library where they can get assistance from library staff."

—**Margaret Hodgins, academic librarian, Okanagan College**

WEB USAGE. Through the internet, you can search files, databases, and homepages. You can participate in various newsgroups, subscribe to discussion lists, send email inquiries, and gain access to publications that exist only in electronic form. Using a browser, you can explore sites on the web, locate experts in all types of specialties, read the latest articles in journals, such as *Nature* or *Science*, or review the latest listings of jobs in your specialty. For online databases in your field, ask your librarian.

Search engines such as Google, Bing, and Yahoo provide nearly instant results when keywords are plugged into their search boxes. For example, a recent Google search for "technical writer" netted nearly 57 million results in 0.20 seconds.

Google's "real-time" search function

Not satisfied with providing access to traditional websites, in December 2009 Google introduced what it calls "real-time search," listing new results as its engine picks them up. By doing so, Google can instantly link to social media posts. Thus, internet surfers will be able to see observers' posted notes as events happen, and surfers will be able to follow conversations and the latest information on Twitter, MySpace, and Facebook.

CAUTION

Assume that any material obtained from the internet is protected by copyright. Before using such material any place other than in a college or university paper (properly documented), obtain written permission from its owner.

ONLINE STARTING POINTS. A pair of collaborative online encyclopedias—Wikipedia and Annotum—provide an excellent beginning for many searches. Wikipedia, the larger and better known of the pair, claims a stable of 75 000 volunteer authors and well over 300 million users. It has wide coverage, but the uneven quality and depth of its articles leads commentators like Mathew Ingram to conclude that "Wikipedia is a starting point for research, not an ending point."

Annotum tries to address the issue of reliability that plagues some Wikipedia articles. Like Wikipedia, Annotum's sites are collaborative, but unlike Wikipedia, they name their authors and exercise editorial control over all entries.

Annotum's self-introduction, at **http://annotum.org/about/,** talks about developing "a simple, robust, easy-to-use authoring system to create and edit scholarly articles." Its developers also want to "deliver an editorial review and publishing system that can be used to submit, review, and publish scholarly articles."

A third site, Citizendium (**http://en.citizendium.org**) also has high ideals. It says that it represents "an endeavor to provide free knowledge with the highest standards of writing, reliability, and comprehensiveness." Unfortunately, it has not yet been able to fully realize its ambitions.

Another alternative is Quora, a question-and-answer site. It operates as a club that requires contributors and users to use their real names, not pseudonyms, when they sign up as members. Users may log in with their Facebook or Twitter accounts. Quora uses a complicated system of upvoting, downvoting, and contributor popularity to determine which "answers" get top billing on the site. Facebook, which has developed an app for Quora, is bullish about the site: "People—real doctors, economists, screenwriters, police officers, and military veterans—bring a wide range of expertise and first-hand experiences to make the content on Quora high quality and enlightening."

In response to criticism about its vetting system, in January 2013 Quora added a blogging component and mobile Rich Text editing so that "anyone with something brilliant to blog, even first-timers, can find a readership" (Constine).

ONLINE NEWSLETTERS, JOURNALS, AND RSS FEEDS.   Individuals can subscribe to specialized online newsletters and journals that come straight to subscribers' email inboxes. These sources of relevant content help people stay well informed about news and activities in their field.

Benefits of RSS feeds

An even more user-friendly way to stay up to date on specific information is through Really Simple Syndication (RSS) feeds. Such feeds use XML files to deliver content straight to a subscriber's computer via an RSS reader. Subscribers choose the kind of information they wish to receive; content-specific articles are taken from a variety of websites, news services, special-interest blogs, and technical updates.

Subscribers can choose RSS feeds that summarize articles and link to the original articles or RSS feeds that link to full articles directly. RSS readers are built into Internet Explorer and Firefox, and users have several other choices, most of which are free.

RSS feeds have an advantage over email delivery because email is often clogged with spam and many other messages competing for attention. However, "the most compelling use of RSS is that it lets users read dozens of websites, all on the same page. The sites can be scanned in seconds rather than having to be laboriously loaded individually" (*BBC Magazine* qtd. in PRESSfeed).

DEALING WITH THE DELUGE.   Proliferating web-based information has made research more accessible, but also more challenging. A 2008 Lexis Nexis survey found that seven out of ten American office workers "feel overwhelmed by information in the workplace, and more than two in five say they are headed for a breaking point" (Tahmincioglu 1). A 2010 follow-up survey confirmed the 2008 results. "The bad news is that wherever you find knowledge workers around the world, you'll also find information overload," said Michael Walsh, CEO of U.S. Legal Markets, Lexis Nexis.

Information purveyors are responding to the deluge of information with new strategies:

◆ RSS feeds go directly to the consumer, so researchers save search time and are not plagued by irrelevant data.

◆ ClearContext Corporation has developed applications that prioritize and organize emails so that receivers can quickly find the emails they want.

◆ Information repositories are used to store data that's not needed immediately. Also, "instead of pushing all information to employees' inboxes," some companies are creating information repositories where employees can find the data they need (Tahmincioglu 2).

For your own information search, see the following guidelines and heed Ross Tyner's advice on page 111 about researching electronic sources.

GUIDELINES   **FOR RESEARCHING ONLINE RESOURCES**

1. *Try to focus your search beforehand.* The more precisely you identify the information you seek, the lower your chance of wandering aimlessly through cyberspace.

2. *Select keywords or search phrases that are varied and technical, rather than general.* Some search terms generate better hits than others. In addition to *electromagnetic radiation*, for example, try *electromagnetic fields*, *power lines and health*, or *electrical fields*. Specialized terms (e.g., *vertigo* versus *dizziness*) offer the best access to sites that are reliable, professional, and specific. Always check your spelling.

3. *Use sites that are most specific and relevant to your needs and interests.* If, for instance, you're looking for news about international reaction to climate change, the news aggregator Google News is a good place to start. The latest news about financial markets can be found at an aggregator, such as Google Finance. Also, specialized newsletters and trade publications are good sources for site listings. For example, the Transportation Association of Canada has searchable online databases, including full-text articles.

4. *Expect limited results from any online search.* Each search engine (Google, AltaVista, Excite, HotBot, GO.com, WebCrawler, Yahoo, etc.) has its own strengths and weaknesses. Some are faster and more thorough, while others yield more targeted and updated hits. Some search titles only—instead of the full text—for keywords. In addition, studies show that "Web content is increasing so rapidly that no single search engine indexes more than about one-third of it" (Peterson). Broaden your coverage by using multiple search engines.

5. *Use Boolean logic to reduce the massive number of responses to a query.* Let's say that you're looking for information about the impact of power lines on health. *EMFs and health* results in over 5 million hits, but *EMFs AND health AND power lines* brings that list down to 317 000 hits. Then, *EMFs AND health AND power lines NOT USA* reduces the list to 245 000 hits. If you want to find sites that discuss the Harper government's views on the topic, add *AND "Stephen Harper"* (i.e., add quotation marks to the name) to the query string, and the results will yield just 6230 sites.

6. *Use natural language searches for general issues rather than specific topics.* Instead of using quotation marks to indicate specific terms and instead of word strings that include AND or NOT, enter general descriptive terms in any order. The results are often useful in supplementing a terms and connectors search, to get a more comprehensive set of resources. Here's an example: *effects of power lines on cancer rates, scientific studies.*

(continued)

7. *Use bookmarks and hot lists for quick access to favourite websites.* Mark a useful site with a bookmark, which you then add to your hot list.

8. *Expect material on the internet to have a brief lifespan.* Site addresses can change overnight; material is rapidly updated or discarded. If you find something of special value, save or print it before it changes or disappears.

9. *Get written authorization from the copyright holder before downloading copyrighted material.* The Copyright Policy Branch of the Department of Canadian Heritage says, "the same copyright rules which apply to conventional media also apply to new media such as the Internet. The fact that certain works may be accessible on the Internet does not imply that the copyright owner of these works authorizes reproduction or any other use." Only material in the public domain is exempted.

Before downloading *anything* from the internet, ask yourself, "Am I violating someone's privacy (as in forwarding an email or a newsgroup entry)?" or "Am I decreasing, in any way, the value of this material for the person who owns it?" Obtain permission beforehand and cite the source.

The following websites provide tips and tutorials for successful, efficient internet searches: **www.internettutorials.net**, **www.mta.ca/library/search_strategy.html**, and **www.mta.ca/library/gov_info.html**. Also, your college or university library likely has online help for internet research.

Source: Adapted from Baker 57; Branscrum; Busiel and Maeglin 39–40, 76; Fugate; Kawasaki, "Get Your Facts"; Matson.

Access Tools for Government Publications

The Canadian and U.S. governments publish maps, periodicals, books, pamphlets, manuals, monographs, annual reports, research reports, and a bewildering array of other information. The Canadian Research Index lists a wide range of Canadian federal and provincial government publications.

Types of Canadian information available to the public include ministerial and government proclamations, government bills and reports, judiciary rulings, and publications from all other government agencies. Much of this information can be searched online as well as in printed volumes. Your best bet for tapping this valuable but complex resource is to request assistance from the librarian in charge of government documents.

Here are some basic access tools for documents issued or published at government expense, as well as some privately issued documents. Fees are charged for most of these tools, but you can use many of them for free at your university or college library.

◆ *Micromedia ProQuest.* It provides access to Canadian news and periodical references, the Canadian Business and Current Affairs Index (CBCA), Canadian Newsstand, CBCA Education, and the Canadian Research Index and its microlog collection.

◆ *Canadian Research Index.* This comprehensive research guide includes all depository publications of research value issued by federal and provincial government agencies and departments. It also indexes scientific and technical reports issued by research institutes and government laboratories. The guide even lists policy, social, economic, and political reports, and theses and dissertations from Canadian universities.

A growing body of government information is posted to the internet. One starting point is the Government of Canada's federal departments and agencies homepage: **http://canada.gc.ca**; then choose "Departments and Agencies" from the side menu.

Examples of government agency postings include the following:

1. The Canadian Food Inspection Agency (**www.inspection.gc.ca**) provides links to information on a wide variety of topics, from avian influenza to plant imports. The agency also highlights current information on its homepage, such as food recalls and safety tips to avoid *E. coli.*

2. The U.S. Food and Drug Administration's electronic bulletin board lists information on experimental drugs to fight AIDS, drug and device approvals, recalls and litigations involving drugs or devices, health fraud, and a host of related items.

EXPLORING PRIMARY SOURCES

Work-related research is often based on primary research, an original, first-hand study of the topic, involving sources like those in Figure 7.5.

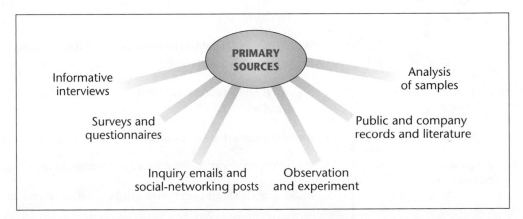

Figure 7.5 Sources for Primary Research

Informative Interviews

An excellent primary source of information unavailable in any publication is the personal interview. Much of what an expert knows may never be published (Pugliano 6). Also, a respondent might refer you to other experts or sources of information.

GUIDELINES   **FOR INFORMATIVE INTERVIEWS**

Planning the Interview

1. *Focus on your purpose.* Determine exactly what you hope to learn from this interview. Write out your purpose.

Purpose statement

> I will interview Anne Hector, chief engineer at Northport Electric, to ask her about the company's approaches to EMF risk avoidance—within the company as well as in the community.

2. *Do your homework.* Learn all you can about the topic beforehand. The more you know, the better your chance of getting the facts straight. If the respondent has published anything relevant, read it before the interview. Be sure the information this person might provide is unavailable in print.

3. *Contact the intended respondent.* Do this by telephone, letter, or email, and be sure to introduce yourself and your purpose.

4. *Request the interview at your respondent's convenience.* Give the respondent ample notice and time to prepare, and ask whether she or he objects to being quoted or recorded. If you use a digital recorder, insert fresh batteries and set the recording volume loud enough.

Preparing the Questions

5. *Make each question clear and specific.* Vague, unspecific questions elicit vague, unspecific answers.

A vague question

> How is this utility company dealing with the problem of electromagnetic fields?

> Which problem—public relations, potential liability, danger to electrical workers, to the community, or what?

A clear and specific question

> What safety procedures have you developed for risk avoidance among electrical work crews?

6. *Avoid questions that can be answered with a mere "yes" or "no."*

An unproductive question

> In your opinion, can technology find ways to decrease EMF hazards?

> Instead, phrase your question to elicit a detailed response:

A productive question

> Of the various technological solutions being proposed or considered, which do you consider most promising?

> This is one instance in which your earlier homework pays off.

7. *Avoid loaded questions.* A loaded question invites a particular bias:

A loaded question

> Wouldn't you agree that EMF hazards have been overstated?

(continued)

An impartial question does not lead the interviewee to respond in a certain way.

An impartial question

> In your opinion, have EMF hazards been accurately stated, overstated, or understated?

8. *Save the most difficult, complex, or sensitive questions for last.* Leading off with your toughest question might annoy respondents, making them uncooperative.

Conducting the Interview

1. *Make a good start.* Dress appropriately and arrive on time. Thank your respondent; restate your purpose; explain why you believe she or he can be helpful; explain exactly how you will use the information.
2. *Be sensitive to cultural differences.* If the respondent belongs to a culture different from your own, then consider the level of formality, politeness, directness, relationship building, and other behaviours considered appropriate in that culture.
3. *Let the respondent do the most talking.* Keep opinions to yourself.
4. *Be a good listener.* Don't doodle or let your eyes wander. People reveal more when the listener seems genuinely interested.
5. *Stick to your interview plan.* If the respondent wanders, politely nudge the conversation back on track (unless the added information is useful).
6. *Ask for clarification or explanation whenever necessary.* If you don't understand an answer, say so. Request an example, an analogy, or a simplified version—and keep asking until you understand.

Clarifying questions

   ◆ Would you go over that again please?
   ◆ Is there a simpler explanation?

Science writer Ronald Kotulak argues that "no question is dumb if the answer is necessary to help you understand something. . . . Don't pretend to know more than you do" (144).

7. *Keep checking on your understanding.* Repeat major points in your own words and ask if the technical details are accurate and if your interpretation is correct.
8. *Be ready with follow-up questions.* Some answers may lead to additional questions.

Follow-up questions

   ◆ Why is it like that?
   ◆ Could you say something more about that?
   ◆ What more needs to be done?
   ◆ What happened next?

9. *Keep note-taking to a minimum.* Record statistics, dates, names, and other precise data, but not every word. Jot key terms or phrases that later can refresh your memory.

Concluding the Interview

1. *Ask for closing comments.* Perhaps the respondent can lead you to additional information.

Concluding questions

   ◆ Have I missed anything?
   ◆ Would you care to add anything?

(continued)

◆ Is there anything I've neglected to ask?
◆ Is there anyone else I should talk to?
◆ Is there anyone who has a different point of view?

2. *Invite the respondent to review your version.* If the interview will be published, ask the respondent to check your final draft (for misspelled names, inaccurate details, misquotations, and so on) and to approve it. Offer to provide copies of any document in which this information appears.

3. *Thank your respondent and leave promptly.*

4. *As soon as you leave the interview, write a complete summary (or record one verbally).* Do this while responses are fresh in your memory.

Sources: Several guidelines are adapted from Blum 88; Dowd 13–14; Kotulak 147; McDonald 190; Rensberger 15; Hopkins-Tanne 23, 29; Young 114, 115.

Of course, an expert opinion can be just as mistaken or biased as anyone else's. Like patients who seek second opinions about serious medical conditions, researchers seek a balanced range of expert opinions about a complex problem or controversial issue—not only from a company engineer and environmentalist, for example, but also from an independent and presumably more objective third party, such as an academic or journalist who has studied the issue.

SELECTING AN INTERVIEW MEDIUM.   Once you decide whom to interview about what, select your medium carefully:

◆ *In-person interviews* are most productive because they allow human contact (Hopkins-Tanne 24).
◆ *Phone interviews* can be convenient and productive, but they lack the human contact of in-person interviews—especially when the interviewer and respondent have not met.
◆ *Email interviews* are convenient and inexpensive, and they allow plenty of time for respondents to consider their answers.

Whatever your medium, obtain a respondent's approval *beforehand*—instead of alienating the person with an unwanted surprise.

Surveys and Questionnaires

Surveys help us develop profiles and estimates about the concerns, preferences, attitudes, beliefs, or perceptions of a large, identifiable group (a *target population*) by studying representatives of that group (a *sample group*).

Surveys help us make assessments like these:

◆ Do consumers prefer brand A or brand B?
◆ What percentage of students feel safe on our campus?
◆ Is public confidence in technology increasing or decreasing?

The tool for conducting surveys is the questionnaire. While interviews allow for greater clarity and depth, questionnaires offer an inexpensive way to survey a large group. Respondents can answer privately and anonymously—and often in more detail than in an interview.

However, questionnaires carry certain limitations:

Limitations of survey research

- ◆ *A low rate of response (often less than 30 percent).* People refuse to respond to a questionnaire that seems too long, complicated, or in some way threatening. They might be embarrassed by the topic or afraid of how their answers could be used.
- ◆ *Responses that might be non-representative.* A survey will get responses from the people who want to respond, but you will know nothing about the people who didn't respond. Those who responded might have extreme views, a particular stake in the outcome, or some other motive that represents inaccurately the population being surveyed (Plumb and Spyridakis 625–26).
- ◆ *Lack of follow-up.* Survey questions do not allow for the kind of follow-up and clarification possible with interview questions.

Even surveys by professionals carry the potential for error. As consumers of survey research, we need to understand how surveys are designed, administered, and interpreted, and what can go wrong in the process.

DEFINING THE SURVEY'S PURPOSE.   Why is this survey being done? What, exactly, is it measuring? How much background research is needed? How will the survey findings be used?

DEFINING THE TARGET POPULATION.   Who is the exact population being studied ("the chronically unemployed," "part-time students," "computer users")?

IDENTIFYING THE SAMPLE GROUP.   How will intended respondents be selected? How many respondents will there be? Generally, the larger the sample surveyed, the more dependable the results (assuming a well-chosen and representative sample). Will the sample be randomly chosen? In the statistical sense, *random* does not mean "chosen haphazardly": a random sample means that any member of the target population stands an equal chance of being included in the sample group.

Even a sample that is highly representative of the target population carries a measure of *sampling error*.

A type of survey error

The particular sample used in a survey is only one of a large number of possible samples of the same size that could have been selected using the same sampling procedures. Estimates derived from the different samples would, in general, differ from each other. (U.S. Department of Commerce 949)

The larger the sampling error (usually expressed as the *margin of error*), the less dependable the survey findings.

DEFINING THE SURVEY METHOD.   What type of data (opinions, ideas, facts, figures) will be collected? Is timing important? How will the survey be conducted—in person, by mail, or by telephone? How will the data be collected, recorded, analyzed, and reported (Lavin 277)?

Telephone, email, and in-person surveys yield fast results, but respondents consider telephone surveys annoying and, without anonymity, people tend to be less candid. Telephone surveys generate high response rates, but mail surveys are less expensive and more confidential. Computerized surveys create the sense of a video game: the program analyzes each response and automatically designs the next question. Respondents who dislike being quizzed by a human researcher seem more comfortable with this automated format (Perelman 89–90).

GUIDELINES FOR INFORMATIVE INTERVIEWS.

1. *Decide on the types of questions* (Adams and Schvaneveldt 202–12; Velotta 390). Questions can be *open-ended* or *closed-ended*. Open-ended questions allow respondents to express exactly what they're thinking or feeling in a word, phrase, sentence, or short essay, as in these questions about EMFs and health:

Open-ended questions

How much do you know about electromagnetic radiation at our school?

What do you think should be done about electromagnetic fields (EMFs) at our school?

2. Since one never knows what people will say, open-ended questions are a good way to uncover attitudes and obtain unexpected information. But essay-type questions are difficult to answer and tabulate.

3. When you want to measure where people stand on an issue, choose closed-ended questions:

Closed-ended questions

Are you interested in joining a group of concerned parents?

YES _____ NO _____

Rate your degree of concern about EMFs at our school.

HIGH _____ MODERATE _____ LOW _____ NO CONCERN _____

Circle the number that indicates your view about the town's proposal to spend $20 000 to hire its own EMF consultant.

1..........2..........3..........4..........5..........6..........7

Strongly   No   Strongly
Approve   Opinion   Disapprove

Respondents may be asked to *rate* one item on a scale (from high to low, best to worst), to *rank* two or more items (by importance, desirability), or to select items from a list. Other questions measure percentages or frequency.

How often do you … ?

ALWAYS _____ OFTEN _____ SOMETIMES _____ RARELY _____ NEVER _____

Although closed-ended questions are easy to answer, tabulate, measure, and analyze, they might elicit biased responses. Some people, for instance, automatically prefer items near the top of a list or the left side of a rating scale (Plumb and Spyridakis 633). Also, people are prone to agree rather than disagree with assertions in a questionnaire (Sherblom, Sullivan, and Sherblom 61).

4. *Design an engaging introduction and opening questions.* Persuade respondents that the survey relates to their concerns, that their answers matter, and that their

anonymity is assured. Explain how respondents will benefit from your findings, or offer an incentive (such as a copy of your final report).

A survey introduction

Your answers will enable our school board to speak accurately for your views at our next town meeting. Results will appear in our campus newspaper. Thank you.

Researchers often include a cover letter with the questionnaire.

Begin with the easiest questions. Once respondents commit to these, they are likely to complete more difficult questions later.

5. *Make each question unambiguous.* All respondents should be able to interpret the same question identically. An ambiguous question leaves room for misinterpretation.

An ambiguous question

Do you favour weapons for campus police? YES _____ NO _____

6. *Weapons* might mean tear gas, clubs, handguns, all three, or two out of three. Consequently, responses to the above question would produce a misleading statistic, such as "Over 95 percent of students favour handguns for campus police," when the accurate conclusion might be "Over 95 percent of students favour some form of weapon." Moreover, the limited choice ("yes/no") reduces an array of possible opinions to an either/or choice.

A clear, incisive question

Do you favour (check all that apply):

_____ Having campus police carry mace and clubs?
_____ Having campus police carry non-lethal stun guns?
_____ Having campus police store handguns in their cruisers?
_____ Having campus police carry small-calibre handguns?
_____ Having campus police carry large-calibre handguns?
_____ Having campus police carry no weapons?
_____ Don't know

To ensure a full range of possible responses, include options such as "Other _____," "Don't know," "Not applicable," or an "Additional comments" section.

7. *Make each question unbiased.* Avoid *loaded questions* that invite or advocate a particular viewpoint or bias:

A loaded question

Should our campus tolerate the needless endangerment of innocent students by lethal weapons?
YES _____ NO _____

Emotionally loaded and judgmental words (*endangerment, innocent, tolerate, needless, lethal*) in a survey are unethical because their built-in judgments manipulate people's responses (Hayakawa 40).

8. *Make it brief, simple, and inviting.* Try to limit questions and response space to two sides of a single page. Include a stamped, return-addressed envelope, and give a specific return date. Address each respondent by name, sign your letter or your introduction, and give your title.

NOTE *A tool, such as the U.S.-based SurveyMonkey, helps people create surveys. FluidSurveys bills itself as a Canadian alternative. Both enterprises charge for their survey creation and analytics services, but FluidSurveys has a free option that can be attractive to researchers on a limited budget. That option allows an unlimited number of surveys but each survey is limited to 20 questions and 150 responses to the survey. Surveys can be distributed through web links, mobile devices, tablets, social media, or email.*

A SAMPLE QUESTIONNAIRE.   The student-written questionnaire in Figure 7.6 on page 125, sent to presidents of local companies, is designed to elicit responses that can be tabulated easily. Written reports of survey findings usually include an appendix that contains a copy of the questionnaire as well as the tabulated responses.

Inquiries

Letters, phone calls, or email inquiries to experts listed in webpages are handy for obtaining specific information from government agencies, legislators, private companies, university researchers, trade associations, and research foundations.

Office Files

Organization records (reports, memos, computer printouts, etc.) are good primary sources. Most organizations also publish pamphlets, brochures, annual reports, or prospectuses for consumers, employees, investors, or voters. But be alert for bias in company literature. If you were evaluating the safety measures at a local nuclear power plant, you would want the complete picture. Along with the company's literature, you would also want studies and reports from government agencies and publications from environmental groups.

Observation and Experiments

If possible, amplify and verify your findings with a first-hand look. Observation should be your final step because you now know what to look for. Have a plan. Know how, where, and when to look, and jot down observations immediately. You might even take photos or make drawings.

Informed observations can pinpoint real problems. Here is an excerpt from a report investigating low morale at an electronics firm. This researcher's observations and interpretation are crucial in defining the problem:

Direct observation is often essential

> Our on-site communication audit revealed that employees were unaware of any major barriers to communication. Over 75 percent of employees claimed they felt free to talk to their managers, but the managers, in turn, estimated that fewer than 50 percent of employees felt free to talk to them.
>
> The problem involves misinterpretation. Because managers don't ask for complaints, employees are afraid to make them, and because employees never ask for an evaluation, they never get one. Each side has inaccurate perceptions of what the other side expects, and because of ineffective communication, each side fails to realize that its perceptions are wrong.

Communication Questionnaire

1. Describe your type of company (e.g., manufacturing, high tech) _____

   _____

2. Number of employees? (Please check one.)

_____ 0–4	_____ 26–50	_____ 101–150	_____ 301–450
_____ 5–25	_____ 51–100	_____ 151–300	_____ 450+

3. What types of written communication occur in your company? (Label by frequency: never, rarely, sometimes, often.)

_____ memos	_____ letters	_____ advertising
_____ manuals	_____ reports	_____ newsletters
_____ procedures	_____ proposals	_____ other (specify)
_____ email	_____ catalogues	_____

4. Who does most of the writing? (Please give titles.) _____

   _____

5. Please characterize your employees' writing effectiveness.

   _____ good   _____ fair   _____ poor

6. Does your company have formal guidelines for writing?

   _____ no   _____ yes   (Please describe briefly.) _____

   _____

7. Do you offer in-house communication training?

   _____ no   _____ yes   (Please describe briefly.) _____

   _____

8. Please rank the usefulness of the following areas in communication training (from 1 to 10, 1 being the most important).

_____ organization information	_____ audience awareness
_____ summarizing information	_____ persuasive writing
_____ editing for style	_____ grammar
_____ document design	_____ researching
_____ email etiquette	_____ webpage design

   _____ other   (Please specify.) _____

9. Please rank these skills in order of importance (from 1 to 6, 1 being the most important).

_____ reading	_____ listening	_____ speaking to groups
_____ writing	_____ collaborating	_____ speaking face to face

10. Do you provide tuition reimbursement for employees?

    _____ no   _____ yes

11. Would you consider having UMD communication interns work for you part-time?

    _____ no   _____ yes

12. Should UMD offer Saturday seminars in communication?

    _____ no   _____ yes

    Additional comments/suggestions: _____

    _____

Figure 7.6 A Sample Survey

An experiment is a controlled form of observation designed to verify an assumption (e.g., the role of fish oil in preventing heart disease) or to test something untried (the relationship between background music and worker productivity). Each specialty has its own guidelines for experiment design.

Even direct observation is not foolproof. For instance, you might be biased about what you see (focusing on the wrong events or ignoring something important), or instead of behaving normally, people being observed might behave in ways they think are expected of them (Adams and Schvaneveldt 244).

Analysis of Samples

Workplace research can involve collecting and analyzing samples: water, soil, or air, for contamination and pollution; foods, for nutritional value; ore, for mineral value; or plants, for medicinal value. Investigators analyze material samples to find the cause of an airline accident. Engineers analyze samples of steel, concrete, or other building materials to determine their load-bearing capacity. Medical specialists analyze tissue samples for disease.

ON THE JOB...

Social Networking and Research

"The internet provides access to experts around the world. Topic-specific chat rooms, technical discussion threads, and Facebook groups all can facilitate the sharing of information. There are so many collaborative and special interest groups; you can post a question and somebody will answer. If not, someone will advise you where to start looking. Facebook is a valid research tool, although you would cite what someone has posted on Facebook (or on another social networking site) **only** *if that person has established credibility regarding that topic."*

—**Ross Tyner, director of library services, Okanagan College**

Social Networking for Primary Sources

Most of the people you will interview will be identified through your personal networking group, or through your professors, colleagues, friends, family, or fellow students. Your network can be used to find and contact primary information sources.

A variety of social-networking sites now facilitate additional ways of identifying useful sources. You can post a question on Facebook or Twitter. You can join a professional association that has a discussion board. You can use a search engine to find experts in various fields. (Several of the sources quoted in the On the Job boxes in this text were discovered through online searches.)

When students who have questions that emanate from an article, librarian Margaret Hodgins encourages them to contact the author of that article: "Often an author will have a Facebook account or other Web presence," she says. She also says that students in collaborative groups "naturally gravitate toward using social networks as a way of sharing information and sources because they find that social networks are so fast and easy to use."

EXERCISES

1. Begin researching for the analytical report (Chapter 19) due at semester's end. Complete these steps. (Your instructor might establish a timetable.)

   Phase One: Preliminary Steps
   a. Choose a topic of *immediate practical importance*, something that affects you or your community directly.
   b. Identify a specific audience and its intended use of your information. Complete an audience/purpose profile (Chapter 2, page 33).
   c. Narrow your topic, and check with your instructor for approval.
   d. Make a working bibliography to ensure sufficient primary and secondary resources. Don't delay this step!
   e. List things you already know about your topic.
   f. Write a clear statement of purpose and submit it in a proposal memo (Chapter 18, page 329) to your instructor.
   g. Develop a tree chart of possible questions (as on page 108).
   h. Make a working outline.

Phase Two: Collecting Data (read Chapter 8 in preparation for this phase)

a. In your research, move from general to specific; begin with general reference works for an overview.

b. Skim your material, looking for high points.

c. Take selective notes. Don't write everything down! Use notecards.

d. Plan and administer questionnaires, conduct interviews, and send inquiry emails.

e. Whenever possible, conclude your research with direct observation.

f. Evaluate and interpret your findings.

g. Use the checklist on page 148 to reassess your research methods and reasoning.

Phase Three: Organizing Your Data and Writing Your Report

a. Revise and adjust your working outline, as needed.

b. Compose an audience/purpose profile sheet, like the sample in the Feasibility Assessment Case in Chapter 19 on pages 366–367.

c. Fully document all sources of information.

d. Proofread carefully and add all needed supplements (title page, letter of transmittal, abstract, summary, appendix, glossary).

Due Dates: (To be assigned by your instructor)
List of possible topics due:
Final topic due:
Proposal memo due:
Working bibliography and working outline due:
Notecards due:
Copies of questionnaires, interview questions, and inquiry emails due:
Revised outline due:
First draft of report due:
Final draft with supplements and documentation due:

2. Using the printed or electronic card catalogue, locate and record the full bibliographic data for five books in your field or on your semester report topic, all published within the past year.

3. List the title of each of these specialized reference works in your field or on your topic: a bibliography, an encyclopedia, a dictionary, a handbook, an almanac (if available), and a directory.

4. Identify the major periodical index in your field or on your topic. Locate a recent article on a specific topic (e.g., use of artificial intelligence in medical diagnosis). Photocopy the article (get Access Copyright clearance) and write an informative abstract.

5. Revise these questions to make them appropriate for inclusion in a questionnaire:

a. Would a female prime minister do the job as well as a male?

b. Don't you think that euthanasia is a crime?

c. Do you oppose increased government spending?

d. Do you think welfare recipients are too lazy to support themselves?

e. Are teachers responsible for the decline in literacy among students?

f. Aren't humanities studies a waste of time?

g. Do you prefer Rocket Cola to other leading brands?

h. In meetings, do you think men interrupt more than women?

6. Arrange an interview with someone in your field. Decide on general areas for questioning: job opportunities, chances for promotion, salary range, requirements, outlook for the next decade, working conditions, job satisfaction, and so on. Compose specific interview questions; conduct the interview, and summarize your findings in a memo to your instructor.

COLLABORATIVE PROJECT

Divide into small groups, and decide on a campus or community issue or some other topic worthy of research. Elect a group manager to assign and coordinate tasks. At project's end, the manager will provide a performance appraisal by summarizing, in writing, the contribution of each team member. Assigned tasks will include planning, information gathering from primary and secondary sources, document preparation (including visuals) and revision, and classroom presentation.

See pages 60–61 for collaboration guidelines. Conduct the research, write the report, and present your findings to the class

MyWritingLab

MyWritingLab: Technical Communication offers the best multimedia resources for technical communication in one, easy-to-use place. You can find a variety of interactive model documents and case studies. There are also extensive guidelines, tutorials, and exercises for Document Design, Writing and the Writing Process, and Research, and a large bank of diagnostics and practice for grammar review.

8

Recording and Reviewing Research Findings

Explore

Click here in your eText to access a variety of student resources related to this chapter.

LEARNING OBJECTIVES

After reading this chapter, you should be able to

- use a consistent, recoverable method of recording research findings
- carefully record all information sources
- evaluate the quality and value of the researched data by answering pertinent questions about information sources and about the evidence that supports findings
- carefully evaluate in evaluating internet-based material
- interpret the data using effective inductive and deductive reasoning, and avoid fallacies

As you discover material during research, you confront questions like these: *How much is worth keeping? How should I record it? Can I trust this information? What, exactly, does it mean? How will I credit the source?* These latter stages of the research process require the same quality of critical thinking as the earlier stages discussed in Chapter 7.

RECORDING THE FINDINGS

Findings should be recorded in ways that enable you to easily locate, organize, and control the material as you work with it. So record primary research findings with a computer, photographs, drawings, a digital recorder, or any medium that suits your purpose. Record secondary research findings as notes.

Taking Notes

Notecards are convenient because they are easy to organize and reorganize. In place of notecards, many researchers take notes on a laptop or tablet, using information or database management software that allows notes to be filed, rearranged, and retrieved by author, title, topic, date, and so forth.

Follow these suggestions when making notes:

1. Make a separate bibliography card (see Figure 8.1 on page 129) or computer file for each work you plan to consult. Record the complete entry, using the identical citation format that will appear in your document. When searching an online catalogue, you can print or electronically save the full bibliographic record for each work, thereby ensuring accurate citation.

2. Skim the entire work to locate relevant material.
3. Decide what to record. Use a separate card or electronic file for each item.
4. Decide how to record the item: as a quotation or a paraphrase. When quoting others directly, be sure to record words and punctuation accurately. When restating or adapting material in your own words, be sure to preserve its original meaning and emphasis. **Record the sources of all internet-based material.**

Quoting the Work of Others

When you quote the exact words from another person's work—whether the words were written or spoken (as in an interview or presentation) or whether they appeared in electronic form—you must place quotation marks around those words. Even a single borrowed sentence or phrase, or a single word used in a special way, needs quotation marks, with the source fully and properly cited.

If your notes fail to identify quoted material accurately, you may forget to credit the source in your report. Even when this omission is unintentional, writers face the charge of *plagiarism* (misrepresenting as one's own the words or ideas of someone else). Possible consequences of plagiarism include expulsion from school, the loss of your job, or a lawsuit.

In recording a direct quotation, copy the selection word for word (see Figure 8.2 on page 130) and include the page numbers. If your quotation omits parts of a sentence, use an *ellipsis* (three periods: . . .) to indicate each part that you have omitted from the original. If your quotation omits the end of a sentence, the beginning of the subsequent sentence, or whole sentences or paragraphs, use an ellipsis with four periods (. . . .).

Ellipses within and between sentences

If your quotation omits parts . . . use an ellipsis. . . . If your quotation omits the end. . . .

Be sure that your elliptical expression is grammatically correct and that the omitted material in no way distorts the original meaning.

Record each bibliographic citation exactly as it will appear in your final report

Pinsky, Mark A. *The EMF Book: What You Should Know about Electromagnetic Fields, Electromagnetic Radiation, and Your Health.* New York: Warner, 1995.

Figure 8.1 Bibliography Card

Place quotation marks around all directly quoted material

Pinsky, Mark A. pp. 29–30.

"Neither electromagnetic fields nor electromagnetic radiation cause cancer *per se*, most researchers agree. What they may do is promote cancer. Cancer is a multistage process that requires an 'initiator' that makes a cell or group of cells abnormal. Everyone has cancerous cells in his or her body. Cancer—the disease as we think of it—occurs when these cancerous cells grow uncontrollably."

Figure 8.2 Notecard for a Quotation

If you insert your own words within the quotation, place them inside brackets to distinguish them from those of your source:

Brackets setting off personal comments within quoted material

This profession [aircraft ground controller] requires exhaustive attention.

Sentences and paragraphs that include quotations must be clear and understandable. Read your sentences aloud to be sure they make sense and they read smoothly and grammatically. Generally, integrated quotations are introduced by phrases such as, "Wong argues that" or "Dupuis suggests that," so that readers know who said what. More importantly, readers must see the relationship between the quoted idea and the sentence that precedes it. Use a transitional phrase that emphasizes this relationship by looking back as well as ahead:

An introduction that unifies a quotation with the discussion

After you decide to develop a program, "the first step in the programming process . . ."

Besides showing how each quotation helps advance the main idea you are developing, your integrated sentences should also be grammatical:

Quoted material integrated grammatically with the writer's words

"Quora's blogging capabilities," Franson acknowledges, "have been substantially improved by its new blogging platform."

"She has rejuvenated the industrial economy of our region," Smith writes of Berry's term as regional planner.

Use a direct quotation only when precision, clarity, or emphasis requires the exact words from the original. Avoid excessively long quoted passages. Research writing is more a process of independent thinking, in which you work with the ideas of others in order to reach your own conclusions; you should, therefore, paraphrase, instead of quoting much of your borrowed material.

Paraphrasing the Work of Others

We paraphrase not only to preserve the original idea but also to express it in a clear, simple, direct, or emphatic way—without distorting the idea. *Paraphrasing* means more than changing or shuffling a few words; it means restating the original idea in your own words and giving full credit to the source.

To borrow or adapt someone else's ideas or reasoning without properly documenting the source is plagiarism. To offer as a paraphrase an original passage only slightly altered—even when you document the source—also is plagiarism. It is equally unethical to offer a paraphrase, although documented, that distorts the original meaning.

Effective paraphrases display all or most of these elements (Weinstein 3):

Elements of an effective paraphrase

- ◆ reference to the author early in the paraphrase, to indicate the beginning of the borrowed passage
- ◆ keywords retained from the original, to preserve the meaning
- ◆ original sentences restructured and combined, for emphasis and fluency
- ◆ needless words from the original deleted, for conciseness
- ◆ your own words and phrases that help explain the author's ideas, for clarity
- ◆ a citation (in parentheses) of the exact source, to mark the end of the borrowed passage and to give full credit
- ◆ preservation of the author's original intent

Figure 8.3 shows an entry paraphrased from the passage in Figure 8.2. Paraphrased material does not have quotation marks, but you must acknowledge your debt to the source. Failing to acknowledge ideas, findings, judgments, lines of reasoning, opinions, facts, or insights not considered *common knowledge* is plagiarism—even when these are expressed in your own words.

Signal the beginning of the paraphrase by citing the author, and the end by citing the source

Pinsky, Mark A.

Pinsky explains that electromagnetic waves probably do not directly cause cancer. However, they might contribute to the uncontrollable growth of those cancer cells normally present—but controlled—in the human body (29–30).

Figure 8.3 Notecard for a Paraphrase

IN BRIEF

Copyright Protection and Fair Dealing of Printed Information

Copyright Law
A copyright is the exclusive legal right to reproduce, publish, and sell a literary, dramatic, musical, or artistic work. The law grants the copyright owner the exclusive rights to do and to authorize any of the following:

1. to reproduce the copyrighted work
2. to prepare derivative works

3. to distribute copies of the copyrighted work to the public by sale, rental, lease, or lending
4. in certain cases, such as for literary and musical works, to perform the copyrighted work publicly
5. in certain cases, such as for graphics, images, or other audiovisuals, to display the copyrighted work publicly

(continued)

You must obtain written permission to use all copyrighted material. Works are copyrighted for the author's life plus 50 years. If the author is unknown, the copyright lasts for 50 years from the document's publication date.

Public Domain

Public domain refers to material on which copyright has expired or material that is not protected by copyright. Commonplace pieces of information, such as height and weight charts, are in the public domain. These works occasionally contain copyrighted material used with permission and properly acknowledged. **However, a new translation or version of a work in the public domain can be protected by copyright**; if you are not sure whether something is in the public domain, obtain permission. In Canada, government publications are *not* in the public domain.

Fair Dealing

Fair dealing allows quotes from, or reproduction of, minor excerpts of a copyrighted work, if the quotation or reproduction is for *bona fide* private study, research, criticism, or newspaper summary. But there's a problem: it's difficult to tell the difference between fair dealing and copyright infringement. The limits of fair dealing copying have been broadly defined by Bill C-11, the *Canadian Copyright Modernization Act*, which received Royal Assent on June 29, 2012. For details of how the new law affects educational institutions and its students and educators, see the University of British Columbia's overview at **http://copyright.ubc.ca/copyright-legislation/bill-c-11-the-copyright-modernization-act/.**

Copyright law provides the following criteria to be considered in the determination of fair dealing:

1. the purpose and character of the use, including whether such use is of a commercial nature or is for non-profit educational purposes
2. the nature of the copyrighted work
3. the amount and substantiality of the portion used in relation to the whole
4. the effect of the use upon the potential market for or value of the copyrighted work

When the quoted material forms the core, distinguishable creative effort of the work being cited, use of the material without permission isn't considered fair.

Fair dealing ordinarily does not apply to use of poetry, musical lyrics, dialogue of a play, entries in a diary, case studies, charts and graphs, author's notes, private letters, testing materials, or quotations for use as epigraphs.

Copying under the Access Copyright Licence

Many Canadian college, university, high school, and public libraries have signed licence agreements with Access Copyright, a non-profit organization representing Canadian and foreign authors and publishers. Under such licences, library users can copy certain portions of certain kinds of documents without permission. **However, each licence imposes clear legal restrictions**, so ask your librarian for details.

Source: Based on Canada's *Copyright Modernization Act*, on UBC's summary of Bill C-11, and on material from Access Copyright, **www.accesscopyright.ca.**

IN BRIEF

Copyright Protection and Fair Dealing of Electronic Information

The Problem

Copyright and fair dealing law is quite specific for printed works or works in other tangible form (e.g., paintings, photographs, music, patents, trademarks, integrated circuit designs). But it's harder to define fair dealing of intellectual property in electronic form. Also, movies, music, pictures, technical information, and many other forms of

data and entertainment are freely available on the internet, or on mobile phones, or through digital copying (Makarenko).

Information obtained via email or discussion groups presents additional problems: sources often do not wish to be quoted or named or to have early drafts made public. How do we protect source confidentiality? How do we avoid infringing on works in

(continued)

progress that have not yet been published? How do we quote and cite this material without violating ownership and privacy rights (Howard)?

Solutions Attempted through Electronic Copyright Law

Subscribers to commercial online databases pay fees, and copyholders in turn receive royalties (Communication Concepts, Inc. 13). The Canadian federal government has succumbed to pressure from lobby groups in the United States to enact the kind of tough legislation found in the *Digital Millennium Copyright Act* (DMCA), which expressly forbids any form of unauthorized copying of electronic files. In addition, the *Stop Online Piracy Act* (SOPA) aims to stop online piracy, particularly from foreign copyright infringers. The United States, in an attempt to coerce Canada to enact similar legislation, has placed Canada on a "Watch List" of jurisdictions that harbour potential content pirates. Also, the United States has raised the issue of stringent copyright laws as part of negotiations concerning trade and other matters of common interest to the two nations (Geist).

Responding to that pressure, the government has produced and passed Bill C-11, which has broadened the scope of the fair dealing exception to specifically include education, research, private study, criticism, and review. However, consumers are strongly targeted by the legalization and protection of digital locks. These "locks" are software or hardware mechanisms that prevent unauthorized copying of digital files.

There's a real problem with that part of Bill C-11: "the digital lock prohibitions in the Act could potentially 'trump' or prevail over various exceptions in the Copyright Act, e.g., the fair dealing or educational exceptions" (UBC).

The following examples continue to be considered violations of copyright (Communication Concepts, Inc. 13; Templeton):

- downloading a work from the internet and forwarding copies to other people
- editing, altering, or incorporating an original work as part of your own document or multimedia file
- putting someone else's printed work online without the author's written permission
- copying and forwarding an email message without the sender's authorization

Penalties for Copyright Infringement

Individuals, businesses, or organizations that hold copyright may sue those who infringe copyright. Canadian legislation stipulates minimum penalties for copyright infringements. Violations of copyright on printed or electronic works may exceed the boundaries of civil law and may be prosecuted as summary convictions under criminal law. When in doubt, assume that the work is copyrighted and obtain written permission for its use.

EVALUATING AND INTERPRETING INFORMATION

Not all information is equal. Not all interpretations are equal. Whether you work with your own findings or the findings of other researchers, you need to decide whether the information is valid and reliable. Then you need to decide what your information means. Figure 8.4 on page 134 outlines this challenge.

Evaluating the Sources

Not all data sources are equally dependable. A source might offer information that is out of date, inaccurate, incomplete, mistaken, or biased.

How current is the information?

IS THE SOURCE UP TO DATE?   Newly published books contain information that can be more than a year old, and journal articles often undergo a lengthy process of peer review.

Certain types of information become outdated more quickly than others. Topics that focus on *technology* (superconductivity, internet censorship, alternative cancer treatments) may be outdated. Except for historical or background research, sources in those areas generally should offer the most recent information available.

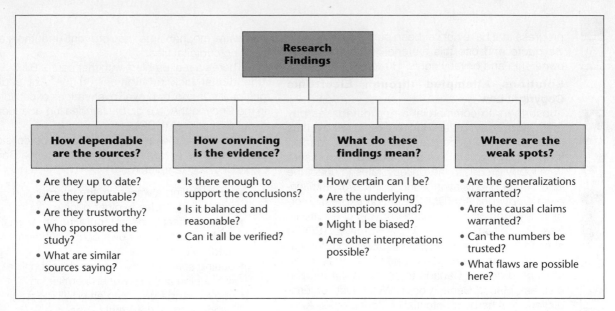

Figure 8.4 Decisions in Reviewing Research Findings

Critical Research Skills

"The most important skill for post-secondary students is critical thinking, so that they can answer questions like, 'What do I really need to find? What are my most likely sources? Which of them are relevant, useful, and credible? Does the search engine's listing really indicate the relevance and importance of the listed sources, or have commercial interests skewed the order of listings?'"

—**Ross Tyner, director of library services, Okanagan College**

What is the source's reputation?

Information on topics that focus on *people* (e.g., business ethics, management practices, workplace gender equality, employee motivation) might offer valuable perspective on present situations even if the information is several decades old.

IS THE SOURCE REPUTABLE?   Some sources enjoy better reputations than others. For research on alternative cancer treatments, you could depend more on reports in the *Canadian Medical Association Journal* or *Scientific American* than on those in scandal sheets or movie magazines. Even researchers with expert credentials, however, can disagree or be mistaken.

One way to assess a publication's reputation is to check its copyright page. Is the work published by a university, professional society, museum, or respected news organization? Is the publication *refereed* (all submissions reviewed by experts prior to acceptance)?

One way to assess an author's reputation is to check citation indexes to see what others have said about this research. Many periodicals also provide brief biographies or descriptions of authors' earlier publications and other achievements.

Websites may initially be evaluated by their domain type and sponsor. A typical URL places the domain type immediately after the name of the website's sponsor. Here are some common domain types:

.com	=	business/commercial organization
.edu	=	educational institution
.gov	=	government organization
.mil	=	military organization
.net	=	anyone with simple software and internet access
.org	=	non-profit organization

Canadian-based sites are not so easy to distinguish. Although some use *.org* and *.com* designators, many use the *.ca* domain label. The problem is that one cannot immediately distinguish among a university (**www.ubc.ca**), a hockey organization (**www.chl.ca**), a car dealer organization (**www.tada.ca**), a think tank (**www.fraserinstitute.ca**), and an online community newspaper (**www.vernoncentral.ca**).

However, Canadian government ministries and agencies usually insert *gc, gov,* or *gouv* into their URLs:

- **www.inspection.gc.ca** (homepage for the Canadian Food Inspection Agency)
- **www.msss.gouv.qc.ca** (Quebec's Ministère de Santé et Services sociaux)
- **www.health.gov.on.ca** (Ontario's Ministry of Health and Long-Term Care)
- **www.gov.bc.ca/health** (British Columbia's Ministry of Health Services)

Provincial government sites also insert the province's initials (e.g., ON, QC).

Can the source be trusted?

IS THE SOURCE TRUSTWORTHY? The internet offers information that may never appear in other sources, such as listservs, blogs, and newsgroups. But much of this information may reflect the bias of the special-interest groups that provide it. Moreover, anyone can publish almost anything on the internet—including a great deal of misinformation—without having it verified, edited, or reviewed for accuracy (Snyder).

Even in a commercial database, decisions about what to include and what to leave out depend on the biases, priorities, or interests of those who assemble that database. In general, try not to rely on any single information source.

Who sponsored the study?

IS THE INFORMATION BIASED? Much of today's research is paid for by private companies or special-interest groups that have their own social, political, or economic agendas (Crossen 14, 19). Medical research may be sponsored by drug or tobacco companies; nutritional research by food manufacturers; and environmental research by oil or chemical companies. Public policy research (e.g., on gun control, seatbelt laws, endangered species) may be sponsored by opposing groups (environmentalists versus the logging industry), producing opposing results. Instead of a neutral and balanced inquiry, this kind of "strategic research" is designed to support one special-interest group or another. Those who pay for strategic research are not likely to publicize findings that contradict their original claims, opinions, or beliefs (profits lower than expected, losses or risks greater than expected). As consumers of research, we should try to determine exactly what the sponsors of a particular study stand to gain or lose from the results.

The following hints may further help you evaluate internet sites:

How to evaluate the validity of internet sites

1. *Identify the purpose of the page or message.* Decide whether the message is intended to merely relay information, to sell something, or to promote a particular ideology or agenda.
2. *Don't be satisfied with generalities.* Look for specific facts, examples, or statistics. Then, check to see which of these facts can be verified by other sources.
3. *Compare the site with other sources.* Check related sites, publications, and other sources to compare the quality of information and to discover what others might have said about this site or author. Comparing many similar sites helps you create a *benchmark*, a standard for evaluating any particular site (based on the criteria in these guidelines). Ask a librarian for help.
4. *Look at your own attitudes and beliefs* to see if you're predisposed to automatically accept certain things that you read.

5. *Exercise extreme caution* in using anything picked up from online discussion groups.
6. *When considering material provided in an academic discussion group, look at the author's credentials.* Does the author have a solid track record of publications or other contributions to the field discussed by that group?
7. *Decide whether the assertions/claims make sense.* How well is each assertion supported? Never accept any claim that seems extreme without verifying it through other sources, such as a professor, a librarian, or a specialist in the field.

CASE	**The Wikipedia Challenge**

Wikipedia articles present a particular challenge to readers. The site's strength—its wide-ranging coverage supplied by tens of thousands of volunteer and often anonymous contributors—is also its weakness. Wikipedia's open source philosophy and its millions of articles mean that the site must basically rely on continuous updating and honest editing to keep its material accurate and reliable.

Clearly, not all Wikipedia articles can be fully trusted. Two rival sites, Citizendium and Knol, are trying to overcome Wikipedia's shortcomings by requiring the authors of all their articles to be named and by asking experts in given fields to check articles for accuracy.

Meanwhile, Wikipedia is starting to work toward a more dependable model. It now requires all articles about people to be subjected to editorial scrutiny. Also, researchers at a Wiki lab have developed a feature called "WikiTrust," which colour-codes newly edited text articles "using an algorithm that calculates author reputation from the lifespan of their previous contributions" (Leggett). The brighter the colour, the less the text's accuracy should be trusted.

Ross Tyner, director of library services at Okanagan College, suggests another approach. "The good Wikipedia articles cite their sources, which can help a researcher determine an article's validity. Other articles provide good ideas for profitably directing one's research; when you're not an expert in a given field, the first time you check into a topic related to that field, Wikipedia can be a gold mine."

Evaluating the Evidence

Evidence is any finding used to support or refute a particular conclusion. While evidence can serve the truth, it also can create distortion, misinformation, and deception. For example, *how much money, material, or energy does recycling really save? How good for your heart is oat bran? How well are public schools educating our children? Which investments or automobiles are safest?* Conclusions about such matters are based on evidence that often can be manipulated in support of one view or another. As consumers of research, we have to assess for ourselves the quality of the evidence presented.

We assess the quality of evidence by examining it critically to understand its limitations; to see if findings conflict; to discover connections, similarities, trends, or relationships; to determine the need for further inquiry; and to raise new questions.

Is there enough evidence?

IS THE EVIDENCE SUFFICIENT? Evidence is sufficient when it enables us to reach an accurate judgment or conclusion. A study of the stress-reducing benefits of low-impact aerobics, for example, would require a broad survey sample: people who have practised aerobics for a long time; people of different genders, different ages, different occupations, and different lifestyles before they began aerobics; and so forth. Even responses from hundreds of practitioners might constitute insufficient evidence unless those responses were supported by laboratory measurements of metabolism, heart rates, and blood pressure.

Personal experience usually offers insufficient evidence from which to generalize. You cannot tell whether your experience is representative, no matter how long you might have practised aerobics. Although anecdotal evidence ("This worked well for me!") might offer a good starting point for an investigation, personal experience should be evaluated within the broader context of *all* available evidence.

Is the evidence hard or soft?

CAN THE EVIDENCE BE VERIFIED? *Hard evidence* consists of factual statements, expert opinion, or statistics that can be verified. *Soft evidence* consists of uninformed opinion or speculation, data obtained or analyzed unscientifically, and findings that have not been replicated or reviewed by experts. Reputable news organizations employ fact-checkers to verify information before it appears in print.

Evidence that seems scientific can turn out to be soft. For example, information obtained from polling often is reported in fancy charts, graphs, and impressive statistics—but it is based on public opinion, which is almost always changing (Crossen 104).

Base your conclusions on hard evidence. For example, suppose an article makes positive claims about low-impact aerobics but provides no data on measurements of pulse, blood pressure, or metabolic rates. Although these claims might coincide with your own experience, your evidence so far consists of only two opinions: yours and the author's—without scientific support. Any conclusion at this point would rest on soft evidence. Only after carefully assessing dependable sources can you decide which conclusions are supported by the bulk of the evidence.

Interpreting the Evidence

Interpreting means trying to reach the truth of the matter: an overall judgment about what the evidence means and what conclusion or action it suggests. Unfortunately, research does not always yield answers that are conclusive or about which we can be certain. Instead of settling for the most *convenient* answer, we should pursue the most *reasonable* answer by examining critically a full range of possible meanings.

WHAT LEVEL OF CERTAINTY IS WARRANTED? As possible outcomes of research, we can identify three distinct and very different levels of certainty:

A practical definition of "truth"

1. The definitive truth—the *conclusive answer*:

   Truth is . . . the reality of the matter, as distinguished from what people wish were so, believe to be so, or assert to be so. From another perspective, in the words of Harvard philosopher Israel Scheffler, truth is the view "which is fated to be ultimately agreed to by all who investigate." The word *ultimately* is important. Investigation may produce a wrong answer for years, even for centuries. . . . Does the truth ever change? No. . . . One easy way to spare yourself any further confusion about truth is to reserve the word *truth* for the final answer to an issue. Get in the habit of using the words *belief*, *theory*, and *present understanding* more often. (Ruggiero 21–22)

ON THE JOB...

Certainty in Environmental Reports

"There's always some uncertainty. Everything we do is constrained by budget and the time frame. With more money and time, we could always generate more data that would bring us closer to certainty. So, the key is to be clear about your level of confidence in the data and inferences, and very clear about the level of uncertainty in your conclusions. There's no definite benchmark for the required level of certainty—it depends on the identified scope of the project. For example, an early Phase 1 contamination assessment project would use a broad approach of enquiries and general observations that may suggest potential areas of ground contamination that should be tested. The wording of the conclusions reflects the inherent degree of uncertainty at that level of analysis. A Phase 2 approach would cost 5 to 10 times more because expensive data gathering and lab analysis would be involved; the resulting report would be able to say with more certainty whether the ground and water are contaminated. A subsequent phase might involve more extensive assessment and remediation, which would cost many times more than a Phase 1 study, but would be able to say with considerably more certainty that the area is free of contaminants and fit for, say, a housing development. . . ."

—Dr. Brian Guy, Vice-president and General Manager, Summit Environmental Consultants

We are often mistaken in our certainty about the *truth*. For example, in the 2nd century A.D., Ptolemy's view of the Universe concluded that Earth was its centre. Though untrue, this judgment was based on the best information available at that time. Ptolemy's view survived for 13 centuries, even after new information had discredited this belief. When Copernicus and Galileo proposed more truthful views in the 15th century, they were labelled heretics.

Conclusive answers are the research outcome we seek, but often we have to settle for answers that are less than certain.

2. The *probable answer*: the answer that stands the best chance of being true or accurate—given the most we can know at this particular time. Probable answers are subject to revision in the light of new information.
3. The *inconclusive answer*: the realization that the truth of the matter is far more elusive, ambiguous, or complex than we expected.

Exactly how certain are we?

To ensure an accurate outcome, we must decide what level of certainty the findings warrant. For example, we are *highly certain* about the perils of smoking or sunburn, *reasonably certain* about the benefits of fruits and vegetables and moderate exercise, but *far less certain* about the perils of coffee drinking, or electromagnetic waves, or the benefits of vitamin supplements.

ARE THE UNDERLYING ASSUMPTIONS SOUND?   *Assumptions* are notions we take for granted, things we accept without proof. The research process rests on assumptions like these: that a sample group accurately represents a larger target group, that survey respondents remember certain facts accurately, and that mice and humans share enough biological similarities for meaningful research. For a particular study to be valid, the underlying assumptions must be accurate.

Consider this example: you are an education consultant evaluating the accuracy of IQ testing as a predictor of academic performance. Reviewing the evidence, you perceive an association between low IQ scores and low achievers. You then check your statistics by examining a cross-section of reliable sources. Can you then conclude that IQ tests do predict performance accurately? This conclusion might be invalid unless you could verify the following assumptions:

1. that neither parents, nor teachers, nor children had seen individual test scores, which could produce biased expectations
2. that, regardless of score, each child had completed an identical curriculum at an identical pace, instead of being "tracked" on the basis of his or her score

DO I HAVE A PERSONAL BIAS? To support a particular version of the truth, our own bias might cause us to overestimate (or deny) the certainty of our findings.

Personal bias is a fact of life

> Expect yourself to be biased, and expect your bias to affect your efforts to construct arguments. Unless you are perfectly neutral about the issue, an unlikely circumstance, at the very outset . . . you will believe one side of the issue to be right, and that belief will incline you to . . . present more and better arguments for the side of the issue you prefer. (Ruggiero 134)

Because personal bias is hard to transcend, *rationalizing* often becomes a substitute for *reasoning*:

Reasoning versus rationalizing

> You are reasoning if your belief follows the evidence—that is, if you examine the evidence first and then make up your mind. You are rationalizing if the evidence follows your belief—if you first decide what you believe and then select and interpret evidence to justify it. (Ruggiero 44)

Personal bias often is unconscious until we examine our own value systems, attitudes long held but never analyzed, notions we've inherited from our own backgrounds, and so on. Recognizing our own biases is a crucial first step in managing them.

What else could this mean?

ARE OTHER INTERPRETATIONS POSSIBLE? Perhaps other researchers would disagree with the meaning of these findings. Some controversial issues (the need for defence spending or causes of inflation) will never be resolved. Although we can get verifiable data and can reason persuasively on some subjects, no close reasoning by any expert and no supporting statistical analysis will prove anything about a controversial subject to everyone's satisfaction. For instance, one could only *argue* (more or less effectively) that federal funds will or will not alleviate poverty or unemployment.

Using Effective Reasoning

Interpreting the evidence involves inductive and/or deductive reasoning, which can later be used to persuasively present interpreted data.

Induction draws conclusions from related data

INDUCTIVE REASONING. This type of reasoning argues from specific cases to general principles. In other words, induction develops and tests hypotheses that explain available data. The more data, and the more similarities observed among the data, the better chance of formulating an accurate hypothesis.

Here's an example of induction. Forty students in a technical writing class wrote two-case study exams. The first exam occurred two weeks before an intensive series of instruction in language basics—grammar, spelling, punctuation, and sentence structure. The students wrote the second exam shortly after the language series. Their results were dramatically different:

1. On average, the second exam grades were 8.3 percent higher than the first exam grades.
2. Of the 40 students, 31 improved their grade, four earned the same grade, and five received a lower grade. The range was from –10 percent to +21 percent.

Cause–effect analysis could lead to the conclusion that the language instruction caused the improved grades, although the results might be partly due to increased familiarity with case-study exams.

Inductive reasoning depends on adequate data

The sample size and group composition (group members came from similar backgrounds) might be too limited to allow the conclusion that intensive language instruction will help all technical writing students improve their grades. A larger study with more students and more diversity of backgrounds would help establish a wider application for the hypothesis that improved language skills acquired through intensive instruction lead to improved grades.

The above example includes two of four common forms of inductive reasoning:

Several forms of induction

1. *The IMRAD pattern.* This reporting format includes introduction of a problem, initiating the research, materials and methods used to explore the topic, quantitative results, and a discussion that uses analogies, examples, and causal analysis to prove a hypothesis that explains the gathered data.
2. *Analogy.* What happens in one environment will work in another, if it's clear that the conditions in both environments are sufficiently similar (analogous). For example, new drugs are tested on laboratory animals and then on small groups of humans, with the understanding that a larger human population will experience the same results as the small groups.
3. *Cause–effect analysis.* What cause will effect a certain outcome. Does exposure to electromagnetic fields created by power lines cause cancer in humans? In a given geographical area, does loss of fish habitat constitute the major cause of declining fish populations? Does the design of SUVs explain their much greater chance of being involved in rollover accidents?
4. *Examples.* After presenting the pertinent data and arriving at a hypothesis, the writer can provide an actual example that further demonstrates the asserted hypothesis.

Deduction is based on widely accepted principles

DEDUCTIVE REASONING. This type of reasoning moves in the opposite direction from inductive reasoning, as it applies a general principle to specific cases. That general, widely accepted principle is called the *major premise.* The second statement, called a *minor premise*, represents a particular instance of the class identified in the major premise. The following example was reported on CBC Radio One: "Insurance carriers in Ontario automatically pay compensation claims made by veteran firefighters who have contracted leukemia (after 15 years' service) or brain tumours (after 20 years' service)."

Major premise: (Based on *inductive* reasoning: several studies have shown that firefighters have a much higher risk of getting certain kinds of cancers than the general population.) Exposure to a variety of chemicals encountered in fighting fires over an extended period of time can lead to brain tumours (after 20 years of firefighting) or to leukemia (after 15 years of firefighting).

Minor premise: Firefighter A has leukemia.
Minor premise (or observation): Firefighter A has worked fighting fires for 15 years or more.
Conclusion: Firefighter A's leukemia is likely due to his or her occupation. The disease is classed as a workplace disease, and firefighter A is entitled to full compensation.

Premises must be
accurate

Deductions work if, and only if, premises are accurate. Consider this example: "Don't expect Bill to write well—he's an engineer."

Major premise: (Implied) Engineers don't write well.
Minor premise: Bill is an engineer.
Conclusion: Bill doesn't write very well.

Avoiding Errors in Reasoning

Finding the truth, especially in a complex issue or problem, often is a process of elimination, of ruling out or avoiding errors in reasoning. As we interpret, we make *inferences*: we derive conclusions about what we don't know by reasoning from what we do know (Hayakawa 37). For example, we might infer that a drug that boosts immunity in laboratory mice will boost immunity in humans, or that a rise in campus crime statistics is caused by the fact that young people have become more violent. Whether a particular inference is accurate or inaccurate depends largely on our answers to one or more of these questions:

◆ To what extent can these findings be generalized?
◆ Is Y really caused by X?
◆ How much can the numbers be trusted, and what do they mean?

Following are four major reasoning errors that can distort our interpretations.

How much can we
generalize?

FAULTY GENERALIZATIONS.   When we accept research findings uncritically and jump to conclusions about their meaning, we commit the error of *hasty generalization*. For example, "We didn't see any of the expected endangered vegetation species during our April 2 field survey, so we conclude that this area doesn't contain such species and thus there's no threat from the proposed development." (This report conclusion doesn't consider that most of the plant species are annuals that don't usually appear until late April.)

When we overestimate the extent to which the findings reveal some larger truth, we commit the error of *overstated generalization*. For example, "At three points of observation along the 3 kilometres of creek, there's no apparent erosion, so we conclude that last fall's erosion control methods have been successful." (This conclusion has not considered that the following spring runoff levels were lower than usual, nor has the writer admitted that three observation points over 3 kilometres are not adequate to draw such a definitive conclusion.)

Are we limiting our
perspective?

LIMITED THINKING.   Our biases and limited experience may restrict our ability to make accurate inferences. *Either/or thinking* may limit the possible hypotheses for a set of phenomena. For example, "If you're not at your desk, you're not working," and "Either the East Coast cod stocks will revitalize in 10 years or the fish will disappear entirely."

One-valued thinking: when our thinking is based on only one set of values, we fail to realize how people from different cultures or different value systems might form a different conclusion from a body of information. For example, it is likely that real estate developers and Aboriginal groups will have conflicting responses to the discovery of ancient Aboriginal artifacts in an excavated basement in a new city subdivision.

FAULTY CAUSAL REASONING.   Causal reasoning tries to explain *why* something happened or *what* will happen: often very complex questions. Faulty causal reasoning oversimplifies or distorts the cause–effect relationship through errors like these:

Ignoring other causes

Investment builds wealth. [Ignores the role of knowledge, wisdom, timing, and luck in successful investing.]

Ignoring other effects

Running improves health. [Ignores the fact that many runners get injured, and that some even drop dead while running.]

Inventing a cause

Right after buying a rabbit's foot, Felix won the 6/49 lottery. [Posits an unwarranted causal relationship merely because one event follows another.]

Confusing correlation with causation

Poverty causes disease. [Ignores the fact that disease, while highly associated with poverty, has many causes unrelated to poverty.]

Rationalizing

My grades were poor because my exams were unfair. [Denies the possible causes of one's failures.]

Because of bias or impatience, we can be tempted to settle for a hasty cause or to confuse possible, probable, and definite causes.

Did X possibly, probably, or definitely cause Y?

Sometimes, a definite cause is apparent (e.g., "The engine's overheating is caused by a faulty radiator cap"), but usually much analysis is needed to isolate a specific cause. Suppose you want to answer this question: why does our local university not have daycare facilities? Brainstorming yields these possible causes:

◆ lack of need among students
◆ lack of interest among students, faculty, and staff
◆ high cost of liability insurance
◆ lack of space and facilities on campus
◆ lack of trained personnel
◆ prohibition by law
◆ lack of government funding for such a project

Suppose you proceed with interviews, questionnaires, and research into provincial laws, insurance rates, and availability of personnel. You begin to rule out some items, and others appear as probable causes. Specifically, you find a need among students, high campus interest, an abundance of qualified people for staffing, and no provincial laws prohibiting such a project. Three probable causes remain: lack of funding, high insurance rates, and lack of space. Further inquiry shows that lack of funding and high insurance rates *are* issues. These obstacles, however, could be eliminated through new sources of revenue: charging a fee for each child, soliciting donations, or diverting funds from other campus organizations.

NOTE   *Anything but the simplest effect is likely to have more than one cause. You have to make the case that the cause you have isolated is the right one. In the daycare scenario, for example, you might argue that the lack of space and facilities is somehow related to funding. And the university's inability to find funds or space might be related to student need or interest, which is not high enough to exert real pressure. Lack of space and facilities, however, appears to be the "immediate" cause.*

When you report on your research, be sure the audience can draw conclusions identical to your own on the basis of the evidence. The process might be shown like this:

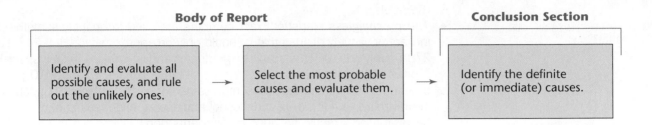

Body of Report   **Conclusion Section**

| Identify and evaluate all possible causes, and rule out the unlikely ones. | → | Select the most probable causes and evaluate them. | → | Identify the definite (or immediate) causes. |

Initially, you might have based your conclusions hastily on soft evidence (an opinion—buttressed by a newspaper editorial—that the campus was apathetic). Now you base your conclusions on solid, factual evidence. You have moved from a wide range of possible causes to a narrow range of probable causes, and finally to one definite cause.

Sometimes, finding a single cause is impossible, but this reasoning process can be tailored to most problem-solving analyses. Anything but the simplest effect is likely to have more than one cause. By narrowing the field, you can focus on the real issues.

FAULTY STATISTICAL REASONING.   The purpose of statistical analysis is to determine the meaning of a collected set of numbers. In primary research, our surveys and questionnaires often lead to some kind of numerical interpretation ("What percentage of respondents prefer X?" "How often does Y happen?"). In secondary research, we rely on numbers collected by primary researchers. Numbers seem more precise, more objective, more scientific, and less ambiguous than words. They are easier to summarize, measure, compare, and analyze. But numbers can be misleading. For example, radio or television phone-in surveys produce grossly distorted data: although 90 percent of callers might express support for a particular viewpoint, callers tend to be those with the greatest anger or the strongest feelings about the issue—representing only a fraction of overall attitudes (Fineman 24). Mail-in surveys can produce similar distortion because only people with certain attitudes might choose to respond.

Before relying on any set of numbers, we need to know exactly where they came from, how they were collected, and how they were analyzed (Lavin 275–76). Can the numbers be trusted, and if so, what do they mean?

Faulty statistical reasoning produces conclusions that are unwarranted, inaccurate, or downright deceptive. The following are some common statistical fallacies:

"Exactly how well are we doing?"

◆ *The sanitized statistic.* Numbers are manipulated to obscure the facts. For instance, a recently revised formula enables the government to exclude from its unemployment figures an estimated 5 million people who remain unemployed after one year—thus creating a far rosier economic picture than the facts warrant. Similar formulas allow for all sorts of sugarcoating in reports of wages, economic growth, inflation, and other statistics that affect the political climate (Morgenson 54). In the United States, the College Board's recentring of SAT scores has raised the average math score from 478 to 500 and the average verbal score from 424 to 500 (a boost of almost 5 and 18 percent, respectively) although actual student performance remains unchanged (Samuelson).

"How many rats was that?"

◆ *The meaningless statistic.* Exact numbers are used to quantify something so inexact or vaguely defined that it should only be approximated (Huff 247; Lavin 278): "Only 38.2 percent of college graduates end up working in their specialty." "Toronto has 3 247 561 rats." "Zappo detergent makes laundry 10 percent brighter." An exact number looks impressive, but certain subjects (e.g., cheating in university, virginity, drug and alcohol abuse on the job, eating habits) cannot be quantified exactly, because respondents don't always tell the truth (e.g., because of denial, embarrassment, or mere guesswork). Or they respond in ways they think the researcher expects.

"Why is everybody griping?"

◆ *The undefined average.* The mean, median, and mode are confused in determining an average (Huff 244; Lavin 279). The *mean* is the result of adding up the value of each item in a set of numbers, then dividing by the number of items. The *median* is the result of ranking all the values from high to low, then choosing the middle value (or the 50th percentile, as in calculating SAT scores). The *mode* is the value that occurs most often in a set of numbers.

Each of these three measurements represents some kind of average, but unless we know which average is being presented, we cannot interpret the figures accurately. Assume that we are computing the average salary among female vice-presidents at XYZ Corporation:

Vice-President	Salary
A	$90 000
B	90 000
C	80 000
D	65 000
E	60 000
F	55 000
G	50 000

In the above example, the mean salary (total salaries divided by people) equals $70 000; the median salary (middle value) equals $65 000; the mode (most frequent value) equals $90 000. Each is legitimately an average, and each could be used to support or refute a particular assertion (e.g., "Women vice-presidents are paid too little" or "Women vice-presidents are paid too much").

Research expert Michael Lavin sums up the potential for bias in reporting averages:

Depending on the circumstances, any one of these measurements [mean, median, or mode] may describe a group of numbers better than the other two. . . . [But] people typically choose the value which best presents their case, whether or not it is the most appropriate to use. (279)

Although the mean is the most commonly computed average, this measurement can be misleading when values on either end of the scale are extremely high or low.

Suppose, for instance, that vice-president A received a $200 000 salary. Because this figure deviates so far from the normal range of salary figures for B through G, it distorts the average for the whole group—increasing the mean salary by more than 20 percent (Plumb and Spyridakis 636).

"Is 51 percent really a majority?"

◆ *The distorted percentage figure.* Percentages are reported without explanation of the original numbers used in the calculation (Adams and Schvaneveldt 359; Lavin 280): "Seventy-five percent of respondents prefer our brand over the competing brand"—without mention that only four people were surveyed. Or "Sixty-six percent of employees we hired this year are women and minorities, compared with the national average of 40 percent"—without mention that only three people have been hired this year, by a company that employs 300 (mostly white males).

Another fallacy in reporting percentages occurs when the *margin of error* is ignored. This is the margin within which the true figure lies, based on estimated sampling errors in a survey. For example, a claim that most people surveyed prefer brand X might be based on the fact that 51 percent of respondents expressed this preference; but if the survey carried a 2 percent margin of error, the true figure could be as low as 49 percent or as high as 53 percent. In a survey with a high margin of error, the true figure may be so uncertain that no definite conclusion can be drawn.

"Which car should we buy?"

◆ *The bogus ranking.* Items are compared on the basis of ill-defined criteria (Adams and Schvaneveldt 212; Lavin 284): "Last year, the Carmobile was the number-one selling car in Canada"—without mention that some competing car makers actually sold *more* cars to private individuals, and that the Carmobile figures were inflated by hefty sales to rental-car companies and corporate fleets. Unless we know how the ranked items were chosen and how they were compared (the criteria), a ranking can produce a scientific-seeming number based on a completely unscientific method.

"Garbage in, garbage out."

◆ *The fallible computer model.* Computer models process complex assumptions to produce impressive, but often inaccurate, statistical estimates about costs, benefits, risks, or probable outcomes.

Computer models to predict global-warming levels, for instance, are based on differing assumptions about wind and weather patterns, cloud formations, ozone levels, carbon dioxide concentrations, sea levels, or airborne sediment from volcanic eruptions. Despite their seemingly scientific precision, different global-warming models generate 50-year predictions of sea-level rises that range from a few inches to several feet (Barbour 121). Other models suggest that warming effects could be offset by evaporation of ocean water and by clouds reflecting sunlight back to outer space (Monastersky, "Do Clouds"). Still other models suggest that the 0.56°C (1°F) warming over the last 100 years may not be the result of the greenhouse effect at all, but of "random fluctuations in global temperatures" (Stone 38). The estimates produced by any model depend on the assumptions (and data) programmed in.

Choice of assumptions might be influenced by researcher bias or the sponsor's agenda. For example, a prediction of human fatalities from a nuclear plant meltdown might rest on assumptions about availability of safe shelter, evacuation routes, time of day, season, wind direction, and structural integrity of the containment unit. But the assumptions could be manipulated to produce an overstated or understated estimate of risk (Barbour 228). For computer-modelled estimates of accident risk (e.g., oil spill, plane crash) or of the costs and benefits

"Does a beer a day keep the doctor away?"

of a proposed project or policy (e.g., a space station, welfare reform), consumers rarely know the assumptions behind the numbers. We wonder, for example, about the assumptions underlying NASA's pre-*Challenger* risk assessment, in which a 1985 computer model reportedly showed an accident risk of less than 1 in 100 000 shuttle flights (Crossen 54).

◆ *Confusion of correlation with causation. Correlation* is the measure of association between two variables (between smoking and increased lung cancer risk, or between education and income). *Causation* is the demonstrable production of a specific effect (smoking causes lung cancer). Correlations between smoking and lung cancer or education and income signal a causal relationship that has been proven by studies of all kinds. But not every correlation implies causation. For example, a recently discovered correlation between moderate alcohol consumption and decreased heart disease risk offers insufficient proof that moderate drinking causes less heart disease.

Many highly publicized correlations are the product of *data mining*. In this process, computers randomly compare one set of variables (various eating habits) with another set (a range of diseases). From these countless comparisons, certain relationships are revealed (say between coffee drinking and pancreatic cancer risk). As dramatic as such isolated correlations may be, they constitute no proof of causation and often lead to hasty conclusions (P.E. Ross 135).

NOTE   *Despite its limitations, data mining is invaluable for "uncovering correlations that require computers to perceive but that thinking humans can evaluate and research further" (Maeglin).*

"Is this good news or bad news?"

◆ *Misleading terminology.* The terms used to interpret statistics sometimes hide their real meaning. For instance, the widely publicized figure that people treated for cancer have a "50 percent survival rate" is misleading in three ways: (1) *survival* to laypersons means "staying alive," but to medical experts, staying alive for only five years after diagnosis qualifies as survival; (2) the "50 percent" survival figure covers *all* cancers, including certain skin or thyroid cancers that have extremely high cure rates, as well as other cancers (such as lung or ovarian) that are rarely curable and have extremely low survival rates; (3) more than 55 percent of all cancers are skin cancers with a nearly 100 percent survival rate, thereby greatly inflating survival statistics for other types of cancer ("Are We" 5; Dana-Farber Cancer Institute 2).

Even the most valid and reliable statistics require that we interpret the reality behind the numbers. For example, the overall cancer rate today is higher than it was in 1910. This may mean that people are living longer and thus are more likely to die of cancer and that cancer today rarely is misdiagnosed—or mislabelled because of stigma ("Are We" 4). The finding that rates for certain cancers double after prolonged exposure to electromagnetic waves may really mean that cancer risk actually increases from 1 in 10 000 to 2 in 10 000.

These are only a few examples of statistics and interpretations that seem highly persuasive but that in fact cannot always be trusted. Any interpretation of statistical data carries the possibility that other, more accurate, interpretations have been overlooked or deliberately excluded (Barnett 45).

Considering Standards of Proof

How much evidence is enough to "prove" a particular claim? The answer often depends on whether the inquiry occurs in the science lab, the courtroom, or the boardroom.

◆ The scientist demands evidence that indicates at least 95 percent certainty. A scientific finding must be evaluated and replicated by other experts. Good science looks at the entire picture. Findings are reviewed before they are reported. Inquiries and answers in science are never "final," but open-ended and ongoing: what seems probable today may be shown improbable by tomorrow's research.

◆ The juror demands evidence that indicates only 51 percent certainty (a "preponderance of the evidence"). Jurors are not scientists. Instead of the entire picture, jurors get only the information made available by lawyers and witnesses. A jury bases its opinion on evidence that exceeds "reasonable doubt" (Monastersky, "Courting" 249; Powell 32+). Courts have to make decisions that are final.

◆ The corporate executive demands immediate (even if insufficient) evidence. In a global business climate of overnight developments (in world markets, political strife, military conflicts, natural disasters), important business decisions often are made on the spur of the moment, and often on the basis of incomplete or unverified information—or even hunches—in order to react to crises and capitalize on opportunities (Seglin 54).

The above examples illustrate the need to consider the specific setting in which a persuasive evidence is presented. But each setting operates within a national or regional culture. Each culture may have its own standards for authentic, reliable, and persuasive evidence. "For example, African cultures rely on storytelling for authenticity. Arabic persuasion is dependent on universally accepted truths. And Chinese value ancient authorities over recent empiricism" (Byrd and Reid 109).

ASSESSING THE ENTIRE RESEARCH PROCESS

The research process is a minefield of potential errors, in what we do and how we reason: we might ask the wrong questions; we might rely on the wrong sources; we might collect or record data incorrectly; we might analyze or document data incorrectly. Therefore, we need to critically examine our methods and our reasoning before reporting findings and conclusions. The following research checklist helps guide our assessment.

CHECKLISTS

for the Research Process

Use this checklist to assess your research process.

Reasoning

◆ Am I reasonably certain about the meaning of these findings?

◆ Does my final answer seem definitive, only probable, or inconclusive?

◆ Am I reasoning instead of rationalizing?

◆ Is this the most reasonable conclusion (or merely the most convenient)?

◆ Can I rule out other possible interpretations or conclusions?

◆ Have I accounted for all sources of bias, including my own?

◆ Are my generalizations warranted by the evidence?

◆ Am I confident that my causal reasoning is correct?

◆ Can all of the numbers and statistics and interpretations be trusted?

◆ Have I resolved (or at least acknowledged) any conflicts among my findings?

◆ Should the evidence be reconsidered?

Method

◆ Did I ask the right questions?

◆ Are the sources appropriately up to date?

◆ Is each source reputable, trustworthy, and relatively unbiased?

◆ Does the evidence clearly support all of the conclusions?

◆ Can all of the evidence be verified?

◆ Is a fair balance of viewpoints presented?

◆ Has the research achieved adequate depth?

◆ Has the entire research process been valid and reliable?

◆ Is all quoted material clearly marked throughout the text?

◆ Are direct quotations used sparingly and appropriately?

◆ Are all quotations accurate and integrated using proper grammar?

◆ Are all paraphrases accurate and clear?

◆ Have I documented all sources not considered common knowledge?

◆ Is the documentation consistent, complete, and correct?

EXERCISES OR COLLABORATIVE PROJECTS

1. Assume that you are an assistant communication manager for a new organization that prepares research reports for decision makers worldwide. (A sample topic: "What is the expected long-term impact of the North American Free Trade Agreement on the Canadian computer industry?") These clients expect answers based on the best available evidence and reasoning.

   Although your recently hired co-workers are technical specialists, few have experience in the kind of wide-ranging research required by your clients. Training programs in the research process are being developed by your communication division but will not be ready for several weeks.

   Meanwhile, your manager directs you to prepare a one- or two-page memo that introduces employees to major procedural and reasoning errors that affect validity and reliability in the research process. Your

manager wants this memo to be comprehensive but not vague.

   Use the four criteria in the above section "Evaluating the Sources," to evaluate the following statements about global warming:

   ◆ "Well, maybe not so alarming. Global temperatures have now held steady for 16 years. They levelled off around 1997." Quoted at **http://www.theglobe andmail.com/commentary/whatever-happened-to-global-warming/article7725145/.**

   ◆ ". . . climate change occurs when long-term weather patterns are altered—for example, through human activity." David Suzuki at **http://www.davidsuzuki.org/issues/climate-change/science/climate-change-basics/climate-change-101-1/?gclid=ClbBsuqujLUCFe5xQgodOGEAEA.**

2. Refer to the scenario in Exercise 1. In an email to colleagues, offer guidelines for avoiding unintentional plagiarism in quoting, paraphrasing, and citing the work of

others. Explain what to document and how, using MLA style for a parenthetical reference and a works-cited entry. (Illustrate with examples, but not those from this text!)

3. From print or broadcast media or from personal experience, identify an example of each of the following sources of distortion or of interpretive error:

- a study with questionable sponsorship or motives
- reliance on soft evidence
- overestimating the level of certainty
- biased interpretation
- rationalizing
- faulty causal reasoning

- hasty generalization
- overstated generalization
- sanitized statistic
- meaningless statistic
- undefined average
- distorted percentage figure
- bogus ranking
- fallible computer model
- misinterpreted statistic

Submit your examples to your instructor, along with a memo explaining each error, or be prepared to discuss your material in class.

MyWritingLab

MyWritingLab: Technical Communication offers the best multimedia resources for technical communication in one, easy-to-use place. You can find a variety of interactive model documents and case studies. There are also extensive guidelines, tutorials, and exercises for Document Design, Writing and the Writing Process, and Research, and a large bank of diagnostics and practice for grammar review.

9

Documenting Research Findings

✳ Explore

Click here in your eText to access a variety of student resources related to this chapter.

LEARNING OBJECTIVES

After reading this chapter, you should be able to

- understand the value of documenting information sources
- understand the basic principles of parenthetical documentation used by MLA and APA documentation styles
- understand the basic principles of numerical citation-sequence systems, such as CSE and IEEE documentation styles

Documenting research findings acknowledges each information source. Proper citation meets professional requirements for ethics, efficiency, and authority.

WHY YOU SHOULD DOCUMENT

Documentation is *ethical* in that the originator of borrowed material deserves full credit and recognition. Moreover, all published material is protected by copyright law. Failure to obtain permission to reproduce published material could make you liable to legal action, even if you have documented the source.

Documentation also is *efficient*. It provides a network for organizing and locating the world's recorded knowledge. If you cite a particular source correctly, your reference will enable interested readers to locate that source themselves.

Finally, documentation *provides authority*. In making any claim—for example, "A Honda Accord is more reliable than a Ford Fusion"—you invite challenge: "Says who?" Data on road tests, frequency of repairs, resale value, workmanship, and owner comments can help validate your claim by showing its basis in *fact*. A claim's credibility increases when expert references support it. For a controversial topic, you may need to cite several authorities who hold various views—as in this next example—instead of forcing a simplistic conclusion on your material:

Citing a balance of views

Opinion is mixed as to whether a marketable quantity of oil rests beneath Great Slave Lake. Edmonton geologist Mo Rajabully feels that extensive reserves are improbable ("Geologist Dampens Hopes" 3). Oil geologist Marta Silverlaug is uncertain about the existence of any oil in quantity at this location ("Northern Oil Drilling" 2). But the Canadian Geological Survey reports that the lake bed may overlay 3.5 billion barrels of oil (Ruston 8).

WHAT YOU SHOULD DOCUMENT

Document any insight, assertion, fact, finding, interpretation, judgment, or other "appropriated material that readers might otherwise mistake for your own" (Gibaldi and Achtert 155)—whether the material appears in published form or not. Specifically, you must document

Sources that require documentation

- any source from which you use exact wording
- any source from which you adapt material in your own words
- any visual illustration: chart, graph, drawing, or the like

Internet and other electronic sources must be documented

The above list includes any form of *intellectual property*, whether it appears in print, online, or in any other electronic form.

You don't need to document anything considered *common knowledge*: material that appears repeatedly in general sources. In medicine, for instance, it is common knowledge that foods high in fat correlate with higher incidences of cancer, so in a report on fatty diets and cancer, you probably would not need to document that well-known fact. But you would document information about how the fat–cancer connection was discovered, subsequent studies (e.g., the role of saturated versus unsaturated fats), and any information for which some other person could claim specific credit. If the borrowed material can be found in only one specific source and not in multiple sources, document it. When in doubt, document the source.

HOW YOU SHOULD DOCUMENT

Borrowed material has to be cited twice: at the exact place that you use the material and at the end of your document. Documentation practices vary in their minor details, but all systems use a two-stage process: a brief reference in the text names the source and refers readers to the complete citation, which enables the source to be retrieved. Four styles are widely used for documenting sources in their respective disciplines:

- Modern Language Association (MLA) style, for the humanities
- American Psychological Association (APA) style, for the social sciences
- Council of Science Editors (CSE) style, for the natural and applied sciences
- Institute of Electrical and Electronics Engineers (IEEE) style, for engineering and electronic applications

Where to find examples of MLA, APA, CSE, and IEEE documentation styles

This chapter introduces the MLA, APA, CSE, and IEEE styles. Complete lists of examples appear in this text's companion resource, **MyWritingLab: Technical Communication**. Other styles have been adopted by various disciplines—a list of the respective style manuals resides at **www.dianahacker.com/resdoc/manual.html**.

Chicago and Turabian documentation styles

Two variant styles are the Chicago and Turabian systems, both of which use an author–date system with notes and parenthetical references. However, both allow footnotes to be used to document sources. The online *Chicago-Style Citation Quick Guide* may be found at **www.chicagomanualofstyle.org/tools_citationguide.html**. If you're looking for help with Turabian style, go to McMaster University's online guide at **http://library.mcmaster.ca/guides/turabian.pdf** or visit the University of Wisconsin–Madison Writing Center's website at **www.wisc.edu/writing/Handbook/DocChicago.html**.

MLA DOCUMENTATION STYLE

As described in the *MLA Handbook for Writers of Research Papers*, MLA style uses parenthetical references that usually include the author's last name and the exact page number of the borrowed material:

Parenthetical reference in the text

> Researchers at the University of Guelph report that the antibody IgY, found in egg yolk, is effective as an oral passive immunization against various gastrointestinal pathogens (Kovacs-Nolan and Mine 39).

Readers seeking the complete citation for Kovacs-Nolan and Mine can easily find it in the Works Cited list at the end of the document, which is arranged alphabetically by author:

Full citation at the end of the document

> Kovacs-Nolan, Jennifer, and Yoshinori Mine. "Passive Immunization through Avian Antibodies." *Food Biotechnology* 18.1 (2004): 39–62. Print.

This complete citation includes page numbers for the entire article and the medium of publication consulted ("Print"). See Figure 9.1, page 158, for other examples.

MLA style also uses footnotes to comment on or explain content in the text or for bibliographic notes.

MLA Parenthetical References

For clear and informative parenthetical references, observe these guidelines:

Citing page numbers only

♦ If your discussion names the author, do not repeat the name in your parenthetical reference; simply list the page number:

> Bowman et al. explain how their recent study indicates an elevated risk of leukemia for children exposed to certain types of electromagnetic fields (59).

Three works in a single reference

♦ If you cite two or more works in a single parenthetical reference, separate the citations with semicolons:

> (Jones 32; Leduc 41; Gomez 293–94)

Two authors with identical last names

♦ If you cite two or more authors with the same last name, include the first initial in your parenthetical reference for each author:

> (R. Jones 32)

> (S. Jones 14–15)

Two works by one author

♦ If you cite two or more works by the same author in your document, add the title of the work (if brief) or a shortened version:

> (Lamont, *Diagnostic Tests* 81)

> (Lamont, *Biophysics* 100–01)

Institutional, corporate, or anonymous author

♦ If you cite a work by an institutional or corporate author, use the author's name; you may shorten terms that are commonly abbreviated. If the work is unsigned (i.e., the author is unknown), use the title of the work (if brief) or a shortened version:

> (American Medical Assn. 2)

> ("Distribution Systems" 18)

To avoid distracting your readers, keep each parenthetical reference as brief as possible. One method is to name the source in your discussion and to place only the page number in parentheses.

For a paraphrase, place the parenthetical reference *before* the closing punctuation mark. For a quotation that runs into the text, place the reference *between* the final quotation mark and the closing punctuation mark. For a quotation set off (indented) from the text, place the parenthetical reference *after* the closing punctuation mark.

The MLA Handbook recommends direct references to electronic sources

As the examples found at **MyWritingLab: Technical Communication** show, documenting electronic sources is often trickier than documenting printed sources. More often than not, internet sources do not name authors, so special identification techniques are required. Also, electronic documents seldom number pages or provide other types of reference numbers. Therefore, if you cite an entire web publication, the *MLA Handbook* recommends including the name of the author in the text rather than in a parenthetical reference, as in this example:

> John Chow's enthusiasm for blogvertising is quite evident in his website, **www.johnchow. com**, where he lists the amount he has earned each month since he began to accept advertisers' dollars for his "reviews."

In-text direct references to electronic sources do not necessarily need to include a URL

This example includes a specific URL, but providing a descriptor would be just as useful for a reader who wishes to locate the online source named in the citation. After all, the reader just has to search for the source name in order to locate the source. A reader of the following example would need only to use the search phrase "SEC's *A Plain English Handbook*" to easily locate a PDF version of the handbook at **www.sec.gov/pdf/handbook.pdf**.

> The SEC's online *A Plain English Handbook* includes sensible advice and convincing examples designed to help writers create readable documents.

MLA Works Cited Entries

How to space and indent entries

The Works Cited list includes sources you have paraphrased or quoted. Double-space the list for academic papers in the humanities. For all other disciplines, single-space within each entry and double-space between entries. Place the first line of each entry at the left margin. Indent the second and subsequent lines five spaces (1.25 cm [½″]). Use a one-character space after any period, comma, or colon.

Online examples of MLA documentation style

Visit **MyWritingLab: Technical Communication** for examples of complete citations as they would appear in the Works Cited section of your document. That site also includes an annotated Works Cited list, which for your convenience is also listed as Figure 9.1 on page 158.

Further examples of MLA style appear at the following websites:

> **http://writing.wisc.edu/Handbook/DocMLA.html**
> **http://owl.english.purdue.edu/owl/resource/747/01**
> **http://dianahacker.com/pdfs/Hacker-Daly-MLA-WC.pdf**

NOTE

If your sources include non-recoverable sources, such as phone conversations and in-person interviews, you may use the title "Sources Cited"—as shown in Figure 9.1 on page 158—rather than "Works Cited."

APA DOCUMENTATION STYLE

An alternative to MLA style appears in the *Publication Manual of the American Psychological Association*. APA style is useful when writers wish to emphasize the publication dates of their references. A parenthetical reference in the text briefly identifies the source, date, and page number(s):

Reference cited in the text

> In a recent study, mice continuously exposed to an electromagnetic field tended to die earlier than mice in the control group (de Jager & de Brun, 1994, p. 224).

The full citation then appears in the alphabetical list of References at the end of the document:

Full citation at the end of the document

> de Jager, L., & de Brun, L. (1994). Long term effects of a 50 Hz electric field on the life-expectancy of mice. *Review of Environmental Health, 10*(3–4), 221–224.

APA style also uses footnotes to add to or explain content in the text or to present copyright permission.

APA Parenthetical References

APA's parenthetical references differ from MLA's as follows: the APA citation includes the publication date; a comma separates the items in the parenthetical reference; and "p." or "pp." precedes the page number(s). When a subsequent reference to a work follows closely after the initial reference, the date need not be included.

Here are specific guidelines:

Author named in the text

◆ If your discussion names the author, do not repeat the name in your parenthetical reference; simply give the date and page number:

> Researchers de Jager and de Brun explain that experimental mice exposed to an electromagnetic field tend to die earlier than mice in a control group (1994, p. 224).

When two authors of a work are named in your text, their names are connected by "and," but in a parenthetical reference their names are connected by an ampersand (&).

Two or more works in the same parentheses

◆ If you cite two or more works in the same parentheses, list the authors in alphabetical order and separate the citations with semicolons:

> (Gomez, 1999; Jones, 2003; Leduc, 2004)

A work with three to five authors

◆ If you cite a work with three to five authors, try to name them in your text the first time the work is mentioned, to avoid an excessively long parenthetical reference:

> Franks, Oblesky, Ryan, Jablar, and Perkins (1993) studied the role of electromagnetic fields in tumour formation.

In any subsequent references of this work, name only the first author followed by "et al." For a work with six or more authors, name only the first author followed by "et al." in all references.

Two or more works by the same author in the same year

◆ If you cite two or more works by the same author published in the same year, assign a different letter to each work:

> (Lamont, 2002a, p. 135)
>
> (Lamont, 2002b, pp. 67–68)

For other examples of APA parenthetical references with their corresponding References entries, visit **MyWritingLab: Technical Communication**.

APA Reference List Entries

How to space and indent entries

The APA reference list includes each source that you cited in your document. In preparing the list, key the first line of each entry flush with the left margin. Indent the second and subsequent lines five spaces (1.25 cm [½″]). Use one character space after any period, comma, or colon.

For examples of complete citations as they would appear in the References section of your document, and for a sample References page, visit **MyWritingLab: Technical Communication**. Shown immediately below each reference entry is its corresponding parenthetical reference as it would appear in the text. Note the capitalization, abbreviation, spacing, and punctuation in the sample entries.

Also see Figure 9.2, page 159, and visit these websites:

Online examples of APA documentation style

www.wisc.edu/writing/Handbook/DocAPA.html
http://owl.english.purdue.edu/owl/resource/560/01
http://apastyle.apa.org
http://library.austincc.edu/help/APA

CSE NUMERICAL DOCUMENTATION STYLE

The Council of Science Editors (CSE; formerly the Council of Biology Editors) accepts a name–year system similar to the APA system; the name–year system is often used by biologists. However, CSE prefers a numerical citation-sequence system in which each work is assigned a number the first time it is cited. This same number then is used for any subsequent reference to that work. Numerical documentation is often used in the physical sciences (astronomy, chemistry, geology, physics) and the applied sciences (mathematics, medicine, engineering, computer science).

The CSE's *Scientific Style and Format: The CSE Manual for Authors, Editors, and Publishers*, 7th edition, 2006, is widely consulted for its guidelines on numerical documentation. This current edition of the manual has added a third acceptable system of citing sources: the citation–name system, which lists references alphabetically by author's name and also by number. In this third option, references are cited by number in the text itself. (All three systems are illustrated at **MyWritingLab: Technical Communication**.)

CSE Citation-Sequence System

In the most common version of CSE style, a citation in the text appears as a superscript number immediately following the source to which it refers:

Numbered citations in the text

A recent study[1] indicates an elevated leukemia risk among children exposed to certain types of electromagnetic fields. Related studies[2-3] tend to confirm the EMF/cancer hypothesis.

When referring to two or more sources in a single superscripted note (as in "[2-3]" above), separate the numbers with a hyphen if they are consecutive and by commas but no space if they are not consecutive: ("[2,6,9]").

The full citation for each source then appears in the numerical listing of references at the end of the document:

REFERENCES

1. Bowman JD. Hypothesis: the risk of childhood leukemia is related to combinations of power-frequency and static magnetic fields. Bioelectromagnetics 1995; 16(1): 48–59.

2. Feychting M, Ahlbom A. Electromagnetic fields and childhood cancer: meta-analysis. Cancer Causes Control 1995 May; 6(3):275–277.

3. Bosanac SD. Dynamics of particles and the electromagnetic field. World Scientific Series in Contemporary Chemical Physics 2005; 24: 325–334.

4. Brodeur P. Currents of death. New York: Simon and Schuster; 2000. 336p.

To refer again to a particular source later in your document, use the same number. For example, all textual references to the Brodeur book would use the number "4" to identify the source for readers.

For more examples of CSE citations, see Figure 9.3, page 160 and visit **MyWritingLab: Technical Communication**. Also see the following websites:

> http://www.library.ualberta.ca/uploads/augustana/cse_qg_march_23_2010.pdf
> www.nlm.nih.gov/pubs/formats/recommendedformats.html
> http://library.duke.edu/research/citing/workscited
> http://library.austincc.edu/help/CSE/CSE-cs.php
> www.monroecc.edu/depts/library/cse.htm

The latest edition of the *CSE Manual* includes extensive guidance on citing print and internet sources, more than can be included in this text or in the websites listed above. The CSE's expanded coverage now conforms to standards set by the National Information Standards Organization (NISO), which lists 67 different kinds of citations.

IEEE DOCUMENTATION STYLE

The Institute of Electrical and Electronics Engineers (IEEE) provides format and stylistic advice for practitioners in electrical, electronics, and computer fields. (Note, though, that CSE is also used by computer science writers.) IEEE documentation style uses a numerical citation-sequence system similar to the system preferred by the CSE. However, as the following examples illustrate, IEEE style differs somewhat from CSE style.

As in CSE style, IEEE in-text citations place numbers at appropriate points, but IEEE format presents citation numbers enclosed in brackets—"[1]"—rather than as superscripts—"[1]":

> A recent study [1] indicates an elevated risk among children exposed to certain types of electromagnetic fields. Related studies [2], [3] tend to confirm the EMF/cancer hypothesis.

As in CSE style, the numbered citations appear in sequence at the end of the document. See the following examples and those in Figure 9.4, page 161.

REFERENCES

Full citations at the end
of the document

[1] J.D. Bowman, "Hypothesis: the risk of childhood leukemia is related to combinations of power-frequency and static magnetic fields," *Bioelectromagnetics*, vol. 16, no. 1, pp. 48-59, 1995.

[2] M. Feychting and A. Ahlbom, "Electromagnetic fields and childhood cancer: meta-analysis," *Cancer Causes Control*, vol. 6, no. 3, pp. 275-277, May 1995.

Subsequent textual references to the Bowman source would continue to use the [1] designation.

References for online journals follow this format: Author. (year, month). Title. *Journal*. [Type of medium]. *volume (issue)*, pages. Available: URL. Here's an example of a cited online journal article:

[3] J.H. Kordower. (2009, Sept.). Animal rights terrorists: what every neuroscientist should know. *J. Neuroscience*. [Online]. *7 (29)*, 11419-11420. Available: http://www.jneurosci.org/cgi/content/full/29/37/11419

References for online magazines follow this format: Author. (year, month, day). Title. *Magazine* [Type of medium]. Available: URL. Here's an example of a cited online magazine article:

[4] P. McDougall. (2009, Nov. 19). New York tests Xbox-based alert system. *Information Week* [Online]. Available: http://www.informationweek.com/story/showArticle. jhtml?articleID=221900336

For more examples of IEEE format, visit **MyWritingLab: Technical Communication**. Also see the following websites:

Online examples of IEEE
documentation style

http://standards.ieee.org/guides/style/2007_Style_Manual.pdf
www.ecf.toronto.edu/~writing/handbook-docum1b.html
www.ieee.org/portal/cms_docs/pubs/transactions/auinfo03.pdf
http://http-server.carleton.ca/~nartemev/IEEE_style.html
www.lib.monash.edu.au/tutorials/citing/ieee.html
http://www.ieee.org/documents/ieeecitationref.pdf

EXERCISE

1. Locate the style manual for your discipline (ask the faculty in your major or a librarian) or see the style guidelines presented in **MyWritingLab: Technical Communication**.
2. If your discipline uses MLA format, use the style manual to learn whether the MLA references in Figure 9.1 (page 158) are correctly formatted.
3. If your discipline uses APA format, use the style manual to learn whether the APA references in Figure 9.2 (page 159) are correctly formatted.
4. If your discipline uses CSE format, use the style manual to learn whether the CSE references in Figure 9.3 (page 160) are correctly formatted.
5. If your discipline uses IEEE format, use the style manual to learn whether the IEEE references in Figure 9.4 (page 161) are correctly formatted.
6. Identify the three following types of sources for a topic that you're exploring: (1) book or report; (2) website

secondary page or article found at a website; and (3) an article in a periodical or magazine. Then, create a References page including those three sources, using the documentation system common to your field of study.

7. Convert the above References page to the other three documentation styles described in this chapter.
8. Which of the four documentation systems described in this chapter is the easiest to format? Which one is the most flexible? Which one is the easiest to read?

COLLABORATIVE PROJECT

Work in groups of three. Examine the documentation style that you have used in your major report assignment to confirm that you have correctly used the documentation system stipulated by your project supervisor.

SOURCES CITED

AllStarJobs.ca. "Sales and Marketing Jobs." *AllStarJobs.* Copyright 1999-2013. Web. 9 Feb 2013.

Amarjit, Deepak. "Preparation for Marketing." Email to Phil Larkin, 20 Dec. 2012.

Carto, Elaine. Personal Communication. 25 Jan. 2013.

Cavalier, Lise. Phone Interview, 8 Feb. 2013.

Cayer, Roger and Bill Collins. Personal Communication. 2 Feb. 2013.

HRDC. "6221 Technical Sales Specialists—Wholesale Trade." *National Occupational Classification.* 1 Jan. 2013. Ottawa: Human Resources Development Canada. Web. 5 Feb. 2013.

———. "6221 Technical Sales Specialists—Wholesale Trade." *Ontario JobFutures.* 2 Feb. 2013. Ottawa: Human Resources Development Canada. Web. 5 Feb. 2013.

Meaux, Anjoline. "The New Tech Sales Gang" *Report on Business* Jan. 2011: 55–58. Print.

Stewart, Mary. Personal Communication 11 Feb. 2013.

Monravia, Ken. *Specialized Sales and Marketing.* New York: Enterprise Books, 2009. Print.

Monster.ca. "Technical Sales and Marketing." *Monster.ca. 9* Feb 2013. Web. 9 Feb 2013.

Nelson, Joseph. "Sales Engineers." *Occupational Outlook Quarterly,* Fall (2001): 44(3), 20–24. Web. 5 Feb., 2013. Print.

Nezarde, Tao. "The New Tech Careers." *The Globe and Mail.* 17 Dec. 2012, p. C3. Print.

St. Croix, Anne. Personal Communication. January 21, 2013.

Tostenson, Andrew. "Alternate Careers in Marketing." *Electro '12 Conference,* Halifax, 29 Mar. 2012.

Working in Canada. *Explore the World of Work.* 2 Feb. 2013. Ottawa: Government of

Canada. Web. 7 Feb. 2013. <http://www.workingincanada.gc.ca/home-eng.do?lang=eng>

Viskovich, Julio. "What's Social Selling and Can It Increase Sales?" *Social Media Today.* 15 Dec. 2012. Web 5 Feb. 2013. <http://socialmediatoday.com/juliovisko/1082726/>.

NOTES:

MLA no longer requires URLs for online sources; however, your professor may require you to include URLs. Therefore, the last two entries in this sample list of sources contain examples of URL placement.

MLA format stipulates that double spacing should be used within each entry, not just between entries. However, technical documents usually single-space within entries and double-space between entries.

Figure 9.1 MLA Sample List of References

REFERENCES

AllStarJobs.ca. (Copyright 1999–2013). Sales and marketing jobs. *AllStarJobs.* Retrieved from www.allstarjobs.ca/marketing/

Monravia, K. (2009). *Specialized Sales and Marketing.* New York: Enterprise Books.

HRDC. (2013, 1 January). 6221 technical sales specialists—wholesale trade. *National occupational classification.* Ottawa: Human Resources Development Canada. Retrieved from www23.hrdc-drhc.gc.ca/2013/e/groups/6221.shtml

———. (2013, 2 February). 6221 technical sales specialists—wholesale trade. *Ontario Job Futures.* Ottawa: Human Resources Development Canada. Retrieved from http://www1.on.hrdc-drhc.gc.ca/ojf/ojfjsp?lang=e&section=Profile&noc=6221

Meaux, A. (2011, January). The new tech sales gang. *Report on Business, 16,* 55–58.

Monster.ca. (2013, February 9). Technical sales and marketing. *Search Jobs.* Retrieved from http://jobsearch.monster.ca

Nelson, J. (2001, Fall). Sales engineers. *Occupational Outlook Quarterly, 44*(3), 20–24. Retrieved from www.bls.gov/opub/ooq/ooqhome.htm

Nezarde, T. (2012 December 17). The new tech careers. *Globe and Mail.* C3.

Tostenson, A. (2012, March 29). Alternate careers in marketing. Presentation at Electro '12 Conference in Halifax, NS.

Viskovich, J. (2012, December 15). What's social selling and does it increase sales? *Social Media Today.* Retrieved from http://socialmediatoday.com/juliovisko/1082726

Working in Canada. (2013, February 2). Explore the world of work. Ottawa: Government of Canada. Retrieved January 30, 2010, http://www.workingincanada.gc.ca/home-eng.do?lang=eng

NOTES:

APA format stipulates that double spacing should be used within each entry, not just between entries. However, technical documents usually single-space within entries and double-space between entries.

Unlike the MLA format illustrated in Figure 9.1, APA does not include personal communications and other non-recoverable sources in the References section.

Figure 9.2 APA Sample List of References

REFERENCES

1. Working in Canada. Explore the world of work. [Internet]. Ottawa: Government of Canada. [Updated 2013 Feb 2; Cited 2013 Feb 5]. Available from http://www.workingin-canada.gc.ca/home-eng.do?lang=eng

2. Nelson, J. Sales engineers. [Internet]. Occupational Outlook Quarterly 2001, Fall; 44: 20–24. [Cited 2013 Feb 5]. Available from www.bls.gov/opub/ooq/ooqhome.htm

3. Nezarde T. 2012 Dec 17. The new tech careers. Globe and Mail. Sect C:3 (col 6).

4. HRDC. 6221 technical sales specialists—wholesale trade. Ontario Job Futures. [Internet]. Ottawa: Human Resources Development Canada. [Updated 2013 Jan 1; Cited 2013 Feb 5]. Available from http://www1.on.hrdc-drhc.gc.ca/ojf/ojfjsp?lang=e&section=Profile&noc=6221

5. Monravia K. Specialized sales and marketing. 2nd ed. New York: Enterprise Books; 2009. 293p.

6. HRDC. 6221 technical sales specialists—wholesale trade. National occupational classification. [Internet]. Ottawa: Human Resources Development Canada. [Updated 2013 Jan 1; Cited 2013 Feb 5]. Available from www23.hrdc-drhc.gc.ca/2013/e/groups/6221.shtml

7. Monster.ca. Technical sales and marketing. [Internet]. Search Jobs. [Updated 2013 Feb 9; Cited 2013 Feb 9]. Available from http://jobsearch.monster.ca

8. AllStarJobs. Sales and marketing jobs. [Internet]. AllStarJobs. [c1999–2013; Cited 2013 Feb 4]. Available from http://www.allstarjobs.ca/marketing/

9. Tostenson A. 2012 Mar. Alternate careers in marketing. [Internet]. Presentation at Electro '12 Conference in Halifax, NS. Available from http://www.electro12.ca/papers/tost.htm

10. Viskovich J. What's Social Selling and Can It Increase Sales? [Internet]. Social Media Today. Available: http://socialmediatoday.com/juliovisko/1082726/

11. Meaux, A. 2011 Jan. The new tech sales gang. Report on Business; 16: 55–58.

NOTE:

Non-recoverable sources such as personal interviews, phone conversations, and oral presentations are cited in the text, but not listed in the References.

Figure 9.3 Sample CSE References (Citation-Sequence Format)

REFERENCES

[1] Working in Canada. (2013 Feb. 2). *Explore the world of work.* [Online]. Ottawa: Government of Canada. Available: http://www.workingincanada.gc.ca/home-eng.do?lang=eng

[2] J. Nelson, (2001 Fall). "Sales engineers." [Online]. *Occupational Outlook Quarterly,* vol. 44, pp. 20–24. Available: www.bls.gov/opub/ooq/ooqhome.htm

[3] T. Nezarde, "The new tech careers." *The Globe and Mail,* 2012 Dec. 17, p. C3.

[4] HRDC, (2013 Jan. 1). 6221 technical sales specialists—wholesale trade. *Ontario Job Futures.* [Online]. Ottawa: Human Resources Development Canada. Available: http://www1.on.hrdc-drhc.gc.ca/ojf/ojf.jsp?lang=e&section=Profile&noc=6221

[5] K. Monravia, *Specialized Sales and Marketing.* New York: Enterprise Books, 2009.

[6] HRDC, (2013 Jan. 1). "6221 technical sales specialists—wholesale trade." [Online]. *National occupational classification.* Ottawa: Human Resources Development Canada. Available: www23.hrdc-drhc.gc.ca/2013/e/groups/6221.shtml

[7] Monster.ca, (2013 Feb. 9). "Technical sales and marketing." [Online]. *Search Jobs.* Available: http://jobsearch.monster.ca

[8] AllStarJobs. (c1999–2009). "Sales and marketing jobs." [Online]. *AllStarJobs.* Retrieved 2013 February 4 from www.allstarjobs.ca/marketing

[9] A. Tostenson, (2012 Mar. 29). "Alternate careers in marketing." [Online]. Presentation at Electro '09 Conference in Halifax, NS. Available: http://www.electro12.ca/papers/tost.htm

[10] J. Viskovich. (2012 Dec. 15). "What's Social Selling and Can It Increase Sales?" [Online]. *Social Media Today.* Available: http://socialmediatoday.com/juliovisko/1082726/

[11] A. Meaux, "The new tech sales gang." *Report on Business,* vol. 16, pp. 55–58.

NOTES:
Non-recoverable sources such as personal interviews, phone conversations, and oral presentations are cited in the text but not listed in the References unless a recoverable record is available.

IEEE citation guidelines provided by Canadian colleges and universities vary somewhat from the IEEE guidelines listed at <http://www.ieee.org/documents/ieeecitationref.pdf>. For example, some schools recommend that online sources should include the date the source was accessed. Therefore, consult your professor for the version required at your school.

Figure 9.4 Sample IEEE References

MyWritingLab

MyWritingLab: Technical Communication offers the best multimedia resources for technical communication in one, easy-to-use place. You can find a variety of interactive model documents and case studies. There are also extensive guidelines, tutorials, and exercises for Document Design, Writing and the Writing Process, and Research, and a large bank of diagnostics and practice for grammar review.

10

Summarizing Information

✱ Explore

Click here in your eText to access a variety of student resources related to this chapter.

LEARNING OBJECTIVES

After reading this chapter, you should be able to

- understand the purpose and value of summarizing information
- understand the elements of a useful summary
- employ the suggested stages of effectively and efficiently summarizing a document
- distinguish among *closing summary*, *informative abstract*, and *descriptive abstract*

A *summary* is a short version of a longer document. An economical way to communicate, a summary saves time, space, and energy.

PURPOSE OF SUMMARIES

ON THE JOB...

The Importance of Summaries

"The importance of a report summary in our line of work cannot be understated. Often, we prepare lengthy, complex reports for clients who have limited time, or a limited understanding of the technical aspects of the project, or both. A well-crafted summary gives the reader a snapshot of the most important points of the report in plain language. It should be able to stand alone, but it should also try to pull the reader forward so that he or she feels compelled to read the rest of the report."

—**Kellie Garcia Bunting, Senior Technical Writer/Editor, Summit Environmental Consultants Inc.**

As we record our research findings, we summarize and paraphrase to capture and compress the main ideas. In addition to this role as a research aid, summarized information is vital in day-to-day workplace transactions.

On the job, you have to write concisely about your work. You might report on meetings or conferences, describe your progress on a project, or propose a money-saving idea. Many employees provide superiors (decision makers) with summaries of the latest developments in their field.

Given today's constant flow and volume of information, summaries are more vital than ever. Some reports and proposals can be hundreds of pages long. Those who must act on this information need to rapidly identify what is most important in a document. From a good summary, busy people can get enough information to decide whether they should read the entire document, parts of it, or none of it.

Whether you summarize someone else's document or your own, your job is to communicate the *essential message* accurately and in the fewest words, as in the following passage and summary.

The following passage provides background geological information in the introduction to an environmental assessment report.

The original passage

Local bedrock materials are characterized by granite and alkali feldspar granite intrusive rocks with biotite and muscovite granite, quartz monzonite, and pegmatite. The site is found in a contemporary river valley bottom surrounded by terraces composed of glacial lacustrine deposits from Ancient Lake Penticton. The terraces consist of uniformly bedded, very fine silty-sand. Late glacial river channel deposits are also prevalent, but they are likely found only at significant depths.

Groundwater beneath the Subject Site is inferred to flow northwest, towards Wood Lake because of regional topography and the location of surface water features (Figure 1, Appendix A). The inferred groundwater flow direction under the site is a good approximation, but the actual direction would require field verification because localized variations in groundwater flow direction are possible.

An executive summary of the above passage would not need to include the geological details, just the two main ideas:

A summarized version

1. The site being assessed in this report has fine clay and sand deposited over bedrock.
2. Local surface water features and regional topography indicate that groundwater under the site likely flows northwest towards Wood Lake, although testing is required to verify that inference.

This summary is about 35 percent of the original length because the original itself is short. With a longer original, a summary might be 5 percent or less. But length is less important than informative value: an effective summary gives readers only what they need.

ELEMENTS OF A SUMMARY

Effective summaries include the following elements:

◆ *The essential message.* The essential message is the significant material from the original: controlling ideas (thesis and topic sentences); major findings; important names, dates, statistics, and measurements; and conclusions or recommendations. Significant material does not include background; the author's personal comments or conjectures; introductions; long explanations, examples, or definitions; visuals; or data of questionable accuracy.

◆ *Non-technical style.* More people generally read the summary than any other part of a document, so write at the lowest level of technicality. Translate technical data into plain English. "The patient's serum glucose measured 240 mg%" can be translated as "The patient's blood sugar remained critically high." But if you know all your readers are experts, you don't need to simplify.

◆ *Independent meaning.* In meaning as well as style, your summary should stand alone as a self-contained message. Readers should have to read the original only for more detail—not to make sense of the basic ideas.

◆ *No personal assessment.* Avoid personal comments ("This interesting report" or "The author is correct in assuming").

◆ *Conciseness.* Conciseness is vital, but never at the expense of clarity and accuracy. Make the summary economical, but clear and comprehensive.

CRITICAL THINKING IN THE SUMMARY PROCESS

Follow these guidelines for summarizing your own writing or another's.

1. *Read the entire original.* When summarizing another's work, get a complete picture before writing a word.
2. *Reread and underline.* Reread the original, underlining essential material. Focus on the thesis and topic sentences.
3. *Edit the underlined information.* Reread the underlined material, and cross out whatever does not advance the meaning.
4. *Rewrite in your own words.* Include all essential material in the first draft, even if it's too long; you can trim later.
5. *Edit your own version.* When you have everything readers need, edit for conciseness.
   a. Cross out all needless words without harming clarity or grammar. Use complete sentences.
   b. Cross out needless prefaces, such as "The writer argues . . ." or "Also discussed is. . . ."
   c. Use numerals for numbers, except to begin a sentence.
   d. Combine related ideas in order to emphasize relationships.
6. *Check your version against the original.* Verify that you have preserved the essential message and added no comments.
7. *Rewrite your edited version.* Add transitional expressions to reinforce the connection between related ideas.
8. *Document your source.* If summarizing another's work, cite the source immediately below the summary, and place directly quoted statements within quotation marks.

CASE	**Summarizing Information for a Non-technical Audience**

Imagine that you work in the information office of your province's ministry of the environment. In a coming election, citizens will vote on a referendum proposal for constructing municipal trash incinerators. Referendum supporters argue that incinerators would help solve the growing problem of waste disposal in highly populated parts of the province. Opponents argue that incinerators cause air pollution.

To clarify the issues for voters, the ministry will mail a newsletter to each registered voter. You have been assigned the task of researching the recent data and summarizing it. Here is one of the articles. You have underlined key phrases. (The margin notes reflect your critical thinking as you prepare to summarize the article.)

Incinerating Trash: A Hot Issue Getting Hotter

Combine as orienting sentence (controlling idea)

Alarmed by the tendency of landfills to contaminate the environment, both public officials and citizens are vocally seeking alternatives. The most commonly discussed alternative is something called a resource recovery facility. Nearly 40 Canadian cities have built such in the last 15 years, and another 50 or so are in various stages of planning.

Include definition

These recovery facilities are a new form of an old technology. <u>Basically</u>, they're <u>incinerators</u>. But, unlike the incinerators of old, they don't just burn waste. <u>They also recover energy</u>. The energy is <u>sold</u> as steam to an industrial customer, or it is converted to electricity and sold to the local utility. (A few facilities, but not many, also recover metals or other materials before using the waste as fuel.)

Include major fact

<u>A tonne of trash</u> possesses the <u>energy</u> content of <u>a barrel and a half of oil</u>. This is not a trivial amount. Canada discards <u>50 million tonnes</u> of <u>municipal waste</u> a year. If all

Include major statistics

of it were <u>converted to energy</u>, we <u>could replace</u> the equivalent of <u>12 percent of our oil imports</u>.

Include major fact

At the local level, <u>selling energy</u> or materials not only replaces non-renewable resources, it also provides a source of income that <u>partly offsets</u> the <u>cost of operating</u> the facility.

Include major fact

The <u>new facilities are</u>, on average, <u>much cleaner than the</u> municipal incinerators of <u>old</u>. Many have two-stage combustion units, in which the second stage burns exhaust

Delete explanation

gases at high temperature, converting many potential organic pollutants to less harmful emissions, such as carbon dioxide. Some, especially the larger and newer facilities, also come equipped with the latest in pollution-control devices.

Delete questionable point

The environmental community is uneasy with this new technology. Environmentalists have argued for many years that the best method of handling municipal trash is to recycle it—i.e., to separate the glass, metal, paper, and other materials and use them again, either without reprocessing or as raw materials in producing new products. The thought of the potential resources in municipal solid waste simply being burned, even with energy recovery, has made many environmentalists opponents of resource recovery.

Include key finding and explanation

More recently, opponents have found a stronger reason to oppose <u>burning waste: dioxin in the plants' emissions</u>. The <u>amounts</u> present are <u>extremely small</u>, measured in trillionths of a gram per cubic metre of air, <u>but dioxin can be deadly, at least to animals, at very low levels</u>.

What Is Dioxin?

Delete technical details

<u>Dioxin</u> is a <u>generic term for any of 75 chemical compounds</u>, the technical name for which is poly-chlorinated dibenzo-p-dioxins (PCDDs). A related group of 135 chemicals, the PCDFs, or furans, are often found in association with PCDDs.

Delete long explanation

<u>The most infamous of these substances</u>, 2,3,7,8-TCDD, is often <u>referred to as the</u> "<u>most toxic chemical known</u>." This judgment is <u>based on animal test data</u>. In laboratory tests, 2,3,7,8-TCDD is lethal to guinea pigs at a concentration of *500 parts per trillion*. A part per trillion is roughly equivalent to the thickness of a human hair compared to the distance across Canada.

Include major point

<u>The effects on humans are less certain</u>, for many reasons: it is difficult to measure the amounts to which humans have been exposed and difficult to isolate the effects of dioxin from the effects of other toxic substances on the same population; and the latency period for many potential effects, such as cancer, may be as long as 20 to 30 years.

Include continuation of major point

<u>Nevertheless</u>, because of the <u>extreme effects</u> of this substance <u>on animals</u>, known releases of dioxin have generated <u>considerable public alarm</u>. One of the most publicized releases occurred at Seveso, Italy, in July 1976, where a pharmaceutical plant explosion resulted in the contamination of at least 700 acres [283 hectares] of fields and affected more than 5000 people. Dioxin was found in the soil in concentrations of 20 to 55 parts per billion.

Delete long example

The immediate effects on humans were nausea, headaches, dizziness, diarrhea, and an acute skin condition called chloracne, which causes burn-like sores. The effects on animals were more severe: birds, rabbits, mice, chickens, and cats died by the hundreds, within days of the explosion. In response to the explosion, the Italian provincial authorities evacuated 730 people from the zone nearest the plant, and sealed off an area containing another 5000 people from contact with non-residents.

Delete long example

In North America, perhaps the best known dioxin contamination incident occurred at Times Beach, Missouri, where used oil, contaminated with dioxin, was sprayed on roads as a dust suppressant. Soil samples showed dioxin at levels exceeding 100 parts per billion. While no human health effects were documented at Times Beach, a flood in December 1982 led to widespread dispersal of the contamination, as a result of which the entire town was condemned, the population evacuated, and over $30 million of the U.S.'s Superfund money used to purchase the condemned property.

Include the most striking and familiar example

Dioxin was among the substances of concern at Love Canal. And it was the major contaminant in the chemical defoliant Agent Orange, the subject of a lawsuit by 15 000 Vietnam veterans and dependants and an out-of-court settlement of those complaints valued at $180 million.

Include key findings

As early as 1978, trace amounts of dioxin were found in the routine emissions of a municipal incinerator. Virtually every incinerator tested since that date has shown traces of dioxin.

Source: Adapted from McCarthy.

Assume that in two early drafts of your summary, you revised the content and edited the phrasing. Then, for coherence and emphasis, you inserted transitions and combined related ideas. Here is your final draft.

Incinerating Trash: A Hot Issue Getting Hotter (A Summary)

Because landfills often contaminate the environment, trash incinerators (resource-recovery facilities) are becoming a popular alternative. Nearly 40 are operating in Canadian cities, and 50 more are planned. Besides their relatively clean burning of waste, these incinerators recover energy, which can be sold to offset operating costs. One tonne of trash has roughly the energy content of 1.5 barrels of oil. Converting all Canadian refuse to energy could reduce oil imports by 12 percent.

However, incinerator emissions contain very small amounts of dioxin (a generic name for any of 75 related chemicals). Even low dioxin levels can be deadly to animals. In fact, animal tests have helped label one dioxin substance "the most toxic chemical known." Although effects on human beings are less certain, news of dioxin in the environment creates public alarm, as evidenced at Love Canal and by the successful Agent Orange lawsuit by 15 000 Vietnam veterans and dependants. Almost every municipal incinerator tested since 1978 has shown traces of dioxin.

The version above is trimmed, tightened, and edited: word count is reduced to roughly 20 percent of the original.

A summary this long serves well in many situations, but other audiences might want a briefer and more compressed summary—say, 10 to 15 percent of the original:

A More Compressed Summary

Because landfills often contaminate the environment, trash incinerators (resource-recovery facilities) are becoming a popular alternative across Canada. Besides their relatively clean burning of waste, these incinerators recover energy, which can be sold to offset operating costs.

However, incinerator emissions contain very small amounts of dioxin, a chemical proven so deadly to animals, even at low levels, that it has been labelled the most toxic chemical known. Although its effects on human beings are less certain, news of dioxin in the environment creates public alarm. Almost every municipal incinerator tested since 1978 has shown traces of dioxin.

Notice that the essential message is still intact. Related ideas are again combined, and fewer supporting details are included. Clearly, length is adjustable according to your audience and purpose.

FORMS OF SUMMARIZED INFORMATION

In preparing a report, proposal, or other document, you might summarize others' material as part of your presentation. But you will often summarize your own material as well. For instance, if your document extends to several pages, it might include three forms of summary, in different locations, with different levels of detail: *closing summary*, *informative abstract*, or *descriptive abstract*.[1] Figure 10.1 depicts these forms.

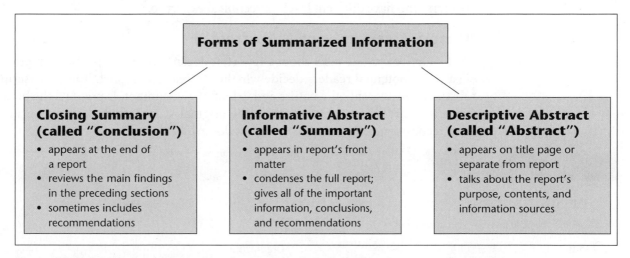

Figure 10.1 Summarized Information Assumes Various Forms

Closing Summary ("Conclusion")

A *closing summary* refers to summarized information at the outset of a conclusion section. It helps readers review and remember the preceding major findings. This look back at the "big picture" helps readers appreciate the conclusions and recommendations that typically follow the closing summary.

---

1. Adapted from Vaughan. Although we take liberties with his classification, Vaughan's insightful article helped clarify our thinking about the overlapping terminology that perennially seems to confound discussions of these distinctions.

Informative Abstract ("Summary")

Readers often appreciate condensed versions of reports. Some of these readers like to see a capsule version of a report before reading the complete document; others simply want to know basically what a report says without having to read the full document.

To meet reader needs, the *informative abstract* appears just after the title page. This summary tells essentially what the full version says: it identifies the need or issue that prompted the report; it describes the report's analytical method; it reviews the key facts and findings; and it condenses the report's conclusions and recommendations.

Actually, the title "Informative Abstract" is not used much these days. You are more likely to encounter "Summary" or "Synopsis." The title "Executive Summary" is used for material summarized for managers who may not understand all the technical jargon a report might contain. By contrast, a "Technical Summary" is aimed at readers at the same technical level as the author of the report. You may need two or three levels of summary for report readers who have different levels of technical expertise.

Descriptive Abstract ("Abstract")

A *descriptive abstract* talks about a report; it doesn't give the report's main points. It just helps potential readers decide whether to read the report. Thus, a descriptive abstract conveys only the nature and extent of a document. It presents the broadest view and offers no major facts from the original. It indicates whether conclusions and recommendations are included, but doesn't list them.

CHECKLIST

for Summaries

Use this checklist to refine your summaries.

Content

◆ Does the summary contain only the essential message?
◆ Does the summary make sense as an independent piece?
◆ Is the summary accurate when checked against the original?
◆ Is the summary free of any additions to the original?
◆ Is the summary free of needless details?
◆ Is the summary economical yet clear and comprehensive?
◆ Is the source documented?
◆ Does the descriptive abstract tell what the original is about?

Organization

◆ Is the summary coherent?
◆ Are there enough transitions to reveal the line of thought?

Style

◆ Is the summary's level of technicality appropriate for its audience?
◆ Is the summary free of needless words?
◆ Are all sentences clear, concise, and fluent?
◆ Is the summary grammatically correct?

Descriptive abstracts usually appear in special publications or in electronic databases, both of which name and briefly describe hundreds of reports; one example of the latter is the widely used Aquatic Sciences and Fisheries Abstracts. However, some reports place a one- to three-sentence abstract on the report's title page. In all of these instances, the title "Abstract" signals a brief description, not all of a report's highlights.

EXERCISES

1. Read each of these two paragraphs, and then list the significant ideas contained in each essential message. Write a summary of each paragraph.

   In recent years, ski-binding manufacturers, in line with consumer demand, have redesigned their bindings several times in an effort to achieve a non-compromising synthesis between performance and safety. Such a synthesis depends on what appear to be divergent goals. Performance, in essence, is a function of the binding's ability to hold the boot firmly to the ski, thus enabling the skier to change rapidly the position of his or her skis without being hampered by a loose or wobbling connection. Safety, on the other hand, is a function of the binding's ability both to release the boot when the skier falls, and to retain the boot when subjected to the normal shocks of skiing. If achieved, this synthesis of performance and safety will greatly increase skiing pleasure while decreasing accidents.

   Contrary to public belief, sewage-treatment plants do not fully purify sewage. The product that leaves the plant to be dumped into the leaching (sieve-like drainage) fields is secondary sewage containing toxic contaminants, such as phosphates, nitrates, chloride, and heavy metals. As the secondary sewage filters into the ground, this conglomeration is carried along. Under the leaching area develops a contaminated mound through which groundwater flows, spreading the waste products over great distances. If this leachate reaches the outer limits of a well's drawing radius, the water supply becomes polluted. And because all water flows essentially toward the sea, more pollution is added to the coastal regions by this secondary sewage.

2. Attend a campus lecture on a topic of interest and take notes on the significant points. Write a summary of the lecture's essential message.

3. Find an article about your major field or area of interest and write both an informative abstract and a descriptive abstract of the article.

COLLABORATIVE PROJECT

Select a long paper that your project group has written for one of your courses. In a work session, combine to write an informative abstract and then a descriptive abstract of the paper.

MyWritingLab

MyWritingLab: Technical Communication offers the best multimedia resources for technical communication in one, easy-to-use place. You can find a variety of interactive model documents and case studies. There are also extensive guidelines, tutorials, and exercises for Document Design, Writing and the Writing Process, and Research, and a large bank of diagnostics and practice for grammar review.

11

Organizing for Readers

✷—Explore

Click here in your eText to access a variety of student resources related to this chapter.

LEARNING OBJECTIVES

After reading this chapter, you should be able to

- create an outline that best organizes a given document according to its content, purpose, and intended audience
- distinguish between standard paragraphs and paragraphs that place the topic sentence in an alternate position
- choose appropriate paragraph structures, from among general-to-specific, specific-to-general, and chronological patterns

---

One of your biggest writing challenges is to transform your material into manageable form. First, you need to unscramble information to make sense of it for yourself; then, you need to shape it for your readers' understanding.

To follow your thinking, readers need a message organized in a way that makes sense to *them*. But data rarely materialize or thinking rarely occurs in neat, predictable sequences. So, you must *shape* random ideas and data into an organized pattern of meaning. As you organize, you face questions, such as the following:

TYPICAL QUESTIONS IN ORGANIZING FOR READERS

- ◆ What major question am I answering for my reader(s)?
- ◆ What secondary questions will help answer the main question?
- ◆ What relationships and answers do the gathered data suggest?

- ◆ What should I emphasize?
- ◆ What belongs where?
- ◆ What do I place first? Why?
- ◆ What comes next?
- ◆ How do I end?

Writers rely on the following strategies for organizing material: topical arrangement, outlining, paragraphing, and sequencing.

TOPICAL ARRANGEMENT

Whenever you analyze something, you break it down to discover constituents, connections, similarities, trends, associations, correlations, relationships, and perspectives.

You break the topic into subtopics (commonly called *partitioning*). You also identify which things share similarities and should therefore be discussed at the same level in a certain category (called *classifying*). The following example explains how topical arrangement can be achieved.

An example of partitioning

A discussion of Olympic Games events could be initially divided (*partitioned*) into two main types: Summer and Winter Games. Then, the winter events could be listed as the 15 sports that took place at the 2010 Vancouver Games, which included ice sports, alpine sports, and Nordic events. Sports could be further partitioned: for example, "figure skating" could be divided into men's and women's singles, pairs, and ice dancing. "Skiing" could be partitioned into several events: alpine, cross-country, ski jumping, freestyle skiing, and biathlon, which could be further differentiated by gender. A partial partition of the 2010 Winter Games sports appears in Figure 11.1 below.

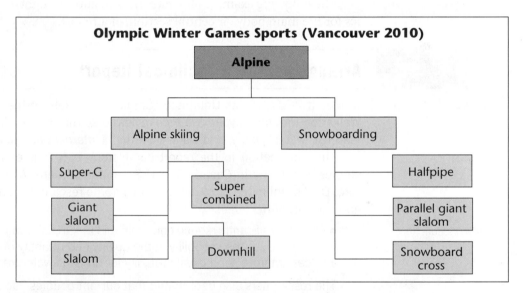

Figure 11.1 Partition of Topics (Partial)

Classifying according to purpose

However, someone might choose to *classify* the 15 official events differently. For example, a writer might want to discuss "team games" and group together hockey, curling, short-track speed skating, bobsled, and cross-country relay. Or a writer might want to focus on the events in which Canadians won gold medals (hockey, curling, bobsled, freestyle skiing, short-track speed skating, snowboarding, speed skating, skeleton). A focus on race events could include alpine skiing, cross-country skiing, speed skating, snowboarding, and biathlon, while judged events could include figure skating and freestyle skiing. The classification depends on the writer's purpose. Figure 11.2 on page 172 illustrates a complete classification of the three main categories and 15 subcategories of Winter Olympic sports.

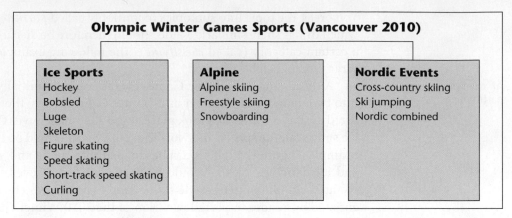

Figure 11.2 Classification of Main Categories

The following example illustrates how an author might arrange and develop topics for the main body, or central section, of a report.

CASE	**Arrangement in a Technical Report**

The reason for the report

In March 2001, Ducks Unlimited submitted a "State of the Science" report to the Walkerton Inquiry. *Beyond the Pipe* reviewed studies of the "link between watershed features, water quality, and water quantity" (*International Scientific Review*).

The information in the report came from 165 articles and reports, based on studies conducted in Canada, the U.S., and elsewhere. After reviewing that literature, Ducks Unlimited and University of Toronto researchers concluded that wetlands and riparian buffer zones

The report's conclusions

- ◆ "can significantly reduce contaminants in surface and groundwater,"
- ◆ "can reduce the variability in the quality and quantity of water sources," and
- ◆ "can improve source water quality for drinking water treatment."

The researchers also determined that current policies "do not sufficiently protect wetlands and riparian areas" and that protection and restoration costs are "small compared to the costs of in-pipe treatment." However, where land costs are high, restoration may not be economically feasible (Gabor et al. 40–41).

The report recommended that the province should

The report's recommendations

1. create integrated, comprehensive water management policy
2. enhance wetland protection
3. encourage and enhance wetland restoration
4. encourage riparian-area protection programs
5. improve its understanding of watershed management (Gabor et al. 41–43)

The authors' main challenge

Beyond the Pipe's purpose was to persuade the Walkerton Inquiry that the Ducks Unlimited goal of preserving wetlands is consistent with the need to protect drinking water sources. The gathered evidence supports that contention, but the report could fulfill its persuasive purpose *only* if it used a clear, logical structure. After a 1.5-page introduction that provides a context and describes the report's four main objectives, the report's 31 pages of findings are organized into five smoothly integrated sections.

Table 11.1 below shows the report's main topics and subtopics.

Table 11.1 Relationships within a Report's Topical Arrangement (Based on a Ducks Unlimited Report)

Wetlands	Permanent Cover	Buffer Strips	Wetland Loss	Policy

Wetlands of   Southern Ontario
Hydrological functions
   Water storage and
     flood reduction
   Groundwater recharge
Water-quality functions
   Nutrient assimilation
   Sediments
   Pathogens
   Contaminants
Summary
	Upland
	   conservation
	   programs
		Sediment removal
		   and erosion control
		Nutrient assimilation
		   Nitrogen
		   Phosphorus
		Pathogens
		Pesticides
		Contaminants
		Summary
				Wetland protection
				   in Ontario
				   Federal policies
				   Provincial policies
				   Other instruments
				Riparian area
				   management
				Why protect and
				   restore?
				   Water quality
				   Costs of natural vs.
				     in-pipe treatment

OUTLINING

The value of outlining

The arrangement of related topics and subtopics in Table 11.1 illustrates the technique of *outlining*, an important skill developed by efficient and effective technical writers. Outlines use partition and classification to guide readers through the material in a pattern that readers find logical and ultimately persuasive.

How should you organize the document to make it logical for your audience? Begin with the basics. Useful writing of any length—a book, report, chapter, letter, or memo—typically has an introduction, central section, and conclusion.

◆ The *introduction* provides orientation by doing any of these things: explaining the topic's origin and significance and the document's purpose; identifying briefly your intended audience and your information sources; defining specialized terms or general terms that have special meanings in your document; accounting for limitations, such as incomplete or questionable data; previewing the major topics to be discussed in the central section.

    Some introductions need to be long and involved; others, short and to the point. Reports too often waste readers' time with needless background information. If you don't know your readers well enough to give them only what they need, use subheadings so that they can choose what they want to read.

◆ The *central section* delivers on the promise implied in your introduction ("Show me!"). Some reports may have one central section, while others may have several. Here you present your data, discuss your evidence, lay out your case, or tell readers what to do and how to do it. Central sections come in all different sizes, depending on how much readers need and expect.

    Central sections are titled to reflect their specific purpose: "Description and Function of Parts," for a mechanical description; "Required Steps," for a set of instructions; "Collected Data," for a feasibility analysis; "Operating Instructions," for a user's manual.

◆ The *conclusion* of a document has assorted purposes: it might evaluate the significance of the report, re-emphasize key points, take a position, predict an outcome, offer a solution, or suggest further study. If the issue is straightforward, the conclusion might be brief and definite. If the issue is complex or controversial, the conclusion might be lengthy and open-ended. Whatever the conclusion's specific purpose, readers expect a clear perspective on the whole document.

Conclusions vary with the type of document. You might conclude a mechanical description by reviewing the mechanism's major parts and then briefly describing one operating cycle. You might conclude a comparison or feasibility report by offering judgments about the facts you've presented and then recommending a course of action.

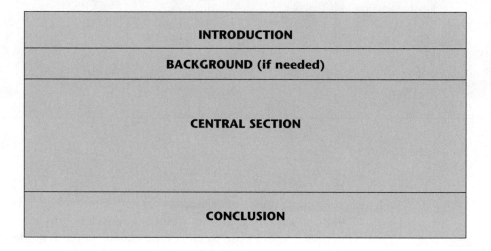

A suitable beginning, middle, and ending are essential, but alter your outline as you see fit. No single form of outline should be followed slavishly. **The organization of any document ultimately is determined by its audience's needs and expectations**. In many cases, specific requirements about a document's organization and style are spelled out in a company's style guide. Structures for specific types of documents are provided in various sections of this text.

The computer is especially useful for rearranging outlines until they reflect the sequence in which you expect readers to approach your message.

The Formal Outline

A simple list usually suffices for organizing short documents. However, long or complex documents call for a systematic, formal outline to mark divisions and to show how categories relate.

The section in Chapter 19 on writing formal reports shows three levels of outlines: *planning outline*, which guides the writer's research and initial organization; *working outline*, a lengthy formal device that acts as a blueprint for a document; and *paragraph outline*, which is a short, informal overview of the structure of a paragraph or series of related paragraphs.

Here is a formal working outline for the report illustrated in Table 11.1 on the previous page.

BEYOND THE PIPE—WETLANDS, RIPARIAN BUFFER STRIPS, AND WATER QUALITY

A formal outline whose topical arrangement uses an appropriate pattern:

◆ nature of need (problem)
◆ nature of wetlands and their restorative functions (solution)
◆ nature of natural cover (solution)
◆ riparian buffer strips (solution)
◆ wetland loss (a reminder of the extent of the problem)
◆ inadequate protection policies (cause of problem)
◆ conclusions and recommendations (why and how to implement a solution)

INTRODUCTION
◆ The need for fresh water
◆ Wetlands and watershed management
    Objectives

WETLANDS
    Wetlands of Southern Ontario
    Hydrological Functions of Wetlands
        Water storage and flood reduction
        Groundwater recharge
    Water-Quality Functions
        Nutrient assimilation
            Nitrogen
            Phosphorus
        Sediments
        Pathogens
        Contaminants
    Summary

PERMANENT COVER
    Upland Conservation Programs

BUFFER STRIPS
    Sediment Removal and Erosion Control
    Nutrient Assimilation
        Nitrogen
        Phosphorus
    Pathogens
    Pesticides
    Contaminants
    Summary

WETLAND LOSS
◆ Percentage of loss in Ontario

POLICY
    Wetland Protection in Ontario
        Federal policies and legislation
        Provincial policies
        Other instruments for protection
    Riparian-Area Management
    A Case for the Protection and Restoration of Wetlands and Riparian Areas
        Wetlands and riparian areas as water-quality support systems
        Funding natural versus in-pipe water treatment—a comparison

CONCLUSIONS AND RECOMMENDATIONS

Technical documents often use decimal notation:

Decimal notation in a
technical document

2.0 WETLANDS
    2.1 Wetlands of Southern Ontario
    2.2 Hydrological Functions of Wetlands
        2.2.1 Water storage and flood reduction
        2.2.2 Groundwater recharge
    2.3 Water Quality Functions
        2.3.1 Nutrient assimilation
        2.3.1.1 Nitrogen
        2.3.1.2 Phosphorus

Decimal notation, which makes it easier to refer readers to specific sections of a document, is usually preferred in engineering, government, and industry.

The Importance of Being Organized

The neat and ordered outline shown earlier represents the *product* of outlining, not the *process*. Beneath any finished outline (or document) lies *planning*. Whether you work alone or as part of a team, planning is an essential element of the writing process. Initially, planning may take the form of general discussions and brainstorming. The ideas generated are often written on a whiteboard or displayed on a computer screen as notes or flow charts.

The second stage of planning *organizes* the ideas into specific topics. Ideas may be discarded because they are not within the scope of the project; other ideas may be generated as the topics are being identified. When you know the main topics you need to cover for your project, you are ready to put them into an outline. If you have used a computer during your planning and organizing, you already have a draft outline. The next step is to rearrange your outline so that it flows logically (has a sequence). Various types of sequencing are covered later in this chapter, but first here are two rules of effective and efficient writing:

1. **Never** start writing until you have thoroughly planned the topics to be in your document and you have created an organized, logical outline.
2. When you start writing, follow your outline closely; don't start planning again!

Organizing for Cross-cultural Audiences

Different cultures often have different expectations as to how information should be organized. A document considered well organized by one culture may confuse or offend another. For instance, a paragraph in English typically begins with a main idea directly expressed as a topic or orienting sentence, followed by specific support; any digression from this main idea is considered harmful to the paragraph's *unity*. Some cultures, however, consider digression a sign of intelligence or politeness. To native readers of English, the long introductions and digressions in certain Spanish or Russian documents might seem tedious and confusing, but a Spanish or Russian reader might view the more direct organization of English as overly abrupt and simplistic (Leki 151).

Expectations can differ even among same-language cultures. British correspondence, for instance, typically expresses the bad news directly up front, instead of the indirect approach preferred in North America. A bad-news letter or memo appropriate for North American readers could be considered evasive by British readers (Scott and Green 19).

Despite all our electronic communication tools, connecting with readers—especially in a global context—requires, above all, human sensitivity and awareness of audience.

PARAGRAPHING

Readers look for orientation—for shapes they can recognize. Beyond its larger shape (introduction, central section, conclusion), a document depends on the smaller shapes of each paragraph.

Although paragraphs can have various structures and purposes (paragraphs of introduction, conclusion, or transition), our focus here is on *standard support paragraphs*. Although it is part of the document's larger design, each support paragraph usually can stand alone in meaning.

The Standard Paragraph

All the sentences in a standard paragraph relate to the main point, which is expressed as the *topic sentence*:

Topic sentences

> Computer literacy has become a requirement for all "educated" people.
>
> A video display terminal can endanger the operator's health.
>
> Chemical pesticides and herbicides are both ineffective and hazardous.

Each topic sentence introduces an idea, judgment, or opinion. But in order to grasp the writer's exact meaning, readers need explanation. Consider the third statement:

> Chemical pesticides and herbicides are both ineffective and hazardous.

Imagine you are a researcher for the Eastern Power Utility and have been asked to determine whether the company should (1) begin spraying pesticides and herbicides under its power lines or (2) continue with its manual (and non-polluting) ways of minimizing foliage and insect damage to lines and poles. If you simply responded with the preceding assertion, your employer would have questions:

- Why, exactly, are these methods ineffective and hazardous?
- What are the problems?
- Can you explain?

To answer these questions and to support your assertion, you need a fully developed paragraph:

Introduction *or* topic sentence (1)
Body (2–6)

> [1]**Chemical pesticides and herbicides are both ineffective and hazardous**. [2]Because none of these chemicals have permanent effects, pest populations invariably recover and need to be re-sprayed. [3]Repeated applications cause pests to develop immunity to the chemicals. [4]Furthermore, most of these products attack species other than the intended pest, killing off its natural predators, thus actually increasing the pest population. [5]Above all, chemical residues survive in the environment (and living tissue) for years, often carried hundreds of kilometres by wind and water. [6]This toxic legacy includes such

biological effects as birth deformities, reproductive failures, brain damage, and cancer. [7]Although intended to control pest populations, these chemicals, ironically, threaten to make the human population their ultimate victims. [8]We therefore recommend continuing our present control methods.

Conclusion (7–8)

Most standard paragraphs in technical writing have an introduction–central section–conclusion structure. They begin with a clear topic (or orienting) sentence, stating a generalization. Details in the central section support the generalization.

The Topic Sentence

Readers look to a paragraph's opening sentences for a framework. When they don't know exactly what the paragraph is about, readers struggle to grasp your meaning. Read this next paragraph once only, and then try answering the questions that follow.

A paragraph with its topic sentence omitted

Besides containing several toxic metals, it percolates through the soil, leaching out naturally present metals. Pollutants, such as mercury, invade surface water, accumulating in fish tissues. Any organism eating the fish—or drinking the water—in turn faces the risk of heavy metal poisoning. Moreover, acidified water can release heavy concentrations of lead, copper, and aluminum from metal plumbing, making ordinary tap water hazardous.

Can you identify the paragraph's main idea? Probably not. Without a clear topic sentence, you have no framework for understanding this information in its larger meaning. And you don't know where to place the emphasis: on polluted fish? on metal poisoning? on tap water?

Now, insert the following opening sentence and reread the paragraph:

The missing topic sentence

Acid rain indirectly threatens human health.

With this orientation, the exact meaning becomes obvious.

The topic sentence should appear *first* in the paragraph, unless you have a good reason to place it elsewhere. Most of the time, it will be a single sentence. However, in some instances a paragraph's main idea may require two or more sentences (a topic statement):

A topic statement can have two or more sentences

The most common strip-mining methods are open-pit mining, contour mining, and auger mining. The specific method employed will depend on the type of terrain that covers the coal.

The topic sentence should immediately tell readers what to expect. Don't write *Some pesticides are less hazardous and often more effective than others* when you mean *Organic pesticides are less hazardous and often more effective than their chemical counterparts*. The first topic sentence leads everywhere and nowhere; the second helps focus the reader on what to expect from the paragraph. Don't write *Acid rain poses a danger*, leaving readers to decipher your meaning of *danger*. If you mean *Acid rain is killing our lakes and polluting our water supplies*, say so. Uninformative topic sentences keep readers guessing.

ALTERNATIVE TOPIC SENTENCE PLACEMENT.   Nearly all paragraphs in technical and business writing open with a topic sentence, but sometimes the opening sentence will provide a transition, which is followed by the topic in the second sentence and then by the rest of the paragraph. Here's an effective paragraph in a report's conclusion section:

Transition sentence (1)
Topic sentence (2)
Supporting reasons (3–5)

[1]As this report has demonstrated, erosion control blankets have a wide range of uses. [2]**Therefore, these blankets should be considered for controlling run-offs on the Summit Highway**. [3]The blankets provide a warm, moist environment for the hydro-seeded wild grass to germinate. [4]Also, the blankets prevent erosion until the grass can provide a stable subsoil–root system to prevent erosion. [5]Finally, the straw and the coconut fibre in the blankets gradually decay into the soil to make the soil more fibrous and resistant to run-off erosion.

Some paragraphs open with a transition that continues the direction established by the previous paragraph(s) and then switch to the real topic of the new paragraph.

Transition sentence (1)
Switch to the real
topic (2) Descriptive
explanation (3–5)

[1]We recommend erosion control blankets for the majority of road banks on the Summit Highway. [2]**However, for banks that are 2:1 or steeper, we recommend heavy-duty geogrids**. [3]Skalned Engineering's studies have shown that steep banks require geogrids of 10-centimetre side wall mesh that can be filled with small-scale blast rock (Appendix 2). [4]The geogrid can also be partially filled with soil, which can be hydro-seeded. [5]Skalned's study revealed that geogrid covers of this type prevent erosion of banks as steep as 1.5 to 1.

Other paragraphs, such as the following, effectively place the main idea at the end of the paragraph.

Sentence 1 introduces
the topic area, but not
the paragraph's main
idea

List of details (2 and 3)
Main idea (4)

[1]Worried about heavy erosion, we used Gabion baskets to retain the 2:1 banks of Tumbledown Creek below LeBihan Falls. [2]We observed that the baskets retained their geometric shapes despite the heavy aggregate that filled them. [3]Also, because we placed geotextile fabric beneath the Gabion structure, the strong 2010 spring run-off didn't scour or undercut the Gabion system. [4]**Overall, the entire system remained intact and stable despite the powerful erosive conditions that ran from early March to late May**.

Paragraph Unity

A paragraph is unified when all its material belongs there—when every word, phrase, and sentence directly support the topic sentence.

A unified paragraph

Solar power offers an efficient, economical, and safe solution to Eastern Canada's energy problems. To begin with, solar power is highly efficient. Solar collectors installed on fewer than 30 percent of roofs in Eastern Canada would provide more than 70 percent of the area's heating and air-conditioning needs. Moreover, solar heat collectors are economical, operating for up to 20 years with little or no maintenance. These savings recoup the initial cost of installation in only 10 years. Most important, solar power is safe. It can be transformed into electricity through photovoltaic cells (a type of storage battery) in a noiseless process that produces no air pollution—unlike coal, oil, and wood combustion. In sharp contrast to its nuclear counterpart, solar power produces no toxic waste and poses no catastrophic danger of meltdown. Thus, massive conversion to solar power would ensure abundant energy and a safe, clean environment for future generations.

One way to damage unity in the paragraph above would be to discuss the differences between active and passive solar heating, or manufacturers of solar technology, or the advantages of solar power over wind power. Although these matters do broadly relate to the general issue of solar energy, none directly advances the meaning of *efficient*, *economical*, or *safe*.

Every topic sentence has a keyword or phrase that carries the meaning. In the pesticide/herbicide paragraph (pages 177–178), the keywords are *ineffective* and *hazardous*. Anything that fails to advance their meaning throws the paragraph—and readers—off track.

Paragraph Coherence

In a unified paragraph, everything belongs. In a coherent paragraph, everything sticks together: the topic sentence and support form a connected line of thought, like the links in a chain. To convey precise meaning, a paragraph must be unified. To be readable, a paragraph must also be coherent.

Paragraph coherence can be damaged by (1) short, choppy sentences, (2) sentences in the wrong sequence, or (3) insufficient transitions and connectors for linking related ideas. Here is how the solar-energy paragraph might become incoherent:

An incoherent paragraph

> Solar power offers an efficient, economical, and safe solution to Eastern Canada's energy problems. Unlike nuclear power, solar power produces no toxic waste and poses no danger of meltdown. Solar power is efficient. Solar collectors could be installed on fewer than 30 percent of roofs in Eastern Canada. These collectors would provide more than 70 percent of the area's heating and air-conditioning needs. Solar power is safe. It can be transformed into electricity. This transformation is made possible by photovoltaic cells (a type of storage battery). Solar heat collectors are economical. The photovoltaic process produces no air pollution.

Here, in contrast, is the original, coherent paragraph with sentences numbered for later discussion and with transitions and connectors shown in boldface. Notice how this version reveals a clear line of thought:

A coherent paragraph

> [1]Solar power offers an efficient, economical, and safe solution to Eastern Canada's energy problems. [2]**To begin with**, solar power is highly efficient. [3]Solar collectors installed on fewer than 30 percent of roofs in Eastern Canada would provide more than 70 percent of the area's heating and air-conditioning needs. [4]**Moreover**, solar heat collectors are economical, operating for up to 20 years with little or no maintenance. [5]**These savings** recoup the initial cost of installation within only 10 years. [6]**Most important**, solar power is safe. [7]**It** can be transformed into electricity through photovoltaic cells (a type of storage battery) in a noiseless process that produces no air pollution—unlike coal, oil, and wood combustion. [8]**In sharp contrast** to its nuclear counterpart, solar power produces no toxic waste and poses no danger of catastrophic meltdown. [9]**Thus**, massive conversion to solar power would ensure abundant energy and a safe, clean environment for future generations.

We can easily trace the sequence of thoughts in this paragraph.

1.   The topic sentence establishes a clear direction.
2–3.   The first reason is given and then explained.
4–5.   The second reason is given and explained.
6–8.   The third and major reason is given and explained.
9.   The conclusion sums up and re-emphasizes the main point.

Within this line of thinking, each sentence follows logically from the one before it. Readers know where they are at any place in the paragraph. To reinforce the logical sequence, related ideas are combined in individual sentences, and transitions and connectors signal clear relationships. The whole paragraph holds together.

Paragraph Length

Paragraph length depends on the writer's purpose and the reader's capacity for understanding. Actual word count is less important than how thoroughly the paragraph makes your point. Consider these guidelines:

- In writing that carries highly technical information or complex instructions, short paragraphs (perhaps in a vertically displayed list) give the reader plenty of breathing space. A clump of short paragraphs, however, can make a document seem choppy and poorly organized.
- In writing that explains concepts, attitudes, or viewpoints, support paragraphs generally run from 100 to 300 words. But long paragraphs can be tiring and hard to follow, especially if important ideas get buried in the middle. On average, report paragraphs should be kept to 100 words, while paragraphs in letters and memos should average about 60 words.
- Long paragraphs can be broken into parts—using bullets, for example—to make the information more accessible to the reader.
- In letters, memos, or news articles, paragraphs of only one or two sentences focus the reader's attention. A short paragraph (even a single-sentence paragraph) can highlight an important idea in any document.
- Avoid long paragraphs at the beginning or end of a document because they can discourage the reader or obscure the emphasis.

SEQUENCING

Research demonstrates that readers more easily understand and remember material that is organized in a logical sequence (Felker et al. 11). In practice, these logical sequences tend to be one of three main types: *general to specific*, *specific to general*, and *chronological* (see Figure 11.3). A single paragraph usually follows one particular sequence. A longer document may use one particular sequence or a combination of sequences.

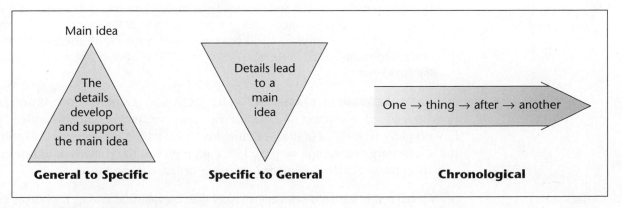

Figure 11.3 Three Main Types of Paragraph Patterns

General to Specific

Most technical and business writing relies heavily on general-to-specific patterns, such as

- spatial sequence (or other type of descriptive sequence)
- statement plus illustration
- emphatic sequence (statement plus detailed evidence or arguments)
- extended definition
- classification
- comparison–contrast sequence

SPATIAL SEQUENCE (OR OTHER TYPE OF DESCRIPTIVE SEQUENCE).   The most common *descriptive* pattern uses a spatial sequence, which begins at one location and ends at another. Such a sequence is most useful in describing a physical item or a mechanism; its parts appear in the sequence in which readers would actually view the parts or in the order in which each part functions (left to right, inside to outside, top to bottom). This description of a hypodermic needle proceeds from the needle's base (hub) to its point:

"What does it look like?"

> **A hypodermic needle is a slender, hollow steel instrument used to inject medication into the body (usually through a vein or muscle). The instrument has three parts, all considered sterile: the hub, the cannula, and the point**. The hub is the lower, larger part of the needle that attaches to the necklike opening on the syringe barrel. Next is the cannula (stem), the smooth and slender central portion. Last is the point, which consists of a bevelled (slanted) opening, ending in a sharp tip. The diameter of a needle's cannula is indicated by a gauge number; commonly, a 24–25 gauge needle is used for subcutaneous injections. Needle lengths are varied to suit individual needs. Common lengths used for subcutaneous injections include 0.85 cm, 1.25 cm, 1.5 cm, and 1.85 cm. Regardless of length and diameter, all needles have the same functional design.

Product and mechanism descriptions almost always have some type of visual to support or amplify the verbal description.

STATEMENT PLUS ILLUSTRATION.   Often, the best way to prove a point is to illustrate it with an example or story. The main point appears in the topic sentence, and the illustration forms the remainder of the paragraph, as in the following example:

"Why should I believe you?"

> **Sometimes, steep erosion-prone slopes can be stabilized simply and inexpensively**. For example, our client at 458 Summit Road was unwilling to spend an estimated $38 000 for a recommended reinforced concrete retaining wall at the rear of his property. Our low-budget solution involved C350 erosion control blankets, hydro-seeded wild grass cover, and multi-flow drainage pipe, all for only $8546. For 27 months, the slope has withstood erosion.

EMPHATIC SEQUENCE (STATEMENT PLUS DETAILED EVIDENCE OR ARGUMENTS).   Reasons offered in support of a specific viewpoint or recommendation often appear in workplace writing, as in the pesticide–herbicide paragraph on pages 177 to 178 or the solar-energy paragraph on page 179. For emphasis, the reasons or examples usually are arranged in decreasing or increasing order of importance.

"What should I remember about this?"

> **Although strip mining is safer and cheaper than conventional mining, it damages the surrounding landscape**. Among its effects are scarred mountains, ruined land, and polluted

waterways. Strip operations are altering our country's land at the rate of 506 hectares (1250 acres) per week. An estimated 7085 kilometres (4400 miles) of streams have been poisoned by silt drainage in British Columbia alone. If strip mining continues at its present rate, 14 900 square kilometres (5750 square miles) of Canadian land eventually will be stripped barren.

EXTENDED DEFINITION. Another descriptive pattern used to organize a paragraph is found in an extended definition (or *expanded definition*). Such a paragraph piles up the details to help a reader understand a term.

"What does this term really mean?"

Open source software code develops collaboratively; it may be copied, modified, and distributed by anyone, without fee charges. The Open Source Initiative further stipulates that all source code must remain clear and that the code licence must state that open source software can be freely distributed, although each modified version should have a different name or version number. Also, the code licence must accept use by any person or group for all suitable applications. To make that last condition feasible, the code licence must be available to any product, compatible with other software, and usable in a variety of electronic formats.

CLASSIFICATION. Close examination of any complex issue requires both partition and classification. (See the previous discussion on page 170.) The following paragraph shows both partition and classification at work.

"How are the components of the issue related?"

While researching the health effects of electromagnetic fields, you'll encounter information about

♦ various radiation sources
♦ the ratio of the level of risk to the level of exposure
♦ workplace studies of effects on workers
♦ lab studies of cell physiology, biochemistry, and behaviour
♦ statistical studies of disease in certain populations
♦ conflicting expert views, and
♦ local authorities' views

COMPARISON–CONTRAST SEQUENCE. An evaluation of two or more items on the basis of their similarities or differences often appears in job-related writing. As in the following example, point-by-point comparison is quite adequate when general features are being compared.

"How do these items compare?"

Point-by-point comparison

The BlackBerry Bold 9000 and iPhone 3GS smartphones each have competitive advantages. Users who like physical keyboards will likely prefer the BlackBerry, while iPhone users will be drawn to its large display and multitouch controls, including a virtual keyboard. Both offer good resolution, excellent speed, and adequate storage, although the iPhone's 32 GB capability is much greater than the BlackBerry's (128 MB onboard memory, 1 GB internal media memory). Another iPhone advantage is its display size, which is 50 percent larger than the BlackBerry's because the latter's QWERTY keyboard covers about half of the device's face. While some users prefer the BlackBerry's digital camera that has a built-in LED flash, the iPhone's camera offers 3.15 megapixels compared with the BlackBerry's 2.0 megapixels, and BlackBerry lacks the iPhone camera's autofocus.

For comparing and contrasting specific data on these smartphones, two lists would be more effective.

Block comparison

The **BlackBerry Bold 9000** has the following features and specifications:

1. It is 127 mm long, 66 mm wide, and 12.7 mm thick.
2. Its 66 mm diagonal touch display has 480 × 320 resolution.
3. Its operations are controlled by a trackball and a built-in QWERTY keyboard.
4. It has 802.11a, 802.11b, and 802.11g wireless connectivity, Bluetooth access, and built-in GPS.
5. Its 2.0 MP digital camera has a built-in LED flash and 5x digital zoom, and on-board Roxio software helps manage and edit photos and video.

Apple's iPhone 3GS has similar features and specifications, but it does not employ a keyboard or trackball:

1. It is slightly smaller at 115.5 mm long, 62 mm wide, and 12.5 mm thick.
2. Its 88.9 mm diagonal display has 480 × 320 resolution.
3. Its operations are manipulated through a multitouch screen display that includes a virtual keyboard.
4. It has 802.11a, 802.11b, and 802.11g wireless connectivity, Bluetooth access, and GPS functionality.
5. Its 3.15 megapixel camera has touch focus and autofocus, but no flash or zoom.

Instead of this block structure (in which one phone is discussed and then the other), the writer might have chosen a point-by-point structure in which points common to both items are listed together (such as megapixel ratings in the above combined paragraph). The point-by-point comparison is favoured in feasibility and recommendations reports because it offers readers a meaningful comparison between common points.

Specific to General

Much less common than the general-to-specific patterns just described, *specific-to-general* patterns finish with the paragraph's main point. One such paragraph, the Gabion baskets example on page 179, builds to a climax. The following paragraph leaves its main point until the end because the writer wants the reader to first see the writer's justification for an unpopular bottom line:

The unpopular bottom line needs preceding justification

Gendron Road Services has attempted several unsuccessful methods to repair the deteriorating approach apron at the north end of the LaSalle Bridge. First, Gendron tried four types of standard surface repair methods, but in each case, potholes reappeared within two months. Then, Gendron engineers tried deep surface patch techniques, but the patched pieces broke up during the spring thaw. Gendron has even experimented with a subsurface heating grid, with poor results. **Gendron therefore recommends that the city authorize a complete reconstruction of the bridge approach, at a cost of $1.9 million above Gendron's annual maintenance contract**.

In the next specific-to-general paragraph, several examples lead to a logical bottom-line conclusion:

The bottom line

Integrating a shop area's radiant heating system with an office area's forced air system appeals to fabrication shops and garages. Other enthusiastic users include nursing homes, hotels, and motels. In all these settings, radiant exchange systems can heat one area of a building while they cool another, with increased efficiency and economy. **Thus, these systems are becoming more popular**.

Chronological

Time-based patterns include past-tense narration sequence, instructions, process description, and causal analysis patterns.

NARRATIVE SEQUENCE.   Past-tense narration is easy to read in the following brief description of how a golf putting green was prepared before it was contoured and seeded. Notice that the paragraph uses transitions to keep the progression of events clear. (The transitions are shown in boldface.)

"What happened?"

In preparing the sub-base soil for the new fifth green, the crew **first** removed all soil and rock to a depth of 1.5 metres. **At that level**, they discovered a small seeping spring, so they **then** established the green's subterranean catch basin at the spring's rocky exit point. **Next**, with excavation complete, the crew laid a cross pattern of drainage pipe leading from the catch basin, and they covered the drain piping with 0.5 metres of pea gravel and 0.75 metres of fine sand. The green's surface **was thus ready** to be contoured.

INSTRUCTIONS.   Explanations of how to do something or how something happened generally are arranged according to a strict time sequence: first step, second step, and so on.

"How is it done?"

Instead of breaking into a jog too quickly and risking injury, take a relaxed and deliberate approach. Before taking a step, spend at least 10 minutes stretching and warming up, using any exercises you find comfortable. (Consult a jogging book for specialized exercises.) When you've completed your warm-up, set a brisk walking pace. Exaggerate the distance between steps, taking long strides and swinging your arms briskly and loosely. After roughly 100 metres at this brisk pace, you should feel ready to jog. So break into a very slow trot: lean your torso forward and let one foot fall in front of the other (one foot barely leaving the ground while the other is on the pavement). Keep the slowest pace possible, just above a walk. *Do not bolt out like a sprinter!* While jogging, relax your body. Keep your shoulders straight and your head up, and enjoy the scenery.

PROCESS DESCRIPTION.   A process description, told in present tense, can be used to organize a single paragraph, as in the following explanation of a vapour compression cycle in a geothermal heating system.

"How does this process work?"

All heat pumps use a vapour compression cycle to transport heat from one location to another. In heating mode, the cycle starts as the cold liquid refrigerant within the heat pump passes through a heat exchanger (evaporator) and absorbs heat from the low temperature source (fluid circulated through the earth connection). The refrigerant evaporates into a gas as heat is absorbed. The gaseous refrigerant then passes through a compressor, where it is pressurized, raising its temperature to over 80°C. The hot gas then circulates through a refrigerant-to-air heat exchanger, where the heat is removed and sent through the air ducts. When the refrigerant loses the heat, it changes back to a liquid. The liquid refrigerant cools as it passes through an expansion valve, and the process begins again (Groundloop 2001).

PROBLEM–CAUSES–SOLUTION SEQUENCE.   Another form of chronological development is used to explain how a problem was solved. The problem-solving sequence proceeds from description of the problem, through diagnosis, to solution. After outlining the cause of the problem, this next paragraph explains how the problem has been solved:

"How was the problem solved?"

On all waterfront buildings, the unpainted wood exteriors had been severely damaged by the previous winter's high winds and sandstorms. **After repairing the damage, we took protective steps against further storms**. First, all joints, edges, and sashes were treated with water-repellent preservative to protect against water damage. Next, three coats of non-porous primer were applied to all exterior surfaces to prevent paint from blistering and peeling. Finally, two coats of wood-quality latex paint were applied over the non-porous primer. To prevent coats of paint separating, we applied the first coat within two weeks of the priming coats, and the second within two weeks of the first. Now, 14 months later, no blistering, peeling, or separation has occurred.

CAUSE–EFFECT ANALYSIS.   Cause–effect analyses represent a useful variation of *chronological* paragraph sequencing. The direct version starts with causes and proceeds to effects—the sequence follows an action to its results. Below, the topic sentence identifies the causes, and the remainder of the paragraph discusses its effects.

"What will happen if I do this?"

Some of the most serious accidents involving gas water heaters occur when a flammable liquid is used in the vicinity. The heavier-than-air vapours of a flammable liquid, such as gasoline, can flow along the floor—even the length of a basement—and be explosively ignited by the flame of the water heater's pilot light or burner. Because the victim's clothing frequently ignites, the resulting burn injuries are commonly serious and extremely painful. They may require long hospitalization, and can result in disfigurement or death. *Never, under any circumstances, use a flammable liquid near a gas heater or any other open flame* (Consumer Product Safety Commission).

EFFECT-TO-CAUSE SEQUENCE.   The indirect version starts with the effect and then traces its cause(s). This sequence identifies the main "effect" (the problem) in the topic sentence.

"How did this happen?"

Modern whaling techniques have brought the whale population to the threshold of extinction. In the 19th century, the invention of the steamboat increased hunters' speed and mobility. Shortly afterward, the grenade harpoon was invented so whales could be killed quickly and easily from a ship's deck. In 1904, a whaling station opened on Georgia Island, in South America. This station became the gateway to Antarctic whaling for the nations of the world. In 1924, factory ships were designed that enabled round-the-clock whale tracking and processing. These ships could reduce a 30-metre-long whale to its by-products in roughly 30 minutes. After World War II, more powerful boats with remote sensing devices gave a final boost to the whaling industry. The number of kills had now increased far beyond the whales' capacity to reproduce.

EXERCISES

1. Locate, copy, and bring to class a paragraph that has the following features:

    ◆ an orienting topic sentence
    ◆ adequate development
    ◆ unity
    ◆ coherence
    ◆ a recognizable sequence
    ◆ appropriate length for its purpose and audience

Be prepared to identify and explain each of these features in a class discussion.

2. For each of the following documents, indicate the most logical sequence (e.g., a description of a proposed computer lab would follow a spatial sequence).

    ◆ a set of instructions for operating a power tool
    ◆ a report analyzing the weakest parts in a piece of industrial machinery

◆ a report analyzing the desirability of a proposed oil refinery in your area

◆ a detailed breakdown of your monthly budget to trim excess spending

◆ a report investigating the reasons for student apathy on your campus

◆ a report evaluating the effects of the ban on DDT in insect control

◆ a report on any highly technical subject, written for a general audience

◆ a report investigating the success of a no-grade policy at other universities and colleges

◆ a proposal for a no-grade policy at your university or college

COLLABORATIVE PROJECT

Organize into small groups. Choose *one* of the following topics, or one your group settles on, and then brainstorm to develop a formal outline for the central section of a report. One representative from your group can write the final draft outline and display it (using a data-projection unit, overhead projector, or similar equipment) for class revision.

◆ job opportunities in your career field

◆ a physical description of the ideal classroom

◆ how to organize an effective job search

◆ how the quality of your higher educational experience can be improved

◆ arguments for and against a formal grading system

◆ an argument for an improvement you think your university or college needs most

MyWritingLab

MyWritingLab: Technical Communication offers the best multimedia resources for technical communication in one, easy-to-use place. You can find a variety of interactive model documents and case studies. There are also extensive guidelines, tutorials, and exercises for Document Design, Writing and the Writing Process, and Research, and a large bank of diagnostics and practice for grammar review.

12

Designing Visuals

✳ Explore
Click here in your eText to access a variety of student resources related to this chapter.

LEARNING OBJECTIVES

After reading this chapter, you should be able to

- appreciate the value of visuals in business and technical documents
- select appropriate visuals for specific purposes and audiences
- follow guidelines for designing readable, useful tables
- employ appropriate types of graphs and charts for specific dramatic and illustrative purposes
- integrate text and visuals

Visuals clarify concepts, emphasize particular meanings, illustrate points, or analyze ideas or data. Besides saving space and words, visuals help audiences process, understand, and remember information. Visuals offer powerful new ways of looking at data, so they reveal trends, problems, and possibilities that otherwise might remain buried in lists of facts and figures. In printed or online documents, in oral presentations or multimedia programs, visuals are a staple of communication today. This chapter covers four main types: tables, graphs, charts, and illustrations.

WHY VISUALS ARE ESSENTIAL

Readers expect more than just raw information; they want the information processed for their immediate understanding. Visuals help us answer many of the questions people ask as they process information:

TYPICAL AUDIENCE QUESTIONS IN PROCESSING INFORMATION

- ◆ Which information is most important?
- ◆ Where, exactly, should I focus?
- ◆ What do these numbers mean?
- ◆ What should I be thinking or doing?
- ◆ What should I remember about this?
- ◆ What does it look like?
- ◆ How is it organized?
- ◆ How is it done?
- ◆ How does it work?

More receptive to images than to words, today's readers resist pages of mere printed text. Visuals help diminish a reader's resistance in several ways:

◆ Visuals enhance understanding by displaying abstract concepts in concrete, geometric shapes.

◆ Visuals make meaningful comparisons possible. "Which industrial sectors report the largest on-site releases of environmental pollutants?" (Figure 12.1, page 190).

◆ Visuals depict relationships. "How does seasonal change affect the rate of construction in our city?" (Figure 12.9, page 202)

◆ Visuals serve as a universal language. In the global workplace, carefully designed visuals can transcend cultural and language differences, and thus facilitate international communication.

◆ Visuals provide emphasis. To emphasize the change in death rates for heart disease and cancer since 1980, a table (Table 12.1) or bar graph (Figure 12.3 on page 198) would be more vivid than a prose statement.

◆ Visuals condense and organize information, making it easier to remember and interpret. A simple table, for instance, can summarize a long and difficult printed passage, as in Table 12.1.

Technical data in printed form can be hard to interpret

Assume that you are researching recent death rates for heart disease and cancer. From various sources, you collect these data:

1. In 1980, 392.1 males and 213.5 females per 100 000 people died of heart disease; 240.3 males and 148.3 females died of cancer.

2. In 1990, 267.5 males and 150.6 females per 100 000 people died of heart disease; 246.6 males and 153.2 females died of cancer.

3. In 1997 . . .

In the written form above, numerical information is repetitious, tedious, and hard to interpret. As the amount of numerical data increases, so does our difficulty in processing this material. Arranged in Table 12.1, these statistics become easier to compare and comprehend.

Table 12.1 Data Displayed in a Table

Death Rates for Heart Disease and Cancer, 1980–2009				
Number of Deaths per 100 000 Population				
	Heart Disease		Cancer	
Year	Male	Female	Male	Female
1980	392.1	213.5	240.3	148.3
1990	267.5	150.6	246.6	153.2
1997	230.8	129.7	229.7	148.5
2003	178.9	98.2	215.3	148.1
2005	160.1	91.5	207.7	143.8
2009	134.2	74.9	192.5	136.8
Percentage change, 1980–2009	−65.8	−64.9	−19.9	−7.8

Source: Based on Statistics Canada, CANSIM, Table 102-0552, retrieved 19 February 2013 at **http://www5.statcan.gc.ca/cansim/a26**.

Along with your visual, analyze or interpret the important trends or the essential message you want your readers to see:

A caption explaining the numerical relationships

As Table 12.1 indicates, both male and female death rates from heart disease decreased from 1980 to 2009, but males showed a slightly larger decrease. Cancer deaths during this period decreased slightly for both males and females.

Besides their value as presentation devices, visuals help us analyze information. Table 12.1 is one example of how visuals enhance critical thinking by helping us identify and interpret crucial information and discover meaningful connections.

WHEN TO USE A VISUAL

Use visuals in situations like these

Translate your writing into visuals whenever they make your point more clearly than the prose. Use visuals to *clarify* and to enhance your discussion, not to *decorate* it. Use a visual display to direct the audience's focus or to help them remember something, as in the following situations (Dragga and Gong 46–48):

- ◆ when you want to instruct or persuade
- ◆ when you want to draw attention to something immediately important
- ◆ when you expect the document to be consulted randomly or selectively (e.g., a manual or reference work)
- ◆ when you expect the audience to be relatively less educated, less motivated, or less familiar with the topic
- ◆ when you expect the audience to be distracted

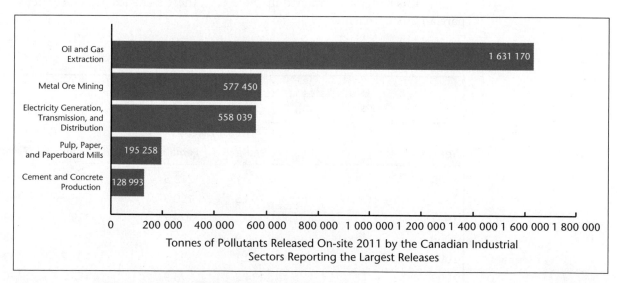

Figure 12.1 A Graph Displaying the "Big Picture"

Source: *Environment Canada's National Pollutant Release Inventory—National Overview 2011.* Available at **http://www.ec.gc.ca/inrp-npri/ default.asp?lang=En&n=0D743E97-1&offset=3&toc=show**.

WHAT TYPES OF VISUALS TO CONSIDER

Different types of visuals serve different functions. The following overview sorts visual displays into four categories: tables, graphs, charts, and graphic illustrations. Each type of visual offers a new way of seeing data—a different perspective.

TABLES DISPLAY ORGANIZED LISTS OF DATA. Tables display data (as numbers or words) in rows and columns for comparison. Use tables to present exact numerical values and to organize data so that readers can sort out relationships for themselves. Complex tables usually are reserved for more specialized readers.

Numerical tables present data for analysis, interpretation, and exact comparison.

TABLE 1 Charting the Lesson			
Lesson	Page	Page	Page
A	1	3	6
B	2	2	5
C	3	1	4

Prose tables organize verbal descriptions, explanations, or instructions.

TROUBLESHOOTING		
Problem	Cause	Solution
• power	• cord	• plug-in
• light	• bulb	• replace
• flicker	• tube	• replace

GRAPHS DISPLAY NUMERICAL RELATIONSHIPS. Graphs translate numbers into shapes, shades, and patterns by plotting two or more data sets on a coordinate system. Use graphs to sort out or emphasize specific numerical relationships. The visual representation helps readers grasp, at a glance, the approximate values, the point being made about those values, or the relationship being emphasized.

Bar graphs often show comparisons.

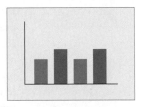

Line graphs often show changes over time.

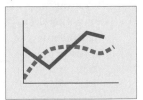

CHARTS DISPLAY THE PARTS OF A WHOLE. Charts depict relationships without the use of a coordinate system by using circles, rectangles, arrows, connecting lines, and other designs.

Pie charts show the parts or percentages of a whole.

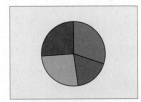

Organization charts show the links among departments, management structures, or other elements of a company.

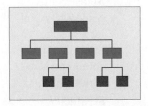

Flow charts trace the steps (or decisions) in a procedure or stages in a process.

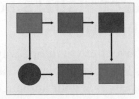

Tree charts show how the parts of an idea or a concept interrelate.

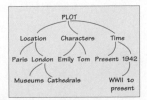

Gantt charts show when each phase of a project is to begin and end.

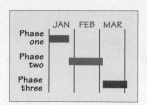

Pictorial charts (pictograms) use icons (or isotypes) to symbolize the items being displayed or measured.

GRAPHIC ILLUSTRATIONS DEPICT ACTUAL OR VIRTUAL VIEWS.   Graphic illustrations are pictorial devices for helping readers visualize what something looks like, how it works, how it's done, how it happens, or where it's located. Certain diagrams present views that could not be captured by photographing or observing the object.

Representational diagrams present a realistic but simplified view, usually with essential parts labelled.

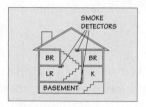

Exploded diagrams show the item pulled apart, to reveal its assembly.

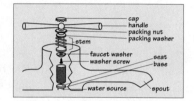

Cutaway diagrams eliminate outer layers to reveal inner parts.

Block or schematic diagrams present the conceptual elements of a principle, process, or system to depict *function* instead of appearance.

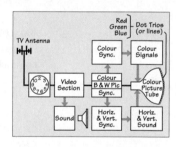

Maps enable readers to visualize a specific location or to comprehend data about a specific geographic region.

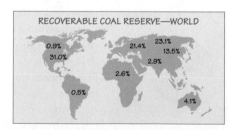

Photographs present an actual picture of the item, process, or procedure.

Source: Marzky Ragsac Jr./Fotolia

HOW TO SELECT VISUALS FOR YOUR PURPOSE AND AUDIENCE

You usually will have more than one way to display information in a visual format. To select the most effective display, consider carefully your specific purpose and the abilities and preferences of your audience.

QUESTIONS ABOUT A VISUAL'S PURPOSE AND INTENDED AUDIENCE

What is my purpose?

◆ What do I want the audience to do or think (know facts and figures, follow directions, make a judgment, understand how something works, perceive a relationship, identify something, see what something looks like, pay attention, other)?

◆ Do I want the audience to focus on one or more exact values, compare two or more values, or synthesize a range of approximate values?

Who is my audience?

◆ What is the audience's technical background on this topic?

◆ What is the audience's level of interest in this topic?

◆ Would the audience prefer the raw data or interpretations of the data?

◆ Is the audience accustomed to interpreting visuals?

Which type of visual might work best in this situation?

◆ What forms of information should this visual depict (numbers, shapes, words, pictures, symbols)?

◆ Which visual display would be most compatible with the type of judgment, action, or understanding I seek from this audience?

◆ Which visual display would this audience find most accessible?

Here are a few examples of the choices you must consider in selecting visuals:

Choices to consider in selecting visuals

◆ To convey facts and figures, a table might be sufficient. But if you want the audience to make a particular judgment about that data, a bar graph, line graph, or pie chart might be preferable.

◆ To show the operating parts of a mechanism, it might be better to use an exploded or cutaway diagram than a photograph.

◆ Expert audiences tend to prefer numerical tables, flow charts, schematics, and complex graphs or diagrams they can interpret for themselves.

◆ General audiences tend to prefer basic tables, graphs, diagrams, and other visuals that direct their focus and that interpret key points extracted from the data.

PREFERRED VISUALS FOR SPECIFIC PURPOSES

Purpose	Preferred Visual
Organize numerical data	Table
Show comparative data	Table, bar graph, line graph
Show a trend	Line graph
Interpret or emphasize data	Bar graph, line graph, pie chart, map
Introduce an unfamiliar object	Photo, representational diagram
Display a project schedule	Gantt chart
Show how parts are assembled	Photo, exploded diagram
Show how something is organized	Organization chart, map
Give instructions	Prose table, photo, diagrams, flow chart
Explain a process	Flow chart, block diagram
Clarify a concept or principle	Block or schematic diagram, tree chart
Describe a mechanism	Photo, representational diagram, or cutaway diagram

Although several alternatives might be possible, one particular type of visual (or a combination) is usually superior for a given purpose and audience. None of the above examples is, of course, immutable. Your particular audience or organization may express its own preferences, or your choices may be limited by lack of equipment (software, scanners, digitizers), insufficient personnel (graphic designers, technical illustrators), or insufficient budget. In any case, your basic task is to enable the intended audience to interpret the visual correctly.

TABLES

Tables can display exact quantities, compare sets of data, and present information systematically and economically. Numerical tables, such as Table 12.1 (see page 189), present *quantitative information* (data that can be measured). In contrast, prose tables present *qualitative information* (brief descriptions, explanations, or instructions). Table 12.2, for example, names Canadian biotechnology companies and lists pertinent basic information about each company.

Table 12.2 A Prose Table

A Cross-section of Canadian Biotechnology Companies		
Company	**Location**	**Overview**
QLT PhotoTherapeutics	Vancouver	Spinoff from UBC; develops drugs that enhance photodynamic therapy. Employs about 400 people.
Alberta Glycomics Centre	Edmonton	A multidisciplinary team of 11 researchers who study the role of carbohydrates and carbohydrate-bearing molecules in biological systems. The study may lead to new drugs, vaccines, and diagnostics.
Bioriginal	Saskatoon	Researches, produces, and distributes high-quality, innovative Essential Fatty Acid (EFA) solutions in Omega-3, Omega-6, and Omega-9 forms. Employed 96 staff as of February 25, 2013.
Generex Biotechnology	Toronto	Researches, develops, and commercializes drug delivery systems and technologies. Generex Oral-lyn™, its flagship product, is an oral insulin that's in worldwide clinical trials. Generex employed 20 in Toronto, as of February 25, 2013.

No table should be too complex for its intended audience. An otherwise impressive-looking table, such as Table 12.3 on page 195, is difficult for non-specialists to interpret because it presents too much information at once. We can see how an unethical writer might use a complex table to bury numbers that are questionable or embarrassing (R. Williams 12). Readers need to understand how the table is organized, where to find what they need, and how to interpret the information they find (Hartley 90).

Tables are constructed with various tools: (1) tab markers and tab keys on a word processor, (2) row-and-column displays in a spreadsheet program, (3) the "Table" command in the better word-processing programs. The third option offers a full range of table editing features: cut and paste, adjust spacing, insert text between rows, add rows or columns, adjust column width, and so on.

Table 12.3 A Complex Table

2011 Air Contaminant Releases By Canadian Industrial Operations								
Total Air Pollutant Emission Inventory								
Province or territory	Sulfur oxides (percent of national emissions)	Nitrogen oxides (percent of national emissions)	Volatile organic compounds (percent of national emissions)	Ammonia (percent of national emissions)	Carbon monoxide (percent of national emissions)	Total particulate matter (percent of national emissions)	Respirable particulate matter (percent of national emissions)	Fine particulate matter (percent of national emissions)
Newfoundland and Labrador	3.0	2.3	1.3	0.3	1.7	3.9	3.2	3.4
Prince Edward Island	0.1	0.3	0.3	0.6	0.6	0.2	0.4	0.5
Nova Scotia	7.2	3.3	2.1	1.0	2.5	3.5	3.9	4.3
New Brunswick	2.2	2.0	2.3	1.0	2.6	3.4	3.3	3.3
Quebec	13	11.9	17.4	17.2	24.9	23.1	25.4	29.1
Ontario	21.4	18.2	23.1	22.2	27.0	22.2	23.2	24.6
Manitoba	10.1	3.0	3.9	10.7	3.9	3.6	3.3	3.4
Saskatchewan	8.5	7.6	11.3	16.3	5.3	8.0	6.3	5.3
Alberta	27.1	38.1	27.1	26.1	16.8	13.2	13.6	12.6
British Columbia	7.3	12.4	10.8	4.6	14.3	17.5	16.7	12.9
Yukon	<0.1	0.1	0.1	<0.1	0.1	0.7	0.2	0.1
Northwest Territories	0.1	0.8	0.2	<0.1	0.2	0.5	0.5	0.5
Nunavut	<0.1	0.1	0.1	<0.1	0.1	<0.1	<0.1	<0.1

Note: Emissions from natural sources (e.g., forest fires) and open sources (e.g., road dust) are not included, except for ammonia, where agricultural sources have been included in the indicator.

*APEI includes emissions reported to the NPRI and Environment Canada estimates.

Source: Environment Canada's *Air Pollution Emissions Data*, accessed at **https://www.ec.gc.ca/indicateurs-indicators/default.asp?lang=en&n=58DE4720-1#prov**

Table Guidelines

Whichever table options you employ, use the following general guidelines:

Guidelines for using tables

◆ Use a table only when you are reasonably sure it will enlighten—rather than frustrate—readers. For non-specialized readers, use fewer tables and keep them simple.

◆ Try to limit the table to one page. Otherwise write "continued" at the bottom, and begin the second page with the full title, "continued," and the original column headings.

◆ If the table is too wide for the page, turn it 90 degrees (landscape) and place its top toward the inside of the binding. Or divide the data into two tables. (Few readers may bother rotating the page to read the table broadside.)

◆ In your discussion, refer to the table by number, and explain what readers should be looking for; or include a prose caption with the table. Specifically, introduce the table, show it, and then interpret it.

For more specific information about creating tables, see the table construction guidelines in Table 12.4.

Table 12.4 Table Construction Guidelines

How to Construct a Table

(1) → **TABLE 1 ■ Science and Engineering Graduates in 2012 and 2013: 2014 Career Status**

COLUMN HEADS

STUB HEAD (2) → DEGREE AND FIELD	Graduates 1999 and 2000 (1000)	2001—PERCENT DISTRIBUTION				Median salary ($1000) (3)
		In school[a]	Employed		Not employed	
			In S&E[b]	In other		
All science fields	(4) 649	(5) 24	11	59	6	34
Computer science/math	61.5	13	32	51	4	51
(6) Life sciences	159.4	37	10	47	5	29
Physical sciences	(7) 32.2	39	27	30	4	34
Social sciences	218.7	(8) → X	5	67	7	30
All engineering fields[c]	109.2	13	64	20	4	49
Civil	16.8	13	67	17	3	42
Electrical/electronics	34.2	11	64	21	4	54
Industrial	6.9	9	59	28	3	49
Mechanical	25.8	12	66	17	4	48

ROW HEADS

NOTE (9) → [a]Full-time grad. students.   [b]Science & engineering.   [c]Other fields not shown. (X) Not available.

SRC (10) → Source: National Science Foundation/SRS, *National Survey of Recent College Graduates: 2001.*
Statistical Abstract of the United States: 2001 (121st edition). Washington: GPO: 611.

1. Number the table in its order of appearance and provide a title that describes exactly what is being compared or measured.

2. Label stub, row, and column heads (*DEGREE AND FIELD*, *Median salary*, *Computer science/math*) to orient readers.

3. Stipulate all units of measurement using familiar symbols and abbreviations ($, hr., no.). Define specialized symbols or abbreviations (Å for angstrom, *dB* for decibel) in a footnote.

4. Compare data vertically (in columns) instead of horizontally (in rows). Columns are easier to compare than rows. Try to include row or column averages or totals, as reference points for comparing individual values.

5. Use horizontal rules to separate headings from data. In a complex table, use vertical rules to separate columns. In a simple table, use as few rules as clarity allows.

6. List the items in a logical order (alphabetical, chronological, decreasing cost). Space listed items so they are not cramped or too far apart for easy comparison. Keep prose entries as brief as clarity allows.

7. Convert fractions to decimals, and align decimals vertically. Keep decimal places for all numbers equal. Round insignificant decimals to the nearest whole number.

8. Use *X*, *NA*, or a dash to signify any omitted entry, and explain the omission in a footnote ("Not available," "Not applicable").

9. Use footnotes to explain entries, abbreviations, or omissions. Label footnotes with lowercase letters so readers do not confuse the notation with the numerical data.

10. Cite data sources beneath any footnotes. When adapting or reproducing a copyrighted table for a work to be published, obtain written permission from the copyright holder.

Tables work well for displaying exact values, but for easier interpretation, readers prefer graphs or charts. Geometric shapes (bars, curves, circles) generally are easier to grasp and remember than lists of numbers (Cochran et al. 25).

Any visual other than a table is usually categorized as a *figure*, and so titled ("Figure 1: Aerial View of the Panhandle Site"). Figures covered in this chapter include graphs, charts, and illustrations.

Like all other components in the document, visuals are designed with audience and purpose in mind (Journet 3). An accountant doing an audit might need a table listing exact amounts, whereas the average public stockholder reading an annual report would prefer an overview in an easily grasped bar graph or pie chart (Van Pelt 1). Similarly, an audience of scientists might find Table 12.3 (page 195) perfectly appropriate, but a less specialized audience (for example, environmental groups) might prefer the clarity and simplicity of Figure 12.1 (page 190).

GRAPHS

Graphs translate numbers into pictures. Plotted as a set of points (a *series*) on a coordinate system, a graph shows the relationship between two variables.

Graphs have a horizontal and a vertical axis. The horizontal axis carries categories (the independent variables) to be compared, such as years within a period (2005). The vertical axis shows the range of values (the dependent variables) for comparing or measuring the categories, such as the number of people who died from heart failure in a specific year. A dependent variable changes according to activity in the independent variable (e.g., a decrease in quantity over a set time, as in Figure 12.2). In the equation $y = f(x)$, x is the independent variable and y is the dependent variable.

Graphs are especially useful for displaying comparisons, changes over time, patterns, or trends. When you decide to use a graph, choose the best type for your purpose: bar graph or line graph.

Bar Graphs

Generally easy to understand, bar graphs show discrete comparisons, as on a year-by-year or month-by-month basis. Each bar represents a specific quantity. Use bar graphs to help readers focus on one value or compare values that change over equal time intervals (expenses calculated at the end of each month, sales figures totalled at yearly intervals). Use a bar graph only to compare values that are noticeably different. Otherwise, all the bars will appear almost identical.

SIMPLE BAR GRAPHS. The simple bar graph in Figure 12.2 displays one relationship taken from the data in Table 12.1, the rate of male deaths from heart disease. To aid interpretation, you can record exact values above each bar—but only if readers need exact numbers.

MULTIPLE-BAR GRAPHS. A bar graph can display two or three relationships simultaneously, each relationship plotted as a separate series. Figure 12.3 displays two comparisons from Table 12.1: the rate of male deaths from both heart disease and cancer.

Whenever a graph shows more than one relationship (or series), each series of numbers is represented by a different pattern, colour, shade, or symbol, and the patterns are identified by a *legend*.

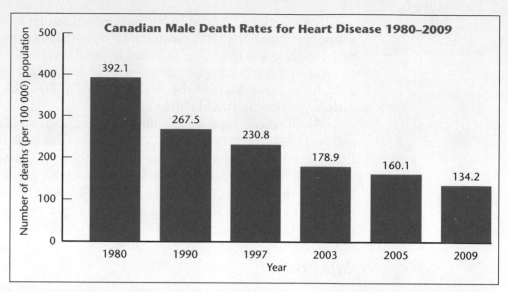

Figure 12.2 A Simple Bar Graph

Source: Based on Statistics Canada, *Cansim,* Table 102-0552, retrieved 19 February, 2013 at **http://www5.statcan.gc.ca/cansim/a26**.

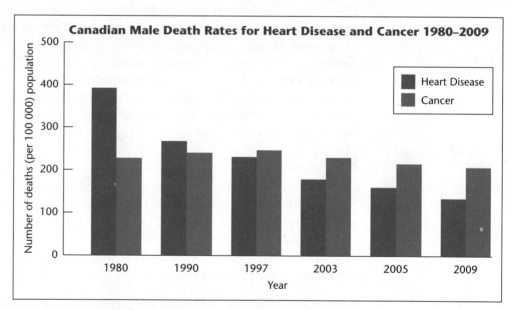

Figure 12.3 A Multiple-bar Graph

Sources: Based on Statistics Canada, *Cansim*, Table 102-0552, retrieved 19 February, 2013 at **http://www5.statcan.gc.ca/cansim/a26**.

The more relationships a graph displays, the harder it is to interpret. As a rule, plot no more than three series of numbers on one graph.

HORIZONTAL BAR GRAPHS.   This type of graph is a good choice for displaying a large series of bars arranged in order of increasing or decreasing value, as in Figure 12.4 on page 199. The horizontal format leaves room for labelling the categories horizontally (*Manufacturing*, etc.). A vertical bar graph leaves no room for horizontal labelling.

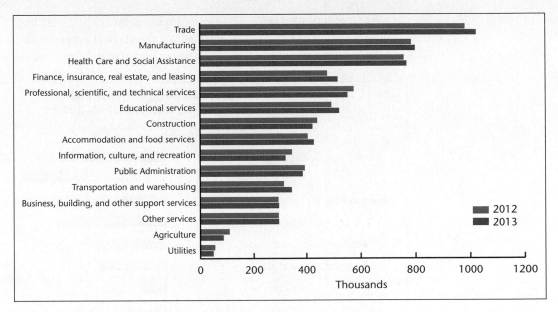

Figure 12.4 A Horizontal Bar Graph

Source: Statistics Canada, CANSIM table 282-0088, reported at **http://www.statcan.gc.ca/tables-tableaux/sum-som/l01/cst01/labr67g-eng.htm**.

STACKED BAR GRAPHS. Instead of side-by-side clusters of bars, you can display multiple relationships by stacking bars. Stacked bar graphs are especially useful for showing how much each item contributes to the whole. Figure 12.5 displays other comparisons from Table 12.1.

Display no more than four or five relationships in a stacked bar graph. Excessive subdivisions and patterns create visual confusion.

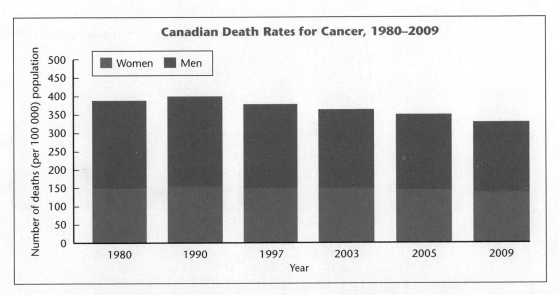

Figure 12.5 A Stacked Bar Graph

Sources: Based on Statistics Canada, *Cansim*, Table 102-0552, available 19 February, 2013 at **http://www5.statcan.gc.ca/cansim/a26**.

DEVIATION BAR GRAPHS.   The deviation bar graph can display both positive and negative values, as in Figure 12.6. Notice how the vertical axis extends to the negative side of the zero baseline, following the same incremental division as above the baseline.

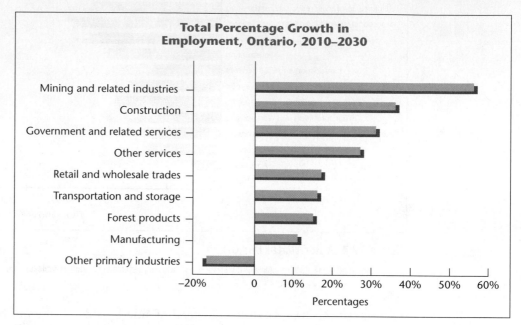

Figure 12.6 A Deviation Bar Graph

Source: The Conference Board of Canada.

3-D BAR GRAPHS.   Graphics software enables you to shade and rotate images and produce three-dimensional views. The 3-D perspectives in Figure 12.7 engage our attention and add visual emphasis to the data.

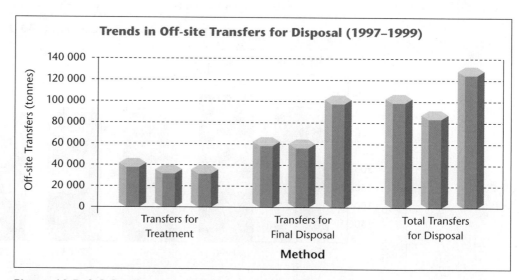

Figure 12.7 3-D Bar Graph

Source: Environment Canada. *1999 National Pollutant Release Inventory.* 19 Apr. 2010. At **http://www.ec.gc.ca/CEPARegistry/documents/subs_list/npri/NO_99/c7.cfm#anchor734**.

Although 3-D graphs can enhance and dramatize a presentation, an overly complex graph can be almost impossible to interpret. Never sacrifice clarity and simplicity for the sake of visual effect.

BAR GRAPH GUIDELINES. Once you decide on a type of bar graph, use the following guidelines for presenting the graph to your audience.

Bar graph guidelines

- Keep the graph simple and easy to read. Avoid plotting more than three types of bars in each cluster. Avoid needless visual details.
- Number your scales in units the audience will find familiar and easy to follow. Units of 1 or multiples of 2, 5, or 10 are best (Lambert 45). Space the numbers equally.
- Label both scales to show what is being measured or compared. If space allows, keep all labels horizontal for easier reading.
- Use *tick marks* to show the points of division on your scale. If the graph has many bars, extend the tick marks into *grid lines* to help readers relate bars to values.

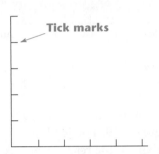
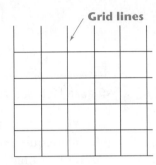

- To avoid confusion, make all bars the same width (unless you are overlapping them). If you must produce your graphs by hand, use graph paper to keep bars and increments evenly spaced.
- In a multiple-bar graph, use a different pattern, colour, or shade for each bar in a cluster. Provide a legend identifying each pattern, colour, or shade.
- If you are trying for emphasis, be aware that darker bars are seen as larger, closer, and more important than lighter bars of the same size (Lambert 93).
- Cite data sources beneath the graph. When adapting or reproducing a copyrighted graph for a work to be published, you must obtain written permission from the copyright holder.
- In your discussion, refer to the graph by number ("Figure 1"), and explain what readers should be looking for; or include a prose caption with the graph.

Many computer graphics programs automatically use most of the design features discussed above, but anyone producing visuals should know all the conventions.

Line Graphs

A line graph can accommodate many more data points than a bar graph (e.g., a 12-month trend, measured monthly). Line graphs help readers synthesize large amounts of information in which exact quantities need not be emphasized. Whereas bar graphs display quantitative differences among items (cities, regions, yearly or monthly intervals), line graphs display data whose value changes over time, as in a trend, forecast, or other change during a specified time (profits, losses, growth). Some line graphs depict cause-and-effect relationships (e.g., how seasonal patterns affect sales or profits).

SIMPLE LINE GRAPHS. A simple line graph, as in Figure 12.8, uses one line to plot time intervals on the horizontal scale and values on the vertical scale. The relationship depicted here would be much harder to express in words alone.

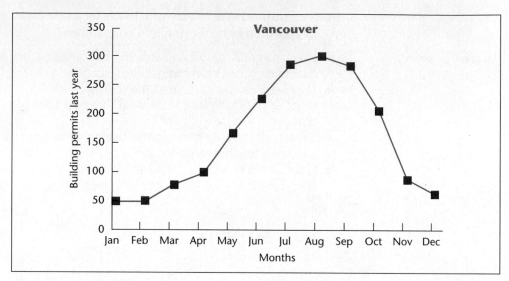

Figure 12.8 A Simple Line Graph

MULTIPLE-LINE GRAPHS. A multiple-line graph displays several relationships simultaneously, as in Figure 12.9. For legibility, use no more than three or four curves in a single graph. Explain the relationships readers are supposed to see.

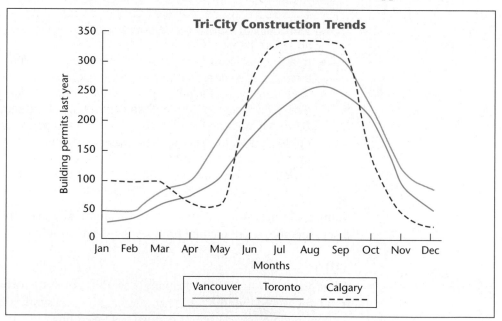

Figure 12.9 A Multiple-Line Graph

A caption explaining the visual relationships

Building permits in all three cities increased steadily as the weather warmed, but Calgary's increase was more erratic. Its permits declined for April–May, but then surpassed Vancouver's and Toronto's for June–September.

BAND OR AREA GRAPHS. For emphasis and appeal, fill the area beneath each plotted line with a pattern. Figure 12.10 is a version of the Figure 12.8 line graph.

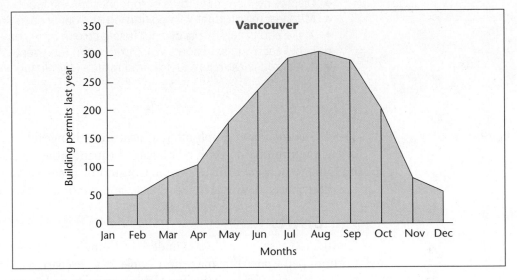

Figure 12.10 A Simple Band Graph

The multiple bands in Figure 12.11 depict relationships among sums instead of the direct comparisons depicted in the equivalent Figure 12.9 line graph.

Despite their visual appeal, multiple-band graphs are easy to misinterpret. In a simple band graph, each line depicts its own distance from the zero baseline. But in a multiple-band graph, the very top line depicts the *total*, each band below it being a part of that total (like stacked bar graph segments). Always clarify these relationships for your audience.

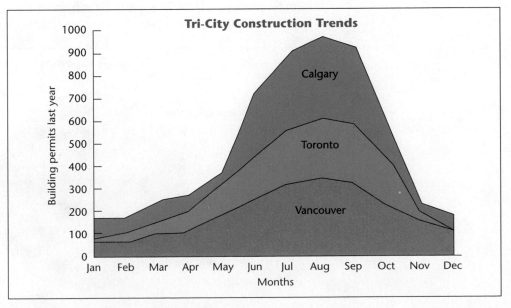

Figure 12.11 A Multiple-Band Graph

LINE GRAPH GUIDELINES.   Follow the bar graph guidelines (see page 201), with these additions:

Line graph guidelines

◆ Display no more than three or four lines on one graph.
◆ Mark each individual data point used in plotting each line.
◆ Make each line visually distinct (using colour, symbols, etc.).
◆ Label each line so readers will know what it represents.
◆ Avoid grid lines that readers could mistake for plotted lines.

CHARTS

The terms *chart* and *graph* often appear interchangeably. But a chart is more precisely a figure that displays relationships (quantitative or cause-and-effect) that are not plotted on a coordinate system. Commonly used charts include pie charts, organization charts, flow charts, tree charts, and pictorial charts (pictograms).

Pie Charts

Considered easy for readers to understand, a pie chart depicts the percentages or proportions of the parts that make up a whole. In a pie chart, readers can compare the parts to each other as well as to the whole (to show how much was spent on what, how much income comes from which sources, and so on). Figure 12.12 shows a simple pie chart.

Figure 12.13 on page 205 shows two other versions of the pie chart in Figure 12.12. Version (a) displays dollar amounts and version (b), the percentage relationships among these dollar amounts.

Pie chart guidelines

PIE CHART GUIDELINES.   Use the following guidelines for constructing a pie chart:

◆ Be sure the parts add up to 100 percent.
◆ If you produce your pie chart by hand, use a compass and protractor for precise segments. Each 3.6-degree segment equals 1 percent. Include any number from two to eight segments. A pie chart with more than eight segments is difficult to interpret, especially if segments are small (Hartley 96).

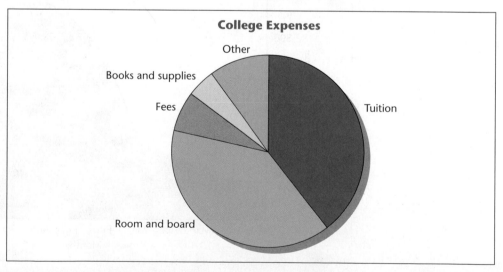

Figure 12.12 A Simple Pie Chart

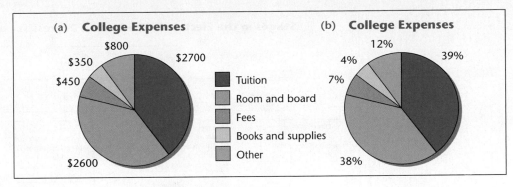

Figure 12.13 Two Other Versions of Figure 12.12

- ◆ Combine small segments under the heading "Other."
- ◆ Locate your first radial line at 12 o'clock and then move clockwise from large to small (except for "Other," usually the final segment).
- ◆ For easy reading, keep all labels horizontal.

Organization Charts

An organization chart divides an organization into its administrative or managerial parts. Each part is ranked according to its authority and responsibility in relation to other parts and to the whole, as in Figure 12.14.

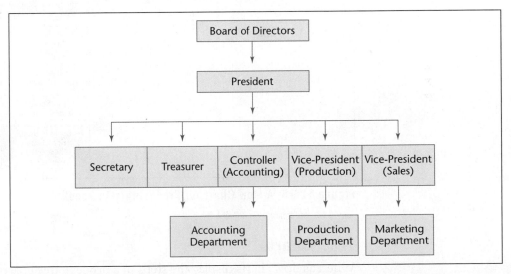

Figure 12.14 An Organization Chart for One Corporation

Flow Charts

A flow chart traces a procedure or process from beginning to end. In displaying the steps in a manufacturing process, the flow chart would begin at the raw materials and proceed to the finished product. Figure 12.15 traces the procedure for producing a textbook.

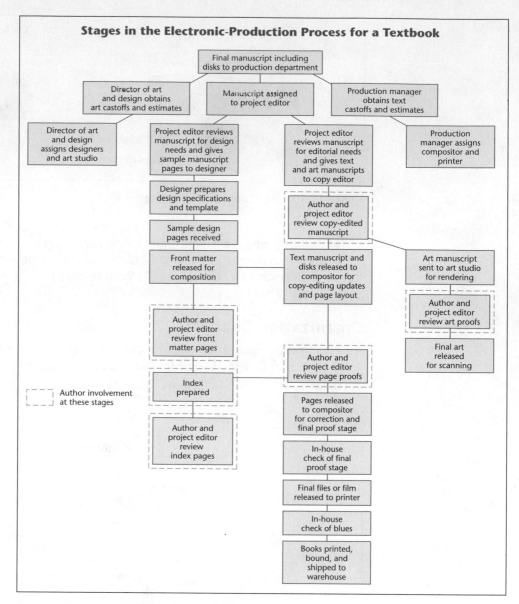

Figure 12.15 A Flow Chart for Producing a Textbook

Source: Adapted from *Harper & Row Author's Guide.*

Tree Charts

Whereas flow charts display the steps in a process, tree charts show how the parts of an idea or concept relate. Figure 12.16 on page 207 displays the parts of an outline for this chapter so that readers can better visualize relationships. The tree version seems clearer and more interesting than the prose listing.

Pictograms

Pictograms depict numerical relationships with icons or symbols (cars, houses, smokestacks) of the items being measured, instead of using bars or lines. Each symbol represents a stipulated quantity, as in Figure 12.17 on page 207. Many graphics programs provide an assortment of pre-drawn symbols.

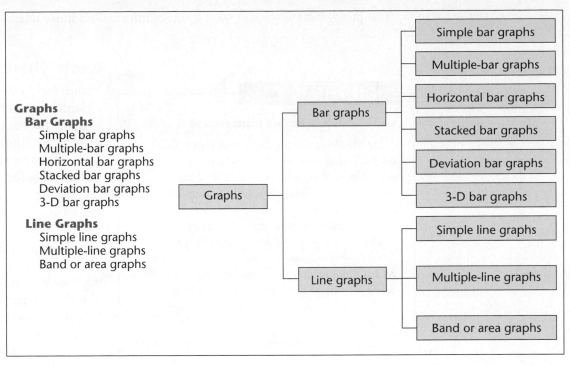

Figure 12.16 An Outline Converted to a Tree Chart

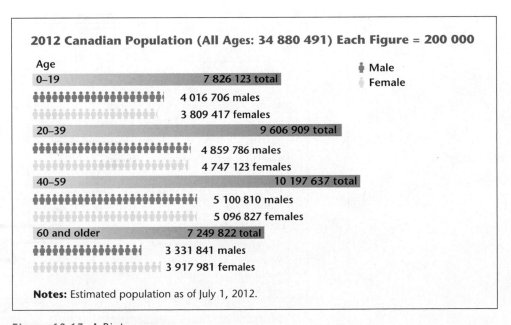

Figure 12.17 A Pictogram

Source: Statistics Canada. *Population by sex and age group.* CANSIM, table 051-0001 Modified: 2012-09-27. Accessed 20 Feb. 2013 at **http://www.statcan.gc.ca/daily-quotidien/120927/t120927b004-eng.htm**.

Use pictograms when you want to make information more interesting for non-technical audiences.

ON THE JOB...

Gantt Charts in Project Management

"I use MS Project to produce Gantt charts for all my projects. A chart helps present the initial proposal (emailed to the client as a PDF); I use a Gantt chart to build and present cost estimates and time estimates; it helps me get quotes from suppliers and subcontractors; and I use the chart to schedule the project. Everybody on the project gets a copy of the complete chart, and, as a result, I get total buy-in from labourers, operators, and the developer. On a big project, I update the chart weekly, and the chart becomes a kind of progress report. . . ."

—**Ken Langedyk, consulting civil engineer**

Gantt Charts

Named for engineer H.L. Gantt, a Gantt chart depicts progress as a function of time. A series of bars or lines (timelines) indicates start-up and completion dates for each phase or task in a project, relative to the other phases or tasks. Gantt charts are especially useful for planning a project (as in a proposal) and tracking it (as in a progress report). The Gantt chart illustrated in Figure 12.18 below shows tasks whose timelines can be simultaneous, overlapping, or consecutive.

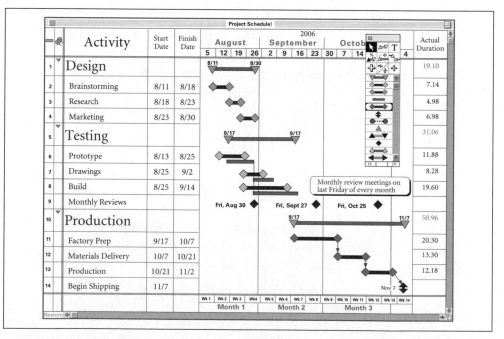

Figure 12.18 A Gantt Chart

Source: Chart created in FastTrack Schedule™. Reprinted by permission from AEC Software. Learn more about FastTrack Schedule at **http://www.aecsoftware.com**.

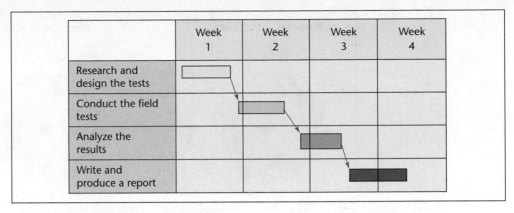

Figure 12.19 A Simple Gantt Chart

Note: Jon Peltier provides advice on using Microsoft Excel to produce Gantt charts at **http://pubs.logicalexpressions.com/pub0009/LPMArticle.asp?ID=343**.

Not all Gantt charts need to be as complex as that illustrated in Figure 12.18. For example, Figure 12.19 above shows a simple Gantt chart for a project that tests potentially contaminated soils.

Other forms of Gantt charts have been developed. One variation—described by Jon Peltier in *TechTrax* online—uses Microsoft *Excel* spreadsheets to produce task and time charts. Still, all Gantt charts have a weakness: they don't account for the problems caused by task areas that fall behind schedule, so PERT charts (Performance Evaluation Review Technique) and CPM charts (Critical Path Method) were developed to show how one delayed task can affect others. Actually, PERT and CPM charts identify two kinds of tasks: *dependent tasks*, which have to be completed in sequence, and *parallel* or *concurrent tasks*, which do not depend on other tasks being completed. A number of software programs, such as Microsoft *Visio* and *PERT Chart Expert*, can routinely create PERT and CPM charts.

GRAPHIC ILLUSTRATIONS

Illustrations consist of diagrams, maps, drawings, and photographs that depict physical relationships rather than numerical ones. Good illustrations help readers understand and remember the material (Hartley 82). Consider this information from a government pamphlet, explaining the operating principle of the seat belt:

> The safety-belt apparatus includes a tiny pendulum attached to a lever, or locking mechanism. Upon sudden deceleration, the pendulum swings forward, activating the locking device to keep passengers from pitching into the dashboard.

Without the illustration in Figure 12.20 on page 210, the mechanism would be difficult to visualize. Clear and uncluttered, a good diagram eliminates unnecessary details and focuses only on material useful to the reader. The following pages illustrate some commonly used diagrams.

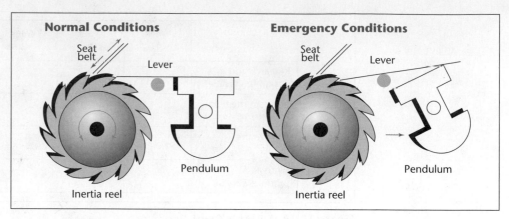

Figure 12.20 A Diagram of a Safety-Belt Locking Mechanism

Source: *Safety Belts*. U.S. Department of Transportation.

Diagrams

Exploded diagrams, like the view of an engineered stand in Figure 12.21 (page 211), show how the parts of an item are assembled. Exploded diagrams often appear in repair, maintenance, or installation manuals. Notice how all parts are clearly separated for easy identification and for reference to the instructions.

Cutaway diagrams show the item with its exterior layers removed in order to reveal interior sections, as in Figure 12.22 on page 212. Unless the specific viewing perspective is immediately recognizable (as in Figure 12.22), name the angle of vision: "top view," "side view," and so on.

Block diagrams are simplified sketches that represent the relationship between the parts of an item, principle, system, or process. Because block diagrams are designed to illustrate concepts, such as current flow in a circuit, the parts are represented as symbols or shapes. The block diagram in Figure 12.23 on page 212 illustrates how any process can be controlled automatically through a feedback mechanism.

Figure 12.24 on page 213 shows the feedback concept applied as the cruise-control mechanism on a motor vehicle.

Increasingly available are electronic drawing programs, clip-art programs, image banks, and other resources for creating visuals or downloading pre-drawn images. Specialized diagrams, however, often require the services of graphic artists or technical illustrators. The client requesting or commissioning the visual provides the art professional with an *art brief* (often prepared by writers and editors) that spells out the visual's purpose and specifications for the visual.

For example, part of the brief addressed to the medical illustrator for Figure 12.22 on page 212 might read as follows:

An art brief for
Figure 12.22

Purpose: to illustrate transsphenoidal adenomectomy for laypersons

◆ View: full cutaway, axial
◆ Range: descending from cranial apex to a horizontal plane immediately below the upper jaw and second cervical vertebra

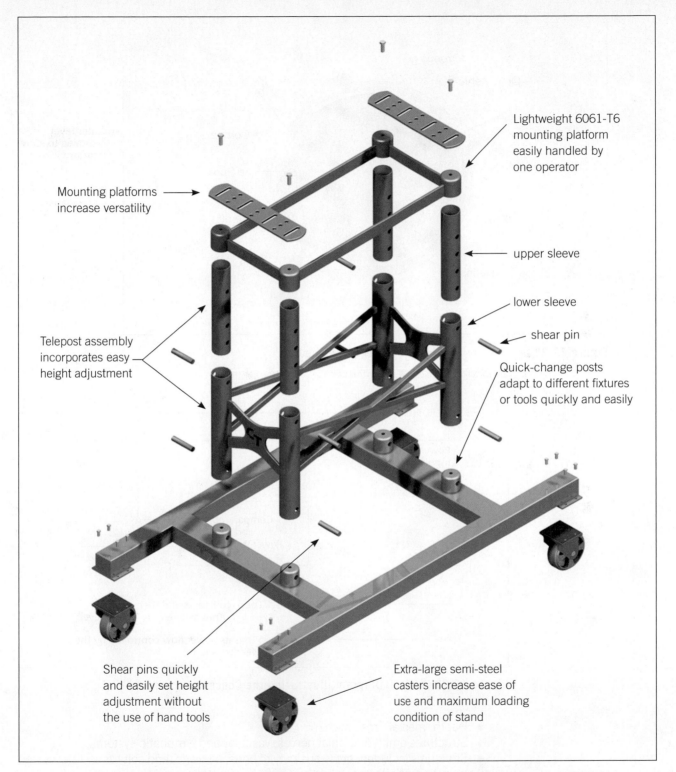

Lightweight 6061-T6 mounting platform easily handled by one operator

Mounting platforms increase versatility

upper sleeve

lower sleeve

shear pin

Telepost assembly incorporates easy height adjustment

Quick-change posts adapt to different fixtures or tools quickly and easily

Shear pins quickly and easily set height adjustment without the use of hand tools

Extra-large semi-steel casters increase ease of use and maximum loading condition of stand

Figure 12.21 Exploded View of the CT04 Engineered Stand's Design Improvements

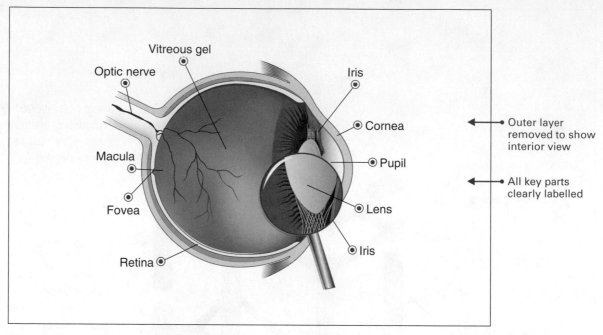

Figure 12.22 A Cutaway Diagram of a Surgical Procedure

Source: *Transsphenoidal Approach for Pituitary Tumor.* © 1986 by the Ludann Co., Grand Rapids, MI.

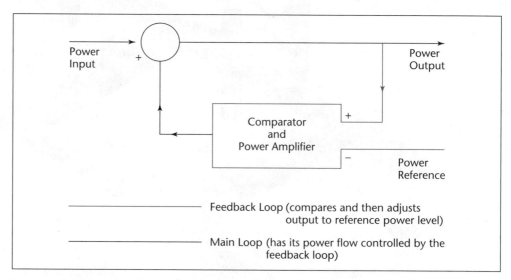

Figure 12.23 A Block Diagram Illustrating the Concept of Feedback

◆ Depth: medial cross-section
◆ Structures omitted: cranial nerves, vascular and lymphatic systems
◆ Structures included: gross anatomy of bone, cartilage, and soft tissue—delineated by colour, shading, and texture
◆ Structures highlighted: nasal septum, sphenoid sinus, and sella turcica, showing the pituitary embedded in a 1.5 cm tumour invading the sphenoid sinus via an area of erosion at the base of the sella

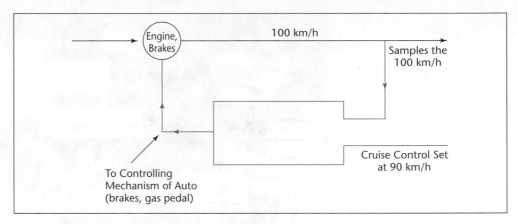

Figure 12.24 A Block Diagram Illustrating a Cruise-Control Mechanism

Photographs

Photographs are especially useful for showing what something looks like (Figure 12.25) or how something is done (Figure 12.26).

No matter how visually engaging, a photograph is difficult to interpret if it includes needless details or fails to identify or emphasize the important material. One graphic design expert offers this practical advice for technical documents:

> To use pictures as tools for communication, pick them for their capacity to carry meaning, not just for their prettiness as photographs . . . [but] for their inherent significance to the [document]. (White, *Great Pages* 110, 122)

Specialized photographs often require the services of a professional who knows how to use angles, lighting, and special film or software to obtain the desired focus and emphasis.

Figure 12.25 A Photograph That Shows the Appearance of Something

Source: Courtesy of Sparkling Hill

Figure 12.26 A Photograph That Shows How Something Is Done

Source: Courtesy of tekmar Control Systems

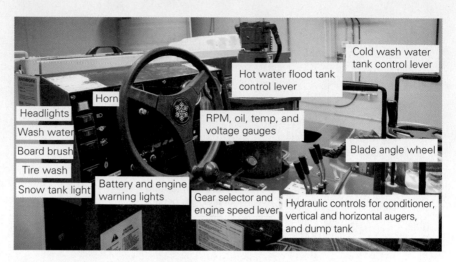

Figure 12.27 Photograph with Essential Features Labelled

Source: Greg Barteluk

PHOTOGRAPH GUIDELINES. Whenever you plan to include photographs in a document or a presentation, use the following guidelines:

Guidelines for using photographs

♦ Try to simulate the approximate angle of vision readers would have in identifying or viewing the item or, for instructions, in doing the procedure (see Figure 12.27).

♦ Trim (or crop) the photograph to eliminate needless details (see Figures 12.28 and 12.29).

♦ To emphasize selected features of a complex mechanism or procedure, consider using diagrams in place of photographs or as a supplement.

♦ For an image unfamiliar to readers, provide a sense of scale by including a person, ruler, or familiar object (such as a hand) in the photo.

♦ If your document will be published, obtain a signed release from any person in the photograph and written permission from the copyright holder. Beneath the photograph, cite the photographer and the copyright holder.

♦ In your discussion, refer to the photograph by figure number and explain what readers should look for or use a caption.

Figure 12.28 A Photograph That Needs to be Cropped

Source: Courtesy of tekmar Control Systems

Figure 12.29 The Cropped Version of Figure 12.28

Source: Courtesy of tekmar Control Systems

Digital-imaging technology allows photographs to be scanned and stored electronically. These stored images then can be retrieved, edited, and altered. Such capacity for altering photographic content creates unlimited potential for distortion and raises questions about the ethics of digital manipulation (Callahan 64–65).

COMPUTER GRAPHICS

Computer technology transfers many of the tasks formerly performed by graphic designers and technical illustrators to individuals with little formal training in graphic design. Whatever your career, you probably will be expected to produce high-quality graphics for conferences, presentations, and in-house publications.

Today's computer systems create sophisticated, multicolour graphic displays and multimedia presentations. Among the virtually countless types of computer-generated visuals are these examples:

◆ With an electronic stylus (a pen with an electronic signal), you can draw pictures on a graphics tablet to be displayed on the monitor, stored, or sent to other computers.
◆ You can create 3-D effects, showing an object from different angles through the use of shading, shadows, on-screen rotation, background lighting, or other techniques.
◆ You can recreate the visual effect of a mathematical model, as in writing equations to explain what happens when high winds strike a tall building. (As the wind deforms the structure, the equations change. Then you can take those new equations and represent them visually.)
◆ You can create a design, build a model, simulate the physical environment, and let the computer forecast what will happen with different variables.
◆ You can integrate computer-assisted design (CAD) with computer-assisted manufacturing (CAM), so that the design will direct the machinery that makes the parts themselves (CAD/CAM).
◆ You can create animations, to see how bodies move (as in a car crash or in athletics).
◆ You can practise dealing with toxic chemicals, operating sophisticated machines, or making other rapid decisions in medical or technical environments, without the cost of or danger in actual situations.
◆ Through various types of scientific visualization, you can do "what if" projections and explore countless ways of conceptualizing and understanding your data. Because the computer can generate and evaluate many possibilities rapidly, it enables you to test hypotheses without doing the calculations.

Computer graphics systems allow you to experiment with scales, formats, colours, perspectives, and patterns. By testing design options on-screen, you can revise and enhance your visual repeatedly until it achieves your exact purpose.

Using Colour or Shading

Colour or shading often makes a presentation more visually interesting. But colour and shading serve purposes beyond visual appeal. Used effectively in a visual, they draw and direct readers' attention to various elements. Colour or shading can help clarify a concept or dramatize how something works, as shown in Figure 12.30: the bright colours against a darker, duller background enable readers to *visualize* concepts, as does the use of dark shading against light.

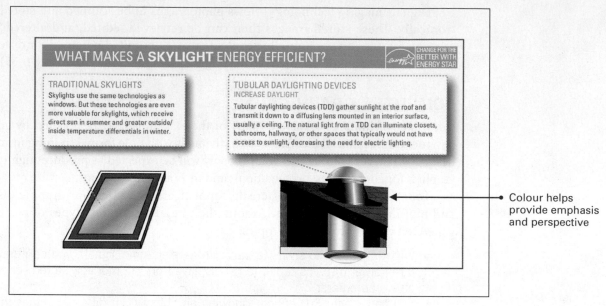

Figure 12.30 Colour Used as a Visualizing Tool

Source: Courtesy of EnergyStar.

Along with shape, type style, and position of elements on a printed page, colour or shading can guide readers through the material. Used effectively on a printed page, colour and shading help organize the reader's understanding, provide orientation, and emphasize important material.

USE COLOUR AND SHADING TO ORGANIZE. Readers look for ways of organizing their understanding of a document (Figure 12.31). Colour or shading can reveal structure and break up material into discrete blocks that are easier to locate, process, and digest:

- ◆ A colour or shaded background screen can set off elements, such as checklists, instructions, or examples.
- ◆ Horizontal rules can separate blocks of text, such as sections of a report or areas of a page.
- ◆ Vertical rules can set off examples, quotations, captions, and so on.

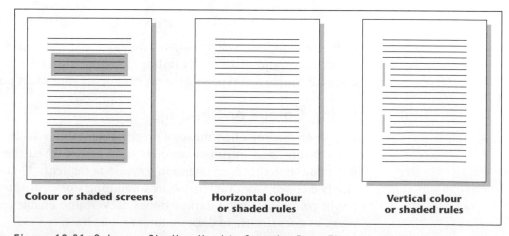

Figure 12.31 Colour or Shading Used to Organize Page Elements

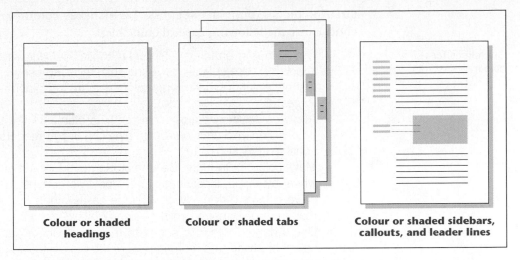

Figure 12.32 Colour or Shading Used as an Orientation Device

USE COLOUR OR SHADING TO ORIENT. Readers look for clear bearings and signposts that help them find their place and find what they need (see Figure 12.32):

- Colour or shading can help headings stand out from the text and differentiate major from minor headings.
- Coloured or shaded tabs and boxes can serve as location markers.
- Coloured or shaded sidebars (for marginal comments), callouts (for labels), and leader lines (for connecting a label to its referent) can guide the eyes.

USE COLOUR OR SHADING TO EMPHASIZE. Readers look for places to focus their attention in a document (see Figure 12.33):

- Colour or shaded typefaces can highlight key words or ideas.
- Colour or shading can call attention to cross-references.
- A coloured or shaded ruled box can frame a warning, caution, note, or hint.

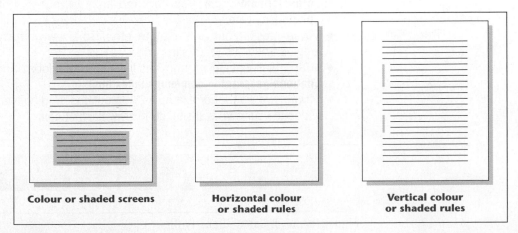

Figure 12.33 Colour or Shading Used for Emphasis

COLOUR OR SHADING GUIDELINES. Whichever colour or shading options you choose, use the following general guidelines:

Guidelines for using colour or shading

- Use colour or shading selectively. It loses impact when it is overused (*Aldus Guide* 39). Use colour or shading sparingly, and use no more than three or four distinct colours when using colour—including black and white (White, *Great Pages* 76).
- Apply colour or shading consistently to elements throughout your document. Inconsistent use of colour or shading can distort readers' perception of the relationships (Wickens 117).
- Make colour redundant. Be sure all elements are first differentiated in black and white: by shape, location, texture, type style, or type size. Different readers perceive colours differently or, in some cases, not at all. A sizable percentage of readers have impaired colour vision (White, *Great Pages* 76).
- Use a darker colour or shade to make a stronger statement. The darker the colour or shade, the more important the material. Darker items can seem larger and closer than lighter objects of identical size.
- Make coloured or shaded type larger than body type. Try to avoid colour or light shading for body type, or use a high-contrast colour or shade (dark against a light background). Colour is less visible on the page than black ink on a white background. The smaller the image or the thinner the rule, the stronger or brighter the colour should be (White, *Editing* 229, 237).
- For contrast in a colour screen, use a very dark type against a very light background, say, a 10- to 20-percent screen (Gribbons 70). The bigger the screen area, the paler the background colour should be (see Figure 12.34).
- Be aware of the colour connotations in different cultures. In North America, for example, red signifies danger and green traditionally signifies safety. But in Ireland, green and orange carry political connotations in certain contexts. In Muslim cultures, green is a holy colour (Cotton 169).

Using Websites for Graphics Support

The internet offers a growing array of visual resources. Following is a sample of useful websites and gateways (Martin).

- *Clip art.* For a comprehensive, updated directory, go to **www.clipart.com**.
- *Photographs.* Vintage photos of people, places, and things can be found at **www.classicphotos.com**. You can find links to many photography sites at **www.photolinks.com**.
- *Art images* (paintings, sculpture, etc.). Go to **www.artresources.com** for a search engine and links to art sites worldwide.
- *Maps.* National Geographic has links to local, global, and political maps at **www.nationalgeographic.com/maps/index.html**.
- *Audio and video.* For examples, instructions, and software for adding audio and video to your own website, go to **www.flashkit.com**.

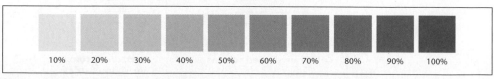

Figure 12.34 A Colour-Density Chart

Be extremely cautious about downloading visuals (or any material, for that matter) from the internet and then using it. Be aware of copyright law. Originators of *any* work on the internet own the work and the copyright.

HOW TO AVOID VISUAL DISTORTION

Although you are perfectly justified in presenting data in its best light, you are ethically responsible for avoiding misrepresentation. Any set of data can support contradictory conclusions. Even though your numbers may be accurate, their visual display could be misleading.

Present the Real Picture

Visual relationships in a graph should accurately portray the numerical relationships they represent. Begin the vertical scale at zero. Never compress the scales to reinforce your point. Notice how visual relationships in Figure 12.35 become distorted when the value scale is compressed or fails to begin at zero.

In version A, the bars accurately depict the numerical relationships measured from the value scale. In version B, item Z (400) is depicted as three times X (200). In version C, the scale is overly compressed, causing the shortened bars to understate the quantitative differences.

Deliberate distortions are unethical because they imply conclusions contradicted by the actual data.

Present the Complete Picture

Without bogging down in needless detail, an accurate visual includes all essential data. Figure 12.36 shows how distortion occurs when data that would provide a complete picture are selectively omitted. Version A accurately depicts the numerical relationships measured from the value scale. In version B, too few points are plotted. Decide carefully what to include and what to leave out of your visual display.

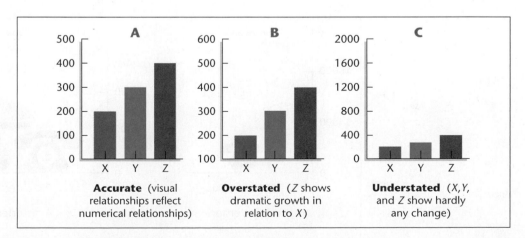

Figure 12.35 An Accurate Bar Graph and Two Distorted Versions

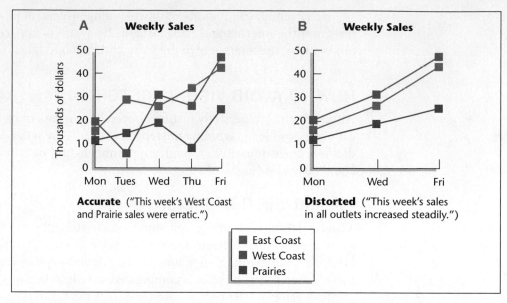

Figure 12.36 An Accurate Line Graph and a Distorted Version

Never Mistake Distortion for Emphasis

When emphasizing a point (a sales increase, a safety record, etc.), be sure your data support the conclusion implied by your visual. Don't use inordinately large visuals to emphasize good news or small ones to downplay bad news (R. Williams 11). When using pictograms or drawings to dramatize a comparison, make the relative size of the images or icons reflect the quantities being compared.

A visual accurately depicting a 100 percent increase in phone sales at your company might look like version A in Figure 12.37 below. Version B overstates the good news by depicting the larger image four times the size, instead of twice the size, of the smaller. Although the larger image is twice the height, it is also twice the *width*, so the total area conveys the visual impression that sales have *quadrupled*.

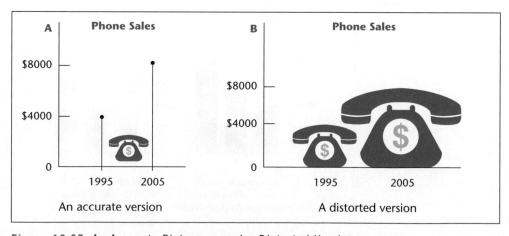

Figure 12.37 An Accurate Pictogram and a Distorted Version

Visuals have their own rhetoric and persuasive force, which we can use to advantage—for positive or negative purposes, for the reader's benefit or detriment (Van Pelt 2). Avoiding visual distortion is ultimately a matter of ethics.

HOW TO INCORPORATE VISUALS WITH THE TEXT

An effective visual enables readers to locate and extract the information they need. To simplify the reader's task, visual and verbal elements in a document should complement each other. For example, a visual should be able to stand alone in meaning, even when isolated from related text.

VISUAL AND VERBAL ELEMENT GUIDELINES.   To incorporate visual and verbal elements effectively, use the following guidelines:

Guidelines for fitting
visuals with text

- ◆ Place the visual where it will best serve your readers. If it is central to your discussion, place the visual as close as possible to the material it clarifies. (Achieving proximity often requires that you ignore the traditional "top or bottom" design rule for placement of visuals on a page.) If the visual is peripheral to your discussion or of interest to only a few readers, place it in an appendix so that interested readers can refer to it as they wish. Tell readers when to consult the visual and where to find it.
- ◆ Never refer to a visual that readers cannot easily locate. In a long document, don't be afraid to repeat a visual when you discuss it again later.
- ◆ Never crowd a visual into a cramped space. Set your visual off by framing it with plenty of white space, and position it on the page for balance. To save space and to achieve proportion with the surrounding text, consider carefully the size of each visual and the amount of space it will occupy.
- ◆ Number the visual and give it a clear title and clear labels. Your title should tell readers what they are seeing. Label all of the important material.
- ◆ Match the visual to your audience. Don't make it too elementary for specialists or too complex for non-specialists. Be sure your intended audience will be able to interpret the visual correctly.
- ◆ Introduce and interpret the visual.

Informative   Table 2 shows that operating costs have increased 7 percent annually since 1999.

Uninformative   See Table 2.

Visuals alone make ambiguous statements (Girill, *Technical Communication and Art*); pictures need to be interpreted. Instead of leaving readers to struggle with a page of raw data, explain the relationships displayed. Follow the visual with a discussion of its important features:

Informative   This cost increase means that . . .

- ◆ Use captions to explain important points made by the visual. Captions help readers interpret a visual. When possible, use a smaller type size so that captions don't compete with text type (*Aldus Guide* 35).
- ◆ Never include excessive information in one visual. Any visual that contains too many lines, bars, numbers, colours, or patterns will overwhelm readers, causing them to ignore the display. In place of one complicated visual, use two or more straightforward ones.

♦ Be sure the visual's meaning can stand alone. Even though it repeats or augments information already in the text, the visual should contain everything readers will need to interpret it correctly.

The following checklist will help ensure that your visuals enhance your meaning.

CHECKLIST

for Evaluating Visuals

Use this checklist to evaluate your visuals.

Content

♦ Does the visual serve a legitimate purpose (clarification, not mere ornamentation) in the document?
♦ Is the visual titled and numbered?
♦ Is the level of complexity appropriate for the audience?
♦ Are all patterns in the visual identified by label or legend?
♦ Are all values or units of measurement specified?
♦ Are the numbers accurate and exact?
♦ Do the visual relationships represent the numerical relationships accurately?
♦ Are explanatory notes added as needed?
♦ Are all data sources cited?
♦ Has written permission been obtained for reproducing or adapting a visual from a copyrighted source in any type of work to be published?
♦ Is the visual introduced, discussed, interpreted, integrated with the text, and referred to by number?
♦ Can the visual itself stand alone in meaning?

Placement

♦ Is the visual easy to locate?
♦ Are all design elements (title, line thickness, legends, notes, borders, white space) positioned for balance?
♦ Is the visual positioned on the page to achieve balance?
♦ Is the visual set off by adequate white space or borders?
♦ Is the top of a wide visual toward the inside binding?
♦ Is the visual in the best report location?

Style

♦ Is this the best type of visual for your purpose and audience?
♦ Are all decimal points in each column vertically aligned?
♦ Is the visual uncrowded and uncluttered?
♦ Is the visual engaging (patterns, colours, shapes), without being too busy?
♦ Is the visual in good taste?
♦ Is the visual ethically acceptable?

EXERCISES

1. The following statistics are based on data from three Canadian colleges. They give the number of applicants to each college over six years.

    ♦ In 2006, X College received 2341 applications for admission, Y College received 3116, and Z College 1807.
    ♦ In 2007, X College received 2410 applications for admission, Y College received 3224, and Z College 1784.
    ♦ In 2008, X College received 2689 applications for admission, Y College received 2976, and Z College 1929.
    ♦ In 2009, X College received 2714 applications for admission, Y College received 2840, and Z College 1992.
    ♦ In 2010, X College received 2872 applications for admission, Y College received 2615, and Z College 2112.
    ♦ In 2011, X College received 2868 applications for admission, Y College received 2421, and Z College 2267.

    Illustrate these data in a line graph, a bar graph, and a table. Which version seems most effective for a reader who (1) wants exact figures, (2) wonders how overall enrollments are changing, or (3) wants to compare enrollments at each college in a certain year? Include a caption interpreting each of these versions.

2. Devise a flow chart for a process in your field or area of interest. Include a title and a brief discussion.

3. Devise a pie chart to depict your yearly expenses. Title and discuss the chart.

4. Obtain enrollment figures at your university or college for the past five years by sex, age, race, or any other pertinent category. Construct a stacked bar graph to illustrate one of these relationships over the five years.

5. Keep track of your pulse and respiration at 30-minute intervals over a four-hour period of changing activities. Record your findings in a line graph, noting the times and specific activities below your horizontal coordinate. Write a prose interpretation of your graph and give it a title.

6. In textbooks or professional journal articles, locate each of these visuals: a table, a multiple-bar graph, a multiple-line graph, a diagram, and a photograph. Evaluate each according to the evaluation checklist, and discuss the most effective visual in class.

7. Anywhere on campus or at work, locate at least one visual that needs revision for accuracy, clarity, appearance, or appropriateness. Look in computer manuals; lab manuals; newsletters; financial aid or admissions or placement brochures; student, faculty, or employee handbooks; newspapers; or textbooks. Use the Checklist for Evaluating Visuals (see page 222) as a guide to revise and enhance the visual. Submit to your instructor a copy of the original, along with a memo explaining your improvements. Be prepared to discuss your revision in class.

COLLABORATIVE PROJECT

Compile a list of 12 websites that offer graphics support by way of advice, image banks, design ideas, artwork catalogues, and the like. Provide the address for each site, along with a description of the resources offered and the cost. Report your findings in a format stipulated by your instructor.

MyWritingLab

MyWritingLab: Technical Communication offers the best multimedia resources for technical communication in one, easy-to-use place. You can find a variety of interactive model documents and case studies. There are also extensive guidelines, tutorials, and exercises for Document Design, Writing and the Writing Process, and Research, and a large bank of diagnostics and practice for grammar review.

13

Designing Pages and Documents

✳ Explore

Click here in your eText to access a variety of student resources related to this chapter.

LEARNING OBJECTIVES

After reading this chapter, you should be able to

- apply the basics of effective page design: page balance, open space and textual spacing, graphic highlighting, font choice, headings, headers and footers, and suitable, integrated visuals
- master the basics of logical, internally consistent headings systems
- adapt design guidelines for varying media, document types, and audiences

Page design—the layout of words and graphics—determines the look of a page. Well-designed pages attract readers, invite them in, guide them through the material, and help them understand and remember it. An audience's *first* impression of a document tends to involve a purely visual, aesthetic judgment.

PAGE DESIGN IN WORKPLACE WRITING

Page design becomes especially significant when we consider these realities about writing and reading in the workplace:

1. *Technical information must be designed differently from material in novels, news stories, and other forms of writing.* Accessible technical documents require more than just unbroken sequences of paragraphs. Navigating through complex material, readers may need the help of charts, diagrams, lists, various type sizes and typefaces, different headings, and other page-design elements.
2. *Technical documents rarely get readers' undivided attention.* Readers may skim documents while they jot down ideas, talk on the phone, or drink coffee. Or they may refer to sections of the document during a meeting. Following distractions, readers must be able to return easily to documents.
3. *People read work-related documents only because they have to.* Novels, general newspapers, and magazines are read for relaxation, but work-related documents (proposals, reports, newsletters) often require the reader's *labour*. The more complex the document, the harder readers have to work.
4. *These days, every document is forced to compete for readers' attention.* Suffering from information overload, readers resist documents that appear overwhelming. They want formats that help them find the information they really need. A user-friendly document has an accessible format: at a glance, readers can see the document's organization, where they are in the document, which items matter most, and how items relate.

5. *Word-processing software contains tools that allow easy production of sophisti-cated page design.* As a result, people today are used to seeing high-quality design, and they resist reading poorly designed, dense documents.

Notice how the information in Figure 13.1 resists our attention. Without design cues, we have no way of grouping this information into organized units of meaning. Now look at Figure 13.2 on page 226, which shows the same information after a design overhaul. Notice also that the material in Figure 13.2 appears in a slightly different order than in Figure 13.1.

Types of Geothermal Heat Pump Systems

The energy crisis has intensified research into geothermal heating. As a result, several types of geothermal heat pump (GHP) systems have been developed. This report considers two main types of GHP systems, open loop and closed loop. Each of the types can be installed with vertical or horizontal piping.

In an open loop system, groundwater is drawn from an aquifer through a well and is passed through the pump's heat exchanger. Then, the water is pumped back to the aquifer through a second well, at some distance from the first. An open loop system can therefore present problems—local groundwater chemicals can foul the heat pump's exchanger and distributed water.

Therefore, closed loop systems have become more common; when properly installed, closed loops are economical, efficient, and reliable (Rafferty 10). Such systems circulate a water/anti-freeze solution through a continuous buried pipe. The length of piping depends on ground temperature, the ground's thermal conductivity, soil moisture, and system design.

Variations of vertical and horizontal pipe are used. Horizontal closed loop installations are cost-effective for small installations, if sufficient land is available. Pipes are buried in trenches; up to six pipes are placed in each trench, with 3 to 4.5 metres between trenches. By contrast, vertical loops are used for large installations, or for situations where the soil is too shallow for trenching, or for installations where not enough land is available for horizontal trenches. In closed loop systems, a U-tube is installed in a drill hole 30 to 120 metres deep. Most installations require several drill holes; the pipes are joined in parallel or series-parallel patterns.

Two variations of the horizontal type are gaining favour. Pond closed loop systems place loops on the bottom of a pond or stream; others have also appeared in lakes or off shore in the ocean. A special variation is called "slinky" loops—overlapping coils of polyethylene pipe increase the heat exchange per foot of trench. Slinky coil systems can be placed in earth trenches or in water.

Figure 13.1 Ineffective Page Design

TYPES OF GEOTHERMAL HEAT PUMP SYSTEMS

The energy crisis has intensified research into geothermal heating. As a result, several types of geothermal heat pump (GHP) systems have been developed. This report considers two main types of GHP systems, open loop and closed loop. Each of the types can be installed with vertical or horizontal piping.

Open Loop Systems

In an open loop system, groundwater is drawn from an aquifer through a well and is passed through the pump's heat exchanger. Then, the water is pumped back to the aquifer through a second well, at some distance from the first. An open loop system can therefore present problems—local groundwater chemicals can foul the heat pump's exchanger and distributed water.

Closed Loop Systems

The problems with open loops mean that closed loop systems have become more common; when properly installed, closed loops are economical, efficient, and reliable (Rafferty 10). Such systems circulate a water/anti-freeze solution through a continuous buried pipe. The length of piping depends on

- ground temperature,
- the ground's thermal conductivity,
- soil moisture, and
- system design.

Horizontal Closed Loop

Horizontal closed loop installations are cost-effective for small installations, if sufficient land is available. Pipes are buried in trenches; up to six pipes are placed in each trench, with 3 to 4.5 metres between trenches.

Two variations of the horizontal type are gaining favour.

1. Pond closed loop systems place loops on the bottom of a pond or stream; others have also appeared in lakes or off shore in the ocean.

2. "Slinky" loops (overlapping coils of polyethylene pipe) increase the heat exchange per foot of trench. Slinky coil systems can be placed in earth trenches or in water.

Vertical Closed Loop

Unlike the horizontal systems, vertical loops are used for large installations, for situations where the soil is too shallow for trenching, or for installations where not enough land is available for horizontal trenches. In closed loop systems, a U-tube is installed in a drill hole 30 to 120 metres deep. Most installations require several drill holes; the pipes are joined in parallel or series-parallel patterns.

Figure 13.2 Effective Page Design

PAGE-DESIGN GUIDELINES

Approach your design decisions from the top down. First, consider the overall look of your pages; next, the shape of each paragraph; and finally, the size and style of individual letters and words (Kirsh 112). Figure 13.3 on page 227 depicts how design

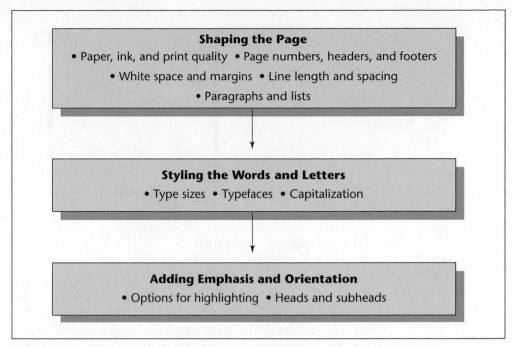

Figure 13.3 A Flow Chart for Decisions in Page Design

considerations follow a top-down sequence, moving from large matters to small. (All design considerations are influenced by the size of the budget for a publication. For instance, adding a single colour to major heads can double the printing cost.)

If your organization prescribes no specific guidelines, the following design principles should satisfy most audiences' expectations.

Shaping the Page

In shaping a page, we consider its look, feel, and overall layout. The following suggestions will help you shape appealing and usable pages.

USE THE RIGHT PAPER AND INK. For routine documents (memos, letters, in-house reports), key or print in black ink, on 21.5 cm by 28 cm (8½" by 11") low-gloss, white paper. Use rag-bond paper (20-pound or heavier) with a high fibre content (25 percent minimum). Shiny paper produces glare and tires the eyes. Flimsy or waxy paper feels inferior.

For documents that will be published (manuals, marketing literature, etc.), the grade and quality of paper are important considerations. Paper varies in weight, grain, and finish—from low-cost newsprint, with noticeable wood fibre, to wood-free, specially coated paper with custom finishes. Choice of paper finally depends on the artwork to be included, the type of printing, and the intended aesthetic effect.

USE HIGH-QUALITY TYPE OR PRINT. Print hard copy on an inkjet or laser printer. If your inkjet's output is blurry, consider purchasing special inkjet paper for the best output quality.

USE CONSISTENT PAGE NUMBERS, HEADERS, AND FOOTERS. For a long document, count your title page as page i, without numbering it, and number all front-matter

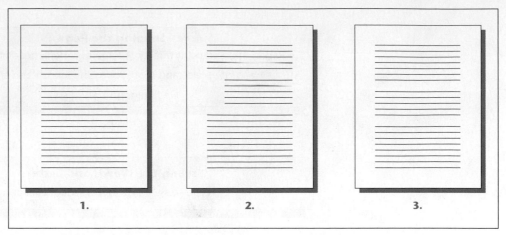

Figure 13.4 Using White Space

pages, including the table of contents and abstract, with lowercase Roman numerals (ii, iii, iv). Number the first text page and subsequent pages with Arabic numerals (1, 2, 3). Along with page numbers, *headers* or *footers* appear in the top or bottom page margins, respectively. These provide chapter or article titles, authors' names, dates, or other publication information. (See, for example, the headers at the tops of the pages in this text.)

USE ADEQUATE WHITE SPACE. White space (all of the space not filled by text and visuals) divides printed areas into small, digestible chunks. It separates sections in a document, headings and visuals from text, and paragraphs from each other (see Figure 13.4). White space enhances a document's appearance, clarity, and emphasis.

Well-designed white space imparts a shape to the whole document, a shape that orients readers and lends a distinctive visual form to the printed matter by

- ◆ keeping related elements together
- ◆ isolating and emphasizing important elements
- ◆ providing breathing room between blocks of information

Pages that look uncluttered, inviting, and easy to follow convey an immediate sense of user-friendliness.

Use white space to orient readers

PROVIDE AMPLE AND APPROPRIATE MARGINS. Small margins make a page look crowded and difficult. On your 21.5 cm by 28 cm (8½" by 11") page, leave margins of at least 2.5 cm (1"). For a document that will be bound, widen the left margin an extra 1.25 cm (½").

Choose between *unjustified* text (uneven or "ragged" right margins) and *justified* text (even right margins). Each arrangement creates its own "feel."

Justified lines

In right-justified text, the spaces vary between words and letters on a line, sometimes creating channels or rivers of white space. The reader's eyes are then forced to adjust continually to these space variations within a line or paragraph. Because each line ends at an identical vertical space, the eyes must work harder to differentiate one line from another (Felker et al. 85). Moreover, to preserve the even margin, words at ends of lines are often hyphenated, and frequently hyphenated line endings can be distracting.

Unjustified lines

Unjustified text, on the other hand, has equal spacing between letters and words on a line, and an uneven right margin. For some readers, a ragged right margin makes reading easier. These differing line lengths can prompt the eye to move from one line to the next line (Pinelli et al. 77). In contrast to justified text, an unjustified page seems to look less formal, less distant, and less official.

Justified text seems preferable for books, annual reports, newsletters, and other publications that use two columns. Unjustified text seems preferable for more personal forms of communication, such as letters, memos, and in-house reports.

KEEP LINE LENGTH REASONABLE.   Long lines tire the eyes. The longer the line, the harder it is for readers' eyes to return to the left margin and locate the next line (White, *Visual Design* 25).

Notice in this three-line example how your eye labours to follow the apparently endless message that seems to stretch in lines that continue long after your eye was prepared to move down to the next line. After reading more than a few of these lines, you begin to feel tired and bored and annoyed, without hope of ever reaching the end.

Short lines force the eyes to move back and forth (Felker et al. 79). "Too-short lines disrupt the normal horizontal rhythm of reading" (White, *Visual Design* 25).

Lines that are too
short cause your eye
to stumble from one
fragment to another
at a pace that too
soon becomes
annoying, if not
nauseating.

A reasonable line length is 70 to 80 characters (or 12 to 15 words) per line for a 21.5 cm by 28 cm (8½" by 11") single-column page. The number of characters will depend on print size. Longer lines call for larger type and wider spacing between them (White, *Great Pages* 70).

Line length, of course, is affected by the number of columns (vertical blocks of print) on your page. Two-column pages often appear in newsletters and brochures, but research indicates that single-column pages work best for complex, specialized information (Hartley 148).

KEEP LINE SPACING CONSISTENT.   For any document likely to be read completely (letters, memos, instructions), single-space within paragraphs and double-space between them. Instead of indenting the first line of single-spaced paragraphs, separate them with a line of space.

TAILOR EACH PARAGRAPH TO ITS PURPOSE.   Readers often skim a long document to find what they want. Most paragraphs therefore begin with a topic sentence forecasting the content.

Shape each paragraph

Use a long paragraph (no more than 15 lines) for clustering material that is closely related (such as history and background, or any body of information best understood in one block).

Use short paragraphs for making complex material more digestible, for giving step-by-step instructions, or for emphasizing vital information.

Avoid *widow* and *orphan* lines. The last line of a paragraph printed at the top of a page is called a *widow*. The first line of a paragraph printed on the bottom of a page is called an *orphan*.

MAKE LISTS FOR EASY READING.   Readers prefer information in list form rather than in continuous prose paragraphs (Hartley 51).

Types of items you might list include advice or examples, conclusions and recommendations, criteria for evaluation, errors to avoid, materials and equipment for a procedure, parts of a mechanism, or steps or events in a sequence. Notice how the preceding information becomes easier to grasp and remember when displayed in the list below.

Types of items you might list:

- advice or examples
- conclusions and recommendations
- criteria for evaluation
- errors to avoid
- materials and equipment for a procedure
- parts of a mechanism
- steps or events in a sequence

A list of brief items usually needs no punctuation at the end of each line. One that is full of sentences or questions requires appropriate punctuation after each item.

Depending on the list's contents, set off each item with some kind of visual or verbal signal. If the items require a strict sequence (as in a series of steps, or parts of a mechanism), use Arabic numbers (1, 2, 3) or the words *first*, *second*, *third*, and so on. If the items require no strict sequence (as in the bulleted list above), use dashes, asterisks, or bullets. For a checklist, use open boxes. Figure 13.5 illustrates different types of lists.

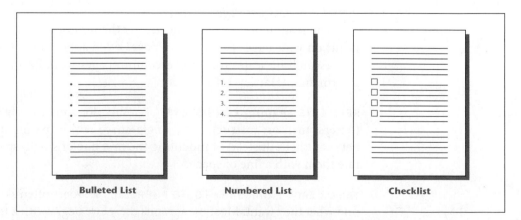

Bulleted List   **Numbered List**   **Checklist**

Figure 13.5 Types of Lists

Introduce your list with an explanation. Phrase all listed items in parallel grammatical form. If the items suggest no strict sequence, try to impose some logical ranking (most important to least important, alphabetical). Set off the list with extra white space above and below.

Styling the Words and Letters

In styling words and letters, consider typographic choices that will make the text easy to read.

USE STANDARD TYPE SIZES.   Word-processing programs offer a wide variety of type sizes:

Select the appropriate type size

9 point
10 point
12 point
14 point
18 point
24 point

The standard type size for most documents is 10- to 12-point. Use larger or smaller sizes for headings, titles, captions (brief explanation of a visual), sidebars (marginal comments), or special emphasis. Use a consistent type size for similar elements throughout the document.

SELECT APPROPRIATE FONTS.   A *font* (also known as a *typeface*) is the style of individual letters and characters. Each font has its own personality, and thus affects the ways readers respond. Also, particular fonts can influence reading speed by as much as 30 percent (Chauncey 26).

Word-processing programs offer a variety of fonts, such as the ones displayed here in 11-point type:

Select a font for its personality

Times New Roman

Brush Script

Mistral

Impact

Arial

Garamond

Palatino

Verdana

Bodoni MT Black

Script MT Bold

Can you assign a personality to each of the above fonts? Thousands more fonts are available on the internet—simply key in "fonts" or "free fonts" in a search engine. (But expect to run the gauntlet of advertising!)

For greater visual unity, try to use different sizes and versions (**bold**, *italic*, **expanded**, condensed) of the same typeface throughout your document—with the possible exception of headings, captions, sidebars, or visuals.

All typefaces divide into two broad categories: *serif* and *sans serif*. Serifs are the fine lines that extend horizontally from the main strokes of a letter.

Decide between serif and sans serif type

Serif type makes body copy more readable, because the horizontal lines "bind the individual letters" and thereby guide the reader's eyes from letter to letter—as in the type you are now reading (White, *Visual Design* 14).

In contrast, sans serif type is purely vertical (as in this Arial font). Clean-looking and "businesslike," sans serif is considered ideal for technical material (numbers, equations, etc.), marginal comments, headings, examples, tables, captions to pictures and visuals, and any other material set off from the body copy (White, *Visual Design* 16).

Be aware of cultural variations

European readers generally prefer sans serif type throughout their documents, and other cultures have their own preferences as well. Learn all you can about the design conventions of the culture you are addressing.

Except for special emphasis, stick to the more conservative styles and avoid ornate ones altogether.

AVOID SENTENCES IN FULL CAPS.   Sentences or long passages in full capitals (uppercase letters) are difficult to read, because all uppercase letters and words have the same visual outline (Felker et al. 87). The longer the passage, the harder readers work to grasp your emphasis.

MY DOG HAS MANY FLEAS.

My dog has many fleas.

Full caps are good for emphasis but hard to read

Hard   ACCORDING TO THE NATIONAL COUNCIL ON RADIATION PROTECTION, YOUR MAXIMUM ALLOWABLE DOSE OF LOW-LEVEL RADIATION IS 500 MILLIREMS PER YEAR.

Easier   According to the National Council on Radiation Protection, your MAXIMUM allowable dose of low-level radiation is 500 millirems per year.

Lowercase letters take up less space, and the distinctive shapes make each word easier to recognize and remember (Benson 37).

Use full caps as section headings (INTRODUCTION) or to highlight a word or phrase (WARNING: NEVER TEASE THE ALLIGATOR). As with other highlighting options discussed below, use full caps sparingly in your document.

Highlighting for Emphasis

Effective highlighting helps readers distinguish the important from the less important elements. Highlighting options include typefaces, type sizes, white space, and other graphic devices that

- ◆ emphasize key points
- ◆ make headings prominent
- ◆ separate sections of a long document
- ◆ set off examples, warnings, and notes

You can highlight with <u>underlining</u>, FULL CAPS, dashes, parentheses, and asterisks.

You can indent to set off examples, explanations, or any material that should be distinguished from other elements in your document.

Using ruled (or keyed) horizontal lines, you can separate sections in a long document:

Using ruled lines, broken lines, or ruled boxes, you can set off crucial information, such as a warning or a caution:

- -

Caution: A document with too many highlights can appear confusing, disorienting, and unprofessional.

- -

In adding emphasis and orientation, consider design elements that will direct readers' focus and help them navigate the text.

Word-processing software offers highlighting options that might include **bold-face**, *italics*, SMALL CAPS, varying type sizes and typefaces, and colour. For specific highlighted items, some options are better than others:

Not all highlighting is equal

Boldface works well for emphasizing a single sentence or brief statement, and is perceived by readers as being "authoritative" (*Aldus Guide* 42).

Italics suggest a more subtle or "refined" emphasis than boldface (Aldus Guide *42*). *They can highlight words, phrases, book titles, and so on. But multiple lines (like these) of italic type are difficult to read.*

SMALL CAPS WORK FOR HEADINGS AND SHORT PHRASES. BUT ANY LONG STATEMENT ALL IN CAPS IS DIFFICULT TO READ.

Small type sizes (usually sans serif) work well for captions, labels for visuals, or to set off other material from the body copy.

Large type sizes and dramatic typefaces are both difficult to miss and difficult to digest. Be conservative—unless you really need to convey forcefulness.

Colour is appropriate only in some documents, and only when used sparingly.

Whichever highlights you select for a document, be consistent. Make sure that all headings at one level are highlighted identically, that all warnings and cautions are set off identically, and so on. Use the "styles" feature of your word processor to ensure this consistency.

Never combine too many highlights.

Using Headings for Access and Orientation

Readers of a long document often look back or jump ahead to sections that interest them most.

Headings announce how a document is organized, point readers to what they need, and divide the document into accessible blocks or "chunks." An informative heading can help a reader decide whether a section is worth reading (Felker et al. 17). Besides cutting down on reading and retrieval time, headings help readers remember information (Hartley 15).

MAKE HEADINGS INFORMATIVE. A heading should be informative but not wordy. Informative headings orient readers, showing them what to expect. Vague or general headings can be more misleading than no headings at all (Redish et al. 144). Whether your heading takes the form of a phrase, a statement, or a question, be sure it advances thought.

What should we expect in the following example—specific instructions, an illustration, a discussion of formatting policy in general? We can't tell.

| **Uninformative heading** | Document Formatting |

The following improved versions each tell the reader what to expect.

| **Informative versions** | How to Format Your Document
Format Your Document in This Way
How Do I Format My Document? |

When you use a question as a heading, phrase it in the same way as readers might ask it.

MAKE HEADINGS SPECIFIC AS WELL AS COMPREHENSIVE.   Focus the heading on a specific topic. Do not preface a discussion of the effects of acid rain on lake trout with a broad heading, such as "Acid Rain." Use instead "Effects of Acid Rain on Lake Trout."

Also, provide enough headings to contain each discussion section. If chemical, bacterial, and nuclear wastes are three *separate* discussion items, provide a heading for each. Do not simply lump them under the sweeping heading "Hazardous Wastes." If you have prepared an outline for your document, adapt major and minor headings from it.

MAKE HEADINGS GRAMMATICALLY CONSISTENT.   All topics at a given level in a document share equal rank; to emphasize this equality, express topics at the same level in identical—or parallel—grammatical form.

| **Non-parallel headings** | **How to Care For Your Car:**
1. Check Tires Weekly
2. Change the Oil Regularly
3. Windows and Windshield Wipers
4. It Is Crucial That All Fluid Levels Are Maintained
5. Belts and Batteries
6. Keep the Brakes In Good Shape |

In items 3, 4, and 5 above, the lack of parallelism (no verbs in the imperative mood like the other three) obscures the relationship between individual steps and is confusing. The following edited headings emphasize the equal rank of all six listed items.

| **Parallel headings** | 3. Clean Windows and Inspect Windshield Wipers
4. Maintain All Fluid Levels
5. Inspect and Replace Belts and Batteries |

Parallelism helps make a document readable and accessible.

MAKE HEADINGS VISUALLY CONSISTENT.   "Wherever heads are of equal importance, they should be given similar visual expression, because the regularity itself becomes an understandable symbol" (White, *Visual Design* 104). Use identical type size and typeface for all headings at a given rank. (Use the "styles" feature of your word-processing software.)

LAY OUT HEADINGS BY RANK. Like a road map, your headings should announce clearly the large and small segments in your document. (Use the logical divisions from your outline as a model for heading layout.) Think of each heading at a particular rank as an "event in a sequence" (White, *Visual Design* 95).

Figure 13.6 shows how headings vary in positioning and highlighting, depending on their rank. However, because of space considerations, Figure 13.6 does not show that each higher-level heading yields at least two lower-level headings.

Many variations of the format that appears in Figure 13.6 are used successfully. One such variation is shown in Figure 13.7.

SECTION HEADING

In formal reports, always centre section headings at the top of a new page. Use a type size roughly 4 points larger than body copy (e.g., 16-point section heads for 12-point body copy). Avoid overly large heads, and use no other highlights. Fully capitalize the heading. (Some documents use colour for section headings and capitalize just the first letter of each word.) Leave a full line space above the various headings (as in this example).

Major Topic Heading
Place major topic headings at the left margin (flush left), and begin each word with an uppercase letter. Use a type size roughly 2 points larger than body copy, in boldface. Start the copy immediately below the heading (as shown), or leave one space below the heading.

Minor Topic Heading
Use boldface and the same type size as the body copy, with no other highlights. Start the copy immediately below the heading (as shown), or leave one space below the heading.

Subtopic Heading. Incorporate subtopic headings into the body copy they head. Place subtopic heads flush left and set them off with a period. Use boldface and the same type size as in the body copy, with no other highlights.

1. ***Alternative subtopic heading.*** If numbering is appropriate, place the subtopics in a list, with the numbers flush left and the body copy indented. Use italics and boldface if you want to draw particular attention to this fourth level of heading.

- **Bulleted variation.** When the sequence of items in a list is not important, use bullets to precede the indented subtopic headings.

Figure 13.6 Recommended Format for Headings

The layouts in Figures 13.6 and 13.7 embody the following guidelines:

◆ Ordinarily, use no more than four levels of heading (section, major topic, minor topic, subtopic). Excessive heads and subheads can make a document seem cluttered or fragmented.

◆ Make sure that the headings system has a consistent, logical progression. Each succeeding heading level in Figure 13.6 and Figure 13.7 has smaller type size than the preceding level.

◆ Never begin the sentence right after the heading with *this*, *it*, or some other pronoun referring to the heading. Make the sentence's meaning independent of the heading.

◆ Use boldface for all headings.

◆ Never leave a heading floating on the final line of a page. Unless at least two lines of text can fit below the heading, carry it over to the top of the next page.

1.0 SECTION HEADING

Use a type size roughly 4 points larger than body copy (e.g., 16-point section heads for 12-point body copy). Use colour to draw attention to the main heading.

1.1 Major Topic Heading

Place major topic headings at the left margin (flush left), and begin each word with an uppercase letter. Use a type size roughly 2 points larger than body copy, in boldface. Start the text on the line below the heading.

1.1.1 Minor Topic Heading

Indent minor topic headings. Use boldface and the same type size as the body copy, with no other highlights. Start the body copy immediately below the heading (as shown).

1.1.1.1 Subtopic heading. Incorporate subtopic headings into the body copy they head. Indent subtopic heads two tabs and set them off with a period. Use boldface and the same type size as in the body copy, with no other highlights.

Figure 13.7 Alternative Format for Headings

AUDIENCE CONSIDERATIONS IN PAGE DESIGN

Like any writing decisions, page-design choices are by no means random. An effective writer designs a document for specific use by a specific audience.

In deciding on a format, work from a detailed audience/purpose profile (Wight 11). Know who your readers are and how they will use your information. Design a document to meet their particular needs and expectations, as in these examples:

Topical headings compared with talking headings

- ◆ If readers will use your document for reference only (as in a repair manual), make sure you have plenty of headings.
- ◆ If your relationship with your readers is formal, use *topical headings* ("Advantages of Treated Pipe"); however, if that relationship is more direct and informal, consider using *talking headings* ("We Should Buy Treated Pipe").
- ◆ If readers will follow a sequence of steps, show that sequence in a numbered list.
- ◆ If readers will need to evaluate something, give them a checklist of criteria (as in this text at the end of most chapters).
- ◆ If readers need a warning, highlight the warning so that it cannot possibly be overlooked.
- ◆ If readers have asked for your one-page résumé, save space by using a 10-point type size.
- ◆ If readers will be facing complex information or difficult steps, widen the margins, increase all white space, and shorten the paragraphs.

Consider cultural variations

Consider also your audience's specific cultural expectations. For instance, Arabic and Persian text is written from right to left instead of left to right (Leki 149). In other cultures, readers move up and down the page, instead of across. Be aware that a particular culture might be offended by certain icons or by a typeface that seems too plain or too fancy (Weymouth 144). Ignoring a culture's design conventions can be interpreted as a sign of disrespect.

CHECKLIST

for Page Design

Use this checklist to review the usability of your page design.

- ◆ Is the paper white, low-gloss, rag bond, with black ink?
- ◆ Is all type neat and legible?
- ◆ Does the white space adequately orient readers?
- ◆ Are the margins ample?
- ◆ Is the line length reasonable?
- ◆ Is the right margin unjustified?
- ◆ Is the line spacing appropriate and consistent?
- ◆ Does each paragraph begin with a topic sentence? If not, why not?
- ◆ Does the length of each paragraph suit its subject and purpose?
- ◆ Are all pages free of widow and orphan lines?
- ◆ Do parallel items in strict sequence appear in a numbered list?
- ◆ Do parallel items of any kind appear in a list whenever a list is appropriate?

- ◆ Are pages numbered consistently?
- ◆ Is the body type size 10 to 12 points?
- ◆ Do full caps highlight only words or short phrases?
- ◆ Is the highlighting consistent, tasteful, and subdued?
- ◆ Are all format patterns distinct enough so that readers will find what they need?
- ◆ Are there enough headings for readers to know where they are in the document?
- ◆ Are headings informative, comprehensive, specific, parallel, and visually consistent?
- ◆ Are headings clearly differentiated according to rank?
- ◆ Is the overall design inviting without being overwhelming?
- ◆ Does this design respect the cultural conventions of the audience?

EXERCISES

1. Find an example of effective page design, in a textbook or elsewhere. Photocopy a selection (two or three pages), and attach a memo explaining to your instructor and colleagues why this design is effective. Be specific in your evaluation. Now do the same for an example of poor formatting. Bring your examples and explanations to class, and be prepared to discuss why you chose them.

2. Rewrite the following headings to make them parallel. (The headings are part of a set of instructions for effective listening.)

   ◆ You Must Focus on the Message
   ◆ Paying Attention to Non-verbal Communication
   ◆ Your Biases Should Be Suppressed
   ◆ Listen Critically
   ◆ Listening for Main Ideas
   ◆ Distractions Should Be Avoided
   ◆ Provide Verbal and Non-verbal Feedback
   ◆ Making Use of Silent Periods
   ◆ Are You Allowing the Speaker Time to Make His or Her Point?

COLLABORATIVE PROJECT

Working in small groups, redesign a document you select or your instructor provides. Prepare a detailed explanation of your group's revision. Appoint a group member to present your revision to the class, using an overhead projector, a large-screen computer monitor, data projection unit, or photocopies.

MyWritingLab

MyWritingLab: Technical Communication offers the best multimedia resources for technical communication in one, easy-to-use place. You can find a variety of interactive model documents and case studies. There are also extensive guidelines, tutorials, and exercises for Document Design, Writing and the Writing Process, and Research, and a large bank of diagnostics and practice for grammar review.

14

Definitions

LEARNING OBJECTIVES

After reading this chapter, you should be able to

- identify situations in which definitions are useful
- use (and recognize when to use) parenthetical definitions, sentence definitions, and expanded definitions
- distinguish between categorical and operational definitions
- apply techniques for expanding and developing technical definitions

Explore
Click here in your eText to access a variety of student resources related to this chapter.

When you define a term, you explain the precise meaning you intend when using that term. Clear writing depends on definitions that both reader and writer understand. Unless you are sure your audience knows the exact meaning you intend, always define terms on first use.

PURPOSE OF DEFINITIONS

Every specialty has its own technical language. Engineers, architects, or programmers talk about *torque*, *tolerances*, or *microprocessors*; lawyers, real estate brokers, and investment counsellors discuss *easements*, *liens*, *amortization*, or *escrow accounts*. Whenever such terms are unfamiliar to an audience, they need defining.

Most of the specialized terms previously mentioned are concrete and specific. Once *microprocessor* has been defined for the reader, its meaning will not differ appreciably in another context. Writers know that highly technical terms should be defined for some readers. However, familiar terms like *disability*, *guarantee*, *tenant*, *lease*, or *mortgage* acquire very specialized meanings in specialized contexts. Here, definition becomes crucial. What *guarantee* means in one situation is not necessarily what it means in another. Contracts are detailed (and legal) definitions of the specific terms of an agreement.

Assume that you're shopping for disability insurance to protect your income in case of injury or illness. Besides comparing prices, you want each company to define *physical disability*. Although company A offers the cheapest policy, it might define *physical disability* as inability to work at *any* job. Therefore, if a neurological disorder prevents you from continuing work as a designer of electronic devices, without disabling you for some menial job, you might not qualify as "disabled." In contrast, company B's policy, although more expensive, might define *physical disability* as inability to work at your *specific* job. Both companies use the term *physical disability*, but each defines it differently. Because they are legally responsible for the documents they prepare, all communicators rely on the technique of clear definition.

ELEMENTS OF DEFINITIONS

For all definitions, use an appropriate level of English, list each defined term's basic properties, and remain objective.

Plain English

Clarify meaning by using language readers understand.

Unclear	A tumour is a neoplasm.
Better	A tumour is a cell growth that occurs independently of surrounding tissue and serves no useful function.
Unclear	A solenoid is an inductance coil that serves as a tractive electromagnet. [A definition appropriate for an engineering manual, but too specialized for general readers.]
Better	A solenoid is an electrically energized coil that converts electrical energy to magnetic energy capable of performing mechanical functions.

Basic Properties

Convey the properties of an item that differentiate it from all others. A thermometer has a singular function: it measures temperature. Without this essential information, a definition would have no real meaning for uninformed readers. Any other data about thermometers (types, special uses, materials used in construction) are secondary. A book, on the other hand, cannot be defined according to functional properties because books have multiple functions. A book can be used to write in or to display pictures, to record financial transactions, to read, and so on. Also, other items (individual sheets of paper, posters, newspapers, picture frames) serve the same functions. The basic property of a book is physical: it is a bound volume of pages. Readers would have to know this *first*, to understand what a book is.

Objectivity

Unless readers understand that your purpose is to persuade, omit your opinions from a definition. *Bomb* is defined as "an explosive weapon detonated by impact, proximity to an object, a timing mechanism, or other predetermined means." If you define a bomb as "an explosive weapon devised and perfected by hawkish idiots to blow up the world," you are editorializing, *and* ignoring a bomb's basic property.

TYPES OF DEFINITIONS

Definitions vary greatly in length and detail—from a few words in parentheses, to one or more complete sentences, to multiple paragraphs or pages.

Your choice of definition type depends on what information readers need, and that, in turn, depends on why they need it. *Fuel injector*, for instance, could be defined in one sentence, briefly telling readers what it is and how it works. But this definition would be expanded for the student mechanic who needs to know the origin of the term, how the device was developed, what it looks like, how it is used, and how its parts interact. Audience needs should guide your choice.

Parenthetical Definition

A parenthetical definition explains the term in a word or phrase, often as a synonym in parentheses following the term:

Parenthetical definitions

The effervescent (bubbling) mixture is highly toxic.

The leaching field (sieve-like drainage area) requires crushed stone.

Clarifying Definition

Another option is to express your definition as a clarifying phrase:

Clarifying definition

The trees on the site are mostly deciduous; that is, they shed their foliage at season's end.

Use parenthetical and clarifying definitions to convey the general meaning of specialized terms so that readers can follow the discussion where these terms are used.

Sentence Definition

Elements of sentence definitions

A definition may require one or more sentences with this structure: (1) the item or term being defined, (2) the class (specific group) to which the term belongs, and (3) the features that differentiate the term from all others in its class.

Term	Class	Distinguishing Features
fuel injector	an electronically controlled valve	in gasoline and diesel engines; squirts pressurized fuel through a tiny nozzle to produce a fine atomized mist that burns easily
transit	a surveying instrument	measures horizontal and vertical angles
diabetes	a metabolic disease	caused by a disorder of the pituitary gland or pancreas and characterized by excessive urination, persistent thirst, and decreased ability to metabolize sugar
stress	an applied force	strains or deforms an object
laser	an electronic device	converts electrical energy to light energy, producing a bright, intensely hot, and narrow beam of light
fibre optics	a technology	uses light energy to transmit voices, video images, and data through hair-thin glass fibres

These elements can be combined into one or more sentences.

A complete sentence
definition

Diabetes is a metabolic disease caused by a disorder of the pituitary gland or pancreas. This disease is characterized by excessive urination, persistent thirst, and decreased ability to metabolize sugar.

A sentence definition is especially useful for stipulating the precise working meaning of a term that has several possible meanings. State your working definitions at the beginning of your report:

A working definition

Throughout this report, the term *disadvantaged student* means . . .

ON THE JOB... ▶

Clarity in the Construction Industry

"When we build a custom-designed home, we draw up a contract that essentially defines the project and our respective responsibilities. Then we prepare a complete flow chart that lists the details of work to be completed and relevant timelines. That chart also provides for notes about each stage of the project. We discuss this description with the client at the beginning and at weekly on-site meetings. The client has to make thousands of decisions during the building process, and we're careful to record them all. Everything goes into a web-based project management tool called BuilderTrend. At any time, the client can review the details of the project, including invoices, client-requested changes, photos, and every other aspect of the documented process. At the lock-up stage, we can usually submit a "Cost to Complete" report. All of this communication serves to define and quantify what's happening. Also, our emails to the client frequently ask if there's anything else we need to explain or quantify. Clarity is what I'm after, every step of the way!"

—**Ken Dahlen, custom home builder**

CLASSIFYING THE TERM.   Be specific and precise in your classification. The narrower your classification, the more specific your meaning. *Transit* is correctly classified as a "surveying instrument," not as a "thing" or an "instrument." *Stress* is classified as "an applied force"; to say it "takes place when . . ." or "is something that . . ." fails to reflect a specific classification. Be sure to select precise terms of classification: *diabetes* is precisely classified as "a metabolic disease," not as "a medical term."

DIFFERENTIATING THE TERM.   Differentiate the term by separating the item it names from every other item in its class. Make these distinguishing features narrow enough to pinpoint the item's unique identity and meaning, yet broad enough to be inclusive. A definition of *brief* as "a legal document introduced in a courtroom" is too broad because the definition doesn't differentiate *brief* from all other legal documents (wills, written confessions, etc.). Conversely, differentiating *fuel injector* as "an electronically controlled valve used in automobile engines" is too narrow because it ignores the fuel injector's use in a variety of other gasoline and diesel engines.

Also, avoid circular definitions (repeating, as part of the distinguishing features, the word you are defining). Thus, *stress* should not be defined as "an applied force that places stress on an object." The class and distinguishing features must express the item's basic property ("an applied force that strains or deforms an object").

CATEGORICAL VERSUS OPERATIONAL DEFINITIONS.   So far, our sample definitions have placed the defined term in a category; we could use the term *categorical definition* for this common method of defining things in sentences. Categorical definitions are static, but a second type of sentence definition, *operational definition*, defines things in active terms. Here's an example:

Operational definition

Technologists translate engineering designs into working plans and then see that these plans are carried out.

In the example, technologists are defined in terms of *what they do*, rather than in terms of *what they are*.

Operational definitions work best in proposals, progress reports, and résumés because the active verbs contribute to the sense of an active and successful person, plan, or activity. Also, operational definitions use fewer words to convey meaning. Compare the above example with its categorical equivalent, which uses 25 percent more words):

Categorical definition

> A technologist is **someone who** translates engineering designs into working plans and then sees that these plans are carried out.

Expanded Definition

An expanded definition can include parenthetical and sentence definitions, but it provides greater detail for readers who need it. The sentence definition of *solenoid* on page 248 is sufficient for a general reader who simply needs to know what a solenoid is. But a manual for mechanics or mechanical engineers would define *solenoid* in detail (as on pages 248–249); these readers need to know also how a solenoid works and how to use it.

> ## ON THE JOB...
>
> ### The Case for Operational Definitions
>
> *"Good technical writing often relies on operative definitions, which help writers and readers understand the real essence of what a product does and how it works. That clear communication adds value to customer transactions."*
>
> **—Don Gibbs, CPO, tekmar Control Systems, which designs and manufactures controls for heating and cooling, solar thermal, and snow and ice melting systems**

The problem with defining an abstract and general word, such as *condominium* or *loan*, is different. *Condominium* is a vaguer term than *solenoid* (solenoid A is pretty much like solenoid B) because the former refers to many types of ownership agreements.

An expanded definition may be a single paragraph (as for a simple tool) or may extend to scores of pages (as for a digital dosimeter—a device for measuring radiation exposure); sometimes, the definition itself *is* the whole report.

EXPANSION METHODS

How you expand a definition depends on the questions you think the audience needs answered, as shown in Figure 14.1. Begin with a sentence definition, and then use only those expansion strategies that serve the audience's needs.

Etymology

A word's origin (its development and changing meanings) can clarify its definition. *Biological control* of insects is derived from the Greek *bio*, meaning "life" or "living organism," and the Latin *contra*, meaning "against" or "opposite." Biological control, then, is the use of living organisms against insects. College dictionaries contain etymological information, but your best bet is *The Oxford English Dictionary* and encyclopedic dictionaries of science, technology, and business.

Some technical terms are acronyms or initialisms, derived from the first letters or parts of several words. *Laser* is an acronym for "light amplification by stimulated emission of radiation."

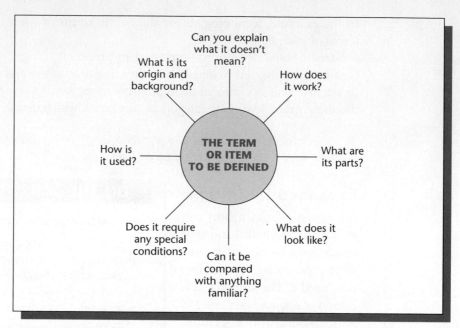

Figure 14.1 Directions in Which a Definition Can Be Expanded

Sometimes, a term's origin can be colourful as well as informative. *Bug* (jargon for "programming error") is said to derive from an early computer at Harvard that malfunctioned because of a dead bug blocking the contacts of an electrical relay. Because programmers, like many of us, were reluctant to acknowledge error, the term became a euphemism for *error*. Correspondingly, *debugging* is the correcting of errors in a program.

History and Background

The meaning of specialized terms, such as *radar, bacteriophage, silicon chips,* or *X-ray* can often be clarified through a background discussion: discovery or history of the concept, development, method of production, applications, and so on. Specialized encyclopedias are good background sources.

"Where did it come from?"

The idea of lasers . . . dates back as far as 212 B.C., when Archimedes used a [magnifying] glass to set fire to Roman ships during the siege of Syracuse. (Gartiganis 22)

"How was it perfected?"

The early researchers in fibre optic communications were hampered by two principal difficulties—the lack of a sufficiently intense source of light and the absence of a medium which could transmit this light free from interference and with a minimum signal loss. Lasers emit a narrow beam of intense light, so their invention in 1960 solved the first problem. The development of a means to convey this signal was longer in coming, but scientists succeeded in developing the first communications-grade optical fibre of almost pure silica glass in 1970. (Stanton 28)

Negation

Readers can grasp some meanings by understanding clearly what the term *does not* mean. For instance, an insurance policy may define coverage for "bodily injury to others" partly by using negation:

"We will *not* pay: (1) For injuries to guest occupants of your auto."

Operating Principle

Most items work according to an operating principle, whose explanation should be part of your definition:

"How does it work?"

A clinical thermometer works on the principle of heat expansion: as the temperature of the bulb increases, the mercury inside expands, forcing a mercury thread up into the hollow stem.

Air-to-air solar heating involves circulating cool air, from inside the home, across a collector plate (heated by sunlight) on the roof. This warmed air is then circulated back into the home.

Basically, a laser [uses electrical energy to produce] coherent light, light in which all the waves are in phase with each other, making the light hotter and more intense. (Gartiganis 23)

Even abstract concepts or processes can be explained on the basis of their operating principle:

Economic inflation is governed by the principle of supply and demand: if an item or service is in short supply, its price increases in proportion to its demand.

Analysis of Parts

When your subject can be divided into parts, identify and explain them:

"What are its parts?"

The standard frame of a pitched-roof wooden dwelling consists of floor joists, wall studs, roof rafters, and collar ties.

In discussing each part, of course, you would further define specialized terms, such as *floor joists*.

Analysis of parts is particularly useful for helping non-technical readers understand a technical subject. This next analysis helps explain the physics of lasing by dividing the process into three discrete parts:

1. [Lasers require] a source of energy, [such as] electric currents or other lasers.
2. A resonant circuit . . . contains the lasing medium and has one fully reflecting end and one partially reflecting end. The medium—which can be a solid, liquid, or gas—absorbs the energy and releases it as a stream of photons [electromagnetic particles that emit light]. The photons . . . vibrate between the fully and partially reflecting ends of the resonant circuit, constantly accumulating energy—that is, they are amplified. After attaining a prescribed level of energy, the photons can pass through the partially reflecting surface as a beam of coherent light and encounter the optical elements.
3. Optical elements—lenses, prisms, and mirrors—modify size, shape, and other characteristics of the laser beam and direct it to its target. (Gartiganis 23)

Figure 1 on page 246 shows the three parts of a laser.

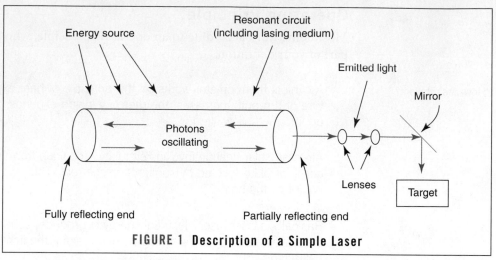

FIGURE 1 Description of a Simple Laser

Source: Gartiganis, Arthur. "Lasers." *Occupational Outlook Quarterly* Winter 1984:23

Visuals

Well-labelled visuals (such as the laser description) are excellent for clarifying definitions. Always introduce your visual and explain it. If your visual is borrowed, credit the source. Unless the visual takes up one whole page or more, do not place it on a separate page. Include the visual near its discussion.

Comparison and Contrast

Comparisons and contrasts help audiences understand a topic. Analogies (a type of comparison) to something familiar can help explain the unfamiliar:

"Does it resemble anything familiar?"

To visualize how a simplified earthquake starts, imagine an enormous block of gelatin with a vertical knife slit through the middle of its lower half. Gigantic hands are slowly pushing the right side forward and pulling the left side back along the slit, creating a strain on the upper half of the block that eventually splits it. When the split reaches the upper surface, the two halves of the block spring apart and jiggle back and forth before settling into a new alignment. Inhabitants on the upper surface would interpret the shaking as an earthquake. ("Earthquake Hazard Analysis")

The average diameter of an optical cable is around two-thousandths of an inch, making it about as fine as a hair on a baby's head. (Stanton 29–30)

Here is a contrast between optical fibre and conventional copper cable:

"How does it differ from comparable things?"

Beams of laser light coursing through optical fibres of the purest glass can transmit many times more information than the present communications systems. . . . A pair of optical fibres has the capacity to carry more than 10 000 times as many signals as conventional copper cable. A 1.25 cm (½") optical cable can carry as much information as a copper cable as thick as a person's arm. . . .

Not only does fibre optics produce a better signal, [but] the signal travels farther as well. All communications signals experience a loss of power, or attenuation, as they move along a cable. This power loss necessitates placement of repeaters at 1.5- or 3.0-kilometre intervals of copper cable in order to regenerate the signal. With fibre, repeaters are necessary about every 50 or 65 kilometres, and this distance is increasing with every generation of fibre. (Stanton 27–28)

Here is a combined comparison and contrast:

"How is it both similar and different?"

Fibre optics technology results from the superior capacity of light waves to carry a communications signal. Sound waves, radio waves, and light waves can all carry signals; their capacity increases with their frequency. Voice frequencies carried by telephone operate at 1000 cycles per second, or hertz. Television signals transmit at about 50 million hertz. Light waves, however, operate at frequencies in the hundreds of trillions of hertz. (Stanton 28)

Required Materials or Conditions

Some items or processes need special materials and handling, or they may have other requirements or restrictions. An expanded definition should include this important information.

"What is needed to make it work (or occur)?"

Besides training in engineering, physics, or chemistry, careers in laser technology require a strong background in optics (study of the generation, transmission, and manipulation of light).

Abstract concepts might also be defined in terms of special conditions:

To be held guilty of libel, a person must have defamed someone's character through written or pictorial statements.

Example

Familiar examples showing types or uses of an item can help clarify your definition. This example shows how laser light is used as a heat-generating device:

"How is it used or applied?"

Lasers are increasingly used to treat health problems. Thousands of eye operations involving cataracts and detached retinas are performed every year by ophthalmologists. . . . Dermatologists treat skin problems. . . . Gynecologists treat problems of the reproductive system, and neurosurgeons even perform brain surgery—all using lasers transmitted through optical fibres. (Gartiganis 24–25)

The next example shows how laser light is used to carry information:

Using lasers in the calculating and memory units of computers, for example, permits storage and rapid manipulation of large amounts of data. And audiodisc players use lasers to improve the quality of the sound they reproduce. The use of optical cable to transmit data also relies on lasers. (Gartiganis 25)

Examples are a most powerful communication tool—as long as you tailor the examples to the readers' level of understanding.

Whichever expansion strategies you use, be sure to document your information sources.

SAMPLE SITUATIONS

The following definitions employ expansion strategies appropriate to their audiences' needs. Specific strategies are labelled in the margin. Each definition, like a good essay, is unified and coherent: each paragraph is developed around one main idea and logically connected to other paragraphs. Visuals are incorporated. Transitions emphasize the connection between ideas. Each definition is at a level of technicality that connects with the intended audience.

To illustrate the importance of audience analysis in a writer's decision about "How much is enough?" this example, like many throughout this text, is preceded by an audience/purpose profile.

CASE | **An Expanded Definition for a Semi-technical Audience**

Audience/Purpose Profile. The intended readers of this material are beginning student mechanics. Before they can repair a solenoid, they will need to know where the term comes from, what a solenoid looks like, how it works, how its parts operate, and how it is used. This definition is designed as merely an *introduction*, so it offers only a general (but comprehensive) view of the mechanism.

Because the intended readers are not engineering students, they do not need details about electromagnetic or mechanical theory (e.g., equations or graphs illustrating voltage magnitudes, joules, lines of force).

Expanded Definition: Solenoid

Formal sentence definition

A **solenoid** is an electrically energized coil that forms an electromagnet capable of performing mechanical functions. The term *solenoid* is derived from the word *sole*,

Etymology

which in reference to electrical equipment means "a part of," or "contained inside, or with, other electrical equipment." The Greek word *solenoides* means "channel," or "shaped like a pipe."

Description and analysis of parts

A simple plunger-type solenoid consists of a coil of wire attached to an electrical source, and an iron rod, or plunger, that passes in and out of the coil along the axis of the spiral. A return spring holds the rod outside the coil when the current is de-energized, as shown in Figure 1 on page 249.

Special conditions and operating principle

When the coil receives electric current, it becomes a magnet and thus draws the iron rod inside, along the length of its cylindrical centre. With a lever attached to its end, the rod can transform electrical energy into mechanical force. The amount of mechanical force produced is the product of the number of turns in the coil, the strength of the current, and the magnetic conductivity of the rod.

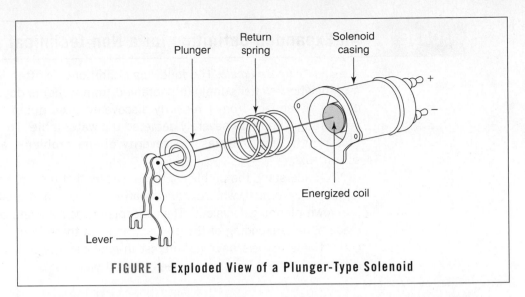

FIGURE 1 Exploded View of a Plunger-Type Solenoid

Example and analysis of parts

Explanation of visual

The plunger-type solenoid in Figure 1 is commonly used in the starter motor of an automobile engine. This type is 11.5 cm (4½″) long and 5 cm (2″) in diameter, with a steel casing attached to the casing of the starter motor. A linkage (pivoting lever) is attached at one end to the iron rod of the solenoid and at the other end to the drive gear of the starter, as shown in Figure 2.

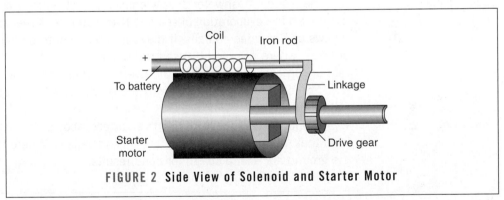

FIGURE 2 Side View of Solenoid and Starter Motor

When the ignition key is turned, current from the battery is supplied to the solenoid coil, and the iron rod is drawn inside the coil, thereby shifting the attached linkage. The linkage, in turn, engages the drive gear, activated by the starter motor, with the flywheel (the main rotating gear of the engine).

Comparison of sizes and applications

Because of the solenoid's many uses, its size varies according to the work it must do. A small solenoid will have a small wire coil, hence a weak magnetic field. The larger the coil, the stronger the magnetic field; in this case, the rod in the solenoid can do harder work. An electronic lock for a standard door would, for instance, require a much smaller solenoid than one for a bank vault.

The audience for the following definition (an entire community) is too diverse to define precisely, so the writer wisely addresses the lowest level of technicality—to ensure that all readers will understand.

CASE   An Expanded Definition for a Non-technical Audience

Audience/Purpose Profile. The following definition is written for members of a community whose water supply (all obtained from wells) is doubly threatened: (1) by chemical seepage from a recently discovered toxic dump site; and (2) by a two-year drought that has severely depleted the water table. This definition forms part of a report that analyzes the severity of the problems and explores possible solutions.

To understand the problems, these readers first need to know what a water table is, how it is formed, what conditions affect its level and quality, and how it figures into town-planning decisions. The concepts of *recharge* and *permeability* are vital to readers' understanding of the problem here, so these terms are defined parenthetically. These readers have no interest in geological or hydrological (study of water resources) theory. They simply need a broad picture.

Expanded Definition: Water Table

Formal sentence definition

The water table is the level below the earth's surface where the ground is saturated with water. Figure 1 shows a typical water table that might be found in Eastern Canada. Wells driven into such a formation will have a water level identical to that of the water table.

Example

Operating principle

The world's freshwater supply comes almost entirely as precipitation that originates with the evaporation of sea and lake water. This precipitation falls to earth and follows one of three courses: it may fall directly onto bodies of water, such as rivers or lakes, where it is directly used by humans; it may fall onto land, and either evaporate or run over the ground to the rivers or other bodies of water; or it may fall onto land, be contained, and seep into the earth. The last precipitation makes up the water table.

Comparison

Similar in contour to the earth's surface above, the water table generally has a level that reflects such features as hills and valleys. Where the water table intersects the ground surface, a stream or pond results.

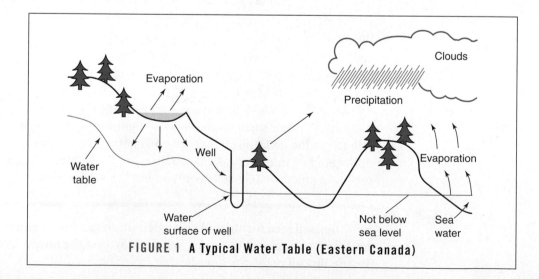

FIGURE 1 A Typical Water Table (Eastern Canada)

A water table's level, however, will vary, depending on the rate of recharge (replacement of water). The recharge rate is affected by rainfall or soil permeability (the ease with which water flows through the soil). A water table, then, is never static; rather, it is the surface of a body of water striving to maintain a balance between the forces that deplete it and those that replenish it. In areas of Nova Scotia and some western provinces where the water table is depleted, the earth caves in, leaving sinkholes.

The water table's depth below ground is vital in water-resources engineering and planning. It determines an area's suitability for wastewater disposal, or a building lot's ability to handle sewage. A high water table could become contaminated by a septic system. Also, bacteria and chemicals seeping into a water table can pollute an entire town's water supply. Another consideration in water-table depth is the cost of drilling wells. These conditions obviously affect an industry's or homeowner's decision on where to locate.

The rising and falling of the water table give an indication of the pumping rate's effect on a water supply (drawn from wells) and of the sufficiency of the recharge rate in meeting demand. This kind of information helps water resources planners decide when new sources of water must be made available.

PLACEMENT OF DEFINITIONS

Poorly placed definitions interrupt the information flow. If you have only a few terms or concepts to define on one page, do so by placing a brief, parenthetical definition immediately after each.

Including more than three or four definitions on one page will be disruptive. Rewrite them as sentence definitions and place them in a "Definitions" section of your introduction, or in a glossary. Definitions of terms in the report's title belong in your introduction.

Place expanded definitions in one of three locations:

1. If the definition is essential to the reader's understanding of the *entire* document, place it in the introduction. A report titled "The Effects of Aerosols on the Earth's Ozone Shield" would require expanded definitions of "aerosols" and "ozone shield" early in the discussion.
2. When the definition clarifies a major part of your discussion, place it in that section of your report. In a report titled "How Advertising Influences Consumer Habits," "operant conditioning" might be defined early in the appropriate section. Too many expanded definitions *within* a report, however, can be disruptive.
3. If the definition serves only as a reference, place it in an appendix. For example, a report on fire safety in a public building might have an expanded definition of carbon monoxide detectors in an appendix.

Electronic documents pose special problems for placement of definitions. In a hypertext document, for instance, each reader explores the material differently. One option for making definitions available when they are needed is the "pop-up note": the term to be defined is highlighted in the text, to indicate that its definition can be called up and displayed in a small window on the actual text screen (Horton, "Is Hypertext" 25).

for Revising and Editing Definitions

Use this checklist to revise your definitions.

Content

◆ Is the type of definition (parenthetical, sentence, expanded) suited to its audience and purpose?
◆ Does the definition convey the basic property of the item?
◆ Is the definition objective?
◆ Is the expanded definition adequately developed?
◆ Are all information sources documented?
◆ Are visuals employed adequately and appropriately?
◆ Does the sentence definition describe features that distinguish the item from all other items in the same class?

Arrangement

◆ Is the expanded definition unified and coherent (like an essay)?
◆ Are transitions between ideas adequate?
◆ Does the definition appear in the appropriate location?

Style and Page Design

◆ Is the definition in plain English?
◆ Will the level of technicality connect with the audience?
◆ Are sentences clear, concise, and fluent?
◆ Is word choice precise?
◆ Is the definition ethically acceptable?
◆ Is the page design inviting and accessible?

EXERCISES

1. Sentence definitions require precise classification and differentiation. Is each of these definitions adequate for a general reader? Rewrite those that seem inadequate. Consult dictionaries and encyclopedias as needed.
   a. A bicycle is a vehicle with two wheels.
   b. A transistor is a device used in transistorized electronic equipment.
   c. Surfing is when one rides a wave to shore while standing on a board specifically designed for buoyancy and balance.
   d. Bubonic plague is caused by an organism known as *Pasteurella pestis*.
   e. Mace is a chemical aerosol spray used by police.
   f. A Geiger counter measures radioactivity.
   g. A cactus is a succulent.
   h. In law, an indictment is a criminal charge against a defendant.
   i. A prune is a kind of plum.
   j. Friction is a force between two bodies.
   k. Luffing is what happens when one sails into the wind.
   l. A frame is an important part of a bicycle.
   m. Hypoglycemia is a medical term.
   n. An hourglass is a device used for measuring intervals of time.
   o. A computer is a machine that handles information with amazing speed.
   p. A Ferrari is the best car in the world.
   q. To meditate is to exercise mental faculties in thought.

2. Standard dictionaries define for the general reader, whereas specialized reference books define for the specialist. Choose an item in your field and copy the definition (1) from a standard dictionary and (2) from a technical reference book. For the technical definition, label each expansion strategy. Rewrite the specialized definition for a general reader.

3. Using reference books as necessary, write sentence definitions for these terms or for terms from your field.

biological insect control	gyroscope
generator	coronary bypass
dewpoint	oil shale
microprocessor	chemotherapy
capitalism	estuary
local area network	Boolean logic
marsh	classical conditioning
artificial intelligence	hypothermia
economic inflation	thermistor
anorexia nervosa	aquaculture
low-impact camping	nuclear fission
hemodialysis	modem

COLLABORATIVE PROJECT

Divide into small groups on the basis of academic majors or interests. Appoint one person as group manager. Decide on an item, concept, or process that would require an expanded definition for a layperson. For example:

◆ From computer science: an algorithm, an applications program, artificial intelligence, binary coding, top-down procedural thinking, or systems analysis

◆ From medicine: a pacemaker, coronary bypass surgery, or natural childbirth

◆ From forestry: forest fragmentation, diameter at breast height, codominant tree

Complete an audience/purpose profile for your layperson reader.

Once your group has decided on the appropriate expansion strategies (etymology, negation, etc.), the group manager will assign each member to work on one or two specific strategies as part of the definition. As a group, edit and incorporate the collected material into an expanded definition, revising as often as needed.

The group manager will assign one member to present the definition in class, using an overhead projection, a large-screen monitor, a data projection unit, or photocopies.

MyWritingLab

MyWritingLab: Technical Communication offers the best multimedia resources for technical communication in one, easy-to-use place. You can find a variety of interactive model documents and case studies. There are also extensive guidelines, tutorials, and exercises for Document Design, Writing and the Writing Process, and Research, and a large bank of diagnostics and practice for grammar review.

15

Descriptions and Specifications

✱⌐Explore
Click here in your eText to access a variety of student resources related to this chapter.

LEARNING OBJECTIVES

After reading this chapter, you should be able to

- incorporate the elements of clear, objective descriptions of mechanisms, places, or objects
- appreciate the range of uses for technical specifications
- apply the principles of effective mechanism description and specifications to technical marketing documents

---

Description (creating a picture with words) is a part of writing. But technical descriptions convey information about a product or mechanism to someone who will use it, buy it, operate it, assemble it, or manufacture it, or to someone who has to know more about it. Any item can be visualized from countless different perspectives. Therefore, *how* you describe—your perspective—depends on your purpose and the needs of your audience.

Two kinds of descriptions are featured in this chapter: mechanism description and specifications. We start with a mechanism description (see Figure 15.1 on page 255).

THE PURPOSE OF DESCRIPTION

Manufacturers use descriptions to sell products; banks require detailed descriptions of any business or construction venture before approving a loan; and medical personnel maintain daily or hourly descriptions of a patient's condition and treatment.

No matter what the subject of description, readers expect answers to as many of these questions as are applicable: *What is it? What does it do? What does it look like? What is it made of? How does it work? How was it put together?* The description in Figure 15.1, part of an installation and operation manual, answers applicable questions for do-it-yourself homeowners.

OBJECTIVITY IN DESCRIPTION

A description is mainly *subjective* or *objective*: based on feelings or fact. Subjective description emphasizes the writer's attitude toward the thing, whereas objective description emphasizes the thing itself.

Essays describing "An Unforgettable Person" or "A Beautiful Moment" express opinions, a personal point of view. Subjective description aims at expressing feelings,

tekmar® Submittal
tekmarNet® 2 House Control 400

C 400
01/11

HVAC Systems   Replaces: New

Job _____   Designer _____   Contact _____

The tekmarNet® 2 House Control 400 is designed to operate all of the mechanical equipment in a residential hydronic heating system, coordinating their operation with network communication. It can be used in applications ranging from a single zone of baseboard with an on/off boiler, to multiple radiant floor zones with a modulating/condensing boiler. This control regulates one space heating water temperature through Outdoor Reset and Indoor Temperature Feedback. It is capable of controlling a single on/off, dual stage, or modulating boiler, Domestic Hot Water, setpoint loads, 4 on-board zone valves, and can connect up to 24 zones in total.

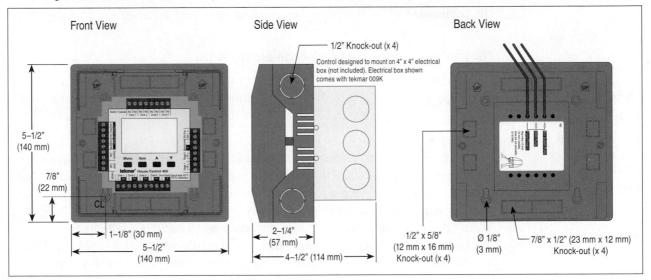

Figure 15.1 A Mechanism Description
Source: Courtesy of tekmar Control Systems.

attitudes, and moods. You create an *impression* of your subject ("The weather was miserable"), more than communicating factual information about it ("All day, we had freezing rain and gale-force winds").

Objective description shows an impartial view, filtering out personal impressions and focusing on observable details.

Except for promotional writing, descriptions on the job should be impartial, if they are to be ethical. Pure objectivity is, of course, humanly impossible. Each writer filters the facts and their meaning through her or his own perspective. Nonetheless, we are expected to communicate the facts as we know them and understand them. One writer offers this useful distinction: "All communication requires us to leave something out, but we must be sure that what is left out is not essential to our [reader's] understanding of what is put in" (Coletta 65).

An ethical writer "is obligated to express her or his opinions of products, as long as these opinions are based on objective and responsible research and observation" (MacKenzie 3). Being "objective" does not mean forsaking personal evaluation in cases in which a product may be unsafe or unsound. Even positive claims made in promotional writing (for example, "reliable," "rugged") should be based on objective and verifiable evidence.

Here are guidelines for remaining impartial.

RECORD THE DETAILS THAT ENABLE READERS TO VISUALIZE THE ITEM. Ask these questions: *What could any observer recognize? What would a camera record?*

Subjective   His office has an *awful* view, *terrible* furniture, and a *depressing* atmosphere.

The italicized words only *tell*; they do not *show*.

Objective   His office has broken windows looking out on a brick wall, a rug with a 15 cm hole in the centre, chairs with bottoms falling out, missing floorboards, and a ceiling with plaster missing in three or four places.

USE PRECISE AND INFORMATIVE LANGUAGE. Use high-information words that enable readers to visualize. Name specific parts without calling them "things," "gadgets," or "doohickeys." Avoid judgmental words (*impressive*, *poor*), unless your judgment is requested and can be supported by facts. Instead of "large," "long," and "near," give exact measurements, weights, dimensions, and ingredients.

Use words that specify location and spatial relationships: *above, oblique, behind, tangential, adjacent, interlocking, abutting,* and *overlapping*. Use position words: *horizontal, vertical, lateral, longitudinal, in cross-section, parallel*.

Indefinite	Precise
a late-model car	a 2011 Acura TL sedan
an inside view	a cross-sectional, cutaway, or exploded view
next to the foundation	along the north side of the foundation
a small red thing	a red activator button with a 2.5 cm diameter and a concave surface

Do not confuse precise language, however, with overly complicated technical terms or needless jargon. Don't say "phlebotomy specimen" instead of "blood," or "thermal attenuation" instead of "insulation," or "proactive neutralization" instead of "damage control." The clearest writing uses precise but plain language. General readers prefer non-technical language, as long as the simpler words do the job. Always think about your specific audience's needs.

ELEMENTS OF MECHANISM DESCRIPTION
Clear and Limiting Title

An effective title promises exactly what the document will deliver—no more and no less. "A Description of a Norco Shore Two Bike" promises a complete description, down to the smallest part. If you intend to describe the brakes only, be sure your title so indicates: "A Description of the Avid Elixir Hydraulic Disc Brakes."

Overall Appearance and Component Parts

Let readers see the big picture before you describe each part. Consider this description of a standard stethoscope:

> The standard stethoscope is roughly 61 cm long and weighs about 140 grams. The instrument consists of a sensitive sound-detecting and amplifying device whose flat surface is pressed against a bodily area. This amplifying device is attached to rubber and metal tubing that transmits the body sound to a listening device inserted in the ear.
>
> Seven interlocking pieces contribute to the stethoscope's Y-shaped appearance: (1) diaphragm contact piece, (2) lower tubing, (3) Y-shaped metal piece, (4) upper tubing, (5) U-shaped metal strip, (6) curved metal tubing, and (7) hollow ear plugs. These parts form a continuous unit (Figure 1).

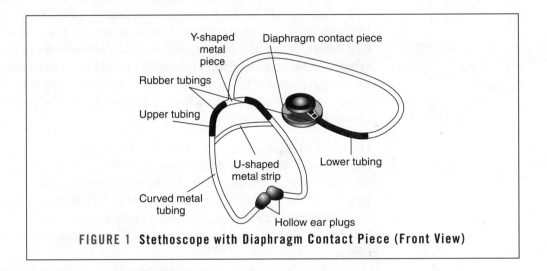

FIGURE 1 Stethoscope with Diaphragm Contact Piece (Front View)

Visuals

Use drawings, diagrams, or photographs generously. Our overall description of the stethoscope is greatly clarified by the above figure.

Function of Each Part

Explain what each part does and how it relates to the whole.

> The diaphragm contact piece is caused to vibrate by body sounds. This part is the heart of the stethoscope that receives, amplifies, and transmits the sound impulse.

Appropriate Level of Detail

Give enough detail for a clear picture, but do not burden readers needlessly. Identify your audience and its reasons for using your description.

The description of the water heater in Figure 15.1 focuses on what this model looks like and what it's made of. Its intended audience of do-it-yourselfers will know already what a hot-water heater is and what it does. That audience needs no background. A description of how it was put together appears with the installation and maintenance instructions later in the manual.

Specifications for readers who will manufacture the water heater would describe each part in exact detail (e.g., the steel tank's required thickness and pressure rating as well as required percentages of iron, carbon, and other constituents in the steel alloy).

Clearest Descriptive Sequence

Any item usually has its own logic of organization, based on (1) the way it appears as a static object, (2) the way its parts operate in order, or (3) the way its parts are assembled. We describe these relationships, respectively, in spatial, functional, or chronological sequence.

SPATIAL SEQUENCE.   Part of all physical descriptions, a spatial sequence answers these questions: *What is it? What does it do? What does it look like? What parts and material is it made of?* Use this sequence when you want readers to visualize the item as a static object or mechanism at rest (a house interior, a document, the CN Tower, a plot of land, a chainsaw, or a computer keyboard). Can readers best visualize this item from front to rear, left to right, top to bottom? (What logical path do the parts create?) A retractable pen would logically be viewed from outside to inside. The specifications in Figure 15.2 on page 264 proceed from the ground upward.

FUNCTIONAL SEQUENCE.   The functional sequence answers, *How does it work?* It is best used in describing a mechanism in action, such as an SLR digital camera, a nuclear warhead, a smoke detector, or a car's cruise-control system. The logic of the item is reflected by the order in which its parts function. Like the heat system controller in Figure 15.1, a mechanism usually has only one functional sequence. The accompanying stethoscope description follows the sequence of parts through which sound travels.

In describing a solar home-heating system, you would begin with the heat collectors on the roof, moving through the pipes, pumping system, and tanks for the heated water, to the radiators or underfloor heating tubes in each room—from source to outlet. After this functional sequence of operating parts, you could describe each part in a spatial sequence.

CHRONOLOGICAL SEQUENCE.   A chronological sequence answers, *How has it been put together?* The chronology follows the sequence in which the parts are assembled.

Use the chronological sequence for an item that is best visualized by its assembly (such as a piece of furniture, an umbrella tent, or a pre-hung window or door unit). Architects might find a spatial sequence best for describing a proposed cottage to clients; however, they would use a chronological sequence (of blueprints) for specifying to the builder the prescribed dimensions, materials, and construction methods at each stage.

COMBINED SEQUENCES.   The "Description of a Standard Bumper Jack" on pages 261–263 alternates among all three sequences: a spatial sequence (bottom to top) for describing the overall mechanism at rest, a chronological sequence for explaining the order in which the parts are assembled, and a functional sequence for describing the order in which the parts operate.

A GENERAL MODEL FOR DESCRIPTION

Description of a complex mechanism almost invariably calls for an outline. This model is adaptable to any description.

    I.   Introduction: General Description
        A. Definition, Function, and Background of the Item
        B. Purpose (and Audience—where applicable)
        C. Overall Description (with general visuals, if applicable)
        D. Principle of Operation (if applicable)
        E. List of Major Parts

    II.   Description and Function of Parts
        A. Part One in Your Descriptive Sequence
            1. Definition
            2. Shape, dimensions, material (with specific visuals)
            3. Subparts (if applicable)
            4. Function
            5. Relation to adjoining parts
            6. Mode of attachment (if applicable)
        B. Part Two in Your Descriptive Sequence (and so on)

    III.   Summary and Operating Description
        A. Summary (used only in a long, complex description)
        B. Interrelation of Parts
        C. One Complete Operating Cycle

This outline is tentative, because you might modify, delete, or combine certain parts to suit your subject, purpose, and audience.

Introduction: General Description

Give readers only as much background as they need to get the picture, as in the case below.

CASE | ## A Description of the Standard Stethoscope

Introduction

Definition and function

The stethoscope is a listening device that amplifies and transmits body sounds to aid in detecting physical abnormalities.

History and background

This instrument has evolved from the original wooden, funnel-shaped instrument invented by a French physician, R.T. Lennaec, in 1819. Because of his female patients' modesty, he found it necessary to develop a device, other than his ear, for auscultation (listening to body sounds).

Purpose and audience

This report explains to the beginning paramedical or nursing student the structure, assembly, and operating principle of the stethoscope.

Finally, give a brief, overall description of the item, discuss its principle of operation, and list its major parts, as in the overall stethoscope description on page 257.

Description and Function of Parts

The body of your text describes each major part. After arranging the parts in sequence, follow the logic of each part. Provide only as much detail as your readers need.

Readers of this description will use a stethoscope daily, so they need to know how it works, how to take it apart for cleaning, and how to replace worn or broken parts. (Specifications for the manufacturer would require many more technical details—dimensions, alloys, curvatures, tolerances, etc.)

Diaphragm Contact Piece

Definition, size, shape, and material

The diaphragm contact piece is a shallow metal bowl, about the size of a dollar coin (and twice its thickness), which is caused to vibrate by various body sounds.

Subparts

Three separate parts make up the piece: hollow steel bowl, plastic diaphragm, and metal frame, as shown in Figure 2.

The stainless steel metal bowl has a concave inner surface with concentric ridges that funnel sound toward an opening in the tapered base, then out through the hollow appendage. Lateral threads ring the outer circumference of the bowl to accommodate the interlocking metal frame. A fitted diaphragm covers the bowl's upper opening.

The diaphragm is a plastic disk, 2 mm thick, 10.2 cm in circumference, with a moulded lip around the edge. It fits flush over the metal bowl and vibrates sound toward the ridges. A metal frame that screws onto the bowl holds the diaphragm in place.

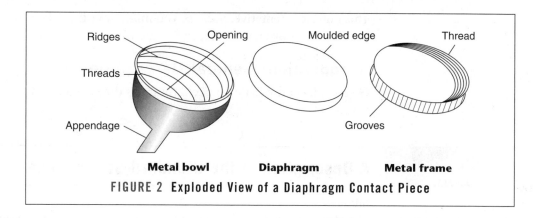

FIGURE 2 Exploded View of a Diaphragm Contact Piece

Function and relation to adjoining parts

A stainless steel frame fits over the disk and metal bowl. A 0.75 cm ridge between the inner and outer edge accommodates threads for screwing the frame to the bowl. The frame's outside circumference is notched with equally spaced, perpendicular grooves—like those on the edge of a dime—to provide a gripping surface.

Mode of attachment

The diaphragm contact piece is the heart of the stethoscope that receives, amplifies, and transmits sound through the system of attached tubing. The piece attaches to the lower tubing by an appendage on its apex (narrow end), which fits inside the tubing.

Each part of the stethoscope, in turn, is described according to its own logic of organization.

Summary and Operating Description

Conclude by explaining how the parts work together to make the whole item function.

Summary and Operating Description

How parts interrelate

The seven major parts of the stethoscope provide support for the instrument, flexibility of movement for the operator, and ease in auscultation.

One complete operating cycle

In an operating cycle, the diaphragm contact piece, placed against the skin, picks up sound impulses from the body surface. These impulses cause the plastic diaphragm to vibrate. The amplified vibrations, in turn, are carried through a tube to a dividing point. From here, the amplified sound is carried through two separate but identical series of tubes to hollow ear plugs.

A SAMPLE SITUATION

The following description of an automobile jack, aimed toward a general audience, follows the previous outline model.

CASE | **A Mechanism Description for a Non-technical Audience**

Audience/Purpose Profile. Some readers of this description (written for an owner's manual) will have no mechanical background. Before they can follow instructions for using the jack safely, they will have to learn what it is, what it looks like, what its parts are, and how, generally, it works. They will not need precise dimensions (e.g., "The rectangular base is 20.25 cm long and 16.5 cm wide, sloping upward 3.75 cm from the front outer edge to form a secondary platform 2.5 cm high and 7.5 cm square"). The engineer who designed the jack might include such data in specifications for the manufacturer. Laypeople, however, need only the dimensions that will help them recognize specific parts and understand their function, for safe use and assembly.

Also, this audience will need only the broadest explanation of how the leverage mechanism operates. Although the physical principles (torque, fulcrum) would interest engineers, they would be of little use to people who simply need to operate the jack safely.

A Description of a Standard Bumper Jack

Introduction—General Description

Definition, purpose, and function

The standard bumper jack is a portable mechanism for raising the front or rear of a car through force applied with a lever. This jack enables even a frail person to lift one corner of a two-ton automobile.

Overall description (spatial sequence)

The jack consists of a moulded steel base supporting a free-standing, perpendicular, notched shaft (Figure 1). Attached to the shaft are a leverage mechanism, a bumper catch, and a cylinder for insertion of the jack handle. Except for the main shaft and leverage mechanism, the jack is made to be dismantled and to fit neatly in the car's trunk.

The jack operates on a leverage principle, with a human hand travelling 46 cm and the car only 1 cm during a normal jacking stroke. Such a device requires many strokes to raise the car off the ground but may prove a lifesaver to a motorist on some deserted road.

Five main parts make up the jack: base, notched shaft, leverage mechanism, bumper catch, and handle.

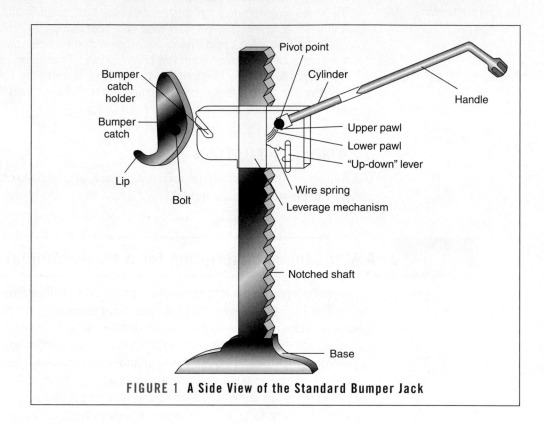

FIGURE 1   A Side View of the Standard Bumper Jack

Description of Parts and Their Function

BASE   The rectangular base is a moulded steel plate that provides support and a point of insertion for the shaft (Figure 2). The base slopes upward to form a platform containing a 1.25 cm depression that provides a stabilizing well for the shaft. Stability is increased by a 1.25 cm cuff around the well. As the base rests on its flat surface, the bottom end of the shaft is inserted into its stabilizing well.

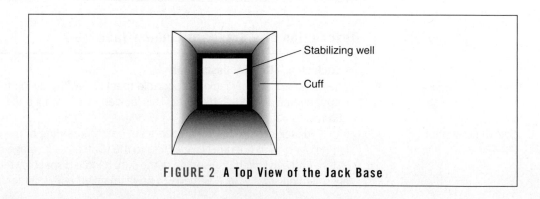

FIGURE 2   A Top View of the Jack Base

NOTCHED SHAFT The notched shaft is a steel bar (80 cm long) that provides a vertical track for the leverage mechanism. The notches, which hold the mechanism in its position on the shaft, face the operator.

The shaft vertically supports the raised automobile, and attached to it is the leverage mechanism, which rests on individual notches.

LEVERAGE MECHANISM The leverage mechanism provides the mechanical advantage needed for its operator to raise the car. It is made to slide up and down the notched shaft. The main body of this pressed-steel mechanism contains two units: one for transferring the leverage and one for holding the bumper catch.

The leverage unit has four major parts: the cylinder, connecting the handle and a pivot point; a lower pawl (a device that fits into the notches to allow forward and prevent backward motion), connected directly to the cylinder; an upper pawl, connected at the pivot point; and an "up-down" lever, which applies or releases pressure on the upper pawl by means of a spring (Figure 1). Moving the cylinder up and down with the handle causes the alternate release of the pawls, and thus movement up or down the shaft— depending on the setting of the "up-down" lever. The movement is transferred by the metal body of the unit to the bumper-catch holder.

The holder consists of a downsloping groove, partially blocked by a wire spring (Figure 1). The spring is mounted in such a way as to keep the bumper catch in place during operation.

BUMPER CATCH The bumper catch is a steel device that attaches the leverage mechanism to the bumper. This 23 cm moulded plate is bent to fit the shape of the bumper. Its outer 0.5 cm is bent up to form a lip (Figure 1), which hooks behind the bumper to hold the catch in place. The two sides of the plate are bent back 90 degrees to leave a 5.0 cm bumper-contact surface, and a bolt is riveted between them. This bolt slips into the groove in the leverage mechanism and provides the attachment between the leverage unit and the car.

HANDLE The jack handle is a steel bar that serves both as lever and lug-bolt remover. This round bar is 56 cm long, 1.5 cm in diameter, and is bent 135 degrees roughly 13 cm from its outer end. Its outer end is a wrench made to fit the wheel's lug bolts. Its inner end is bevelled to form a bladelike point for prying the wheel covers and for insertion into the cylinder on the leverage mechanism.

Conclusion and Operating Description

One quickly assembles the jack by inserting the bottom of the notched shaft into the stabilizing well in the base, the bumper catch into the groove on the leverage mechanism, and the bevelled end of the jack handle into the cylinder. The bumper catch is then attached to the bumper, with the lever set in the "up" position.

As the operator exerts an up-down pumping motion on the jack handle, the leverage mechanism gradually climbs the vertical notched shaft until the car's wheel is raised above the ground. When the lever is in the "down" position, the same pumping motion causes the leverage mechanism to descend the shaft.

SPECIFICATIONS

Airplanes, bridges, smoke detectors, and countless other items are produced according to certain specifications. A particularly exacting type of description, specifications

Ruger, Filstone, and Grant Architects

SPECIFICATIONS FOR THE POWNAL CLINIC BUILDING

Foundation

Footings: 8" x 16" concrete (load-bearing capacity: 3000 lb. per sq. in.)
Frost walls: 8" x 4' @ 3000 psi
Slab: 4" @ 3000 psi, reinforced with wire mesh over vapour barrier

Exterior Walls

Frame: Eastern pine #2 timber frame with exterior partitions set inside posts

Exterior partitions: 2" x 4" kiln-dried spruce set at 16" on centre

Sheathing: ¼" exterior-grade plywood

Siding: #1 red cedar with a ½" x 6' bevel
Trim: Finished-pine boards ranging from 1" x 4" to 1" x 10"
Painting: 2 coats of clear wood finish on siding; trim primed and finished with one coat of bone-white, oil base paint

Roof System

Framing: 2" x 12" kiln-dried spruce set at 24" on centre

Sheathing: ⅝" exterior-grade plywood

Finish: 240 Celotex 20-year fibreglass shingles over #15 impregnated felt roofing paper
Flashing: Copper

Windows

Anderson casement and fixed-over-awning models, with white exterior cladding, insulating glass and screens, and wood interior frames

Landscape

Driveway: Gravel base, with 3" traprock surface
Walks: Timber defined, with traprock surface
Cleared areas: To be rough graded and covered with wood chips
Plantings: 10 assorted lawn plants along the road side of the building

Figure 15.2 Specifications for a Building Project (Partial)

(or "specs") prescribe standards for performance, safety, and quality. For almost any product, specifications spell out

◆ the methods for manufacturing, building, or installing the product
◆ the materials and equipment to be used
◆ the size, shape, and weight of the product

Because specifications define an acceptable level of quality, they have ethical and legal implications. Any product "below" specifications provides grounds for a lawsuit. When injury or death results (as in a bridge collapse caused by inferior reinforcement), the contractor, subcontractor, or supplier who cut corners is criminally liable.

Federal and provincial regulatory agencies routinely issue specifications to ensure safety. Health Canada specifies standards for a wide variety of materials and devices, from the operation of seat belts to the fire-retardant qualities of cloth used for infant pyjamas. Meanwhile, the Canadian Standards Association designates safety and operating specifications for nearly every product sold in this country. Further, provincial and local agencies issue specifications in the form of building codes, electrical codes, and property development requirements, to name just a few.

> ### ON THE JOB...
>
> **Updating Specifications**
>
> *"When the client specifies changes in a project, I cost those changes and send them as soon as possible to the client. In a site meeting, my site superintendent writes site notes about items the client wants changed. They get listed on our web-based project management site, where the client can review them and decide whether to actually go ahead with the changes, partly based on the estimated costs. The site also shows a deadline for the client's decision, which is tied to the timeline chart that tracks our progress. A home construction could entail 50 specified alterations from start to finish. There's no way anyone could remember all those details, so they have to be dealt with one at a time and then documented."*
>
> **—Ken Dahlen, custom home builder**

Government departments (Defence, Environment, etc.) issue specifications for all types of military hardware and other equipment. A set of specifications for replacement parts for a Canadian navy vessel prescribes the standards and 13-digit NATO stock numbers for even the smallest nuts and bolts, down to screw-thread depth and width in millimetres.

The private sector issues specifications for countless products or projects, to help ensure that customers get exactly what they want. Figure 15.2 on page 264 shows partial specifications drawn up by an architect for a building that will house a small medical clinic. This section of the specs covers only the structure's "shell." Other sections detail the requirements for plumbing, wiring, and interior finish work.

The detailed building specifications partially shown in Figure 15.2 provide the basis for the comprehensive agreement between the builder and the client. In addition, the specifications (along with properly drawn building plans) are important in convincing the municipal authority to issue a building permit. Subsequently, building inspectors will use the plans and specifications as part of their criteria when they inspect the clinic in various stages of the building process.

Specifications like those in Figure 15.2 must be clear enough for *identical* interpretation by a broad audience (Glidden 258–59):

◆ *the customer*, who has the big picture of what is needed and who wants the best product at the best price

♦ *the designer* (architect, engineer, computer scientist, etc.), who must translate the customer's wishes into the actual specifications
♦ *a contractor or manufacturer*, who won the job by making the lowest bid and so must preserve profit by doing only what is prescribed
♦ *the supplier*, who must provide the exact materials and equipment
♦ *the workforce*, who will do the actual assembly, construction, or installation (managers, supervisors, subcontractors, and workers—some working on only one part of the product, such as plumbing or electrical)
♦ *the inspectors* (such as building, plumbing, or electrical inspectors), who evaluate how well the product conforms to the specifications

Each of these parties has to understand and agree on exactly *what* is to be done and *how* it is to be done. In the case of a lawsuit over failure to meet specifications, the readership broadens to include judges, lawyers, and jury. Figure 15.3 (below) depicts how a set of clear specifications unifies all users (their various viewpoints, motives, and levels of expertise) in a shared understanding.

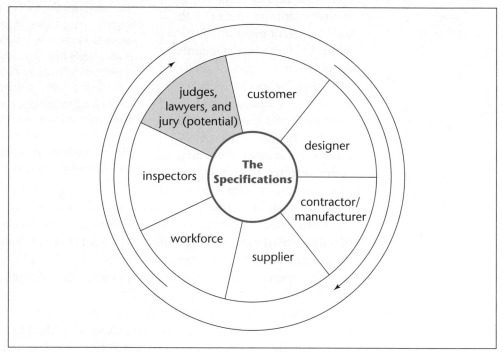

Figure 15.3 Users and Potential Users of Specifications

In addition to guiding a product's design and construction, specifications can facilitate the product's use and maintenance. For instance, specifications in a computer manual include the product's performance limits, or *ratings*: its power requirements, work or processing or storage capacity, environment requirements, the makeup of key parts, and so on. Product support literature for appliances, power tools, and other items routinely contains ratings to help customers select a good operating environment or replace worn or defective parts (Riney 186). The ratings in Figure 15.4 on page 267 are taken from a manufacturer's website.

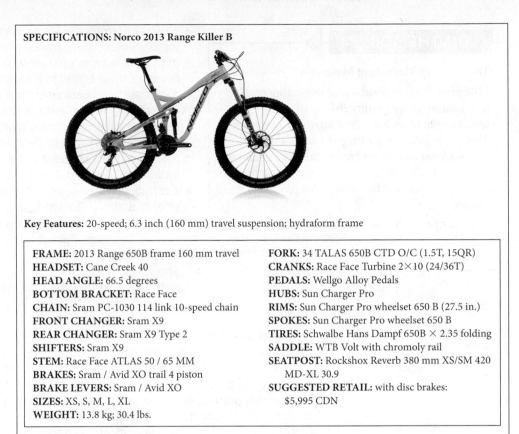

SPECIFICATIONS: Norco 2013 Range Killer B

Key Features: 20-speed; 6.3 inch (160 mm) travel suspension; hydraform frame

FRAME: 2013 Range 650B frame 160 mm travel	**FORK:** 34 TALAS 650B CTD O/C (1.5T, 15QR)
HEADSET: Cane Creek 40	**CRANKS:** Race Face Turbine 2×10 (24/36T)
HEAD ANGLE: 66.5 degrees	**PEDALS:** Wellgo Alloy Pedals
BOTTOM BRACKET: Race Face	**HUBS:** Sun Charger Pro
CHAIN: Sram PC-1030 114 link 10-speed chain	**RIMS:** Sun Charger Pro wheelset 650 B (27.5 in.)
FRONT CHANGER: Sram X9	**SPOKES:** Sun Charger Pro wheelset 650 B
REAR CHANGER: Sram X9 Type 2	**TIRES:** Schwalbe Hans Dampf 650B × 2.35 folding
SHIFTERS: Sram X9	**SADDLE:** WTB Volt with chromoly rail
STEM: Race Face ATLAS 50 / 65 MM	**SEATPOST:** Rockshox Reverb 380 mm XS/SM 420
BRAKES: Sram / Avid XO trail 4 piston	MD-XL 30.9
BRAKE LEVERS: Sram / Avid XO	**SUGGESTED RETAIL:** with disc brakes:
SIZES: XS, S, M, L, XL	$5,995 CDN
WEIGHT: 13.8 kg; 30.4 lbs.	

Figure 15.4 Specifications for the Norco 2013 Range Killer B
Source: Courtesy of Norco Bicycles.

TECHNICAL MARKETING LITERATURE

Technical marketing literature is designed to sell a technical or scientific product or service to audiences that range from novice to highly informed. Descriptions and specifications are essential marketing tools because they help potential customers to visualize the product and to recognize how its special features can fit their exact needs.

Technical marketing audiences expect a factual presentation

Even though technical marketing has persuasion as its main goal, readers dislike a "hard-sell" approach. They expect upbeat performance claims, such as "high-performance components," to be backed up by solid evidence—results of objective product testing, performance ratings, and specific technical data that indicate how the product meets or exceeds industry specifications.

Unlike proposals, which are also used to sell a product or service, technical marketing materials tend to be less formal and more dynamic, colourful, and varied. A typical proposal is tailored to one specific client's needs and follows a fairly standard format. Marketing literature, on the other hand, uses a wide variety of formats for a broad array of audiences and needs:

Common forms of technical marketing documents

◆ *Business letters* are the most personal type of marketing document. See, for example, how Manson Harding's sales letter in chapter 5 on pages 85–86 creates a human connection with a potential customer.

◆ *E-newsletters*, easily sent via email or RSS feeds to consenting, often enthusiastic, consumers, are a recent addition to direct marketing strategy. These newsletters, which can be quickly produced, include human-interest stories and technical product features and usage (Arellano, Email).

◆ *Fact sheets* offer basic data about the product or service in a straightforward, unadorned format, usually on a single 8.5″ by 11″ page, sometimes using both sides of the page. Fact sheets provide reams of technical information, but they are designed to be inviting and navigable, with engaging, easily "read" visuals; concise, readable paragraphs; clear headings; and complex data chunked into clearly labelled lists (Hilligoss 63).

◆ *Webpages* are especially effective for technical marketing. A visitor can explore the links of particular interest, read or download the material, and easily return to the homepage. Webpages offer several advantages: information is easily updated; customers can interact to ask questions or place orders; and, through animation, the product can be shown in operation (Gurak and Lannon 238). Figure 15.5 (on page 269) shows a Norco Bicycles webpage, which opens to hyperlinked specifications and product descriptions, "news" stories, and a "community" of video clips, podcasts, photos, and stories by and about people who ride Norco bikes. These stories are found in blogs, an e-newsletter, and a special section called "Ride Guide." Overall, Norco's site reaches its audience of hardcore riders and other enthusiasts by combining specs, tech talk, and human-interest stories.

◆ *Brochures* remain a popular marketing medium. Good brochures feature panels that provide a logical sequence, with each panel offering its own discrete chunk of information about the product (Hilligoss 63). Large-format brochures, such as Norco's 2013 magazine-style brochure, can run to over 80 pages of material that mimics the look of a magazine. Figure 15.6 on page 269 shows a segment of that brochure, which is also available at **www.norco.com/img/downloads/brochuresmanuals/media/2012December/23.pdf**. Online brochures have special advantages, including hyperlinks to specifications and descriptions of a company's full range of products.

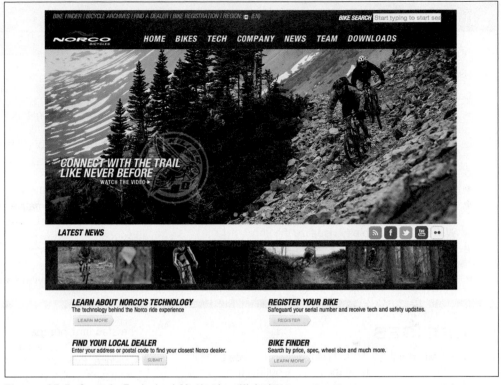

Figure 15.5 Sample Technical Marketing Website
Source: Courtesy of Norco Bicycles.

Figure 15.6 Excerpt from a Large-Format Brochure
Source: Courtesy of Norco Bicycles.

CHECKLIST

for Revising and Editing Descriptions

Use this checklist to check the usability of your description.

Content

◆ Does the title promise exactly what the description delivers?

◆ Are the item's overall features described, as well as each part?

◆ Is each part defined before it is discussed?

◆ Is the function of each part explained?

◆ Do visuals appear whenever they can provide clarification?

◆ Will readers be able to visualize the item?

◆ Are any details missing, needless, or confusing for this audience?

◆ Is the description ethically acceptable?

Arrangement

◆ Does the description follow the clearest possible sequence?

◆ Are relationships among the parts clearly explained?

Style and Page Design

◆ Is the description sufficiently impartial?

◆ Is the language informative and precise?

◆ Will the level of technicality connect with the audience?

◆ Is the description written in plain English?

◆ Is each sentence clear, concise, and fluent?

◆ Is the page design inviting and accessible?

EXERCISES

1. Select an item from the following list or a device used in your field of study. Using the general outline (page 259) as a model, develop an objective description. Include (a) all necessary visuals; (b) a rough diagram for each visual; or (c) a "reference visual" (a copy of a visual published elsewhere) with instructions for adapting your visual from that one. (If you borrow visuals from other sources, provide full documentation.) Write for a specific use by a specified audience. Attach your written audience/purpose profile to your document.

breathalyzer   Skinner box
sphygmomanometer   radio
transit   distilling apparatus
sabre saw   bodily organ
hazardous waste site brand of woodstove
photovoltaic panel   catalytic converter

Remember, you are simply describing the item, its parts, and its function: *do not* provide directions for its assembly or operation.

As an optional assignment, describe a place you know well. You are trying to convey a visual image, not a mood; therefore, your description should be impartial, discussing only the observable details.

2. The standard bumper jack description in this chapter (pages 261–263) is aimed toward a general reading audience. Evaluate it using the Checklist. In one or two paragraphs, discuss your evaluation, and suggest revisions.

3. Locate a description and specifications for a particular brand of automobile or some other consumer product. Evaluate this material for promotional and descriptive value and ethical appropriateness.

COLLABORATIVE PROJECT

Assume that your group is an architectural firm designing buildings at your university or college. Develop a set of specifications for duplicating the interior of the classroom in which this course is held. Focus only on materials, dimensions, and equipment (whiteboard, desk, etc.), and use visuals as appropriate. Your audience includes teachers, school administrators, and the firm that will construct the classroom. Use the same format as in Figure 15.2 (page 264), or design a better one. Appoint one member to present the completed specifications in class. Compare versions from each group for accuracy and clarity.

MyWritingLab

Process Analyses, Instructions, and Procedures

LEARNING OBJECTIVES

After reading this chapter, you should be able to

- understand the similarities and important differences between instructions and process analysis, and between instructions and procedures
- research, plan, and write an effective process analysis
- demonstrate the elements of readable, useful instructions
- apply principles of effective instruction to print and online tutorials

A process is a series of actions or changes leading to a product or result. *Instructions* and *tutorials* describe how to carry out a "process"; a *procedure* is a special kind of instructional set. *Process analysis* identifies, describes, and explains the sequence of events in a repeatable process. This chapter discusses all four types of process-related description.

Although process analysis and instructions both present chronological steps leading to a predicted result, that's where the similarity ends. The reasons for writing and reading a process analysis are quite different from the motivations for writing and reading instructions. These fundamental differences lead to the differences in content, structure, voice, mood, appearance, and style summarized in Table 16.1 on pages 272–273.

PROCESS ANALYSIS

Readers of process analyses want to know *how* and *why* a process occurs, so a writer's first order of business is to divide the process into its parts or principles. Usually, those parts occur in chronological order. Moreover, the first part or step of the process usually leads to the second step and often creates the conditions that allow that second step to occur. Then, the second step leads to the third step, and so on.

> Why readers read process analyses

This chronological development of dependent steps is particularly noticeable to writers who analyze mechanical processes (operation of a piston-driven engine), geological processes (formation of icebergs), or chemical processes (chemical hydration within concrete). Even electronic processes, which occur at blinding speed, can be understood as a series of causative actions: it's possible to know the exact order, duration, and effect caused by each sub-process within an electronic circuit.

The electronic example raises another very interesting point about most processes. They depend on the "conditions" that cause them. The process of an electronic circuit's

Table 16.1 Process Analysis and Instructions Compared

Process Analysis

Purpose	Helps the reader understand how and why the process occurs
Audience	Aimed at people interested in understanding how something works or how it happens
Content	Explanations are essential, in addition to straight chronological description of the stages of the process. Description of the physical environment is part of some descriptions. Illustrations are often very useful. Descriptions are specific and detailed.
Structure	*General idea (lead-in)* ◆ Names and defines process and its special features ◆ Where, when, why, how often the process occurs ◆ Where necessary, gives background theory ◆ Lists the main stages or actions of the process *Individual stages (chronological)* ◆ Each stage is described in detail and related to the stages that precede and follow; the importance of particular stages is noted. ◆ Each stage includes applicable measurements of time, distance, direction, density, volume, etc. *Conclusion (lead out to practical considerations)* ◆ Where applicable, comments about time needed for overall process, cost, process applications, special problems, immediate and long-term results
Voice/Mood	Uses indicative mood: e.g., "The next stage takes three hours . . ." Stays detached, in third person: e.g., "The skier's first move . . ." Active or passive voice: e.g., "The signal travels . . ." or "The signal is next transferred to the filtering stage . . ."
Appearance and Style	Usually looks formal (headings, paragraphs, standard spacing) Reads like a "serious" discussion Uses a mixture of sentence types and lengths Uses precise, accurate vocabulary

(continued)

operation depends on the design of the circuit itself. In a less confining way, perhaps, the process by which a road is washed by heavy spring run-off depends on the physical conditions of water volumes, soil composition, and terrain.

As Table 16.1 illustrates, a process analysis must include enough detail to enable readers to follow the process step by step. That level of detail depends on the reader's needs. For example, a back-country skier who wants to avoid avalanches will be satisfied with a basic description of the forces and conditions that

Table 16.1 Process Analysis and Instructions Compared (*Continued*)

Instructions

Purpose	Helps the reader perform the process that is described
Audience	Aimed at people who need to complete a task or want to improve performance
Content	Provides no more detail than is necessary (Note: Analysis and explanations may be necessary.)
	Features a very careful chronological listing of steps
	Very carefully describes exact steps to take
	Includes frequent visual illustration
Structure	*Introduction*
	◆ Concisely explains the overall actions to be performed
	◆ In some cases, provides background information and, where necessary, lists materials/equipment to be used or the conditions necessary for successful action
	◆ In some cases, cautions reader about safety factors
	Chronological list of steps (plus necessary explanations)
	◆ Where appropriate, combines groups of steps under subheadings: e.g., "Setting the Timer," "Selecting Programs"
	◆ Shows the interrelations and sequence of actions by using numbered steps and sequence transitions: e.g., "next," "then," "10 minutes later," "after the liquid cools"
	◆ Gives reasons for performing certain actions in a specific way or at a specific time
	◆ Uses illustrations to show the results of performed actions, not just the techniques for performing the actions
	Brief practical conclusion
	◆ Reminds the reader of expected results/performance times
Voice/Mood	Uses imperative mood: e.g., "Set the timer by choosing . . ."
	Directly addresses reader: e.g., "Your first task will be to . . ."
	Uses active voice: e.g., "Choose one of three settings . . ."
Appearance and Style	Uses some paragraphs, but mostly uses numbered point form
	Looks user-friendly
	Writes in phrases or short sentences
	Features direct, straightforward vocabulary
	Employs lots of white space

affect the slab-avalanche process. However, a civil engineer studying avalanches will need to know much more about snow compaction forces, changes in crystalline structures, and the forces that cause snow layers to shear apart.

Because it emphasizes the process itself, rather than the reader's role, process analysis is written in the third person. Indeed, all aspects of a process analysis resemble a technical essay:

- ◆ It uses standard paragraphs, most of which follow a chronological pattern.
- ◆ It employs serious, reflective phrasing.
- ◆ It employs precise, accurate vocabulary.
- ◆ It presents a formal appearance, usually with headings, formal illustration format, and formal documentation of sources.

To help the reader fully understand the process, the writer must carefully plan the structure of a process analysis. The writer's first step is to analyze the process itself, at the level of the reader's interests and needs. This analysis will help the writer produce a detailed outline.

For an idea of the components of such an outline, see the structure for process analysis summarized in Table 16.1.

The following process analysis has used a structure like the one outlined in Table 16.1. The document's writer, Bill Kelly, belongs to an environmental group studying the problems of acid rain in its Southern Ontario community. To gain community support, the environmentalists must educate citizens about the problem. Kelly group is publishing and mailing a series of brochures. The first brochure explains how acid rain is formed.

Here is Kelly's audience/purpose profile for the document.

CASE	**A Process Analysis for a Non-technical Audience**

Audience/Purpose Profile. My audience will consist of general readers. Some already will be interested in the problem; others will have no awareness (or interest). Therefore, I'll keep my explanation at the lowest level of technicality (no chemical formulas, equations). But my explanation needs to be vivid enough to appeal to less aware or less interested readers. I'll use visuals to create interest and to illustrate simply. To give an explanation thorough enough for broad understanding, I'll divide the process into three chronological steps: how acid rain develops, spreads, and destroys.

The document resulting from Kelly's analysis of both his subject and his audience is below.

How Acid Rain Develops, Spreads, and Destroys

INTRODUCTION

Definition

Acid rain is environmentally damaging rainfall that occurs after fossil fuels burn, releasing nitrogen and sulphur oxides into the atmosphere. Acid rain, simply stated, increases the acidity level of waterways because these nitrogen and sulphur oxides combine with the air's normal moisture. The resulting rainfall is far more acidic than normal rainfall. Acid rain is a silent threat because its effects, although slow, are cumulative. This analysis explains the cause, the distribution cycle, and the effects of acid rain.

Purpose

Brief description of the process

Most research shows that power plants burning oil or coal are the primary cause of acid rain. The burnt fuel is not completely expended, and some residue enters the atmosphere. Although this residue contains several potentially toxic elements, sulphur oxide and, to a lesser extent, nitrogen oxide are the major problem, because they are transformed when they combine with moisture. This chemical reaction forms sulphur dioxide and nitric acid, which then rain down to Earth.

Preview of stages

The major steps explained here are (1) how acid rain develops, (2) how acid rain spreads, and (3) how acid rain destroys.

THE PROCESS

First stage

How Acid Rain Develops Once fossil fuels have been burned, their usefulness is over. Unfortunately, it is here that the acid rain problem begins.

Fossil fuels contain a number of chemicals that are released during combustion. Two of these, sulphur oxide and nitrogen oxide, combine with normal moisture to produce sulphuric acid and nitric acid. (Figure 1 illustrates how acid rain develops.) The released gases undergo a chemical change as they combine with atmospheric ozone and water vapour. The resulting rain or snowfall is more acidic than normal precipitation.

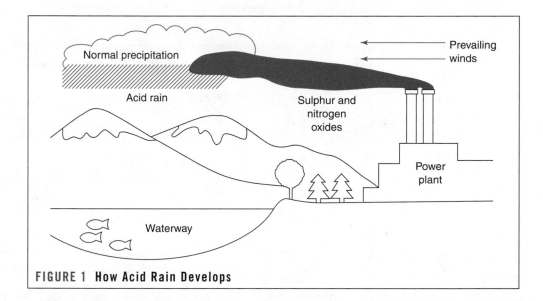

FIGURE 1 How Acid Rain Develops

Definition

Acid level is measured by pH readings. The pH scale runs from 0 through 14—a pH of 7 is considered neutral. (Distilled water has a pH of 7.) Numbers above 7 indicate increasing degrees of alkalinity. (Household ammonia has a pH of 11.) Numbers below 7 indicate increasing acidity. Movement in either direction on the pH scale, however, means multiplying by 10. Lemon juice, which has a pH value of 2, is 10 times more acidic than apples, which have a pH of 3, and is 1000 times more acidic than carrots, which have a pH of 5.

Because of carbon dioxide (an acid substance) normally present in air, unaffected rainfall has a pH of 5.6. At this time, the pH of precipitation in the northeastern United States and Canada is between 4.5 and 4. In Massachusetts, rain and snowfall have an average pH reading of 4.1. A pH reading below 5 is considered to be abnormally acidic, and therefore a threat to aquatic populations.

Second stage

How Acid Rain Spreads Although it might seem that areas containing power plants would be most severely affected, acid rain can in fact travel thousands of kilometres from its source. Stack gases escape and drift with the wind currents. The sulphur and nitrogen oxides are thus able to travel great distances before they return to Earth as acid rain.

For an average of two to five days after emission, the gases follow the prevailing winds far from the point of origin. Estimates show that about 50 percent of the acid rain that affects Canada originates in the United States; at the same time, 15 to 25 percent of the U.S. acid rain problem originates in Canada.

The tendency of stack gases to drift makes acid rain a widespread menace. More than 200 lakes in the Adirondacks, hundreds of kilometres from any industrial centre, are unable to support life because their water has become so acidic.

How Acid Rain Destroys Acid rain causes damage wherever it falls. It erodes various types of building rock, such as limestone, marble, and mortar, which are gradually eaten away by the constant bathing in acid. Damage to buildings, houses, monuments, statues, and cars is widespread. Some priceless monuments and carvings already have been destroyed, and some tree varieties are dying in large numbers.

More important, however, is acid rain damage to waterways in the affected areas. (Figure 2 illustrates how a typical waterway is infiltrated.)

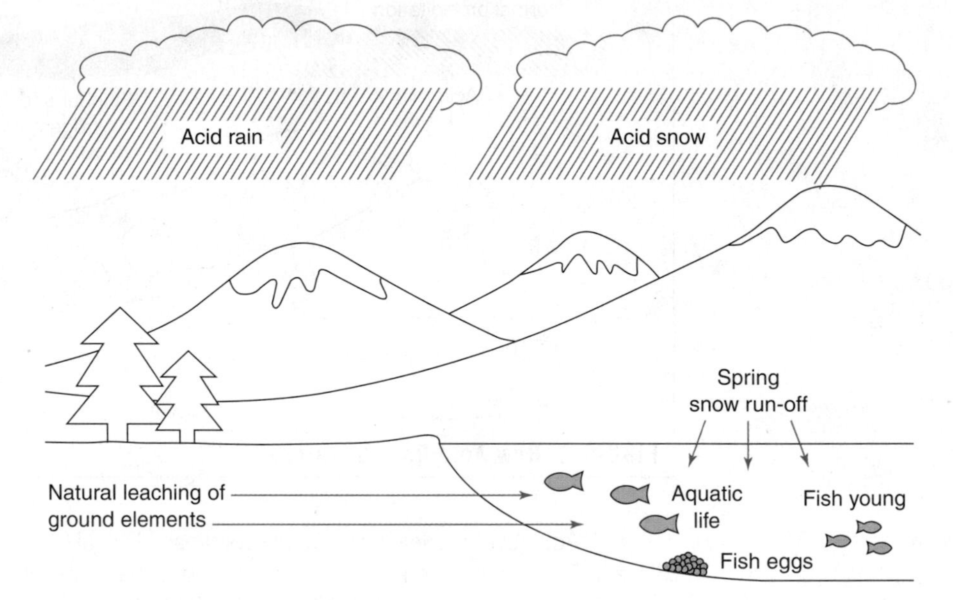

Third stage (margin)

Substage (margin)

FIGURE 2 How Acid Rain Destroys

Because of its high acidity, acid rain dramatically lowers the pH in lakes and streams. Although its effect is not immediate, acid rain eventually can make a waterway so acidic it dies. In areas with natural acid-buffering elements, such as limestone, the diluted acid has less effect. The northeastern United States and Canada, however, lack this natural protection, and so they are continually vulnerable.

The pH level in an affected waterway drops so low that some species cease to reproduce. In fact, a pH level of 5.1 to 5.4 means that fisheries are threatened; once a waterway reaches a pH level of 4.5, no fish reproduction occurs. Because each creature is part of the overall food chain, loss of one element in the chain disrupts the whole cycle.

In the northeastern United States and Canada, the acidity problem is compounded by the run-off from acid snow. During the cold winter months, acid snow sits with little melting, so that by spring thaw, the acid released is greatly concentrated. Aluminum and other heavy metals normally present in soil are also released by acid rain and run-off. These toxic substances leach into waterways in heavy concentrations, affecting fish in all stages of development.

SUMMARY

One complete cycle

Acid rain develops from nitrogen and sulphur oxides emitted by industrial and power plants burning fossil fuels. In the atmosphere, these oxides combine with ozone and water to form acid rain: precipitation with a lower-than-average pH. This acid precipitation returns to earth many kilometres from its source, severely damaging waterways that lack natural buffering agents. The northeastern United States and Canada are the most severely affected areas in North America.

ELEMENTS OF INSTRUCTIONS

Why people need instructions

As consumers, we need instructions to learn how to operate everything from automobiles to smartphones. But we also seek instruction on topics that we know reasonably well. For instance, we may read instructional magazine articles on how to ski steep slopes or how to hit a particular golf shot or how to perform an aerobics sequence, even though we might already be able to perform the activity. Why? To perform the activity better!

Almost anyone with a responsible job writes and reads instructions. The new employee uses instructions for operating office equipment or industrial machinery; the employee going on vacation writes instructions for the replacement person. The person who buys a computer reads the manuals (or documentation) for instructions on connecting a printer or downloading a software program.

Instructions carry profound ethical and legal implications. Each year, as many as 10 percent of workers are injured on the job (Clement 149). Countless injuries also result from misuse of consumer products, such as power tools and car jacks—misuse often caused by defective instructions.

A person injured because of inaccurate, incomplete, or misleading instructions can sue the writer. Courts have ruled that a writing defect in product support literature carries liability, as would a design or manufacturing defect in the product itself (Girill, "Technical Communication and Law" 37). Some legal experts argue that writing defects carry even greater liability than product defects because they are more easily demonstrated to a non-technical jury (Bedford and Stearns 128).

> ## ON THE JOB...
>
> ### The Value of Instructions
>
> *"You may have the greatest software on the planet, but if your on-screen help and user guides are not clear, then your customers will struggle with the software. They'll call your help desk and they'll complain. I help software companies to reduce their customer support costs. How? By writing clear instructions for their software. The result is that customers don't call the help desk as often as they used to. There's a case study on* **www.techscribe.co.uk/techw/cssdl.htm**.
>
> — Mike Unwalla, principal writer, TechScribe (qtd. in Beaumont 18)

To ensure that your own instructions meet professional and legal requirements for accuracy, completeness, and clarity, observe the following guidelines.

Clear and Limiting Title

Make your title promise exactly what your instructions deliver—no more and no less. The title "How to Clean a Laptop Computer's Drive Head" tells readers what to expect: instructions for a specific procedure on a selected part. But the title "The Laptop Computer" gives no such forecast. A reader of a document so titled might think the document contains a history of the laptop, a description of each part, or a wide range of related information.

Informed Content

Never count on ignorance as an excuse

Make sure that you know exactly what you're talking about. Ignorance on your part makes you no less liable for instructions that are faulty or inaccurate:

> If the author of [a car repair] manual . . . provided faulty instructions on the repair of the car's brakes, the home mechanic who was injured when the brakes failed may recover [damages] from the author. (Walter and Marsteller 165)

Do not write instructions unless you know the procedure in detail and you actually have performed it.

Visuals

In addition to showing what to do, instructional visuals attract the attention of today's graphics-oriented readers and help keep words to a minimum. Instructions also often include a persuasive dimension: to motivate interest, commitment, or action.

Types of visuals especially suited to instructions include icons, representational and schematic diagrams, flow charts, photographs, and prose tables.

Illustrate any step that might be difficult for readers to visualize. Show the same angle of vision the reader will have when doing the activity or using the equipment—and name the angle (*side view, top view*) if you think readers will have trouble figuring it out for themselves.

The less specialized your audience, the more visuals they are likely to need. But do not illustrate any action simple enough for readers to visualize on their own, such as "Press ENTER" for any user familiar with a keyboard.

Figure 16.1 on page 279 depicts an array of visuals and their specific instructional functions. Virtually each of these visuals is easily constructed and some could be further enhanced, depending on your production budget and graphics capability. Writers and editors often provide an *art brief* and a rough sketch describing the visual and its purpose for the graphic designer or art department.

Appropriate Level of Technicality

Unless you know your readers have the relevant background and skills, write for a general reader, and do three things:

1. Give them enough background to understand why they need your instructions.
2. Give them enough detail to understand *what* to do.
3. Give them enough examples to visualize the procedure clearly.

HOW TO LOCATE SOMETHING

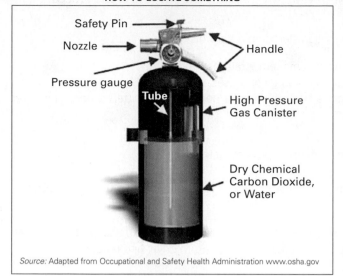

Safety Pin
Nozzle
Handle
Pressure gauge
Tube
High Pressure Gas Canister
Dry Chemical Carbon Dioxide, or Water

Source: Adapted from Occupational and Safety Health Administration www.osha.gov

HOW TO IDENTIFY SAFE OR ACCEPTABLE LIMITS

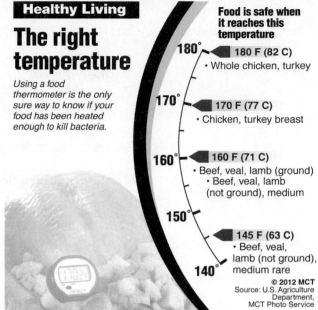

Healthy Living

The right temperature

Using a food thermometer is the only sure way to know if your food has been heated enough to kill bacteria.

Food is safe when it reaches this temperature

180° — **180 F (82 C)** • Whole chicken, turkey

170° — **170 F (77 C)** • Chicken, turkey breast

160° — **160 F (71 C)** • Beef, veal, lamb (ground) • Beef, veal, lamb (not ground), medium

150°

145° — **145 F (63 C)** • Beef, veal, lamb (not ground), medium rare

140°

© 2012 MCT
Source: U.S. Agriculture Department, MCT Photo Service

HOW TO REPAIR SOMETHING

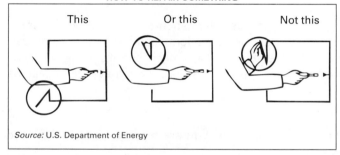

This   Or this   Not this

Source: U.S. Department of Energy

HOW TO POSITION SOMETHING

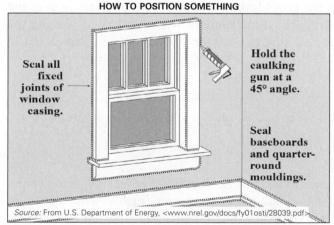

Seal all fixed joints of window casing.

Hold the caulking gun at a 45° angle.

Seal baseboards and quarter-round mouldings.

Source: From U.S. Department of Energy, <www.nrel.gov/docs/fy01osti/28039.pdf>

HOW TO OPERATE SOMETHING

Figure 16.1 Common Types of Instructional Visuals and Their Functions

Sources: **Locate:** Adapted from Occupational and Safety Health Administration, **www.osha.gov**; **repair:** U.S. Department of Energy; **position:** from U.S. Department of Energy, **www.nrel.gov/docs/fy01osti/28039.pdf**; **identify:** Staf/MCT/Newscom; **operate** David Burton/Alamy

BACKGROUND INFORMATION.   Begin by explaining the purpose of the procedure.

Tell readers why they are
doing this

You might easily lose information stored on a hard disk if

1. The disk is damaged by extreme temperature or moisture.
2. The disk is erased by a power surge or a user error.
3. The stored information is scrambled by a nearby magnet (telephone, computer terminal, or the like).

Always make a backup copy of any important material.

Also, state your assumptions about your reader's level of technical understanding.

Tell readers what they
should know already

To follow these instructions, you should be able to identify these parts of a Mac system: computer, monitor, keyboard, mouse, CD-Rom drive, and recordable compact disk.

Define any specialized terms that appear in your instructions.

Tell readers what each
key term means

Initialize: Before you can store or retrieve information on a new disk, you must initialize the blank disk (unless you are using pre-formatted disks). Initializing creates a format the computer can understand—a directory of specific memory spaces (like post office boxes) on the disk, where you can store and retrieve information as needed.

When your reader understands *what* and *why,* you are ready to explain *how* the reader can carry out the procedure.

Adequate detail for
general readers

ADEQUATE DETAIL.   Explain the procedure in enough detail for readers to know exactly what to do. Vague instructions result from the writer's failure to consider the readers' needs, as in these unclear instructions for giving first aid to an electrical shock victim:

1. Check vital signs.
2. Establish an airway.
3. Administer external cardiac massage as needed.
4. Ventilate, if cyanosed.
5. Treat for shock.

These instructions might be clear to medical experts, but they will not be to general readers. Not only are the details inadequate, but terms, such as *vital signs* and *cyanosed,* are too technical for laypeople. Such instructions posted for workers in a high-voltage area would be useless. The instructions need illustrations and explanations, as in the illustrations on the following pages.

The following instruction for Step 2 above, "Establish an airway," provides another example of the value of providing illustrations and explanations:

Step 2: While you maintain the head in a backward tilt position, place your cheek and ear close to the victim's mouth and nose. Look for the chest to rise and fall while you listen and feel for breathing. Check for about 5 seconds.

Step 1   Step 2   Step 3

Source: Reprinted with permission from *New York Public Library Desk Reference*, 3rd ed., copyright © 1998, 1993, 1989 by The New York Public Library and the Stonesong Press, Inc.

It's easy to overestimate what people already know, especially when the procedure is almost automatic for you. (Think about when a relative or friend was teaching you to drive a car, or perhaps you tried to teach someone else.) Always assume that the reader knows less than you. A colleague will know at least a little less; a layperson will know a good deal less—maybe nothing—about this procedure.

Exactly how much information is enough? These suggestions can help you find an answer:

How to provide adequate detail

- ◆ Give everything readers need, so the instructions can stand alone.
- ◆ Give only what readers need. Don't tell them how to build a computer when they need to know only how to copy a CD.
- ◆ Instead of focusing on the *product* ("How does it work?"), focus on the *task* ("How do I use it?" or "How do I do it?") (Grice 132).
- ◆ Omit steps obvious to readers (e.g., "Seat yourself at the computer").
- ◆ Adjust the *information rate* ("the amount of information presented in a given page," Meyer 17) to readers' background and the difficulty of the task. For complex or sensitive steps, slow the information rate. Don't make readers do too much too fast.
- ◆ Reinforce the prose with visuals. Don't be afraid to repeat information if it saves readers from flipping pages.
- ◆ When writing instructions for consumer products, assume "a barely literate reader" (Clement 151). Simplify.
- ◆ Recognize the persuasive dimension of the instructions. You may need to persuade readers that this procedure is necessary or beneficial, or that they can complete this procedure with relative ease and competence.

EXAMPLES. Procedures require specific examples (how to load a program, how to order a part), to help readers follow the steps correctly.

To load your program, key this command:

Give plenty of examples

```
Load "Style Editor"
```

Then press ENTER.

Like visuals, examples *show* readers what to do. Examples in fact often appear as visuals.

Several examples in Figure 16.1 illustrate these important components of effective instruction:

1. Begin each instruction with an action verb.
2. Let the visual repeat, restate, or reinforce the prose.
3. Place the visual close to the step.

Logically Ordered Steps

Instructions not only divide the procedure into steps; they also guide users through the steps in *chronological order*. They organize the facts and explanations in ways that make sense to readers.

Show how steps are connected

> You can splice two wires to make an electrical connection only when you have removed the insulation. To remove the insulation, you will need . . .

Try to keep all information for one step close together.

Warnings, Cautions, and Notes

Here are the only items that should interrupt the steps in a set of instructions (Van Pelt 3):

◆ A *note* clarifies a point, emphasizes vital information, or describes options or alternatives:

Note: The computer will not burn a CD that is scratched or imperfect. If your CD is rejected, try a new CD.

◆ A *caution* prevents possible mistakes that could result in injury or equipment damage:

Caution: A momentary electrical surge or power failure will erase the contents of internal memory. To avoid losing your work, every few minutes, save on a backup disk what you have just typed into the computer.

◆ A *warning* alerts users against potential hazards to life or limb:

Warning: To prevent electrical shock, always disconnect your printer from its power source before cleaning internal parts.

◆ A *danger* notice identifies an immediate hazard to life or limb:

Danger: The red canister contains **DEADLY** radioactive material. **Do not break the safety seal** under any circumstances.

In addition to prose warnings, attract readers' attention and help them identify hazards by using symbols or icons such as the following (Bedford and Stearns 128):

Warning   **Do not enter**   **Radioactivity**   **Fire danger**

Preview the warnings, cautions, and notes in your introduction, and place them, *clearly highlighted*, immediately before the respective steps.

NOTE   *A recent study found that product users were six times more likely to comply with warnings included in the product usage directions than with a separate warning label* ("Notes" 72).

Use notes, warnings, and cautions only when needed; overuse will dull their effect, and readers may overlook their importance.

Appropriate Words, Sentences, and Paragraphs

Of all communications, instructions have the strictest requirements for clarity, because they lead to *immediate action*. Readers are impatient and often will not read the entire instructions before plunging into the first step. Poorly phrased and misleading instructions can be disastrous.

Like descriptions, instructions name parts, locations, and positions, and they state exact measurements, weights, and dimensions. Instructions also require your strict attention to phrasing, sentence structure, and paragraph structure.

TRANSITIONS TO MARK TIME AND SEQUENCE.   Transitional words provide a bridge between related ideas. Some transitions (*in addition*, *next*, *meanwhile*, *finally*, *10 minutes later*, *the next day*, *before*) mark time and sequence. They help readers understand the step-by-step process.

CAREFULLY SHAPED PARAGRAPHS AND SENTENCES.   Much of the introductory and explanatory material in instructions takes the form of standard prose paragraphs, with enough sentence variety to keep readers interested. But the steps themselves have unique paragraph and sentence requirements. Unless the procedure consists of simple steps, separate the steps by using a numbered list—one step for one activity. If the activity is especially complicated, use a new line (not indented) to begin each sentence in the step.

Instructions ordinarily employ short sentences. But brief is not always best, especially when readers have to fill in the information gaps. Never "telegraph" your message by omitting articles (*a*, *an*, *the*).

Unlike other documents, instructions call for very little sentence variety. Use similar structures ("Do this. Then do that") to avoid confusing readers. If a single step covers two related actions, follow the sequence of required actions:

Logical sequence   Switch on the computer, then insert the CD in the drive.

Make your explanations easy to follow by using a familiar-to-unfamiliar sequence:

Difficult   You must initialize a blank CD before you can store information on it.

This sentence is clearer if the familiar material comes first:

Easier   Before you can store information on a blank CD, you must initialize the disk.

Shape every sentence and every paragraph so that they are accessible to readers.

ACTIVE VOICE AND IMPERATIVE MOOD.   Use the active voice and imperative mood ("Insert the CD") to address the reader directly. Otherwise, your instructions can lose authority ("You should insert the CD") or become ambiguous ("The CD is inserted"). In the ambiguous example, we can't tell if the CD is to be inserted or if it already has been inserted.

Indirect or confusing	The user keys in his or her access code.
	You should key in your access code.
	It is important to key in the access code.
	The access code is keyed in.
Direct and clear	**Key in** your access code.

The imperative makes instructions more definite and easier to understand because the action verb—the crucial word that identifies the next action—comes first. Instead of burying your verb in mid-sentence, *begin* with an action verb (*raise, connect, wash, insert, open*) to give readers an immediate signal.

AFFIRMATIVE PHRASING.   Research shows that readers respond more quickly and efficiently to instructions phrased affirmatively rather than negatively (Spyridakis and Wenger 205).

Weaker	Verify that your CD is not contaminated with dust.
Stronger	Examine your CD for dust contamination.

PARALLEL PHRASING.   Like any items in a series, steps should use identical grammatical form. Parallelism is important in all writing, but especially in instructions, because repeating those forms emphasizes the step-by-step organization of the instructions. Parallelism also increases readability and lends continuity to the instructions.

Not parallel	To log on to the VAX 950, follow these steps:
	1. Switch the terminal to "on."
	2. The CONTROL key and C key are pressed simultaneously.
	3. Typing LOGON, and pressing the ESCAPE key.
	4. Type your user number, and then press the ESCAPE key.
Parallel	To log on to the VAX 950, follow these steps:
	1. Switch the terminal to "on."
	2. Press the CONTROL key and C key simultaneously.
	3. Type LOGON, and press the ESCAPE key.
	4. Type your user number, and then press the ESCAPE key.

Effective Page Design

Instructions rarely get undivided attention. The reader usually does two things more or less at once: interpreting the instructions and performing the task. Effective instructional design is *accessible* and *inviting*. An effective design conveys the sense that the task is within a qualified user's range of abilities.

Here are suggestions for designing instructions that help users find, recognize, and remember what they need:

How to design
instructions

- ◆ Use informative headings that tell readers what to expect, that emphasize what is most important, and that serve as an aid to navigation. The heading "How to Initialize Your Compact Disk" is more informative than "Compact Disk Initializing."
- ◆ Arrange all steps in a numbered list.

- Single-space within steps and double-space between, to separate steps visually. (Use the Spacing Before and After Paragraph feature of your word-processing program for maximum editing flexibility.)
- Instead of indenting, double-space to signal a new paragraph.
- Use white space and highlighting to separate discussion from step.

Set off your discussion on a separate line (like this), indented or highlighted or both. You can highlight with underlining, capitals, dashes, parentheses, and asterisks. Alternatively, you can use **boldface**, *italics*, varying type sizes, and different typefaces.

- Set off warnings, cautions, and notes in ruled boxes or use highlighting and plenty of white space.
- Keep the visual and the step close together. If room allows, place the visual right beside the step; if not, right after the step. Set off the visual with plenty of white space.
- Strive for format variety that is appealing but not overwhelming or inconsistent. Readers can be overwhelmed by a page with excessive or inconsistent graphic patterns.

The more accessible and inviting the design, the more likely your readers are to follow the instructions. Don't be afraid to experiment until you find a design that works. Then, as with all aspects of effective written instructions, test the usability of your document.

A SAMPLE SET OF INSTRUCTIONS

Figure 16.2 shows a complete set of instructions written for a non-technical audience. These instructions follow the basic outline by offering an overview, a list of equipment needed, and simple numbered instructions. The design, which uses informative headings, numbered steps, and simple visual diagrams, is easy to use.

CASE **Preparing Instructions for a Non-technical Audience**

The Situation. The owner of your town's local hardware store tells you that he often is asked the same questions by customers who want to make simple home repairs. One common question is about how to replace a worn faucet washer. He decides to hire a technical writing student (you) to help him write and design a simple, yet effective, set of instructions.

Audience/Purpose Profile. These customers come from a wide range of backgrounds, but most of them are not engineers or plumbers. They are just regular homeowners who are confident in their ability to work with basic tools and comfortable trying a new task. They don't want a lot of detail about the history of faucets or the various types of faucets. They just want to know how to fix the problem. These people are busy—they have lots of chores to do on the weekend, and they want instructions that are easy to follow and use

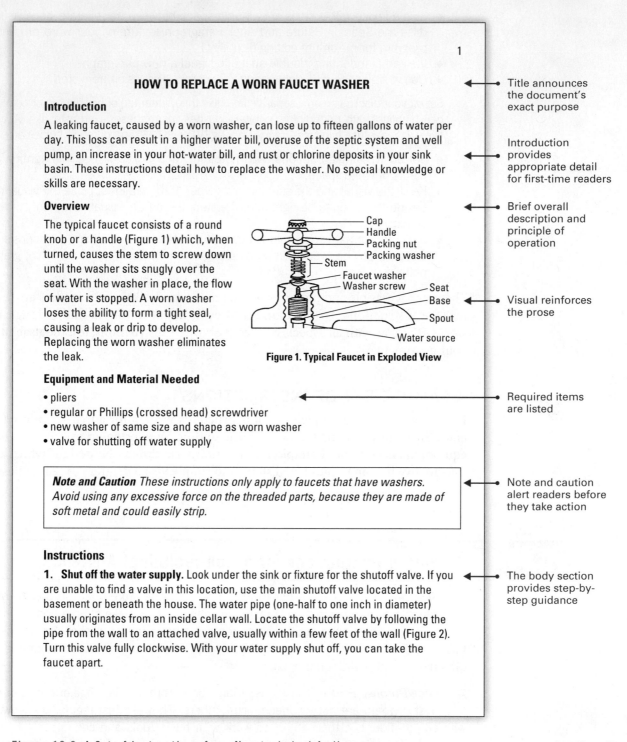

1

Title announces the document's exact purpose

Introduction provides appropriate detail for first-time readers

Brief overall description and principle of operation

Visual reinforces the prose

Required items are listed

Note and caution alert readers before they take action

The body section provides step-by-step guidance

HOW TO REPLACE A WORN FAUCET WASHER

Introduction

A leaking faucet, caused by a worn washer, can lose up to fifteen gallons of water per day. This loss can result in a higher water bill, overuse of the septic system and well pump, an increase in your hot-water bill, and rust or chlorine deposits in your sink basin. These instructions detail how to replace the washer. No special knowledge or skills are necessary.

Overview

The typical faucet consists of a round knob or a handle (Figure 1) which, when turned, causes the stem to screw down until the washer sits snugly over the seat. With the washer in place, the flow of water is stopped. A worn washer loses the ability to form a tight seal, causing a leak or drip to develop. Replacing the worn washer eliminates the leak.

Cap
Handle
Packing nut
Packing washer
Stem
Faucet washer
Washer screw
Seat
Base
Spout
Water source

Figure 1. Typical Faucet in Exploded View

Equipment and Material Needed

• pliers
• regular or Phillips (crossed head) screwdriver
• new washer of same size and shape as worn washer
• valve for shutting off water supply

Note and Caution These instructions only apply to faucets that have washers. Avoid using any excessive force on the threaded parts, because they are made of soft metal and could easily strip.

Instructions

1. **Shut off the water supply.** Look under the sink or fixture for the shutoff valve. If you are unable to find a valve in this location, use the main shutoff valve located in the basement or beneath the house. The water pipe (one-half to one inch in diameter) usually originates from an inside cellar wall. Locate the shutoff valve by following the pipe from the wall to an attached valve, usually within a few feet of the wall (Figure 2). Turn this valve fully clockwise. With your water supply shut off, you can take the faucet apart.

Figure 16.2 A Set of Instructions for a Non-technical Audience

(continued)

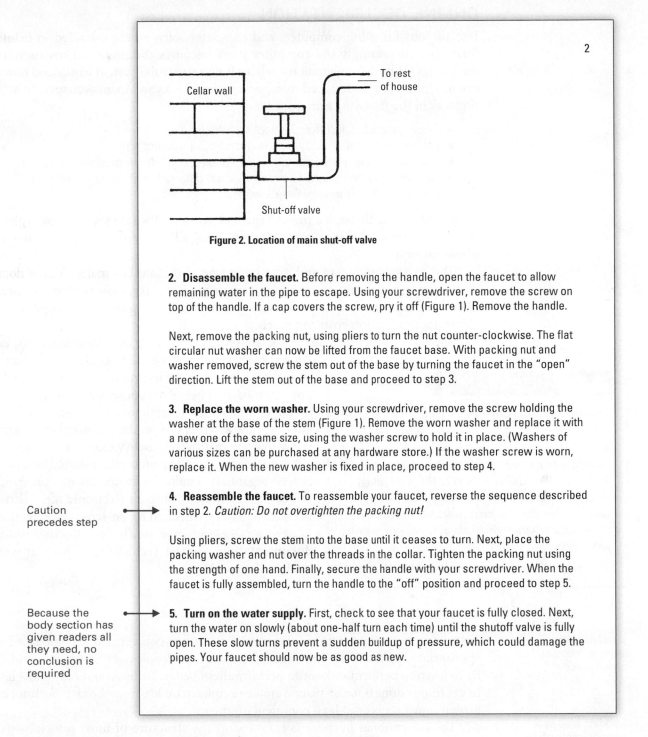

2

Figure 2. Location of main shut-off valve

2. Disassemble the faucet. Before removing the handle, open the faucet to allow remaining water in the pipe to escape. Using your screwdriver, remove the screw on top of the handle. If a cap covers the screw, pry it off (Figure 1). Remove the handle.

Next, remove the packing nut, using pliers to turn the nut counter-clockwise. The flat circular nut washer can now be lifted from the faucet base. With packing nut and washer removed, screw the stem out of the base by turning the faucet in the "open" direction. Lift the stem out of the base and proceed to step 3.

3. Replace the worn washer. Using your screwdriver, remove the screw holding the washer at the base of the stem (Figure 1). Remove the worn washer and replace it with a new one of the same size, using the washer screw to hold it in place. (Washers of various sizes can be purchased at any hardware store.) If the washer screw is worn, replace it. When the new washer is fixed in place, proceed to step 4.

4. Reassemble the faucet. To reassemble your faucet, reverse the sequence described in step 2. *Caution: Do not overtighten the packing nut!*

Using pliers, screw the stem into the base until it ceases to turn. Next, place the packing washer and nut over the threads in the collar. Tighten the packing nut using the strength of one hand. Finally, secure the handle with your screwdriver. When the faucet is fully assembled, turn the handle to the "off" position and proceed to step 5.

5. Turn on the water supply. First, check to see that your faucet is fully closed. Next, turn the water on slowly (about one-half turn each time) until the shutoff valve is fully open. These slow turns prevent a sudden buildup of pressure, which could damage the pipes. Your faucet should now be as good as new.

Caution precedes step

Because the body section has given readers all they need, no conclusion is required

Figure 16.2 A Set of Instructions for a Non-technical Audience

ONLINE DOCUMENTATION

Instructions for using computers and computer software are provided in printed form, but increasingly the computer itself becomes the preferred instructional medium. Online documentation explains how a particular system works and how to use it. Online help is designed to support specific tasks and to answer specific questions, as in the following examples:

◆ error messages and troubleshooting advice
◆ reference guides to additional information or instructions
◆ tutorial lessons that include interactive exercises with immediate feedback
◆ help and review options to accommodate different learning sites
◆ link to software manufacturer's website

Instead of leafing through a printed manual, users find what they need by typing a simple command, clicking a mouse button, using a help menu, or following an electronic prompt.

The cost of producing and distributing printed materials makes online documentation especially attractive to software producers. Also, most business software is sold through subscription, with new versions appearing regularly, so providing paper documentation costs too much.

Special software, such as ComponentOne's *Doc-to-Help* or Adobe *RoboHelp,* can convert print material to online help files that appear as dialogue boxes asking the user to input a response, or as pop-up or balloon help that appears when the user clicks on an icon for more information. Also, recent software advances allow writers to produce instructional documents that automatically adjust to various kinds of devices.

Like webpages, online information should be written in well-organized chunks. Paper documentation should not be merely dumped into an electronic file. Instructions need to be restructured for on-screen presentation.

For advice and case studies of effective online documentation, see the TechScribe website at **www .techscribe.co.uk**.

ON THE JOB...

Documentation across Platforms

"I've cut down the number of RoboHelp *outputs in my project from six or seven down to just one Multiscreen output—HTML5. In Adobe* RoboHelp 10, *Multiscreen HTML5 publishing displays an optimal view of content based on the user's screen size. No need to set up individual layouts for separate channels, screens, or separate job functions."*

—**Matt Sullivan, Independent Adobe Certified Instructor**

TUTORIALS

Tutorials emphasize task learning, not just performance

Tutorials present instructions placed within a given context and accompanied by appropriate explanations, illustrations, and self-test. People use instructions primarily to learn to perform tasks or to perform them better. Tutorials not only show how to get things done but also place a greater emphasis on learning the task. So tutorials explain most steps or at least comment on them.

Tutorials employ a distinctive structure

The educational purpose is reflected in the structure of most tutorials—full introductions, a modular breakdown, summaries, and post-tests. Glen Coulthard, the lead author and series editor of the *Advantage Series for Computer Education*, stresses that "each chapter or 'module' in our manuals is conceived as a tutorial. Each module opens with learning objectives and concludes with an assessment mechanism. We provide more than 'how-to' instructions; we educate the reader."

The order in which modules are presented depends on the topic area:

◆ Many tutorials use a *sequential* breakdown that mirrors the order in which tasks would be completed. Learn2's tutorial for *Group Wise 5.5 Messaging*, for example, starts with basics, such as "Exploring the toolbar" and "Using the mail functions," before teaching 32 other sequential tasks and concluding with "Utilizing dial sender" and "Converting other conversation place features" (**www.tutorials.com**).

◆ Another Learn2 online tutorial, Defensive Driving, uses a *lateral topical break-down*. After an introduction and pre-test, five sessions with titles, such as "What's in it for me?" and "Making choices you can live with," approach the subject from various personal angles. The series ends with a post-test.

◆ Meanwhile, Learn2's *Word 2007* Introduction uses a hierarchical approach: Chapter 1 deals with fundamentals such as opening *Word*, inserting text, and using various keys. Then, after laying this basic groundwork, the tutorial moves on to four specialized chapters at the next level in the hierarchy.

The sample tutorial excerpt in Figure 16.3 shows part of a tutorial's opening module. This excerpt's structure mirrors the tutorial's overall structure. Other tutorials are available at **www.learnthat.com**.

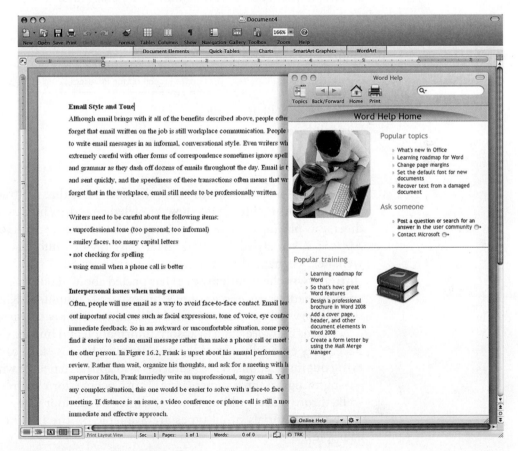

Figure 16.3 An Online Help Screen

Source: Microsoft product screenshot reprinted with permission from Microsoft Corporation.

Print versus Online Tutorials

Print's chief advantage is its convenience—people can use the tutorial any time, anywhere, without a computer. Readers can progress at their own pace, and can easily move back and forth within the tutorial pages. Self-directed tutorials are inexpensive, too; they don't require a trainer and, compared with web-based tutorials, they are cheap to write and require less extensive testing (Price and Korman 190).

Computer-based training (CBT) appeals to many users, however. Online tutorial users don't have to flip from book to monitor when learning computer software, for example. Often, the instructional screen and the corresponding software screen can be open simultaneously, in separate windows. Also, users report that they feel they are actually performing tasks, not just reading about them and imitating them (Price and Korman 191).

CBT tutorials have an additional technical advantage. Although linear structures suit the building of increasingly complex tasks in tutorials, web users often get restless. They would prefer free-flowing hyperlinks through a hierarchical structure. CBT tutorials can offer multipath structures that, for example, can provide "spur nodes" to digress into related topics and then return to the main tutorial (Farkas and Farkas 306, 309–10). "Split-joins," illustrated in Figure 16.4, allow users to choose which path to follow before rejoining the main tutorial.

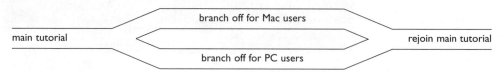

Figure 16.4 Split-Join Path
Source: Based on Farkas and Farkas 310.

Video Tutorials

One of the wonderful things about the internet is that it attracts all kinds of people who have an urge to share their knowledge and to instruct others. On topics as diverse as playing the guitar, basic car maintenance, and mirroring one's iPhone and Apple TV, self-styled "teachers" have posted video tutorials on YouTube and other websites. Perhaps the most viewed teacher is Sal Khan, whose Khan Academy (**http://www.khanacademy.org**) has posted about 4000 YouTube videos on subjects, such as math, science, economics, history, and computers. Khan's appeal is his clear, straightforward approach coupled with an evident joy of learning.

Videos allow learners to absorb information and implement instructions at their pace. They especially appeal to visual learners because they can see the results of carrying out instructions. Video tutorials also appeal to people who have difficulty visualizing things in three dimensions. Those people, for example, much prefer a video walk-through of a building to a two-dimensional floor plan.

Video tutorials of all types are available on the internet. Here are just a few examples:

> **http://about.jstor.org/video-tutorials**
> **http://www.mendeley.com/videos-tutorials**

http://www.pexsupply.com/resources/videos
http://www.teach-ict.com/videohome.htm
http://www.firestonecompleteautocare.com/car-care/videos.jsp

WEBINARS. One hybrid version of video tutorials is known as a *webinar*, which is essentially an online seminar. Webinars combine an audio lecture with a stream of visual illustrations. Most are delivered live, often followed by a discussion period that involves those who were tuned in. Many webinars combine information and ideas with an instructional purpose, such as this example that presents product information and training about a thermostat that controls a heat pump system:

http://www.tekmarcontrols.com/support/training/557introwebinar.html

---

GUIDELINES   **FOR CONDUCTING WEBINARS**

- ◆ If your company doesn't have its own webinar software, use a commercial site, such as *WebEx*, *GoToMeeting*, or Adobe *Acrobat Connect*.
- ◆ Upload and display *PowerPoint* or other presentation slides in real time during the webinar. Video can also help make your points.
- ◆ Or, draw and make notes on digital whiteboards.
- ◆ Use the webinar software's real-time chat tools to engage your audience.
- ◆ Make sure that all webinar participants will be available at the presentation's scheduled time.
- ◆ For those people who were unable to attend the webinar or would like to review its content, post the webinar online or on your company's internal network.
- ◆ Prepare your slides and other supporting materials in advance.
- ◆ If possible, ask one or two colleagues to sign in for a trial run the day before the presentation.

PROCEDURES

Procedures differ from instructions in two major respects:

1. The reader already knows how to perform the tasks outlined in the procedure and thus does not need detailed instructions.
2. The reader does not necessarily know the order in which tasks are to be performed or how and when to coordinate with other members of a work team.

Therefore, procedures are usually aimed at groups of people; the procedures describe how the group members will coordinate their activities or when they will perform their particular functions. Examples include evacuation procedures (in case of a fire or toxic spill), maintenance procedures, and installation procedures.

Not all procedures are aimed at groups, however. A software installation will likely be performed by a single technician who has the skills and general knowledge to do the job but who needs to know the specific variations of the procedure required by a given software program. Other procedures such as a union grievance or a procedure for reporting an on-site accident will assume that the reader needs to know whom to contact, when to contact, and what form to use. In these circumstances, the assumption is that the reader doesn't need to be told *how* to grieve the perceived infraction or *how* to phrase the accident report.

The excerpt from a group procedure in Figure 16.5 (page 293) comes from a Canadian power utility. The procedure demonstrates how workers must coordinate their skills with one another. In the conductor repair procedure, that coordination is essential in preventing the instant death that could result from a botched manoeuvre.

The Figure 16.5 procedure assumes that the crew knows how to "conduct a tailboard session," how to operate "the aerial device," and how to "test the insulators." The procedure is aimed at trained technicians who don't need instruction in performing the tasks. They need to know who does what, when, and with which equipment.

CHECKLIST

for Revising and Editing Instructions

Content

- Does the title promise exactly what the instructions deliver?
- Is the background adequate for the intended audience?
- Do explanations enable readers to understand what to do?
- Do examples enable readers to see how to do it correctly?
- Are the definition and purpose of each step given as needed?
- Is all needless information omitted?
- Are all obvious steps omitted?
- Do notes, cautions, or warnings appear whenever needed, before the step?
- Is the information rate appropriate for readers' abilities and the difficulty of this procedure?
- Do visuals adequately clarify the steps?
- Do visuals repeat prose information whenever necessary?

Page Design

- Does each heading clearly tell readers what to expect?
- Are steps single-spaced within and double-spaced between?
- Do white space and highlights set off discussion from steps?
- Is everything accurate?

- Are notes, cautions, or warnings set off or highlighted?
- Are visuals beside or near the step, and set off by white space?

Organization

- Is the introduction adequate without being excessive?
- Do the instructions follow the exact sequence of steps?
- Is all the information for a particular step close together?
- For a complex step, does each sentence begin on a new line?
- Is a conclusion necessary and, if so, adequate?

Style

- Do introductory sentences have enough variety to maintain interest?
- Does the familiar material appear first in each sentence?
- Do steps generally have short sentences?
- Does each step begin with an action verb?
- Are all steps in the active voice and imperative mood?
- Do all steps have parallel phrasing?
- Are transitions adequate for marking time and sequence?

CONDUCTOR REPAIR ON
72/138 Kv H-FRAME

(INSTALL PRE-FORMED ARMOUR SPLICES USING A SECOND UNIT TO SUPPORT CONDUCTORS)

1. Operate in a crew that has a minimum of four (4) certified Journeymen Power Line Technicians.

2. Conduct a tailboard session to confirm the order in which this procedure is to be performed.

3. Position the aerial device in the best position to maintain safe working clearances while allowing two line technicians to work satisfactorily on the line. Verify safe working clearances with the measuring stick, as outlined in the General Rules.

   NOTE: Position the aerial device directly under the centre phase. Back in with the turret two (2) metres from the centre of the structure. Park the boom truck, which has a wire holder on an insulated jib, on the other side of the structure in such a way that the truck can support each conductor without having to move the truck. Install unit grounding on both vehicles.

4. *Line technicians*—Carry all tools and materials along in the buckets.

5. *Line technicians*—Test the insulators. Install a temporary jumper if it's necessary.

6. *Second unit operator*—Support the outside conductor with the jib and raise the conductor slightly.

7. *Line technician in the insulated aerial bucket*—Move into position while maintaining both phase-to-phase and phase-to-ground clearance. Disconnect the clamp from the insulators with hot sticks.

8. *Second unit operator*—Move the conductor to a position where pre-formed armour rod may be installed without reducing the limits of approach. The line technicians may now bond on and complete the repairs.

9. *Line technicians*—Remove the bonds and back away. Move the conductor back to the insulators and use hot line tools to attach a suspension clamp to the insulators again.

10. Repeat this operation on the other outside phase.

11. *Line technicians*—Connect the centre phase and loosen the suspension clamp with hot sticks. The clamp may be slid out on the conductor to a point where the technician in the insulated aerial device can maintain clearances and bond on to the line. After replacing the suspension clamp, you may remove the bonds and move into position to use hot line tools to apply the armour rod.

Figure 16.5 Sample Technical Procedure

EXERCISES

1. Improve the readability of the following instructions by editing them for a more appropriate voice and design. Use the checklist above to evaluate the usability of instructions.

What to Do Before Jacking Up Your Car

Whenever the misfortune of a flat tire occurs, some basic procedures should be followed before the car is jacked up. If possible, your car should be positioned on as firm and level a surface as is available. The engine has to be turned off; the parking brake should be set; and the automatic transmission shift lever must be placed in "park," or the manual transmission lever in "reverse." The wheel diagonally opposite the one to be removed should have a piece of wood placed beneath it to prevent the wheel from rolling. The spare wheel, jack, and lug wrench should be removed from the luggage compartment.

2. Select a specialized process that you understand well and that has several distinct steps. Using the process analysis on pages 274–77 as a model, explain this process to colleagues who are unfamiliar with it. Begin by completing an audience/purpose profile. Some possible topics: how the body metabolizes alcohol; how economic inflation occurs; how the federal deficit affects our future; how a lake or pond becomes a swamp; how a volcanic eruption occurs.

3. Choose a topic from the following list, your major, or an area of interest. Using the general outline in this chapter as a model (pages 277–85), outline instructions that require at least three major steps. Address a general reader, and begin by completing an audience/purpose profile. Include (1) all necessary visuals; or (2) an art brief and a rough diagram for each visual; or (3) a "reference visual" (a copy of a visual published elsewhere) with instructions for adapting your visual from that one. (If you borrow visuals from other sources, provide full references.)

planting a tree	hitting a golf ball
hot-waxing skis	removing the rear wheel
hanging wallpaper	of a bicycle
filleting a fish	avoiding hypothermia

4. Select any one of the instructional visuals in Figure 16.1 (pages 279) and write a prose version of those instructions—without using visual illustrations or special page design. Bring your version to class and be prepared to discuss the conclusions you've derived from this exercise.

5. Locate an example of five or more visuals from the following list.
   A visual that shows

   ◆ how to locate something
   ◆ how to operate something
   ◆ how to handle something
   ◆ how to assemble something
   ◆ how to position something
   ◆ how to avoid damage or injury
   ◆ how to diagnose and solve a problem
   ◆ how to identify safe or acceptable limits
   ◆ how to proceed systematically

   Bring your examples to class for discussion, evaluation, and comparison.

COLLABORATIVE PROJECT

Any of the above exercises may be a collaborative project.

MyWritingLab

MyWritingLab: Technical Communication offers the best multimedia resources for technical communication in one, easy-to-use place. You can find a variety of interactive model documents and case studies. There are also extensive guidelines, tutorials, and exercises for Document Design, Writing and the Writing Process, and Research, and a large bank of diagnostics and practice for grammar review.

Manuals and Usability Testing

LEARNING OBJECTIVES

After reading this chapter, you should be able to

- understand the purpose of a range of manuals
- choose appropriate content and section titles for an assigned manual
- design a manual to be easily accessible and read
- use an appropriate set of criteria to test a manual's usability

This chapter describes manuals and how to write them, and then discusses aspects of testing manuals for their usability. Manuals consist primarily of instructions, either printed and bound in book form or posted online. Their use has increased, as products become increasingly complex. Even an electric wok comes with a manual. And relatively simple printers often come with 60-page operating manuals.

Why are manuals so common and so detailed? The majority exist to help users get the most out of purchased equipment. Also, well-written manuals endear the company to consumers. Further, companies that purchase equipment and software programs do so in order to improve productivity; that goal is undercut if employees can't properly use the equipment or software.

Equipment and software suppliers have found that clearly presented, "user-friendly" manuals reduce the number of complaints, inquiries, and warranty claims from customers. "Microsoft estimates the total cost of a single call answered by a help desk technician to be more than $20 per call" (Candib). In effect, good manuals pay for themselves. Poor manuals, on the other hand, over-tax the company's customer service staff and alienate users. Besides, badly written manuals require frequent revisions, an expensive process.

Even when a product supplier provides extensive employee training along with its product, user and maintenance manuals refresh employees' memories and reduce the incidence of equipment breakdowns.

ON THE JOB...

The Value of Well-Written Manuals

"Customers get a better experience when they purchase products that are accompanied by well-written, readable technical documentation. Also, there's a long-term overall benefit that comes from improving the precision of thinking by everyone in the organization. Precise thinking has a ripple effect on creativity, too."

—**Don Gibbs, CPO, tekmar Control Systems, which designs and manufactures controls for heating and cooling, solar thermal, and snow and ice melting systems**

TYPES OF MANUALS

There is a manual for every purpose. Table 17.1 shows a partial list.

Table 17.1 Types of Manuals and Their Purpose

Types of Manuals	Purpose
Software manuals	Help users navigate the intricacies of computer software
Operating manuals	Instruct users how to use and care for equipment and operating systems
Repair and service manuals	Provide detailed technical information for technicians (the complete service manual for your automobile is much more detailed and probably 20 times thicker than your operator's manual)
Maintenance and troubleshooting manuals	Provide users with detailed information on how to maintain equipment and troubleshoot common problems. They are vital for many businesses, including large-scale installations. For example, Doppelmayr's maintenance manual for its detachable chair-lift system provides several hundred pages of specific maintenance instructions, covering every aspect of the system's hardware and operation
Test manuals	Describe techniques, equipment, and ideal results for lab and field tests of everything from soil samples to sludge samples to electronic equipment
Installation manuals	Outline how to install complicated equipment or software
Construction manuals	Describe construction procedures, pertinent building codes, and contract requirements
Reference manuals	Allow users to easily find specific information
Tutorial manuals	Take trainees through skill-development exercises and explanations of how a system works
Documentation	Include tutorials and descriptions (instructions) of how to do things and explanations of how certain things work (process analysis)
Business practice manuals	Describe procedures for business operation, advertising policies, customer-service methods, collections and complaint handling, business travel procedures, and a multitude of business guidelines (One of the first, and most famous, business-practice manuals was written by Ray Kroc as he began to franchise the McDonald's restaurants across North America.)
User manuals	Provide detailed descriptions of how to use equipment. These are the most common types of manuals. If you want to program your DVD player, you turn to the manual; if you want to reconfigure your computer's hard drive, you refer to the appropriate manual.

MANUAL WRITING

A complex, collaborative process

Writing manuals is quite complex: many manuals are book-length and cover a variety of topics. Full-length manuals usually require more skills and time than one person can devote to the project. Even full-time manual writers collaborate with product designers, page designers, and graphic artists, all of whom add their specialized skills and perspective to the project.

The main challenge in writing (and designing) a manual is reducing the reader's reluctance to use the document; manuals, especially long ones, can overwhelm readers. So manual writing must incorporate all the elements of "user-friendly" instructions.

Planning

Audience/purpose analysis helps you choose the content and structure

You should begin the writing process by analyzing your manual's purpose. Each type of manual requires a primary purpose that you can clarify by analyzing who will read your manual and how they will use it. That analysis will guide your choice of what to include in the manual. The audience's needs and knowledge will later dictate the technical level and the amount of detail included in each section.

After your preliminary analysis of audience and purpose, you need to talk with the product developers and the project manager to learn more about the product and its uses. This consultative stage may take several meetings. Among other things, you need to determine

> ### ON THE JOB...
>
> **Manual Writing at Industry Canada**
>
> *"I spent a lot of time creating, updating, condensing, or expanding manuals. For example, our office spent $25 000 on a videoconferencing system, but there were no user manuals. I had to write the manuals because I was given the job of setting up the system and learning how to run it. When our office got a really good deal on Symantec's remote access software (we paid only for site licensing), again there were no manuals, so I wrote them. Also, I replaced outdated manuals and condensed some lengthy, complicated ones. Finally, I produced an instruction manual for the next co-op student...."*
>
> —Dave Parsons, electronics co-op student

- the product's features and specifications
- exactly how the product works
- the product's full range of uses and applications
- how to troubleshoot problems and malfunctions
- how to install or set up the product
- how to operate, maintain, and adjust the device
- how to obtain parts and accessories
- how to claim warranty work
- which aspects of the product's use could result in damage to the product itself or to the operator or to the system in which the product is installed

The manual should be organized in the order in which readers will use its various sections. A list of what will go in the manual should be finalized *before* a detailed writing outline is completed, and it's essential to examine (and possibly revise) that outline before composing the first draft. Also, you should show your completed outline to members of the manual production team before starting to write. The outline should have the entire headings system in place, along with the names and locations of all illustrations and a brief description of each paragraph or block of instructions.

Drafting

Writing the actual instructions and technical descriptions will be relatively easy if the outline has been thoroughly prepared. All the paragraphs and instruction blocks may be written by one person, or various sections might be written by their respective developers. In the latter case, one person or a small group of editors will revise and edit all of the sections to ensure consistent style and to avoid duplication. In some cases, the writer will create the manual's graphics, but most large organizations employ graphic artists to create artwork and technical illustrations.

Testing and Revising

Measurements of readability can be very helpful, but the best way to determine audience reaction is to have members of the target market test the manual. Testing is especially important for the instructions sections—ask readers to follow their literal interpretations of the instructions. You'll see which instructions are unclear or incomplete, and you can revise accordingly. Then, after revising the content and editing the phrasing, retest the manual on a new batch of readers to see if the revisions have helped.

The remaining task is to ruthlessly edit all the manual's phrasing to make it as direct, clear, and concise as possible.

PARTS OF MANUALS

Manuals contain body sections and supplementary sections.

Body Sections

Manuals do not strictly follow a beginning/middle/closing sequence. However, an introductory section (or two) is needed to orient readers.

INTRODUCTION.   At its beginning, a manual should include

- the manual's purpose and scope
- an overview of how to use the manual
- key definitions and principles
- introductory descriptions of equipment or procedures
- accompanying illustrations (photographs or diagrams of equipment, flow charts that show the relationships among sets of instructions)

Early on, a manual might also provide

- specifications for equipment and equipment operation
- background information, such as an overview of the equipment or procedures that are being replaced
- a list of accessories and/or related equipment

Use topical headings or talking headings

The heading title for an introductory section could be topical; for example, "Introductions and Specifications" would suit a user manual for a sophisticated piece of electronics test equipment, especially if the manual is aimed at electronics technicians. On the other hand, the opening sections for a CD player manual might use talking headings, such as

- Key Facts about Your CD Player
- How to Install the CD Player in Your Stereo System

REMAINING SECTIONS. The number of remaining sections varies with the manual's purpose and readers. Table 17.2 illustrates this principle.

Table 17.2 Section Variations

380-Page Operating and Service Manual for a Phillips Oscilloscope	60-Page Operating Manual for a Sony DVD Player	52-Page Onkyo Tuner/ Amplifier Instruction Manual
◆ Operating Instructions ◆ Theory of Operations ◆ Maintenance Instructions ◆ Calibration Instructions	◆ Getting Started ◆ Basic Operations ◆ Settings and Adjustments ◆ Additional Information	◆ Features, Safeguards, and Precautions ◆ Before Using This Unit ◆ System Connections ◆ Controls ◆ Operations ◆ Troubleshooting Guide ◆ Specifications

ORDER OF SECTIONS. The order of sections within a manual may be based on chronological development, order of importance, or spatial development, as illustrated in Table 17.3.

Table 17.3 Order of Manual Sections

Order	Example/Description
Chronological development	Sets of instructions for an electronic power supply could be provided in the order in which a user would use the device: ◆ Unpacking and Installing the Power Supply ◆ Connecting the Power Supply to a Load ◆ Operating and Adjusting the Supply ◆ Maintaining and Calibrating the Supply ◆ Troubleshooting Operational Problems ◆ Ordering Parts and Accessories
Order of importance	This structure is often used for reference manuals and test manuals. The most important (and most-used) sections are placed first.
Spatial development	The physical structure of a system can form the basis for organizing that system's corresponding manual. For example, the maintenance manual for a ski lift could be divided into sections, such as the following: ◆ Cable Maintenance ◆ Chairs and Grips ◆ Loading Ramps ◆ Drive Systems ◆ Housings

In organizing the components of a manual, you might consider creating two or more separate, related documents. For example, *Simply Accounting* software comes with three manuals: (1) Getting Started, (2) Workbook, and (3) User Guide.

CONCLUSION.   A separate conclusion section is not usually included at the end of a manual. If conclusion material is included, it is placed at the end of a body section. For example, the end of a set of repair instructions might indicate the standard length of time to perform the repair and might reiterate key steps to check if the repair hasn't solved the identified problem.

Supplementary Sections

Manuals include supplementary matter to help readers use the document. Depending on the type of manual, you should include some or all of the following *front matter*.

COVER AND TITLE PAGE.   Essentially, a cover's purpose is to protect the manual, but the cover also carries the manual's title and, where appropriate, the company name and logo, and equipment model number (Figure 17.1 on page 301). As Figure 17.2 on page 302 illustrates, a manual's corresponding title page carries more information. At a minimum, the title page names the document's intended use (in the title), the manufacturer's name and logo, the date (or year) of publication, and the name and model number of the product. Very seldom do manual writers' names appear on the title page. Well-designed covers carry design principles from the cover to the title page.

TABLE OF CONTENTS AND LIST OF ILLUSTRATIONS.   The longer and more complex a manual is, the more its readers need a table of contents and a list of illustrations to find particular topics within the manual.

WARNINGS AND CAUTIONS.   Manuals often require warnings and cautions, especially for electrically powered equipment. Often, these warnings will appear on the inside front cover. Sometimes, warranty information appears with the warnings and cautions.

GLOSSARY.   A glossary might appear in the front matter, if readers need to know the meaning of certain terms before starting to use the manual. Alternatively, the introductory section could define key terms.

Depending on the type of manual, you may need to include some *back matter*.

BACK MATTER.   A manual's back matter might include some or all of the following:

- information on replacement parts and ordering
- schematics
- contact details for dealers and service depots
- service log
- component specifications tables
- construction codes

There are as many possible combinations of back matter as there are types of manuals. You can determine the best combination of back matter sections for your manual by examining a variety of manuals in your field.

INDEXES.   Indexes are commonly used in long or complex manuals, such as computer user manuals.

CUTTING EDGE TECHNOLOGIES
Operating and Service Manual

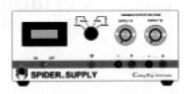

SPIDER™ SUPPLY

Model 001a

Figure 17.1 Sample Manual Cover
Source: Courtesy of Curt Willis.

OPERATING AND SERVICE MANUAL

SPIDER™ SUPPLY

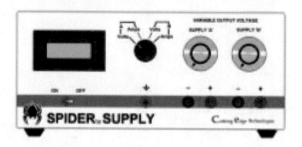

Variable Dual 20-Volt Channel Voltage Regulated Power Supply Model 001a Series 0.1.0-98a

Copyright © Cutting Edge Technologies 1998
4131 Web Crescent, Kelowna, British Columbia V1W 1V8

Figure 17.2 Sample Manual Title Page
Source: Courtesy of Curt Willis.

FORMAT CONSIDERATIONS

In many ways, effective manuals use the same design principles that should be applied to formal reports and books. However, manuals have their own formatting requirements.

Section Identification

In order to constantly orient your manual's readers, use headers (or footers) that combine the page number with a section identifier. (This text uses a version of that system.) Your readers will then have three methods to easily find particular sections in the manual:

1. the table of contents
2. the index
3. the page headers/footers

Figure 17.3 below and Figure 17.4 on the next page show alternative methods of designing section identifiers. (The example illustrated in Figure 17.4 does not show that the page numbers appear at the bottom of the page in that manual.) Some

---

Maintenance **5**

 Caution: If the voltage continues to increase greatly between 80% and 100%, immediately turn off the variac and find the source of the problem before continuing. If you hear or smell anything suspicious, turn off the variac immediately.

7. Once the voltage reaches 100%, turn the Channel A voltage adjustment knob fully counter clockwise. The DMM should read between $1V_{DC}$ and $1.5V_{DC}$.

8. Repeat Steps 1 through 7 for Channel B.

5.3   Ammeter Calibration

5.3.1 Setup procedure

1. Attach the positive output of the channel you are calibrating to the positive channel of 1 DMM. Set this DMM on the $2A_{DC}$ range.

2. Attach the negative output of the "Amps" DMM to one side of the load resistor, and hook the other side of the load resistor to the negative output of the channel you are calibrating.

Figure 17.3 A Section Identifier

Section 6: CALIBRATION

To keep your SPIDER SUPPLY running at an optimal level, you may need to calibrate it. Calibrating the power supply is simple and will not likely need to be performed often.

Several possible circumstances could cause your power supply to drift from its specified settings: extreme temperature changes, over-drawing output current, abuse, or long-term use. The following calibration instructions will help you bring your supply back up to specifications, should the need arise.

6.1 Non-powered Tests

If you plan to calibrate your supply, you should do these quick no-power tests as well. They take little time and require minimal equipment.

To ensure that all connections are indeed *connected*, perform the following checks:

✓ Check every component solder joint and header connection to confirm that they are sound and properly affixed.

✓ Check the following resistances with a DMM:

6.1 Non-powered Tests continued . . .

✓ Confirm the following resistances with a DMM:

Table 6-5 Variable Voltage Resistor (Dials) Checks

Point	Front Panel Voltage Dial Adjustment	Resistance Measurement
Between J1 & J2	SUPPLY 'A' fully CCW	150 Ω
Between J1 & J2	SUPPLY 'A' fully CW	2.60 Ω
Between J3 & J4	SUPPLY 'B' fully CCW	150 Ω
Between J3 & J4	SUPPLY 'B' fully CW	2.60 Ω

Figure 17.4 Variations in Section Identifiers Used for the First and Second Pages of a Section
Source: Courtesy of Curt Willis.

manuals also use sturdy card stock or plastic tabs along the right-hand edge to distinguish one section from another.

Headings System

Headings and subheadings are particularly valuable in lengthy, complex manuals, so make headings stand out by using boldface or colour. Clearly distinguish among levels of headings by using capitalization, type size, and heading placement. You might also use one kind of typeface for the main headings and another kind for the other levels of headings. Figure 17.5 shows colour and subtle shadowing used in an electronics student's manual for a power supply.

Flashy graphics attract attention, but the headings must also be clearly phrased to be usable. Notice how effectively the following heading (from Hewlett-Packard's *HP DeskJet 720 Series Printer User's Guide*) signals the content of its section:

How to Print on Different Paper Sizes

Compare that heading with much less useful headings, such as

Printing and Paper Size Variations

OR

Paper Differentiation Techniques

Page Layout

Chapter 16 discusses effective page design for instructions. Follow that advice, but also remember to use a format that can comfortably accommodate both point-form instructions and paragraphed explanations.

Section 1: INTRODUCTION

We hope you're as excited as we are about the new 2011 version of the industrious SPIDER SUPPLY. The SPIDER SUPPLY Dual 20VDC Voltage Regulated Power Supply is a reliable, user-friendly DC output device.

1.1 Features

The SPIDER SUPPLY comes with many easy-to-use features for your convenience. The new liquid crystal digital display allows you to view the output with more clarity and accuracy than our past analogue meters.

1.1.1 The Basics (batteries are included)

Our new voltage regulated 20 volt DC supplies either a single or dual channel output, ideal for use with OP-Amp circuits. Your power supply also includes a circuit common ground, as well as a true earth ground. And new cutting-edge regulators in accordance with our vented case result in improved heat dissipation, and thus a more durable output.

Figure 17.5 Colour and Subtle Shadowing Used to Distinguish Headings
Source: Courtesy of Curt Willis.

HINTS FOR EFFECTIVE PAGE LAYOUT. Most of the following useful hints for effective page layout are illustrated in Figures 17.1 to 17.4 (pages 301–305):

- ◆ Use a lot of white space.
- ◆ Use pictures, diagrams, or graphs to show the results of the manual user's actions as well as how to achieve those results.
- ◆ Place lengthy explanations in boxes.
- ◆ Use fonts wisely:
  a. boldface, italicize, or change type size before changing font type
  b. use strong fonts (Times New Roman, Verdana, **Impact**)
- ◆ Where appropriate, use colour for headings.
- ◆ Phrase headings clearly and place text after all headings.

Symbols and Design Graphics

Hints, warnings, explanations, and special instructions can be highlighted by placing them in boxes and by using symbols to capture the reader's attention. Think of how you might use some of the following symbols that are found in Microsoft Word's "Insert" pull-down menu. Many more symbols are available on the internet.

DESIGN GRAPHICS BENEFITS. Design graphics, such as those illustrated in Figures 17.3 (page 303), 17.4 (page 304), and 17.5 (page 305), provide three main benefits:

1. They contribute to a professional appearance.
2. They help set up a balanced page layout.
3. They help create an inviting, accessible look that encourages readers to use the manual.

DESIGN TECHNIQUES. The simplest design technique is to use italics, boldface, underlining, and quotation marks for special effects, and to use each of those techniques for one effect. For example, you could use

- ◆ italics for emphasizing key words or phrases
- ◆ boldface for headings and definitions
- ◆ underlining for names of printed publications or for emphasizing important words, warnings, and so on
- ◆ quotation marks for quoted words or for words used in a special way

Remember to use such techniques consistently within a manual. For example, don't use italics to emphasize a phrase in one place and boldface to emphasize a phrase in another place.

Of all the kinds of documents produced by technical writers, a manual is the most likely to be *used*, not just read. Therefore, the document's usability needs to be tested before publication and distribution. A *usable* document is safe, dependable, and easy to use.

Most of the following advice for testing a manual's usability can be applied to other types of documents, such as product brochures, catalogues, stand-alone instruction sheets, and technical reports.

USABILITY TESTING

Companies routinely measure the usability of their products—including the *documentation* that accompanies the product (warnings, explanations, instructions for assembling or operating, and so on). To keep their customers—and to avoid lawsuits—companies go to great lengths to identify flaws or to anticipate all the ways a product might fail or be misused.

What a usable document enables readers to do

The purpose of usability testing is to keep what works in a product or a document and to fix what doesn't. For a document to achieve its objectives—to be *usable*—users must be able to do at least three things (Coe 193; S. Spencer 74):

- ◆ easily locate the information they need
- ◆ understand the information immediately
- ◆ use the information successfully

To measure usability in a set of instructions, for instance, we ask this question: *Do these instructions enable all users to carry out the task safely, efficiently, and accurately?*

Ideally, usability tests for documents occur in a setting that simulates the actual situation, with people who will actually use the document (Redish and Schell 67; Ruhs 8). But even in a classroom setting, we can reasonably assess a document's effectiveness on the basis of the usability criteria discussed below.

How Usability Testing Is Done

Usability testing usually occurs at two levels (Petroski 90): (1) *alpha testing*, by the product's designers or the document's authors and (2) *beta testing*, by the actual users of the product or document. At the beta level, two types of testing can be done: *qualitative* and *quantitative*.

QUALITATIVE TESTING. Qualitative testing assesses what users think about the document by observing how they react to the document or what they say or do. Qualitative testing for usability is done in two ways: through focus groups and protocol analysis ("Testing Your Documents" 1–2):

Two types of qualitative testing

- ◆ *Focus groups*. Based on a list of targeted questions about the document, a small group of readers discuss what information they think is missing or excessive, what they like or dislike, and what they find easy or hard to understand. They might also suggest revisions to the document. When group discussion is not possible, mail surveys can be effective for obtaining responses from a diverse and dispersed group of users.
- ◆ *Protocol analysis*. In a one-on-one interview, a user is asked to read a specific section of a document and then explain what he or she thinks that section means. For long documents, the interviewer also observes how the person actually reads the document: for example, how often she or he flips pages or refers to the index or table of contents to locate information.

Another type of protocol analysis asks users to think out loud as they perform the task; comments are recorded electronically or via the observer's notes (Ostrander 20).

QUANTITATIVE TESTING.   Quantitative testing yields actual numerical data about a document's usability by measuring the performance of a control group: for example, the comparative success rates among people using different versions of a document or the number of people who performed the task accurately ("Testing Your Documents" 2–4). Other items that can be measured include the time required to complete a given task and the types and frequency of user errors (Hughes 489). Tests of this kind are obviously more complicated, time-consuming, and expensive than qualitative tests.

When to Use Which Test

Qualitative testing shows which parts of the document work or don't work

If time, budget, and available users allow, both qualitative and quantitative testing should be done. Quantitative testing is ordinarily done last, as a final check on usability. Each test has its benefits and limitations: "control testing will tell you *if* the new document is a success, but it won't tell you *why* it is or isn't a success" ("Testing Your Documents" 3). In short, to find out if the document succeeds as a whole, use quantitative testing; to find out exactly which parts of the document work or don't work, use qualitative testing.

How Usability Criteria Are Determined

Usability criteria for workplace documents are based on answers to these questions about the task, the users, and the setting:

Questions for assessing usability

- ◆ What specific task is being documented?
- ◆ What do we know about the users' abilities and limitations?
- ◆ In what setting will the document be read/used?

Only after these questions are answered can we decide on criteria for evaluating the effectiveness of a specific document, as shown in Figure 17.6.

Type of Task	**User Characteristics**	**Constraints of the Setting**
• Learn facts	• Motivation/attitude	• Distractions
• Understand concepts	• Prior knowledge/skills	• Interruptions
• Follow directions	• Experience	• Other constraints
• Make judgments	• Limitations	• Ways the document will be read
• Make decisions	• Cultural background	• Things that can go wrong

Specific Usability Criteria for This Document
- • Worthwhile content
- • Sensible organization
- • Readable style
- • Accessible design
- • Ethical, legal, and cultural considerations

Figure 17.6 Defining a Document's Usability Criteria

The Checklist for Usability (see page 311) identifies criteria shared by many technical documents. In addition, specific elements (visuals, page design) and document types (proposals, memos, instructions) have their own usability criteria.

GUIDELINES   **FOR TESTING A DOCUMENT'S USABILITY**

1. *Identify the document's purpose.* Determine how much *learning* versus *performing* is required in the task—and assess the level of difficulty (Mirel, Feinberg, and Allmendinger 79; Wickens 232, 243, 250):
   ◆ *Simply learning facts.* "How rapidly is this virus spreading?"
   ◆ *Mastering concepts or theories.* "What biochemical mechanism enables this virus to mutate?"
   ◆ *Following directions.* "How do I inject the vaccine?"
   ◆ *Navigating a complex activity that requires decisions or judgments.* "Is this diagnosis accurate?" "Which treatment option should I select?"

2. *Identify the human factors.* Determine which characteristics of the user and the work setting (human factors) enhance or limit performance (Wickens 3):
   ◆ *Users' abilities/limitations.* What do these users know already? How experienced are they in this task area? How educated? What cultural differences could create misunderstanding?
   ◆ *Users' attitudes.* How motivated or attentive are they? How anxious or defensive? Do users need persuading to pay attention or be careful?
   ◆ *Users' reading styles.* Will users be scanning the document, studying it, or memorizing it? Will they read it sequentially (page by page) or consult the document periodically and randomly? (For example, expert users often look only for keywords and key concepts in titles, abstracts, or purpose lines, and then determine what sections of the document they will read.)
   ◆ *Workplace constraints.* Under what conditions will the document be read? What distractions or interruptions does the work setting pose? (For example, consider procedures for treating a choking victim, posted in a busy restaurant kitchen.) Will users always have the document in front of them while performing the task?
   ◆ *Possible failures.* How might the document be misinterpreted or misunderstood (Boiarsky 100)? Any potential trouble spots (material too complex for these users, too hard to follow or read, too loaded with information)?
   For a closer look at human factors, review the audience/purpose profile sheet.

3. *Design the usability test.* Ask respondents to focus on specific problems (Hart 53–57; Daugherty 17–18):
   ◆ *Content.* "Where is there too much or too little information?" "Does anything seem inaccurate?"

(continued)

◆ *Organization.* "Does anything seem out of order, or hard to find or follow?"
◆ *Style.* "Is anything hard to understand?" "Are any words inexact or too complex?" "Do any expressions seem wordy?"
◆ *Design.* "Are there any confusing headings, or too many or too few?" "Are any paragraphs, lists, or steps too long?" "Are there any misleading or overly complex visuals?" "Could any material be clarified by a visual?" "Is anything cramped and hard to read?"
◆ *Ethical, legal, and cultural considerations.* "Does anything mislead?" "Could anything create potential legal liability or cross-cultural misunderstanding?"

4. *Administer the test.* Have respondents performed the task under controlled conditions? To ensure dependable responses, look for these qualities (Daugherty 19–20):
◆ *A range of responses.* Select respondents at various levels of expertise with this task.
◆ *Independent responses.* Ask each respondent to work alone when testing the document and recording the findings.
◆ *Reliability.* Have respondents perform the test twice.
◆ *Group consensus.* After individual testing, arrange a meeting for respondents to compare notes and to agree on needed revisions.
◆ *Thoroughness.* Once the document is revised/corrected, repeat this entire testing procedure.

Usability Issues in Online or Multimedia Documents

In contrast to printed documents, online or multimedia documents have unique usability issues (Holler; Humphreys 754–55):

◆ Online documents tend to focus more on "doing" than on detailed explanations. Workplace readers typically use online documents for reference or training rather than for study or memorizing.
◆ Users of online instructions rarely need persuading (say to pay attention or follow instructions) because they are guided interactively through each step of the procedure.
◆ Visuals play an essential role in online instruction.
◆ Online documents are typically organized to be read interactively and selectively rather than in linear sequence. Users move from place to place, depending on their immediate needs. Organization is therefore flexible and modular, with small bits of easily accessible information that can be combined to suit a particular user's needs and interests.

◆ Online users can easily lose their bearings. Unable to shuffle or flip through a stack of printed pages in linear order, they need constant orientation (to retrieve some earlier bit of information or to compare something on one page with something on another). In the absence of page or chapter numbers, an index, or a table of contents, "Find," "Search," and "Help" options need to be plentiful and complete.

CHECKLIST

for Usability

Content

◆ Is all material relevant to this user for this task?
◆ Is all material technically accurate?
◆ Is the level of technicality appropriate for users?
◆ Are warnings and cautions inserted where needed?
◆ Are claims, conclusions, and recommendations supported by evidence?
◆ Is the material free of gaps, unclear passages, or needless details?
◆ Are all key terms clearly defined?
◆ Are all data sources documented?

Organization

◆ Is the structure of the document visible at a glance?
◆ Is there a clear line of reasoning that emphasizes what is most important?
◆ Is material organized in the sequence that users are expected to follow?
◆ Is everything easy to locate?
◆ Is the material "chunked" into easily digestible parts?
◆ Is each sentence understandable the first time it is read?

Style

◆ Is content-rich information expressed in the fewest words possible?
◆ Are sentences put together with enough variety?
◆ Are words chosen for exactness, and not for camouflage?
◆ Is the tone appropriate for the situation and users?

Design

◆ Is page design inviting, accessible, and appropriate for users' needs?
◆ Are there adequate aids to navigation (headings, lists, type styles)?
◆ Are there adequate visuals to clarify, emphasize, or summarize?
◆ Do supplements accommodate the needs of a diverse audience?

Ethical, Legal, and Cultural Considerations

◆ Does the document indicate sound ethical judgment?
◆ Does the document comply with copyright law and other legal standards?
◆ Does the document respect users' cultural diversity?

EXERCISES

1. Select part of a technical manual in your field, or instructions for a general reader, and make a copy of the material. Use the "Checklist for Usability" above to evaluate the sample's usability. In a memo to your instructor, discuss the strong and weak points of the instructions, or be prepared to explain to the class why the sample is effective or ineffective.

2. Create an operating or user manual for a product or system you've developed in one of your classes. Alternatively, write a procedures manual for a series of related techniques and procedures that you've learned in your program of studies. (Examples could include CAD methods, test procedures, fabrication techniques, design principles, or procedures for complying with government regulations.)

COLLABORATIVE PROJECT

Test the usability of a document prepared for your technical communication course:

1. As a basis for your "alpha" test, adapt the guidelines on pages 309–310, the audience/purpose profile (page 33), and the specific parts of the "Checklist for Usability" that apply to this particular document.

2. Revise the document based on your findings.

3. Appoint a group member to explain the usability testing procedure and the results to the class.

MyWritingLab

MyWritingLab: Technical Communication offers the best multimedia resources for technical communication in one, easy-to-use place. You can find a variety of interactive model documents and case studies. There are also extensive guidelines, tutorials, and exercises for Document Design, Writing and the Writing Process, and Research, and a large bank of diagnostics and practice for grammar review.

18

Proposals

LEARNING OBJECTIVES

After reading this chapter, you should be able to

- distinguish among sales proposals, research proposals, and improvement proposals
- understand the process of generating solicited proposals
- follow guidelines in planning, phrasing, and designing informal and formal proposals
- adapt the structure for a standard proposal to a particular situation

✱ Explore
Click here in your eText to access a variety of student resources related to this chapter.

A proposal offers to do something or recommends that something be done. A proposal's general purpose is to improve conditions, authorize work on a project, present a product or service (for payment), or otherwise support a plan for solving a problem or doing a job.

Your proposal may be a letter to your university's dean of engineering to suggest an interdisciplinary program of studies linked with the computer science and arts programs; it may be a memo to your firm's general manager to request funding for computer training; or it may be a 100-page document to the provincial highways ministry to bid for a contract to design a series of highway overpasses. You may write the proposal alone or as part of a team. It may take hours or months.

TYPES OF PROPOSALS

Proposals are as varied as the situations that generate them, but we can identify three main types:

1. *sales proposals*, which promote business
2. *research proposals*, which are used in academic institutions and in business
3. *improvement proposals* (or *planning proposals*), which suggest how to improve situations or ways of doing things, or which present solutions to problems

Proposals may be *internal* (written for members of an organization) or *external* (written for others outside the organization). In either case, the proposal writer has reasons for writing the proposal. But, internal or external, the success or failure of the proposal depends almost entirely on the writer's ability to look at the situation from the *reader's* point of view. Table 18.1 on page 314 shows a variety of proposals whose writers have tried to use supporting arguments that appeal to their readers.

Some of the proposals in Table 18.1 have been solicited by their readers. These proposals may have to be presented differently from unsolicited proposals that have likely not been anticipated by their receivers.

Table 18.1 Types of Proposals

Type	Internal	External
Sales	The design unit in a company that manufactures auto parts solicits proposals for testing new magnesium-alloy brake parts. In keeping with the company's budgetary policies, the company's design unit can choose the company's own research lab's proposal or an external lab's proposal. Either way, the design unit will be billed for the testing. If the company's lab wins the testing contract, it gains credits for salaries and new equipment.	An engineering firm that specializes in environmental studies proposes the methods, timeline, staffing, and budget for assessing the potential environmental impact of a proposed golf course. The proposal responds to a Request for Proposal (RFP) issued by the provincial Ministry of the Environment. The RFP lists several pages of requirements that a successful proposal must meet.
Research	A truck-manufacturing company's engine design team proposes a project to determine methods of reducing engine vibration. In arguing its case, the design team points to negative customer and dealer feedback about excessive engine vibration and noise. The design team also cites examples of how engine vibration has created warranty problems for the manufacturer.	A university physicist proposes a four-year computer modelling study of the properties of surface molecules. The physicist knows that her readers at the Natural Sciences and Engineering Research Council of Canada will understand her highly technical proposal, but for possible public consumption she attaches an executive summary that explains the potential applications of her research on the upper ozone layer.
Improvement	A research chemist in a petroleum firm's product-development branch proposes a new way of blending methanol with gasoline. This new blending method would reduce production costs. In response to a supervisor's request, a fisheries biologist proposes a software package that tracks costs of extended research studies.	A software-development firm, after reading newspaper reports about sailing delays and operating-cost overruns for an East Coast Ferries fleet, proposes a software program that fully coordinates the ferries' sailing schedules, maintenance schedules, staff training, supply loading, and fuelling. Before submitting the proposal, the software firm's partners learn all they can about who screens proposals at the ferry corporation.

A *solicited proposal* is received by a client who is not surprised or annoyed to receive that proposal. However, the client usually states definite requirements and expects to see those requirements met. Also, often, the client indicates the conditions and/or the format for the proposal. In these cases, the proposal writer has to pay strict attention to the reader's expectations.

The writer of an *unsolicited proposal* has a different challenge: often, the reader initially feels reluctant to accept the proposal. Perhaps the reader doesn't see a problem or is happy with the current way of doing things. Perhaps the reader is not aware of the product you have to sell, or has other priorities. However, if you can convince the reader of the need for your proposed solution, you have a chance of persuading the reader. If you can't establish that need in the reader's mind, there's no point in proposing how to meet the need!

THE PROPOSAL PROCESS

The basic proposal process for a solicited proposal can be summarized simply: someone offers a plan for something that needs to be done. In business and government, this process has five stages:

1. Client X needs a service or product.
2. Client X draws up detailed requirements and advertises its need in a Request for Proposal (RFP).
3. Firms A, B, and C research Client X and its needs. (Likely they will request a more detailed version of the RFP.)
4. Firms A, B, and C propose a plan for meeting the need.
5. Client X awards the job to the firm offering the best proposal.

The complexity of each phase will, of course, depend on the situation. In practice, the process could look like the following.

CASE   **Responding to a Request for Proposal**

The Situation. Tony Mutu, a civil engineering technologist with seven years of experience, works for EnviroMax Engineering Consultants in Penticton, in British Columbia's Okanagan Valley. On February 2, 2014, Tony's supervisor, Ken Guenther, assigns him to plan and write a proposal responding to an RFP that appears in a local newspaper and on the District of Beachland's website.

District of Beachland
Request for Proposal
17PW—2014d

The District of Beachland requires restoration of the beach and protection of the beach, shore, and foreshore at 2056 Beach Avenue, Town of Beachland. The proposal must describe a method of repairing the current damage. The firm that wins the contract will design the improvements and supervise the Beachland Public Works crew, who will carry out the restorative measures.

Submitted proposals must
- describe a long-term solution to the erosion problem
- provide a large, protected swimming area and a protected sandy beach
- show how the work will be completed by June 20, 2014, in time for the summer tourist season
- be submitted to the District of Beachland by February 21, 2014
- use Beachland Public Works equipment and employees as much as possible, to control costs

Detailed information and a guided site inspection may be requested from

John Warren, Public Works Manager
District of Beachland Public Works
Box 732
Beachland, British Columbia V9H 1X0
Phone: 250-792-2343   pubworks@beachland.ca

Tony reads reports about Lake Okanagan water levels, stream inlets, prevailing winds, and erosion control measures, and meets John Warren. Tony's site visit and a 2013 report about restoration measures reveal the following information:

1. The shoreline at 2054 Beach Avenue receded 2 metres from 2001 to 2012, an average of 19.3 centimetres per year. However, measurements taken by the owners of 2054 and 2056 show that the shoreline has eroded an average of 33 centimetres per year for the past four years.
2. Beachland businesses noticed a marked decline in business in 2012 and 2013.
3. In September 2010, the District of Beachland placed 21 tonnes of fine sand on the beach at 2054 Beach Avenue. By 2013, few traces of that sand remained.

Next, Tony consults Ken Guenther and other EnviroMax staff to learn who will be available to lend their expertise to the project. Also, Tony gleans information from

◆ a visit to the Beachland Public Works (BPW), to examine BPW heavy equipment and BPW's stockpile of blast rock, gravel, and sand;

- a visit with the District's financial controller, Donna Potter;
- consultations with Bettina Chiu, biologist, and Tom Orman, hydrologist, of the B.C. Ministry of the Environment;
- responses to Tony's email inquiries about heavy equipment rental, underwater video camera rental, and marine construction insurance;
- Ministry of the Environment records—Okanagan water levels, 1946 to 2011;
- an EnviroMax report about the building of a breakwater for Holiday Houseboats' marina on Shuswap Lake; and
- a preliminary study, by Tony and civil technologist Theresa Mucci, regarding current water depths and soil depths offshore in front of 2056 Beach Avenue.

Again, Tony consults Ken Guenther and EnviroMax's Doug Nygren. Together, they reject two short-term solutions and choose a long-term solution described by Tony, even though it's considerably more expensive than the other two. Ken endorses Tony's proposed fee structure and the involvement of several EnviroMax personnel. Tony is nearly ready to write the proposal.

Other firms would also bid for a project such as the one described above. The client would award it to the firm submitting the best proposal, based on the criteria listed in the RFP, and the following criteria, which are generally used to assess proposals:

- understanding of the client's needs
- soundness of the firm's technical approach
- quality of the proposed project's organization and management
- ability to complete the job by the deadline
- ability to control costs
- specialized experience of the firm in this type of work
- qualifications of staff assigned to the project
- the firm's record for similar projects

NOTE　*Sometimes, an RFP will list the client's specific criteria in a point scale, as in the table on the following page, an excerpt from an RFP for harvesting timber on Crown land.*

Some clients hold a pre-proposal conference for the competing firms. During this briefing, the firms are informed of the client's needs, expectations, specific start-up and completion dates, criteria for evaluation, and other details to guide proposal development. Such a conference does not suit John Warren's casual style of doing business.

Nor does the District of Beachland office do what some municipal and government agencies do: restrict a given project to a list of pre-selected firms. Usually, this is done on a rotation basis—if, for example, a city department has 15 firms on its contractor list, it might send invitations to bid to the first five firms on the list, and then to the next five firms for the next project, and so on. Or an agency might require firms to present their qualifications for a certain kind of project before actually soliciting proposals for it. Only those firms that clear the qualifications hurdle will be allowed to submit a proposal.

In the Beachland case, Tony Mutu knows that the merits of his firm's proposed plan will be evaluated solely on the basis of what he puts on paper. Still, it helps to know who will make the final decision about the competing proposals. John Warren is a no-nonsense person who dislikes high-pressure sales tactics. With that in mind, Tony writes a formal proposal that looks very professional and that lays out EnviroMax's full case, but that is as brief as possible. Tony uses businesslike prose that is free from unnecessary jargon.

His proposal appears in this chapter, starting on page 335.

Criteria	Weighting
Employment	30
Proximity	10
Existing plant	10
New capital investment	10
Labour value-added	10
Change in value-added	20
Revenue	10
Total weighting	100

All applicants must submit a proposal that contains a business case for lumber manufacturing or specialty wood products manufacturing and addresses the development objectives of the Crown.

Source: Adapted from a format used by the Ministry of Forests, Province of British Columbia.

ON THE JOB...

Evaluating Proposals

"Many proposals are evaluated by a 'scoring grid' that assigns specific values to each section of the document. The judging criteria are provided in advance, allowing the writer a reference guide to ensure each aspect of the proposal is properly addressed. Several committees will read and score the competing proposals so it is vital to provide all the pertinent technical and written information in a clear and concise format. Readability is critical. . . . The reader must be able to easily understand the content the first time through because the proposal may not get a second chance. . . ."

—**Norm Metcalf, manager of a non-profit agency**

PROPOSAL GUIDELINES

A reader will evaluate your proposal according to how clearly, informatively, and realistically you answer these questions:

- What are you proposing?
- What problem will you solve?
- Why is your plan worthwhile?
- What is unique about your plan?
- What are your (or your firm's) credentials?
- How will the plan be implemented?
- How long will the project take to complete?
- How much will it cost?
- How will we benefit if we accept your plan?

In addition to answering the questions above, successful proposal writers adhere to the following guidelines. This chapter's three sample proposals illustrate the guidelines.

GUIDELINES   **FOR WRITING PERSUASIVE PROPOSALS**

1. *Signal your intent with a clear title.* Decision makers are busy people who have no time for guessing games. The title should clearly signal the proposal's purpose and content. Don't write "Recommended Improvements" when you mean "Recommended Wastewater Treatment." A specific, comprehensive title signals the proposal's intent.

2. *Design an accessible and appealing format.* Format includes such features as
   - the layout of words and graphics, including white space
   - font and font size

(continued)

◆ margins, spacing, headers, and footers
◆ headings and illustration labels
◆ highlights and lists

A poorly designed proposal suggests a careless attitude toward the project. Well-designed proposals help readers quickly find what they need.

3. *Focus on audience needs.* Readers want specific suggestions for filling specific needs. Their biggest question is "What's in this for me?" Show them that you understand their problem and offer a plan for improving their products, sales, or services.

4. *Explain the benefits of implementing your plan.* A persuasive proposal shows readers how they (or their organization) will benefit by adopting your plan. Relate those benefits to the factors that are critical to their business success. Also, *analyze your audience's major concerns* and anticipate likely questions and objections. Such concerns might include personal concerns ("This proposal means a lot of extra work for me!") as well as role concerns ("I like this proposal, but I can't see how we can fit it into this year's budget").

5. *Provide concrete, specific information.* Vagueness in a proposal is a fatal flaw. Be sure to spell things out. *Show* as well as *tell*. Instead of writing "We will install state-of-the-art equipment," write "To meet your automation requirements, we will install 12 Dell PC computers with 400-GB hard drives. The system will be networked for rapid file transfer between offices. We will interconnect four HP laser printers and one HP DeskJet colour printer." To avoid any misunderstanding and to reflect your ethical commitment, a proposal must elicit *one* interpretation only.

6. *Treat contingencies and limitations realistically.* Do not underestimate the project's complexity. Identify contingencies (occurrences subject to chance) that readers might not anticipate, and propose realistic methods for dealing with the unexpected. If the best available solutions have limitations, let readers know. Otherwise, you and your firm could be liable in the case of project failure. Avoid overstatement. If you can't guarantee "eliminate," then write "diminish" or "help avoid."

7. *Propose realistic timetables and budgets.* Complex projects need to be carefully planned and accurately represented, perhaps in a Gantt chart. Also, provide a realistic, accurate budget, with a detailed cost breakdown of all costs that can be anticipated. Indicate variables that will affect final costs, and the predicted range affected by those variables.

8. *Write readable prose.* Avoid language that is overblown or too technical for your audience. Keep paragraphs short, or use lists.

9. *Use convincing language.* Your proposal should move people to action. Keep your tone confident and encouraging, not bossy and critical.

10. *Emphasize key points with effective visuals.* Tables, flow charts, photographs, drawings, and other visuals are sometimes essential to convince readers. Table 18.4, on page 345, provides useful advice.

11. *Tailor supplements for a diverse audience.* A single proposal often addresses a diverse audience: executives, managers, technical experts, lawyers, politicians, and so on. Various reviewers are interested in various parts of your proposal.

(continued)

Experts look for the technical details. Others might be interested in the recommendations, costs, timetable, and expected results, but will need some explanation of technical aspects. Both short and long proposals may include supporting materials (maps, blueprints, specifications, calculations, and so forth). Place supporting material in an appendix to avoid interrupting the discussion.

If the primary audience is expert or informed, keep the proposal text technical and provide a summary, a glossary, and specialized information in appendices for uninformed secondary audiences. If you're unsure as to which supplements to include in an internal proposal, ask the intended audience or study other proposals. For a solicited proposal (one written for an outside agency) follow the agency's instructions *exactly*.

The following advice is abstracted from the March 2013 Complex2Clear newsletter, written by Paul Heron. His firm helps clients develop powerful proposals and other marketing materials.

ON THE JOB...

Respect Your Reader's Needs

"Smart communicators ruthlessly stamp out wordiness and unsupported claims in proposals, websites and collateral. Clients want tight copy that focuses on

1. Measurable claims about how they deliver and perform,
2. Hard proof of those claims, and
3. How they will increase sales and/or reduce cost or risk.

Nearly everyone who reads marketing communications would rather be doing something else. Reading our stuff is work for them, so they naturally crave brevity and clarity. Communicators should respect this need by making copy rewarding and easy to read.

Visit your home page or most recent proposal and stamp out

◆ Copy that's not client-focused,
◆ Unverified claims,
◆ Long sentences and words, and
◆ Use of the passive voice."

—**Paul Heron, President of Complex2Clear**

A GENERAL STRUCTURE FOR PROPOSALS

To be successful, a proposal must clearly explain the proposed actions, and it must present supporting arguments that recognize the reader's needs and priorities. In practice, a proposal also needs to open with an overview of its purpose, method, and benefits. Finally, it needs to close with a request for action—either the reader's authorization or a meeting to discuss the proposal in more detail.

Typical Sections of Proposals

Although a proposal's complexity determines exactly which sections will be included, the sections listed in Table 18.2 on the next page are usually present.

Table 18.2 Typical Sections of Proposals

Section	What to Include
Introduction	◆ Connect with your reader by referring to the reader's request or to a problem that has led to this proposal. ◆ Briefly summarize the proposed plan, service, or product. Perhaps highlight your (or your team's) qualifications. ◆ "Hook" the reader by previewing the main benefit. ◆ Describe the scope of the project and list the topics covered in the proposal. ◆ Match the length of the introduction to the length of the proposal. For example, use a single paragraph for a short letter proposal or one or more pages for a report-length document. Or put the introduction in the accompanying transmittal letter.
Background	◆ Discuss the problem or need that has led to the proposal. This discussion will be more extensive in an unsolicited proposal. In a solicited proposal, show that you understand the problem completely. ◆ Discuss the general situation that has led to the current problem (when appropriate) and discuss the significance of solving that problem. ◆ Talk about the requirements for a solution.
Project Description	◆ Describe your solution to the problem: – what will be done – when and where – by which methods ◆ Include headings such as the following: – Plan of the Work – Schedule – Task Breakdown – Projected Results ◆ Incorporate the project components specified in the client's detailed guidelines when responding to an RFP.
Supporting Material: Facilities/Equipment **Personnel's Past Experience**	◆ Provide evidence of the ability to complete all aspects of the project: – Show that you have access to the required facilities, equipment, and other resources. – Show that the project leaders have the required qualifications and experience to complete the project. ◆ Provide examples of similar work performed by your team. ◆ Include résumés (where appropriate).
Supporting Arguments	◆ Show that the proposed plan is feasible (e.g., that it can indeed be completed in the projected time frame, or that a similar solution has been successfully implemented in a similar situation). ◆ Discuss the benefits of the proposed action and, if necessary, address potential reader concerns.

(continued)

Table 18.2 Typical Sections of Proposals *(continued)*

Section	What to Include
Budget	◆ Complete this section carefully, to avoid problems due to potential increases in your costs—you're legally bound during the time period you specify. Some proposals need detailed cost breakdowns; others need only a bottom-line figure.
Authorization	◆ Use this closing action statement to request your reader to authorize your proposed plan, or request a meeting to present your proposal. ◆ Include the authorization request in the closing paragraph of informal proposals (along with a reminder of the proposal's main benefits). For formal proposals, place the authorization request in the transmittal letter.

Additional Sections in Formal Proposals

In addition to the standard seven sections found in more informal proposals, include the sections presented in Table 18.3 in longer, formal proposals.

Table 18.3 Additional Sections in Formal Proposals

Section	What to Include
Transmittal Document	◆ Use this persuasive letter to address the person who receives the proposal or is responsible for the final decision. This is your only chance to directly address the gatekeeper or decision maker who will decide the fate of the proposal, so word this letter very carefully. ◆ Refer to the RFP. ◆ Describe (briefly) the main points and benefits of the proposal. ◆ Indicate the time limit of your bid. ◆ Ask for the action you desire.
Copy of RFP	◆ Include a copy of the RFP when you know or suspect that the receiving organization has recently issued more than one RFP.
Summary	◆ Summarize the proposal's highlights in one page or less. ◆ Use the headings "Summary" or "Abstract" for technical readers. ◆ Use the heading "Executive Summary" for less technical summaries designed for managers and others. ◆ Include both types of summaries if more than one type of reader may read the proposal.

Title Page	◆ Include, in order,
	  – the title of the proposal
	  – the name of the client organization
	  – the RFP number or other identifier
	  – the author's name (if appropriate) and that of the organization
	  – the date of submission

(continued)

Table 18.3 Additional Sections in Formal Proposals *(continued)*

Section	What to Include
Table of Contents	◆ Include all section headings and the pages where they appear. ◆ Do not list the RFP, transmittal letter, and title page. ◆ List the names of appendices.
Supporting Documents	◆ Include appendices such as the following: – testimonials from satisfied clients – résumés – technical evidence – relevant news or magazine articles, brochures, or photos of previous projects

Special Considerations for Executive Summaries

A proposal's executive summary is probably the one part of the proposal that will be read by all persons who review the document. In particular, busy executives in the receiving organization will want to see that your proposal understands their business needs.

Therefore, as John Clayton argues in an article posted by the Harvard Business School's *Working Knowledge* newsletter, proposal executive summaries must do more than summarize. They must distill the business case that drives the proposal, with three elements highlighted:

1. "Establish the need or problem" that must be addressed in order to fulfill the client's business objectives.
2. "Recommend the solution and explain its value" in such a way that the client is clear about the potential benefits of the proposed solution.
3. "Provide substantiation" in the form of the key reasons why the reader should believe that your firm is the right one to deliver the proposed solution.

Clayton says that the executive summary should be formatted with bullets, headings, and well-chosen graphics to direct attention and to improve readability. Also, Clayton quotes Tom Sant regarding length: "Keep your executive summary short—one to two pages for the first 25 pages of proposal text and an additional page for each 50 pages thereafter."

SAMPLE PROPOSALS

Now, let's see how the identified sections are implemented in three different proposals:

1. an informal improvement proposal
2. an informal research proposal
3. a formal sales proposal

An Informal Improvement Proposal

Here's a situation that requires a planning proposal.

CASE	**An Informal Planning Proposal**

The Situation. The writer, Justine Amarjan, is an engineer in the Moncton office of New Brunswick's Larouche Construction, a multifaceted company that provides engineering services as well as road and bridge construction throughout the Maritimes. She sends an unsolicited proposal to her office manager, Yannick Larose, trying to persuade him to approve the purchase and installation of an anti-virus software package on each of the eight networked computers in the office. In the past, these computers have not needed anti-virus protection because the office intranet has been only recently connected to the internet.

Justine has to meet the following challenges to persuade her reader:

1. This proposal has not been solicited, so Justine must convince her reader (Larose) that there's a problem (or potential problem) that's serious enough to justify the proposed expense.
2. With two months left in the fiscal year, the Moncton office's discretionary budget is nearly exhausted, as Larose has pointedly informed his staff.
3. The reader uses computers because he must do so, in order to do his job. However, he doesn't really like computers, and he doesn't learn any more about them than he absolutely has to. He has essentially delegated computer decisions to Justine and another young staff engineer, but his role requires him to approve all spending decisions, including computer equipment purchases.
4. The reader knows very little computer jargon.

So Justine faces a formidable task. Still, she does have two points in her favour:

- ◆ Yannick Larose is a logical, reasonable person who bases decisions on what is best for his work team and for the Moncton office's operation.
- ◆ This reader makes a point of trusting his staff's decisions, provided that employees present reasonable evidence to support those decisions. In particular, he has told Justine that he values her expertise with computers and computer software.

Let's see in Figure 18.1 (on pages 325–28) how well Justine follows the guidelines listed on pages 318–20.

Larouche Construction   **MEMO**

To::   Yannick Larose, Manager

From:: Justine Amarjan, Planning Engineer

Date:: December 17, 2014

Re: **A Proposal to Prevent Damage to the Office Network**

Connects with reader

You've asked me to keep you informed about how our new computer network is working. It's working well; everyone is pleased, and productivity seems to have improved. However, we have an urgent security issue that I think should be addressed.

The nature of the problem

At last week's company planning meetings in Fredericton, I was reminded of a potential problem that has worried us since we updated our office intranet and connected to the internet two months ago. Colleagues in Fredericton and Dartmouth reported that their offices have been hit by viruses in the past few weeks. Two weeks ago, the Dartmouth office lost most of its archived files in addition to current project files.

An urgent note

We can't afford the same kind of disaster—we have too many ongoing projects, some of which have files dating back to 2005. With thousands of hours invested in engineering plans and project management details, we can't afford to have our files corrupted or erased. It could easily happen—now that we're connected to the internet, computer viruses and worms could enter our network via email or internet downloads at any time. Even if we could retrieve our files, it might take days to do so.

Solutions

Three main kinds of approaches can provide protection:
1. hardware solutions, such as Cisco ASR. 1002
2. server-based software solutions, such as Microsoft's *ISA Server* software
3. any one of several anti-virus software packages designed for individual PCs

Shows her appreciation of the reader's cost concerns

I haven't considered the hardware solution because it's very expensive; when we looked at adding Cisco's hardware to our system in October, we were quoted CDN$20 760 plus tax. Likewise, I have discounted the Microsoft server software because it would cost at least $3500 for the kind of "low-end configuration" we would need.

Figure 18.1 An Informal Improvement Proposal   *(continued)*

An Informal Research Proposal

An informal student research proposal is illustrated in Figure 18.2 on pages 329–332. This proposal memo describes a computer student's idea for a research project required by the Communication 116 class that she's taking. That project will culminate in an analytical report for a "real-world" reader.

Provides useful background for an uninformed reader

Establishes a consistent set of evaluation criteria

Documentation to support opinion

Ethical: The writer admits that another product is also very good

The key reason

So, I've considered five readily available anti-virus software programs that could be loaded on each of our machines. We would pay to load the software on each machine and then would pay an annual subscription fee for each machine, starting one year from the time of initial installation. The five packages include *F-Secure Antivirus Scan 2013, Panda Antivirus Pro, AVG Antivirus 2013, Kaspersky Anti-Virus 2013,* and *BitDefender Antivirus Plus 2013.*

Evaluation Criteria

Each software can be assessed using four criteria:

1. How well does the software work?
   - Does it identify all viruses and worms, whether they come via email, IM apps, or web browsing?
   - Does it effectively clean or isolate infected files?
   - Does it report the result of its scans and what it did with infected files?
   - What kind of scanning engine does it use?
2. Is the software easy to install and use?
3. Is the software compatible with our Windows XP operating system and Windows XP Advanced server?
4. Is it affordable? (Cost includes purchase price, annual subscription, and the cost of technical support.)

In order to consistently and fairly compare the five software programs, I read a number of software reviews. I found the most useful of these reviews at www.anti-virus-software-review.toptenreviews.com. The site's review summary is attached to this memo.

The Best Anti-virus Package

I propose installing *BitDefender 2013 Standard Edition,* although it was a close call: BitDefender and Kaspersky are both ranked very highly in terms of performance and ease of use, and both are compatible with Windows XP and our Windows server (See an attached detailed comparison.)

Although Kaspersky joined *BitDefender* in passing all the tests run by *PC World,* PCMag.com, and TopTenREVIEWS, *BitDefender's* superior user-friendly features and product-leading cost make it the best product under review.

Figure 18.1 An Informal Improvement Proposal   *(continued)*

Benefits and Features

As identified by the independent TopTenREVIEWS "Anti-virus Software Review," *BitDefender* has several features to recommend it:

- The product is well designed and very easy to use, so our staff will require very little training in its use.
- *BitDefender* has a full set of protection and virus-removal features, including automatic or manual virus checks, frequent automatic updates of the viruses and worms detected by the software, and quick check times.
- The package is very easy to install, so I will be able to load it onto all the machines on a weekend morning.
- This software was awarded the VB 100% certification that goes to anti-virus software that detects all viruses thrown at it, while generating no false positives.
- Full-time technical support is available by phone or email.

Installation and Training Plan

As I mentioned, the software is easy to install, so I propose to install it next weekend on each of the seven PCs and the server. The packages can be ordered over the internet and delivered within two working days from Montreal. Then, the following Monday morning, after our staff meeting, I propose to take half an hour to train all staff in how to use *BitDefender*. It's as simple as that.

Personnel

Unlike most of our projects, this proposal involves a one-person band! With your approval, I propose to order the software, install it on the first available weekend morning, and train the staff in its use. Also, I will monitor *BitDefender's* operation and consult Softwin's technical help desk if necessary.

Why am I volunteering my time to implement the proposed plan? I have two motivations, really:

1. I'm very concerned about the possibility of a virus or worm taking down our entire system, so I'm willing to help reduce costs in order to make the proposed solution feasible. (We'll save about $250 in installation and training costs if I perform those tasks instead of an outside technician.)

This schedule is relatively simple

While discussing her motivations, the writer establishes her credentials

Figure 18.1 An Informal Improvement Proposal   *(continued)*

2. As you know, I studied network engineering technology for two years before switching to civil engineering, and I retain a strong interest in all aspects of computer networking. Also, I continue to read networking articles, so my knowledge is current. This project, although relatively straightforward, helps develop my knowledge in the area.

Costs

The total costs are low, compared with the benefits of installing first-rate anti-virus software on our network

BitDefender online purchase	8 × $24.95 = US$199.60
Installation	$0.00
Training	$0.00

Total = US$199.60 (CDN$218.00)

Annual online updates subscription 8 × US$24.50 = US$196.00 (CDN$213.64)

Authorization

May I have your authorization to order the software, install the packages, and train the staff? I believe that our office network remains at great risk without anti-virus protection. Most of the staff are very careful about opening suspicious emails and about researching on the internet, but we could download a devastating virus by simply turning on a computer!

Justine

Justine Amarjan
Attachment: Anti-virus Software Review's detailed rating chart
            (from www.anti-virus-software-review.toptenreviews.com)

An attempt to "close the sale"

Figure 18.1 An Informal Improvement Proposal

Communication 116 Memorandum

Date: February 4, 2014
To: Professor Devon Koenig, English Department
From: Amy Suen, Computer Systems 1st year
Re: **Proposed Research Project for Communication 116**

In response to the proposal assignment that you announced in our January 22 class, this memo outlines my proposal to research firewall products. The research will lead to a recommendation of which product will best suit the needs of 2020 Design, a local electronics company. Its networking specialist, Mark Reuters, has indicated his interest in my findings.

Background

Mark Reuters has recently revamped 2020 Design's total network of computers. He is now in the process of installing new networking software, and will continue fine-tuning the network and training 2020 Design personnel for the next three months. At that time, he will turn his attention to the challenges of partially linking 2020's internal network to the internet.

One of the major problems to solve will be the issue of internal network security. 2020 Design stays competitive by developing new electronic designs, so it doesn't want outsiders tapping into its research and development work. Still, the company wants to market its products on the internet and engage in e-commerce.

Recently, *firewall* has come to mean software products for blocking unwanted access to protected information, but its original meaning included all aspects of network security strategy—software, hardware, and personnel. Mark Reuters has asked me to focus on software products.

Developing appropriate software is very expensive, so most companies purchase rather than develop. Firewall software ranges from a few thousand dollars to about $100 000, depending on performance and user requirements. Each product has advantages and disadvantages that depend on network configurations and the method of implementing the software. Therefore, it's not easy to choose an appropriate firewall product.

Proposed Plan

My initial research into the topic indicates that the following process would be best.

1. **Determine the client's needs.** Subject to your approval of my proposal, Mark Reuters will provide me with details of the network he administers. Those details will include the special features and challenges built into that network (which apparently is unique). With Mark Reuters's help, I'll be able to choose the criteria that I can use to evaluate and compare firewall software products.

2. **Research available software.** As the attached tentative bibliography shows, there seems to be plenty of information available. In addition, I have arranged to possibly interview Marsha Campbell, one of my computer instructors, and Guy Larivière, the network administrator for our college. I'll proceed with those interviews if I receive your authorization for this project.

3. **Evaluate information gathered about firewall products.** Please see the attached outline, which describes the general approach I plan to take. So far, I have identified

Figure 18.2   A Student Research Proposal

(continued)

three products (*AltaVista, Check Point,* and *CyberGuard*), but I'll continue to look for others. I expect to discuss three to five products in the final report; that means I may have to do a preliminary assessment to weed out inappropriate firewall products.

4. **Rigorously apply the assessment criteria and choose the best product for 2020 Design.** Part of this assessment can come from the specifications and product information provided by the manufacturers on their websites. The full assessment will come from actually testing the software.

Feasibility of Project

I've looked at this project quite carefully and I think it's feasible for the following reasons:

- Information is available, and I have access to expert opinion here at the college.
- My strong interest in this subject has already prompted me to read all the sources listed in the attached bibliography, and I've made notes on three of the articles.
- Professor Campbell has agreed to help me assess software on the network in computer lab 218, and she is accepting this project for credit in her Networking class.
- I can get access to firewall products through Mark Reuters, who will request demo software from manufacturers.

Schedule

According to your January comments about task requirements, I have overestimated the time required for the following tasks, but my semester time budget can still accommodate the following:

Activity	Time Required	Dates	Document Produced
Research re: firewalls (types and leading products) and 2020 Design system configuration	10–15 hrs	Feb. 1–12	Refine planning outline
Interview Campbell and Larivière	2–2.5 hrs	Feb. 14–15	Refine planning outline
Evaluate data	5–8 hrs	Feb. 22	Adjusted research plan (?)
Analyze/organize data	6–10 hrs	Feb. 27–28	Working outline (due Mar. 7)
Plan/write progress report	4 hrs	Mar. 5–6	Progress report (due Mar. 7)
Write report from outline	6–10 hrs	Mar. 14–16	Analytical report—draft
Edit report and polish format	5-8 hrs	Mar. 24–27	Analytical report—due Mar. 31

Figure 18.2 A Student Research Proposal

(continued)

Prof. Koenig
February 4, 2014
Page 3

Budget
Because I won't have to buy firewall products, and because I use my home internet connection for many purposes, my budget for this project is minimal:

Photocopy articles	$25
Bus travel for interviews	$20
Report printing and binding	$18
Total:	$63

Authorization
I hope you agree that my proposal topic and approach is appropriate for the Communication 116 project, because I'm committed to doing an excellent job. For one thing, I'm also doing the project for my Networking class, and I can afford to put more effort into a project that fulfills two sets of requirements. Also, I think I may be able to get a student co-op job at 2020 Design this May if I do well on this project.

May I have your authorization to proceed? If you wish to contact me outside of class, please email me at amysuen@silk.net.

AS

Attachments: Tentative Bibliography
Planning Outline

PLANNING OUTLINE: REPORT PROJECT
Communication 116

Topic: Firewalls

Reader (needs/reason for reading/knowledge level):

Mark Reuters, 2020 Design; he's interested in firewall products for his firm's network; he's very knowledgeable, but not about current products

Purpose: Assess firewall products suitable for 2020 Design (maybe recommend best one)

Tentative Structure/Topics:

Reason for report
Background re: current firewall technology

- list of products (*CyberGuard, AltaVista, Check Point,* and others)
- manufacturers
- concepts behind the products

2020 Design's network structure Client's general and special requirements

Figure 18.2 A Student Research Proposal

(continued)

Prof. Koenig
February 4, 2014
Page 4

Product evaluations:

- supporting operating system and services
- performance
- price
- installation and training
- does it meet client's particular requirements?

Conclusion re: which products satisfy the criteria/which *best* satisfies the criteria

Tentative Sources:
Interviews: Marsha Campbell, Guy Larivière, and Mark Reuters.
Check Point Technologies Ltd. "Firewall-1 Products and Solution." *Checkpoint.com.* n.d. Web.
    2 Feb. 2014. <http://www.checkpoint.com.>.
Computer Security Institute. "CSI Firewall Matrix." *Computer Security Insitute.* n.d. Web.
    31 Jan. 2014. <http://www.gocsi.com>.
Daswani, N., C. Kern, and A. Kesavan. *Foundations of Security: What Every Programmer Needs
    to Know.* New York: Apress, 2007.
Digital Equipment Corp. "AltaVista Firewall 10." *AltaVista.com.* n.d. Web. 29 Jan. 2014.
    <http://www.altavista.software.digital.com>.
Hallogram Reviews. n.d. Web. 22 Jan. 2014. <http://hallogram.com/avfirewall>.
Markus, Henry S. *Home PC Firewall Guide.* n.d. Web. 27 Jan. 2014.
    <http://www.firewallguide.com>.
Stein, L.S. *Web Security a Step-by-Step Reference Guide.* 2nd ed. Reading, MA: Addison-
    Wesley, 2006.
Thegild.com. "Firewall Product Overview." *Thegild.com.* n.d. Web. 2 Feb. 2014.
    <http://www.thegild.com/firewall>.

Figure 18.2 A Student Research Proposal

| CASE | **An Informal Research Proposal** |

The Situation. The student, Amy Suen, chooses to examine firewall systems that might provide network security for the networked systems at 2020 Design, an electronics design and manufacturing company. Amy does some preliminary research to confirm that she can find information on firewalls. Also, through a contact at 2020 Design, she learns that her report would interest Mark Reuters, 2020's Design computer networking specialist. Mark has been too busy designing and installing his company's new array of computers to pay much attention to the external security issue. Still, 2020 Design is committed to marketing its products on the internet and therefore risks incursions into its internal communications, including its proprietary designs.

Amy knows that 2020 Design will not make decisions based on a first-year computer student's report, but Mark Reuters assures her that her report could provide a springboard for his own investigation of firewall software. For that reason, he is willing to help Amy with her preliminary research.

In preparing her proposal for her instructor, Devon Koenig, Amy starts by analyzing her reader and her purpose for writing the proposal, as follows:

Audience. Devon Koenig, project supervisor. He will use my proposal to decide whether to approve my proposed approach. If my proposal is rejected, I'll have to prepare another.

> His knowledge: severely limited (so I'll have to provide background)
> His questions: (in the assignment memo dated January 22)

1. What's your topic? Who's the reader for your proposed report?
2. What main question does your reader want answered? (Identify the report's specific analytical purpose.)
3. How will the proposed report achieve that purpose? (Here, an attached detailed outline would help establish that you do indeed have a plan.)
4. Do you have the time, resources, and commitment to complete the project? (You should include a time budget and a resources budget.)
5. Will you find sufficient information to answer the report's overall question? (Attach a tentative bibliography.)

Purpose. Convince Prof. Koenig that the topic has merit and that the project is feasible. (Show that I can research, that I can manage the project, that I can afford it, that I can think, that I can write well enough to handle a report of this complexity.)

Audience Attitude and Temperament. Prof. Koenig has high standards and expectations. Therefore, I'll have to answer all his questions thoroughly. He said he'll be skeptical, so I'll have to support all my statements. This won't be an easy sell! I don't think he cares much about how much the project will cost me, as long as I can prove I can afford it. He's a tough marker, so I'll need to phrase the memo carefully and get it proofread.

Audience Expectations. He doesn't want more than three pages, not including attachments. His comments in class emphasized our need to provide proof of all positive statements. That's the only way I'll get his support. I'll need to use a direct, businesslike tone—he doesn't like extra words or pompous language. (I'd better stay away from words like *utilize*.) And I'd better stick pretty close to the proposal structure he recommended: connect with reader/provide "hook" in the proposal pre-summary/give background/show need for report/describe planned approach/propose schedule/prove that it's feasible—sources/my qualifications/budget/request authorization.

The memo that Amy Suen presents (Figure 18.2) to her project supervisor begins on page 329.

A Formal Sales Proposal

The formal proposal (Figure 18.3) that begins on page 335 presents a commercial sales pitch. Its writer, Tony Mutu, uses a formal report format to impress his reader with a professional-looking document, even though the main body proposal is only six pages. Note, though, that another 12 pages of supplements are included as appendices.

One of the appendices in Figure 18.3 includes a copy of the RFP and accompanying documents that the writer retrieved from the client. To some extent, this inclusion is overkill: as far as the writer knows, the client has just this one RFP outstanding at the moment.

This proposal does not include a separate summary—after all, the proposal report is just six pages and its main points are already summarized in the transmittal letter. Also, the concluding authorization request ("closing the sale") is included in the transmittal letter, so it is not repeated on page 6 of the proposal report.

Notice that this proposal uses imperial measurements, a system often still used in construction projects. The writer has been told that his reader, John Warren, is "old school," and not eager to use the metric system.

Other aspects of the writer's audience and purpose analysis have been discussed in the section "The Proposal Process"; see page 315.

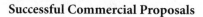

ON THE JOB...

Successful Commercial Proposals

"Over the years, I've ghost-written dozens of proposals for clients, and I write my own proposals to bid for editing and writing contracts. Typically, these proposals range from 15 to 40 pages, but I have produced proposals that exceed 100 pages. The secret of writing successful proposals is to clearly show how the reader will benefit from the proposed product or service. I'm now well established in Ottawa, and on the standing offer lists of several government departments, but I must write clear, comprehensive proposals to win contracts. One of my successful techniques is a 'compliance matrix' that matches client requirements with components of my proposal—this matrix could present several pages that show details of how my proven skills and knowledge will allow me to fulfill each requirement listed in the client's full RFP. . . ."

—**Judith Whitehead, writer and editor, Vankellers Editorial and Writing Services**

EnviroMax Engineering Consultants

103-1025 Ellis Street   Penticton, British Columbia   V5T 2W9
Ph: 250-493-2380   Fax: 250-493-2395   email: emax@shaw.ca

11 February 2014

Mr. John Warren, Public Works Manager
District of Beachland Public Works
Box 732
Beachland, British Columbia
V0H 1X0

Dear Mr. Warren:

Re: RFP 17PW - 2014d

Thank you for the opportunity to bid on the engineering design and supervision of the project to improve the beach and swimming areas fronting 2056 Beach Avenue, Beachland. As your RFP describes, the situation at 2056 Beach Avenue has serious implications for Beachland businesses and requires a long-term solution. EnviroMax believes that the problems outlined in RFP 17PW-2014d can be solved by the right engineering work.

Our enclosed proposal meets the RFP's requirements by building an offshore breakwater and by repairing previous erosion damage.

As our proposal indicates, we have considered all the available information concerning the problems at 2056 Beach Avenue. In addition, we have conducted a preliminary study of the current lake-water levels and the underwater soil depths fronting the property. Further, we have consulted with faculty members in the Water Quality and Environmental Sciences programs at Okanagan College to learn more about the potential environmental impact of our proposed construction. Those faculty have volunteered their expertise in return for the chance of project work for their senior students.

EnviroMax trusts you agree that our proposal is the best long-term (and most cost-effective) solution to the erosion problem and recreation needs at 2056 Beach Avenue. May we have your authorization to proceed? To discuss the proposal, please call me at 493-2380 or email me at emax@shaw.ca.

Sincerely,

Tony Mutu

Tony Mutu, C.E.T.

Figure 18.3 A Formal Sales Proposal

PROPOSAL FOR
BEACH RESTORATION
AND EROSION PROTECTION
RFP 17PW – 2014d

Prepared for

John Warren
Public Works Manager

District of Beachland
Beachland, British Columbia

Prepared by
Tony Mutu, C.E.T.
Project Manager
EnviroMax Engineering Consultants
Penticton, British Columbia

February 11, 2014

Figure 18.3 A Formal Sales Proposal

(continued)

TABLE OF CONTENTS

Figure 18.3 A Formal Sales Proposal   *(continued)*

⠿ *EnviroMax Engineering Consultants* ⠿

INTRODUCTION

EnviroMax is well qualified to supply the network design and implementation outlined in RFP 17PW – 2014d. We are an established business in the Okanagan Valley with a reputation for providing outstanding quality and customer service. We have extensive experience in construction design, in environmental solutions, and in environmental project management. This proposal details our plan to create a long-term, cost-effective solution to the erosion problems at 2056 Beach Avenue. EnviroMax is excited about the possibility of contributing to the beauty and economic viability of Beachland's lakefront.

Statement of Need

Before designing the proposed solution to the problems at 2056 Beach Avenue, EnviroMax considered the following key factors:

- Research conducted from 1998 to 2009 shows the shoreline receding at an average rate of 7.6 inches per year, but measurements taken by the owners of 2054 and 2056 Beach Avenue show that the shoreline has eroded an average of 13 inches per year for the past four years. *In other words, the erosion seems to be worsening.* As a result, the beach sand in front of the 125 feet of shoreline at 2056 Beach Avenue has been eroded away, leaving a pebble and rock beach. This erosion has made the campground at 2056 less desirable for tourist campers, and Beachland businesses have noticed a marked decline in business the last few years.
- Simply replacing the beach sand does not seem to work, as has been shown by the eroded sand that was placed at 2054 Beach Avenue in 2007.
- In addition to restoring the beach, the restorative work should create a large, safe swimming area.
- Beachland Public Works equipment, materials, and employees are available to complete much of the required work, which could result in substantial cost savings to the municipality.
- In winter, the prevailing winds come from the east and northeast, while in the other three seasons the prevailing winds come from the south or southeast.
- Current Okanagan water levels are slightly lower than normal for this time of year.

Potential Solutions

EnviroMax's experience, supplemented by research into erosion scenarios in British Columbia, Alberta, Manitoba, and Washington State, suggested three main types of solutions:

- a compacted bed of river rock covered by sand
- a foreshore retaining wall, with an inner area filled with rock and covered with sand, to reclaim the 11 feet of eroded shoreline and create a larger beach
- an offshore breakwater, supplemented by beach restoration

- 1 -

Figure 18.3 A Formal Sales Proposal

(continued)

The first two options were examined, and then rejected because they do not meet stated requirements:

⠿ *EnviroMax Engineering Consultants* ⠿

1. **A compaction method** would place a bed of rock about 60 feet out into the lake, to a water depth of 6.5 feet. The first layer would consist of 3- to 6-inch rocks at the 6.5-foot depth, to anchor the entire rock bed. Then, progressively smaller rocks would be placed as the water depth decreases, closer and closer to shore. Close to the lakeshore, the underwater rocks would be only about 1.5 inches. On the lakeshore, gravel and sand would be used to reconstruct the surface layer of the beach. Finally, the entire new surface would have to be compacted, right out to the 6.5-foot water level.

    The compaction is usually time-consuming and difficult because of the water depth in which the compaction equipment must operate. Also, an extensive environmental study must be completed before the work can proceed. This method repairs existing damage, but will not prevent subsequent erosion at 2056 Beach Avenue, especially from the waves driven by northeast and southeast winter winds. Also, it would not provide a safe swimming area.

2. **A foreshore retaining wall** would be built parallel to the existing shoreline, about 11 feet out (in order to reclaim the shore lost to erosion). Five steps are required: ① conducting an environmental study; ② preparing the site for the retaining wall; ③ building the wall so that it extends about 6.5 feet above the September low-water mark; ④ filling in the area on the land side of the wall; ⑤ compacting the fill and surfacing the compacted fill with sand, gravel, or grass.

    The retaining wall would cost $79 000 to build—costs include fuel, Public Works employees' wages, Public Works equipment repairs, equipment rentals, and the services of a professional stonemason. The filling and surfacing and cleanup would cost $9500. The environmental study and EnviroMax's engineering fees would cost $34 250.

    This method repairs existing damage and will prevent subsequent erosion. However, it does not provide for a *protected* swimming area, a major consideration for families who vacation at this beach.

    Therefore, we propose an offshore breakwater and a restored beach.

Information Sources
EnviroMax has researched several sources in preparing the proposal:
1. Examination of the 2009 report based on the study Beachland commissioned in 2007;
2. Consultations with Mario Zumbo, an experienced stonemason;
3. Investigation of the operational and maintenance costs of using Beachland Public Works employees and heavy equipment;
4. Enquiries regarding rental costs for specialized trenching and excavating equipment;
5. Probes of the lake bottom where we propose to build the breakwater—this survey has revealed the current water depth (6.5 feet), soil thickness (18 inches to 6 feet), and the lake bed profile (relatively flat); and
6. Ministry of the Environment records of Lake Okanagan water levels, recorded since 1943.

- 2 -

Figure 18.3 A Formal Sales Proposal *(continued)*

::: *EnviroMax Engineering Consultants* :::

THE PROPOSED PLAN

Restorative and Protective Measures

Building a breakwater and repairing existing damage would require a six-stage process, shown in the following table.

Stage	Performed By	Comments
1. *Conduct environmental study* re: impact on aquatic life and impact on water quality.	Instructors and students from Okanagan College's (OC) Water Quality and Environmental programs, supervised by EnviroMax	Required by the Ministry of the Environment, this study may result in modifications to the original project design.
2. *Lay a delivery causeway* from the centre of the beach to the centre of the proposed wall. (Figures 1 and 2)	Beachland Public Works (BPW) employees and equipment, supervised by EnviroMax	Requires trucks to haul dirt and blast rock from the Public Works materials yard.
3. *Prepare the site* for the breakwater wall. An 8-ft.-wide, 150-ft.-long trench must be dug to remove lake bed soil and thus provide a solid base for the wall. (Figure 2)	BPW employees and equipment, supervised by EnviroMax	The soil thickness varies from 1.5 ft. to 6 ft., so the excavator arm will operate in water as deep as 14 ft. Thus, a special excavator must be rented at a cost of $1200 per day.
4. *Build the wall*, in 3 steps: • Use the excavated soil and hauled material to form a causeway parallel to the trench, on the beach side.	BPW employees and equipment, supervised by EnviroMax	For this first stage, BPW's Caterpillar 920C excavator could be used, but the rental machine would have more capacity.
• Place large blast rock in the trench, to a point 2 ft. above the waterline.	BPW employees and equipment, supervised by EnviroMax and Mario Zumbo, stonemason	Here, the rented excavator and rented underwater video monitors would be needed.
• Build the upper wall, with jagged blast rock on the wave side and mortared flat rock on the beach side. (Figure 3)	BPW would deliver the blast rock from its stockpile; Mario Zumbo and his assistant would construct the wall	The rental equipment would not be needed for this step. Special marine mortar would be used.
5. *Remove the causeway.*	BPW employees and equipment, supervised by EnviroMax	The rental excavator would be needed for depths greater than 4 ft.
6. *Clean up and restore the beach.*	BPW employees and equipment, supervised by EnviroMax	After the beach area is graded and cleaned, about 60 t. of of fine sand (from the BPW yards) will be placed on the beach and on the lake bed, 12 ft. out from the shoreline.

- 3 -

Figure 18.3 A Formal Sales Proposal

(continued)

⦂⦂⦂ *EnviroMax Engineering Consultants* ⦂⦂⦂

Proposed Location of Work
The following figures show the location of the proposed work.

Figure 1. Approximate Breakwater Location

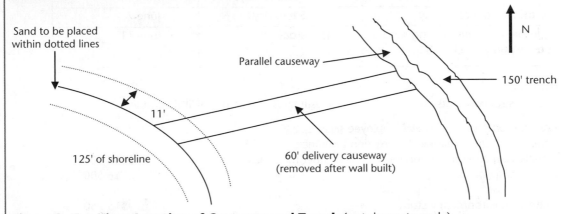

Figure 2. Top View: Location of Causeway and Trench (not drawn to scale)

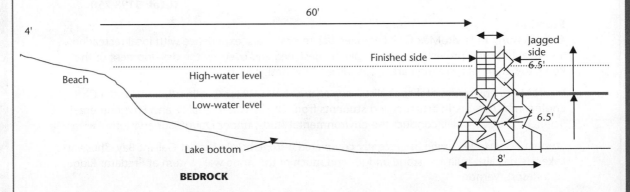

Figure 3. Side View: The Finished Wall (not drawn to scale)

- 4 -

Figure 18.3 A Formal Sales Proposal   *(continued)*

EnviroMax Engineering Consultants

Proposed Schedule

The following schedule allows for unforeseen delays due to bad weather, equipment breakdown, or environmental factors. If begun by March 1, work will be completed before the June 20 deadline.

Task	Time Required	Completion Date
Environmental study	10 to 15 days	March 19
Building a delivery causeway	5 to 8 days	March 30
Preparing the site for the wall	10 to 12 days	April 18
Building the parallel causeway	5 to 6 days	April 26
Building the wall	15 to 18 days	May 14
Removing the causeway	5 to 6 days	June 7
Cleaning the site; placing sand on the existing beach and in the water	4 days	June 11

Budget

The construction costs represent maximum estimates; the work will likely cost less.

Construction costs: Fuel, BPW employee salaries;
equipment rental; marine construction insurance;
stonemason contract; marine mortar, and other materials $158 000
Place sand/clean site $6 500
EnviroMax Fees
Conduct environmental study $15 750
Design wall and supervise construction $18 500

Total: $198 750

Staffing

Gary Stewart, an EnviroMax Civil Engineer, has several years' experience with local recreation and beautification projects. (For example, he designed and oversaw the development of the Penticton channel recreation area.) He will lead the design team.

EnviroMax's environmental specialist *Doug Nygren* has a Water Quality diploma and a Civil Engineering degree. Instructors and students from OUC's Water Quality and Environmental Sciences programs will conduct the environmental study, under Doug's supervision.

Stonemason *Mario Zumbo* has recently completed a stone breakwater at Gallant Bay, Shuswap Lake. He has also built the stone bridges and much of the stone wall system at Predator Ridge Golf Resort, Vernon.

Civil Engineering Technologist *Tony Mutu* has seven years' experience. Skilled in project management, he specializes in slope stability and erosion issues.

- 5 -

Figure 18.3 A Formal Sales Proposal

(continued)

CONCLUDING COMMENTS

Scope of Services
The prices in this proposal are valid until February 28, 2014. The scope of the services offered is only for the provision of a "large, protected swimming area and a protected beach" as specified in RFP 17PW – 2014d.

Benefits
The proposed work offers a long-term solution to the erosion problem. The dual-sided break-water will be functional and attractive:

a. The points of the jagged rocks on the wave side will dissipate wave energy and reduce the impact on the wall (stonemason Mario Zumbo estimates a minimum wall life of 50+ years).
b. The mortared smooth beach-facing wall will feature a classic Italian look, with a flat finished top. Similar breakwaters built on Shuswap Lake and on Washington State's Lake Chelan have added attractive components to the local landscape.

The wall will require no maintenance.

The proposed work repairs existing damage and fully restores the sandy beach. Between the beach and the breakwater, campers will have a large, protected swimming area.

The plan is the most expensive of the three plans considered, but it provides excellent value because BPW workers, materials, and equipment are used. Possibly, the wage costs, fuel costs, and some of the rental costs can be folded into the existing 2014 BPW budget. Also, the environmental study costs are considerably reduced by involving Okanagan College students.

Supporting Documents

Appendix A: RFP and Accompanying Documents — p. 7

Appendix B: Construction Cost Calculations — p. 9

Appendix C: EnviroMax Client List — p. 11

Appendix D: Cross-section of EnviroMax Projects— descriptions, photos, and client comments — p. 12

Appendix E: Information Sources and Proposal Contributors — p. 18

- 6 -

Figure 18.3 A Formal Sales Proposal
Note: To save space in this textbook, the appendices have not been included.

AN INTERPERSONAL PERSPECTIVE

Some of the examples on the preceding pages show external proposals, most of which attempt to win business contracts. Often, however, as employees, we have the urge to improve workplace conditions or ways of doing things. If we try to ignore that need, we can end up feeling bitter and helpless. But if we have methods of convincing others to make necessary changes, we can feel better about ourselves and the place where we work.

The following action gradient (Figure 18.4), which becomes more positive as it moves to the right, illustrates the value of suggesting improvements.

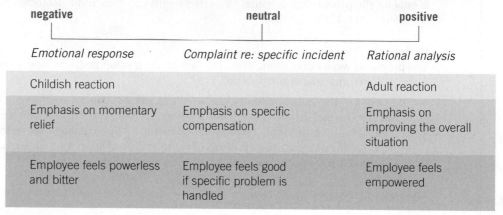

Figure 18.4 The Action Gradient

CASE	**The Action Gradient at Work**

Let's assume that you work as an engineering technologist for a civil engineering firm that specializes in developing city subdivisions. Because you're frequently placed in charge of projects, you essentially act as an assistant manager. Your office manager is technically competent and works hard, but she uses negative feedback as her primary motivational tool; the only time she comments on a person's work is when that person has made a mistake. Usually, she presents her criticisms in a hostile, aggressive manner. Employee morale and productivity are starting to suffer.

By contrast, your co-workers prefer your positive motivational techniques. You give credit for work well done. When you have to comment on incomplete or shoddy work, you take care to focus on the work itself, not on the worker. Lately, though, the manager's attitude and behaviour have been particularly hard to stomach because everyone has been working extra hard to meet a contract deadline.

Avoiding Action

If you operate at the left side of the action gradient, you don't confront the problem directly in the situation described above; instead, you use passive aggression by talking behind your manager's back. This approach allows you and the other employees to take care of your resentment and frustration for a moment, but you don't feel good about yourself. Meanwhile, the interpersonal climate worsens.

Taking Limited Action

If you operate in the middle ground, you may comment about a specific incident and ask for an apology. However, even if you receive that apology, the basic situation hasn't changed because you haven't confronted the underlying problem.

Taking Positive Action

The advantage of operating on the right side of the gradient is that you try to improve the overall situation. This response pattern requires a rational analysis of the problem and the formulation of a workable solution. Once you believe you have a valid solution, you have a choice of presenting it as a proposal or as recommendations. In the proposal, you single-mindedly argue in favour of a definite course of action, while the recommendations report assesses two or more possible solutions and then chooses one of those solutions.

Whether you went the proposal route or opted for the (apparently) more objective recommendations report, you have set in motion a series of productive possibilities. For example, your proposal of an incentive plan for managers and staff may get adopted. But even if it doesn't gain approval, you will feel good about yourself for positively confronting the situation. Moreover, you could be perceived as a positive influence within the organization, which will not hurt your subsequent chances for promotion. And, in the long run, positive actions improve the working climate, which was your goal in the first place.

GRAPHICS IN PROPOSALS

Proposals tend to feature text, not graphics. Still, as Table 18.4 suggests, visuals can help make proposals more persuasive.

Table 18.4 Graphics in Proposals

The Message	The Graphic
We offer high performance at low costs.	◆ line, bar, or pie chart ◆ table
Our plan is logical.	◆ flow chart
Our system or equipment does the job.	◆ schematic diagram ◆ hybrid graphic (such as drawings, photos, and tabular data pasted onto a flow diagram)
The parts are easy to assemble.	◆ exploded view drawing
We can meet the schedule.	◆ timeline with milestones ◆ critical path diagram
We have the resources and experience.	◆ data chart ◆ résumé with experience timelines ◆ photos (people, facilities, and equipment)

Source: Adapted from G. Edward Quimby. "Make Text and Graphics Work Together," *Intercom* [Newsletter of the Society for Technical Communication] Jan. 1996: 34.

CHECKLIST

for Revising and Editing Proposals

Use this checklist as a guide to revising and refining your proposals.

Format

◆ Have you chosen the best format (letter, memo, report) for your purpose and audience?

◆ Does the long proposal include appropriate appendices?

◆ Does the title forecast the proposal's subject and purpose?

Content

◆ Is the problem clearly identified?

◆ Is the objective clearly identified?

◆ Does everything in the proposal support its objective?

◆ Does the proposal show as well as tell?

◆ Does the proposed plan, service, or product benefit the reader's personal or organizational needs?

◆ Are the proposed methods practical and realistic?

◆ Are all foreseeable limitations and contingencies identified?

◆ Is the proposal free of overstatement?

◆ Is the proposal's length appropriate to the subject?

Arrangement

◆ Does the proposal include all relevant sections of the recommended structure?

◆ Does the introduction provide sufficient orientation to the problem and the plan?

◆ Does the plan explain how, where, and how much?

◆ Are there clear transitions between related ideas?

Style

◆ Is the writing style clear, concise, and fluent?

◆ Is the level of technicality appropriate for the primary reader?

◆ Do supplements follow the appropriate style guidelines?

◆ Does the tone connect with the reader?

◆ Is the language convincing and precise?

◆ Is the proposal grammatically correct?

◆ Is the proposal ethically acceptable?

EXERCISES

1. After identifying your primary and secondary audience, compose a short planning proposal for improving an unsatisfactory situation in the classroom, on the job, in your dorm, or in your apartment (e.g., poor lighting, drab atmosphere, health hazards, poor seating arrangements). Choose a problem or situation whose resolution is more a matter of common sense and lucid observation than of intensive research. Be sure to (a) identify the problem clearly, give brief background, and stimulate interest; (b) state clearly the methods proposed to solve the problem; and (c) conclude with a statement designed to gain the audience's support for your proposal.

2. Write a research proposal to your instructor (or an interested third party) requesting approval for the final term project (an analytical report or formal proposal). Identify the subject, background, purpose, and benefits of your planned inquiry, as well as the intended audience, scope of inquiry, data sources, methods of inquiry, and a task timetable. Be certain that adequate primary and secondary sources are available. Convince your reader of the soundness and usefulness of the project.

3. As an alternative term project to the formal analytical report (Chapter 19), develop a long proposal for solving a problem, improving a situation, or satisfying a need in your college, community, or job. Choose a subject sufficiently complex to justify a formal proposal, a topic requiring research (mostly primary). Identify an audience (other than your instructor) who will use your proposal for a specific purpose. Compose an audience/purpose profile, using the sample on pages 333–34 as a model. Here are possible subjects for your proposal:

   ◆ improving living conditions in your dorm
   ◆ creating a daycare centre on campus
   ◆ creating a new business or expanding a business
   ◆ saving labour, materials, or money on the job
   ◆ improving working conditions
   ◆ improving campus facilities for people with disabilities
   ◆ supplying a product or service to clients or customers

- eliminating traffic hazards in your neighbourhood
- reducing energy expenditures on the job
- improving in-house training or job-orientation programs
- improving tutoring in the learning centre
- making the course content in your major more relevant to student needs

- changing the grading system at your school
- establishing more equitable computer use

COLLABORATIVE PROJECT

Exercise 1, 2, or 3 may be used for a collaborative project.

MyWritingLab

MyWritingLab: Technical Communication offers the best multimedia resources for technical communication in one, easy-to-use place. You can find a variety of interactive model documents and case studies. There are also extensive guidelines, tutorials, and exercises for Document Design, Writing and the Writing Process, and Research, and a large bank of diagnostics and practice for grammar review.

19

Formal Analytical Reports

✳ Explore

Click here in your eText to access a variety of student resources related to this chapter.

LEARNING OBJECTIVES

After reading this chapter, you should be able to

- differentiate among four main types of analytical reports: evaluation/assessment, feasibility analysis, causal analysis, and recommendations
- incorporate the structural and format elements of formal analytical reports
- use a planning outline and detailed working outline to efficiently produce a formal report

ON THE JOB...

The Report as "Product"

"Providing consulting services is our business. Our reports are our products, along with maps and other forms of documentation. All our reports provide detailed information, but most present analysis as well. After establishing its scope and objectives, the report describes the methodology and resulting data, interprets the data, draws conclusions, and (usually) recommends actions. A long-term project may require a series of informational and analytical documents. . . ."

—Dr. Brian Guy, vice-president and general manager,
Summit Environmental Consultants

Formal analytical reports are used for lengthy discussions (usually 10 pages or more) or when the topic is important enough to warrant formal presentation (title page, table of contents, formal heading system, formal documentation, and so on).

Less formal reports can use a memo, letter, or semi-formal format.

Analytical reports answer these questions:

1. What data, observations, ideas, and background information can we gather about the topic discussed in this report? (*What do we know?*)
2. What inferences can we draw about the collected data? (*What does it mean?*)
3. What bottom-line conclusions can we draw? (*What does it all mean?*)
4. What recommendations stem from our conclusions? (*What should we do?*)

All readers of analytical reports want the first three questions answered. Many readers want the fourth question answered as well.

"Real-world" analytical reports answer questions for decision makers. Often, such reports provide the main basis for a reader's practical business decision. By contrast, "academic" analytical reports usually employ a more theoretical model although, increasingly, academic researchers are being approached by business firms for answers to difficult real-world questions.

FOUR MAIN TYPES OF ANALYSIS

As an employee, you may be asked to *evaluate* a new assembly technique on the production line. Or, you may be asked to *recommend* the best equipment at the best price. You might have to *identify the cause* of a monthly drop in sales, the reasons for low morale among employees, the causes of an accident, or the reasons for equipment failure. You might need to *assess the feasibility* of a proposal for a company's expansion or investment. There are many varieties of these four main types of analysis, but the procedure remains the same: (1) ask the right questions, (2) search for information, (3) evaluate and interpret your findings, and (4) draw conclusions and possibly recommend actions.

In general, then, your prime responsibility is to answer your reader's questions. In providing those answers, you will need to fulfill several attendant responsibilities:

◆ make the report's purpose clear
◆ use an appropriate structure for that purpose
◆ examine the topic at an appropriate level, and use appropriate language
◆ ensure that the report is readable by evaluating it objectively
◆ write ethically: admit data limitations, and do not suppress contrary evidence
◆ make points clear and well-supported
◆ make the report professional and error-free so that it gets the attention it deserves

Now, let's examine the nature of productive analysis, which allows readers to make informed decisions.

TYPICAL ANALYTICAL PROBLEMS

Far more than an encyclopedic presentation of information, the analytical report shows how you arrived at your conclusions and recommendations. Here are some typical analytical problems.

Will X Work for a Specific Purpose? Analysis can answer practical questions. For example, imagine that your employer is concerned about the effects of stress on employees. She asks you to investigate the claim that low-impact aerobics has therapeutic benefits—with an eye toward such a program for employees. You design your analysis to answer this question: *Do low-impact aerobics programs significantly reduce stress?* The analysis follows a *questions–answers–conclusions* sequence. Because the report could lead to action, you include recommendations based on your conclusions.

The questions posed in such a *feasibility report* are also termed *assessment criteria*. In order to answer the main question of whether low-impact aerobics reduce stress, supporting questions have to be asked (assessment criteria have to be applied):

◆ What causes stress?
◆ How is stress revealed physiologically?
◆ Can the physical manifestations of stress be measured?
◆ What kinds of activities reduce stress? How strenuous do they have to be?
◆ How long do these activities have to be followed before measurable effects are detected?
◆ Do the stress-reducing activities work equally well for all subjects?

Has X Worked as Well as Expected? *Evaluation (assessment) reports*, like feasibility studies, use a series of evaluation criteria to assess the performance or value of equipment, facilities, or programs. Unlike feasibility reports, however, assessment reports apply those criteria after a decision has been made.

Let's imagine, for example, that your engineering firm decided last year to network all of the firm's computers. Now, in assessing that network's performance, your report might use the following criteria to determine if the predicted gains have actually happened:

◆ performance gains, if any (the amount and quality of design work)
◆ communication within the firm (savings in meeting time)
◆ compatibility with the firm's design and communication software
◆ network reliability and downtime
◆ the firm's ability to accept more complex projects

Is X or Y Better for a Specific Purpose? Analysis is essential in comparing machines, processes, business locations, computer systems, or the like. Assume that you manage a ski resort and need to answer this question: *Which of the two most popular snowboard bindings is best for our rental boards?* In a *comparative analysis* of the Burton Cartel and the Flow Five All-Mountain bindings, you would assess the strengths and weaknesses of each binding on the basis of specific criteria (rider control, cost, ease of entry, ease of adjustment, parts availability, and so on), which you would rank in order of importance.

The comparative analysis follows a *questions–answers–conclusions* sequence and is designed to help the reader make a choice. Examples appear in magazines, such as *Consumer Reports* and *Consumers Digest.*

Why Does X Happen? *Causal analysis* is designed to answer questions like this: *Why do small businesses have a high failure rate?* This kind of analysis follows a variation of the *questions–answers–conclusions* structure: namely, *problem–causes–solution.* Such an analysis follows this sequence:

1. Identify the problem.
2. Examine possible and probable causes, and isolate definite ones.
3. Recommend solutions.

An analysis of low morale among employees would investigate causal relationships.

How Can X Be Improved or Avoided? Another form of problem solving focuses on desired results and recommends methods of achieving these results. This type of analysis answers questions like these: *How can we operate our division more efficiently? How can we improve campus security?*

Usually, a *recommendations report* first identifies causes of a problem or components of a desired result. Then, the report presents possible solutions and uses a consistent set of criteria to evaluate each solution in turn. Finally, the report recommends which solution or combination of solutions to implement.

Many readers of solicited recommendations prefer to see the final recommendations first, *before* the full analysis that leads to those recommendations. Chapter 21 describes when a direct recommendations pattern would be more suitable than an indirect pattern.

What Are the Effects of X? An analysis of the consequences of an event or action would answer questions like these: *How has air quality been affected by the local power plant's change from burning oil to coal? Does electromagnetic radiation pose a significant health risk?*

Another kind of problem-solving analysis is done to predict an effect: *What are the consequences of my changing majors?* Here, the sequence is *proposed action– probable effects–conclusions and recommendations.*

Is X Practical in This Situation? The feasibility analysis assesses the practicality of an idea or plan: *Is wood frame housing safe, practical, and affordable in an earthquake zone in the tropics?* In a variation of the *questions–answers–conclusions* structure, a feasibility analysis presents *reasons for–reasons against*, with both sides supported by evidence. Business owners often use this type of analysis.

Combining Types of Analysis. Types of analytical problems overlap considerably. Any one study may in fact require answers to two or more of the previous questions. The sample report in Figure 19.3 (pages 358–365) is both a feasibility analysis and a comparative analysis. It is designed to answer these questions: *Is technical marketing the right career for me? If so, how do I enter the field?*

ELEMENTS OF ANALYSIS

Successful analytical reports feature the following elements.

Clearly Identified Problem or Question

Know what you're looking for. If your car's engine fails to turn over when you switch on the ignition, you would wisely check battery and electrical connections.

Above, a hypothetical employer asked whether a low-impact aerobics program could significantly reduce stress among her employees. The aerobics question obviously requires answers to three other questions: *What are the therapeutic claims for aerobics? Are they valid? Will aerobics work in this situation?* How aerobic exercise got established, how widespread it is, who practises it, and other such questions are not relevant to this problem (although some questions about background might be useful in the report's introduction). Always begin by defining the main questions and thinking through any subordinate questions they may imply. Only then can you determine the data or evidence you need.

With the main questions identified, the writer of the aerobics report can formulate his or her statement of purpose:

Define your goal

This report examines and evaluates claims about the therapeutic benefits of low-impact aerobic exercise.

The writer might have mistakenly begun instead with this statement:

Vague

This report discusses low-impact aerobic exercise.

Words such as *examines* and *evaluates* (or *compares, identifies, determines, measures, describes,* and so on) enable readers to understand the specific analytical activity that is the subject of the report.

Notice how the first version sharpens the focus by expressing the precise subject of the analysis: not aerobics (a huge topic), but the alleged *therapeutic benefits* of aerobics.

Define your purpose by condensing your approach to a basic question: *Does low-impact aerobic exercise have therapeutic benefits?* or *Why have our sales dropped steadily for three months?* Then restate the question as a declarative sentence in your statement of purpose.

Subordination of Personal Bias

Interpret evidence impartially. Throughout your analysis, stick to your evidence. Do not force viewpoints on your material that are not substantiated by dependable evidence.

Accurate and Adequate Data

Do not distort the original data by excluding vital points. Imagine you are asked to recommend the best chainsaw for a logging company. Reviewing test reports, you come across this information.

> Of all six brands tested, the Bomarc chainsaw proved easiest to operate. It also had the fewest safety features, however.

If you cite these data, present *both* findings, not simply the first—even though you may prefer the Bomarc brand. *Then* argue for the feature you think should receive priority.

As space permits, include the full text of interviews or questionnaires in appendices.

Fully Interpreted Data

Explain the significance of your data. Interpretation is the heart of the analytical report. You might interpret the chainsaw data in this way:

> Our cutting crews often work suspended by harness, high above the ground. Also, much work is in remote areas. Safety features therefore should be our first requirement in a chainsaw. Despite its ease of operation, the Bomarc saw does not meet our safety needs.

By saying "therefore," you engage in analysis—not mere information sharing. *Merely listing your findings is not enough.* Tell readers what your findings mean.

Clear and Careful Reasoning

Each stage of your analysis requires decisions about what to record, what to exclude, and where to go next. As you evaluate your data (*Is this reliable and important?*), interpret your evidence (*What does it mean?*), and make recommendations based on your conclusions (*What action is needed?*), you might have to alter your original plan. Remain flexible enough to revise your thinking if contradictory new evidence appears.

Appropriate Visuals

Use visuals generously. Graphs are especially useful in an analysis of trends (rising or falling sales, radiation levels). Tables, charts, photographs, and diagrams work well in comparative analyses.

Valid Conclusions and Recommendations

Along with the informative abstract, conclusions and recommendations are the sections of a long report that receive most attention from readers. The goal of analysis is to reach a valid conclusion—an overall judgment about what all the material means (that X is better than Y, that B failed because of C, that A is a good plan of action).

Here is the conclusion of a report on the feasibility of installing an active solar heating system in a large building:

Offer a final judgment

1. Active solar space heating for our new research building is technically feasible because the site orientation will allow for a sloping roof facing due south, with plenty of unshaded space.
2. It is legally feasible because we are able to obtain an access easement on the adjoining property, to ensure that no buildings or trees will be permitted to shade the solar collectors once they are installed.
3. It is economically feasible because our sunny, cold climate means high fuel savings and faster payback (15 years maximum) with solar heating. The long-term fuel savings justify our short-term installation costs (already minimal because the solar system can be incorporated during the building's construction—without renovations).

Conclusions are valid when they are logically derived from accurate interpretation.

Having explained *what it all means*, you then recommend *what should be done*. Taking into account all possible alternatives, your recommendations urge specific action (to invest in *A* instead of *B*, to replace *C* immediately, to follow plan *A*, or the like). Here are the recommendations based on the previous interpretations:

Tell what should be done

1. I recommend we install an active solar heating system in our new research building.
2. We should arrange an immediate meeting with our architect, building contractor, and solar heating contractor. In this way, we can make all necessary design changes before construction begins in two weeks.
3. We should instruct our legal department to obtain the appropriate permits and easements immediately.

Recommendations are valid when they propose an appropriate response to the problem or question.

Because they culminate your research and analysis, recommendations challenge your imagination, your creativity, and—above all—your critical-thinking skills. Having reached a valid conclusion about *what is*, you now must decide *what ought to be done*. But what strikes one person as a brilliant idea might be seen by others as idiotic or offensive. Depending on whether recommendations are carefully thought out or off the wall, writers earn an audience's respect or its scorn. Figure 19.1 on the next page depicts the types of decisions writers encounter in formulating, evaluating, and refining their recommendations.

Present the report's recommendations directly and clearly

When you do achieve definite conclusions and recommendations, express them with assurance and authority. Unless you have reason to be unsure, avoid noncommittal statements ("It would seem that" or "It looks as if"). Be direct and assertive ("The earthquake danger at the reactor site is acute," or "I recommend an immediate investment"). Let readers know where you stand.

If, however, your analysis yields nothing definite, do not force a simplistic conclusion on your material. Instead, explain your position ("The contradictory responses to our consumer survey prevent a definite conclusion. Before we make any decision about this product, we should conduct a full-scale market analysis"). The wrong recommendation is far worse than no recommendation at all.

Consider All the Details

- What exactly should be done?
- How exactly should it be done?
- When should it begin and be completed?
- Who will do it, and how willing are they?
- Is any equipment, material, or resources needed?
- Are any special conditions required?
- What will this cost, and where will the money come from?
- What consequences are possible?
- Whom do I have to persuade?
- How should I order my list (priority, urgency, etc.)?

Locate the Weak Spots

- Is anything unclear or difficult to follow?
- Is it unrealistic?
- Is it risky or dangerous?
- Is it too complicated or confusing?
- Is anything about it illegal or unethical?
- Will it cost too much?
- Will it take too long?
- Could anything go wrong?
- Who might object or be offended?
- What objections might be raised?

Make Improvements

- Can I rephrase anything?
- Can I change anything?
- Should I consider alternatives?
- Should I reorder my list?
- Can I overcome objections?
- Should I get advice or feedback before I submit this?

Figure 19.1 How to Think Critically about Your Recommendations

Source: Adapted from *The Art of Thinking*, 8th ed. by Vincent R. Ruggiero, copyright © 2006. Reprinted by permission of Pearson Education, Inc.

A GENERAL MODEL FOR ANALYTICAL REPORTS

Every analytical report identifies an issue to be examined or an overall question to be answered for the intended audience. That issue or question is eventually settled in the report's conclusion section. In between, the report employs a series of supporting questions (analytical criteria) to lead to the bottom-line answer. Figure 19.2 illustrates the line of reasoning that might be used for a feasibility report about a proposed downtown location for a multipurpose arena. The city council likes the idea of a downtown location but wants to be certain that the location is practical, so it instructs the consultant to examine traffic flow in the area, potential parking problems, and the ability of existing city services to handle increased demands. The city also asks the consultant to predict the impact on surrounding businesses.

Figure 19.2 shows the relationships among the introduction, the central sections, and the conclusion in one particular type of analytical report. Though other reports will use the same basic *introduction–analysis–conclusion* pattern, no general model can cover all formal analytical reports. Some reports will have one central section; others will have several. Much depends on the scope of the report and on the complexity of the analysis.

Question ⟶ **Answer**

INTRODUCTION	BACKGROUND	ASSESSMENT	CONCLUSION
Report's purpose: *Assess feasibility of proposed arena location*	Explain process by which this site was identified	**Cost of Land** • City property—reserved for public • Equivalent value	Land cost makes location desirable
To be evaluated on basis of ① Cost of land ② Impact on traffic ③ Availability of parking ④ City infrastructure (water, sewer, gas) ⑤ Parking ⑥ Impact on business	Explain why other sites will be assessed later: *priority given to downtown development* Explain information sources	**Impact on Traffic** • Concerts and conventions • Hockey games **Availability of Parking** • Concerts and conventions • Hockey games **City Infrastructure** • Water • Sewer • Gas **Parking** • Requirements for various events • On-site parking • Parking within 4-block radius **Impact on Business** • Restaurants • Shopping	Feasible in terms of traffic flow, parking, existing infrastructure Little positive impact on local business **Recommendation** Retain this site as feasible option, but look at other sites also

Figure 19.2 A Question-to-Answer Development Pattern for Analytical Reports

PARTS OF A FORMAL REPORT

Table 19.1 on the next page presents the sections that usually appear in formal reports, in the order they most often follow. (The numbers in parentheses refer to the suggested order for preparing the sections; following that order of preparation will help you write the report efficiently.) Your supervisor will tell you which sections are required for the specific report you're writing.

Strictly speaking, the transmittal document (letter or memo) does not belong in the front matter (i.e., between the title page and introduction). Transmittal documents *accompany* reports.

Table 19.1 Parts of a Formal Report

Front Matter	Body	Back Matter
Transmittal Document (13) Cover (14) Title Page (8)	Introduction (3)	Sources Cited (6)
Summary (4)	Central Section(s) (1)	Recommendation (2) Consulted (7)
Table of Contents (12)	Conclusion (2)	
List of Illustrations (11)	Recommendation (2)	
Glossary (9) List of Symbols (9) Acknowledgments (10)		Appendices (5)

Introduction

The most important function of an analytical report's introduction is to identify the report's analytical purpose and preview how that purpose will be achieved. Usually that preview includes a list of supporting questions that will be used to answer the report's major question. In other words, the introduction lists the criteria used in the report's assessment or outlines the process used to determine causes, or previews the logical path to be used in arriving at a recommended action. In some cases, you will need to justify your choice of criteria or explain your analytical method.

An introduction may also require some or all of the following elements:

1. The context, situation, or problem prompting this report (background)
2. Type of data on which the report is based and the type of source
3. Other pertinent theoretical or background information
4. Useful illustrations

An introduction also indirectly sets the tone of the report. The phrasing reflects whether the report takes an aggressive stance or uses a more cautious or conciliatory approach. For example, a causal report's direct, no-nonsense approach is signalled as follows:

> The Forestry Ministry assembled an investigation team to determine if
>
> 1. forestry activities in the area contributed to the large destructive debris flow that killed three people, and if
> 2. additional investigation is required to assess the future risk of mass mudslides in the area.

Central Section

Some reports need just one central section. For example, a 10-page causal analysis report might use a central section called "Contributory Causes," with a subsection for each of the factors that may have helped cause the problem or situation.

Other reports may need several central sections. For example, a 35-page assessment of three submitted proposals for a truck-leasing contract might have four central sections, one for each proposal and one entitled "Comparison of Alternatives." Each of four central sections would use the same set of assessment criteria to organize the analysis within each section. The report's conclusion section would identify the best of the three proposals and recommend whether to accept that proposal in its entirety, or to negotiate a modified version.

As you read the central sections of the two sample reports in this chapter, notice how the analytical criteria (supporting or exploratory questions) are presented in the introduction and then used to form logical structures in the report's central section. Notice also that the sample reports use clear, informative headings that identify exactly where you are at any point in the discussion.

ON THE JOB...

Investigation Reports

"We do investigations, which culminate in formal reports that present our technical findings, our explanations, and our recommendations. The reports examine what happened and why. These reports could go to insurance adjustors, lawyers, engineers, or clients, so it would be wise to phrase the summary and recommendations differently for each type of reader. . . ."

—Tom Guenther, consulting structural engineer

Conclusion

The conclusion of an analytical report will interest readers because it answers the questions that sparked the analysis in the first place. Some workplace reports, therefore, place the conclusion *before* the introduction and central sections.

In the conclusion, you summarize, interpret, and (perhaps) recommend. Although you have interpreted evidence at each step in the analysis, your conclusion pulls the strands together in a broader interpretation. This final section must be consistent in three ways:

1. The summary must reflect accurately the main body of the report, and the bottom-line conclusions must be firmly based on information, ideas, and analysis already presented in the report. Do not introduce new material in the conclusion section.
2. Your overall interpretation must be consistent with the findings in your summary and must present an honest and objective appraisal of the material.
3. If you include recommendations, they must be consistent with the purpose of the report, the evidence presented, and the interpretations given.

Some reports require a separate recommendations section

Often, recommendations form the last part of the conclusion section, especially if the report's primary purpose is to assess or to identify causes, but if the report's main purpose is to advise the reader what action to take, create a separate recommendations section. Remember also that not all reports require a set of recommended actions.

A SAMPLE SITUATION

The report in Figure 19.3 (pages 358–365) combines a feasibility analysis with a comparative analysis. The report author's audience/purpose analysis appears on pages 366 and 367.

INTRODUCTION

Technical occupations in environmental industries will experience strong job prospects for the near future, according to Working in Canada (2013). A variety of other technical fields can expect fair to good opportunities in the five-year period ending in 2016 (*Working In Canada*, 2013).

One of the fields with decent job prospects is *sales engineer*, a specially trained professional who markets and sells highly technical products and services. Therefore, recent and impending graduates might consider this alternative career where they could apply their technical training and combine science and engineering expertise with "people" skills. (Nelson, 2001, p. 21; Nezarde, 2012, p. C3)

What specific type of work do technical marketers and sales specialists perform? The Ontario Job Futures website offers this job description:

> [They] sell a range of technical goods and services, such as scientific and industrial products, electricity, telecommunications services and computer services. ... They usually specialize in a particular line of goods or services. ...Sales require constant interaction with clients. ... There is also a growing interdependence between product development and sales. This link expands the specialist's role in providing vital information to product developers about customers' needs. (*Ontario Job Futures,* 2013. para. 1, 3, 6)

(For a more detailed job description, refer to the "Technical Marketing Process," on page 2.)

Undergraduates interested in a technical marketing career need answers to these basic questions:

- Is this the right career for me?
- If so, how do I enter the field?

To help answer these questions, this report analyzes information gathered from professionals as well as from the literature.

After defining *technical marketing*, the following analysis examines the field's employment outlook, required skills and personal qualities, career benefits and drawbacks, and various entry options.

Figure 19.3 A Feasibility Assessment Report   *(continued)*

COLLECTED DATA

Key Factors in a Technical Marketing Career

Anyone considering technical marketing needs to assess whether this career fits his or her interests, abilities, and aspirations.

The technical marketing process. Although the terms "marketing" and "sales" are often used interchangeably, technical marketing traditionally has involved far more than sales work. The process itself (identifying, reaching, and selling to customers) entails six major activities. (Monravia, 2009, p. 38–9):

1. *Market research:* gathering information about the size and character of the target market for a product or service.
2. *Product development and management:* producing the goods to fill a specific market need.
3. *Cost determination and pricing:* measuring every expense in the production, distribution, advertising, and sales of the product, to determine its price.
4. *Advertising and promotion:* developing and implementing all strategies for reaching customers.
5. *Product distribution:* coordinating all elements of a technical product or service, from its conception through its final delivery to the customer.
6. *Sales and technical support:* creating and maintaining customer accounts, and servicing and upgrading products.

Fully engaged in all these activities, the technical marketing professional gains a detailed understanding of the industry, the product, and the customer's needs (Figure 1).

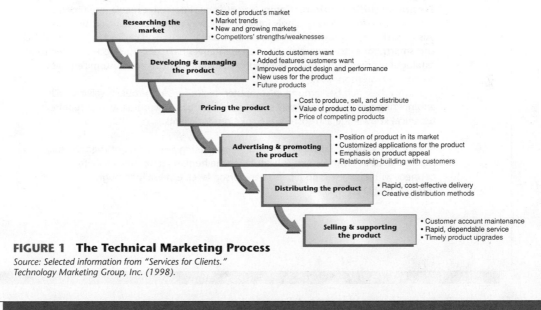

FIGURE 1   The Technical Marketing Process
Source: Selected information from "Services for Clients."
Technology Marketing Group, Inc. (1998).

Figure 19.3 A Feasibility Assessment Report   *(continued)*

Employment outlook. The employment outlook for technical marketing appears excellent for graduates with the right combination of technical and personal qualifications. In 2013, Human Resources and Skills Development Canada predicted strong growth through 2016: "there will be many opportunities for workers with advanced computer skills, knowledge of import and export regulations, and the ability to speak a second language" (HRDC, 2013, para. 5).

As of February 2013, that prediction appears to have been accurate:

- From January 25 to February 4, 2013, Monster.ca posted over 100 Canadian sales and marketing positions in its job listings.
- For the Toronto area alone, on February 4, 2013, AllStarJobs.ca listed 9 technical sales and marketing positions, while on the same day Yahoo Canada's HotJobs site listed 17 such positions in Toronto (2013).
- On February 4, 2013, *eluta.ca,* a Toronto-based job search engine, listed 70 technical sales positions Canada-wide, while Service Canada's online *Job Bank* listed another 54 positions.

Many of the positions listed in early 2013 suggested that speaking a second language was a definite asset.

One especially promising area is environmental technical sales and marketing. Consulting firms, manufacturers of environmental products, recycling programs and equipment providers, purveyors of water purifying equipment, and others all require knowledgeable sales specialists, as all sorts of companies realize the practical values of reducing their so-called "environmental footprint."

Technical skills required. Computer networks, interactive media, multimedia, and social networking will increasingly influence the way products are advertised and sold. Also, marketing representatives often work from a "virtual office," using laptops, PDAs, and smartphones to contact clients. Representatives have real-time access to electronic catalogues, pricing for customized solutions, inventory data, and customized sales contacts (Tostenson, 2012).

With their rich background in computer, technical, and problem-solving skills, engineering graduates are ideally suited for working in environments that require technical knowledge and customized solutions for clients.

Other skills and qualities required. In marketing and sales, not even the most sophisticated information can substitute for the human factor, the ability to connect with customers on a person-to-person level, especially through social media (Viskovich, 2012.)

Figure 19.3 A Feasibility Assessment Report   *(continued)*

One senior sales engineer praises the efficiency of her automated sales system, but thinks that automation will "get in the way" of direct customer contact. Other technical marketing professionals express similar views about the continued importance of human interaction (p. 94).

Besides a strong technical background, marketing requires a generous blend of those traits summarized in Figure 2.

FIGURE 2 Requirements for a Technical Marketing Career

Motivation is essential for marketing work. Professionals must be energetic and able to function with minimal supervision. Career counsellor Anne St. Croix describes the ideal candidates as people who can plan and program their own tasks, who can manage their time, and who have no fear of hard work (personal interview, January 21, 2013). Leadership potential, as demonstrated by extracurricular activities, is an asset.

Motivation alone provides no guarantee of success. Marketing professionals are paid to communicate the virtues of their products or services. This career therefore requires skill in communication, both written and oral. Documents for readers outside the organization include advertising copy, product descriptions, sales proposals, sales letters, and user manuals and online help. In-house writing includes recommendation reports, feasibility studies, progress reports, memos, and email correspondence. Increasingly, marketing professionals also need to incorporate digital images and digital video into their work (Meaux, 2013, p. 57).

Skilled oral presentation is vital to any sales effort, as Anne St. Croix points out. Technical marketing professionals need to speak confidently and persuasively—to represent their products and services in the best possible light (personal communication, January 21, 2013). Sales presentations often involve public speaking at conventions, trade shows, and other, similar forums.

Figure 19.3 A Feasibility Assessment Report   *(continued)*

Beyond motivation and communication skills, interpersonal skills are the ultimate requirement for success in marketing (A. St. Croix, personal communication, January 21, 2013). Consumers are more likely to buy a product or service when they like the person selling it. Marketing professionals are extroverted, friendly, and diplomatic; they can motivate people without alienating them.

Advantages of the career. As shown in Figure 1, technical marketing offers diverse experience in every phase of a company's operation, from a product's design to its sales and service. Such broad exposure provides excellent preparation for countless upper-management positions.

 In fact, sales engineers with solid experience often open their own businesses as "manufacturers' agents" representing a variety of companies. These agents represent products for companies that have no marketing staff of their own. In effect their own bosses, manufacturers' agents are free to choose, from among many offers, the products they wish to represent (Tostenson, 2012).

Another career benefit is the attractive salary—marketing professionals typically receive base pay plus commissions. National figures in March 2012 indicated that Technical Sales Specialists earned, on average, $21.95 per hour, while Sales, Marketing, and Advertising Managers earned an average of $32.97 per hour. The average for all occupations in the federal government's database was $17.81 per hour (Job Futures, 2012). These conservative government figures do not include sales commissions or performance bonuses. Very few of the technical sales positions listed at various websites on February 4, 2013, named base salaries below $40, 000 per year.

Technical marketing is especially attractive for its geographic and job mobility. Companies nationwide seek recent graduates, especially in Ontario and in Saskatchewan. In addition, the interpersonal and communication skills possessed by marketing and sales professionals are highly portable (Meaux, 2012, p. 57).

Drawbacks of the career. Technical marketing is by no means a career for every engineer. Sales engineer Roger Cayer cautions that personnel might spend most of their time travelling to meet potential customers. Success requires hard work over long hours, evenings, and occasional weekends. Above all, the job is stressful because of constant pressure to meet sales quotas (personal communication, February 2, 2013). Anyone considering this career should be able to work and thrive in a highly competitive environment.

A Comparison of Entry Options
Engineers and other technical graduates enter technical marketing through one of four options. Some join small companies and learn their trade directly on the job. Others join companies that offer formal training programs. Some begin by getting experience in their

Figure 19.3 A Feasibility Assessment Report

(continued)

technical specialty. Others earn a graduate degree beforehand. These options are compared below.

Option 1: Entry-level marketing with on-the-job training. Smaller manufacturers offer marketing positions in which people learn on the job. Elaine Carto, president of ABCO Electronics, believes small companies offer a unique opportunity; entry-level salespersons learn about all facets of an organization and have a good possibility for rapid advancement (personal communication, January 26, 2013). Career counsellor Anne St. Croix says, "It's all a matter of whether you prefer to be a big fish in a small pond or a small fish in a big pond" (personal communication, January 21, 2013).

Entry-level marketing offers immediate income and a chance for early promotion. A disadvantage, however, might be the loss of any technical edge one might have acquired in college.

Option 2: A marketing and sales training program. Formal training programs offer the most popular entry into sales and marketing. Mid-size to large companies typically offer two formats: (a) a product-specific program, focused on a particular product or product line, or (b) a rotational program, in which trainees learn about an array of products and develop the various skills outlined in Figure 1. Programs can last from weeks to months.

Former trainees Roger Cayer, of Northland Products, and Bill Collins, of Ontarex, speak of the diversity and satisfaction such programs offer: specifically, solid preparation in all phases of marketing, diverse interaction with company personnel, and broad knowledge of various product lines (phone interviews, February 2, 2013).

Like direct entry, this option offers trainees the advantage of immediate income and early promotion. With no chance to practise in their technical specialty, however, trainees might eventually find their technical expertise compromised.

Option 3: Prior experience in one's technical specialty. Instead of directly entering marketing, some candidates first gain experience in their specialty. This option combines direct exposure to the workplace with the chance to sharpen technical skills in practical applications. In addition, some companies, such as Roger Cayer's, will offer marketing and sales positions to outstanding staff engineers, as a step toward upper management (phone interview, February 2, 2013).

Although this option delays a candidate's entry into technical marketing, industry experts consider direct workplace and technical experience key assets for career growth in any field. Also, work experience becomes an asset for applicants to top MBA programs (Meaux, 2012, p. 58).

Figure 19.3 A Feasibility Assessment Report *(continued)*

Option 4: Graduate program. Instead of direct entry, some people choose to pursue an MSc degree in their specialty or an MBA. According to engineering professor Mary Stewart, MSc degrees are usually unnecessary for technical marketing unless the particular products are highly complex (personal interview, January 29, 2013).

In general, job seekers with an MBA have a distinct competitive advantage. More significantly, new MBA with a technical bachelor's degree and one to two years of experience command salaries from 10 to 30 percent higher than MBAs who lack work experience and a technical bachelor's degree. In fact, no more than 3 percent of job candidates offer a "techno-MBA" specialty, making this unique group highly desirable to employers (Meaux, 2012, p. 58).

A motivated student might combine graduate degrees. Deepak Amarjit, president of Northeast Systems, sees the MSc/MBA combination as ideal preparation for technical marketing (email to Phil Larkin, December 20, 2012).

One disadvantage of a full-time graduate program is lost salary, compounded by school expenses. These costs must be weighed against the prospect of promotion and monetary rewards later in one's career.

An overall comparison by relative advantage. Table 1 compares the four entry options on the basis of three criteria: immediate income, rate of advancement, and long-term potential.

TABLE 1   Relative Advantages among Four Technical-Marketing Entry Options

	Relative Advantages		
Option	Early, immediate income	Greatest advancement in marketing	Long-term potential
Entry level, no experience	yes	yes	no
Training program	yes	yes	no
Practical experience	yes	no	yes
Graduate program	no	no	yes

Figure 19.3 A Feasibility Assessment Report

(continued)

CONCLUSION

Summary of Findings

Technical marketing and sales involves identifying, reaching, and selling the customer a product or service. Besides a solid technical background, the field requires motivation, communication skills, and interpersonal skills. This career offers job diversity and excellent income potential, balanced against hard work and relentless pressure to perform.

College graduates interested in this field confront four entry options: (1) direct entry with on-the-job training, (2) a formal training program, (3) prior experience in a technical specialty, and (4) graduate programs. Each option has benefits and drawbacks based on immediacy of income, rate of advancement, and long-term potential.

Interpretation of Findings

For graduates with a strong technical background and the right skills and motivation, technical marketing offers attractive career prospects. Anyone contemplating this field, however, needs to be able to enjoy customer contact and thrive in a highly competitive environment.

Those who decide that technical marketing is for them can choose among the various entry options:

- For hands-on experience, direct entry is the logical option.
- For sophisticated sales training, a formal program with a large company is best.
- For sharpening technical skills, prior work in one's specialty is invaluable.
- If immediate income is not vital, graduate school is an attractive option.

Recommendations

Those whose interests and abilities match the requirements should consider these suggestions:

1. To get an involved opinion, seek advice from people in the field. A good place to start is by contacting professional associations such as the Canadian Professional Sales Association through its website, www.cpsa.com, or the Direct Sellers Association of Canada, www.dsa.ca.
2. Any career in technical sales and marketing rests on thorough knowledge of the products and services sold, so a good starting point is a technical or scientific diploma or degree.
3. Each of the entry options has its advantages and disadvantages, so the fledgling marketer needs to match the chosen entry option to his/her career goals.

REFERENCES

Notes:

- This report uses APA documentation. References normally start on a fresh page. In this report, the References would be placed on page 9.
- That complete list of references is shown in Figure 9.2, page 159.
- MLA, CSE, and IEEE format variations are appear in Figure 9.2.

Figure 19.3 A Feasibility Assessment Report

CASE	**A Feasibility Assessment**

ON THE JOB...

Components of a Successful Analytical Report

"It must precisely address the scope of the project, nothing omitted, nothing added. It has to be clear and concise. The report must meet all the study's objectives: each objective must get the appropriate amount of coverage. The objectives must be clearly stated, and then it must be clear that they've all been met. The intended audience must find the report readable. For example, we write direct, no-nonsense environmental management plans for construction managers and contractors, but we also present complex environmental assessments for the government review agencies that issue permits. Different readers require different levels of content and readability. . . ."

—**Dr. Brian Guy, vice-president and general manager, Summit Environmental Consultants**

Richard Larkin, author of the report shown in Figure 19.3, has a work-study job 15 hours weekly in his school's placement office. His supervisor, Mimi Lim (placement director), likes to keep abreast of trends in various fields. Richard, an engineering major, has become interested in technical marketing and sales. In need of a report topic for his writing course, Richard offers to analyze the feasibility of a technical marketing and sales career, both for himself and for technical and science graduates in general. Mimi accepts Richard's offer, looking forward to having the final report in her reference file for use by students choosing careers.

Richard wants his report to be useful in three ways: to satisfy a course requirement, to help him in choosing his own career, and to help other students with their career choices.

With his topic approved, Richard begins gathering his primary data, using interviews, letters of inquiry, telephone inquiries, and lecture notes. He supplements these primary sources with articles in recent publications. He will document his findings in APA (author–date) style.

As a guide for designing his final report, Richard completes the following audience/purpose profile.

Audience Identity and Needs

My primary audience consists of Mimi Lim, placement director, and the students who will be referring to my report as they choose careers. The secondary audience is my writing instructor. The data I've uncovered will help me make my own career choice.

Lim is highly interested in this project, and she has promised to study my document carefully and to make copies available to interested students. Because she already knows something about the technical marketing field, Lim will need very little background to understand my report. Many student readers, however, may know little or nothing about technical marketing, and so will need background, definitions, and detailed explanations. Here are the questions I can anticipate from my collective audience:

- What, exactly, is technical marketing and sales?
- What are the requirements for this career?
- What are the pros and cons of this career?
- Could this be the right career for me?
- How do I enter the field?
- Is there more than one option for entering the field? If so, which option would be best for me?

Audience Attitude and Temperament

Readers likely to be most affected by my document are students who will be making career choices. I would expect my readers' attitudes to vary widely.

To connect with this array of readers, I will need to persuade them that my conclusions are based on dependable data and careful reasoning.

Audience Expectations

I know that my readers are busy and impatient, so I'll want to make this report concise enough to be read in no more than 15 or 20 minutes.

Essential information will include an expanded definition of technical marketing and sales, the skills and attitudes needed for success, the career's advantages and drawbacks, and a comparison of various paths for entering the career. Throughout, I'll relate my material to many technical and science majors, not just engineers.

The central section of this report combines a feasibility analysis with a comparative analysis. Therefore, I'll structure the feasibility section with reasons for and reasons against. In the comparison section, I'll use a block structure followed by a table that presents a point-by-point comparison of the four entry paths. Because I want this report to lead to informed decisions, I will include concrete recommendations that are based solidly on my conclusions. An informative abstract will address various readers who may not want to read the entire report.

My tone throughout should be conversational. However, because I am writing for a mixed audience (placement director, students, and writing instructor), I will use a third-person point of view.

This report's front matter (title page and so on) appears in Chapter 20.

THE PROCESS OF WRITING REPORTS

The efficient writing process described in Chapter 3 certainly applies to writing lengthy reports. Also, the research advice in Chapters 7, 8, and 9 applies to the process of gathering, recording, and documenting information for formal reports. To avoid unnecessary effort and to save time in producing a lengthy report, follow the advice in Chapters 3 and 7.

ON THE JOB...

Visual Outlines

"For longer documents, I set up a graphics page with the sections of the report or proposal, almost like a flow chart. Then, I write the sections in turn, referring to my notes and rough paragraph outlines. It's a kind of outline system that's suited to my visual mind. Taking this organized approach has been especially useful in writing the training and user documentation for the software development and quality management systems I've been working on for the past few years. In both of those topic areas, the reader could take more than one path, which further necessitated visual planning of the entire document before I started to write it. . . ."

—Jan Bath, civil engineering technologist and application developer

Project: Assess the CT83 Pit Jack Adapter's design and perhaps recommend improvements in (1) the weight and manoeuvrability of the jack adapter, (2) the choice of component materials, (3) the CT83's range of applications, and (4) the height.

R & D method:

- Consult the internet for competing products—automotive, heavy-duty equipment; transit vehicles.
- With assistance from our professors, assess the current design to evaluate the chosen materials, design components and stress points, and overall dimensions.
- Interview experts in the industry.
- Assess the gathered data and use the assessment as a basis for design modifications.
- Assess the cost of modifications; if necessary, redesign the adapter to decrease costs.
- Create drawings and models for engineering assessments and, later, tech sales material.
- If time allows, build and test a prototype.

Possible report topics:

- history of the CT83's development (?)
- problems/obstacles/design flaws—analysis after industry feedback
- solutions/design improvements
- cost analysis
- final recommendations (if any)

Information sources:

Omega Lift website.

Mott, R.L. (2008). *Applied strength of materials*, 5th ed. New Jersey: Pearson Education.

Southern Tool Company website.

Interviews with Bill Glaus, Bill Sand, Terry Lockhart, Henry Murphy.

Figure 19.4 A Planning Outline

Using Outlines

You can use three types of outlines to write top-quality reports efficiently:

1. Use a *planning outline* to guide your research and initial planning. That outline will change as you gather material, but such an outline will continually remind you of the report's purpose and the analytical criteria to achieve that purpose. Figure 19.4 above shows a planning outline for a recommendations report.

2. When you have chosen, evaluated, and analyzed the material for your report, write a detailed formal *working outline*, including:

   ◆ the report's *working title*
   ◆ a *purpose/audience* statement to remind yourself of the reason for the report

The equivalent of a pre-summary statement in an essay is the *thesis* or main point

- a *pre-summary* statement to further remind yourself that everything in the report contributes to a "bottom-line" answer
- all *headings* and subheadings, named and formatted as they will be in the finished report's body
- a brief description of every *paragraph* in the finished report
- the name and number of each *illustration*, placed where it will appear in the report

This working outline will take some time to write because it forms a complete blueprint for the first draft, but a thorough, well-conceived outline will dramatically decrease the time required to compose, revise, and edit your first draft. Also, *each keystroke that goes into the working outline will appear in that first draft*; the headings and illustration labels will all be in place, and even the paragraph description phrases will likely end up in their respective paragraphs.

Compare the headings and paragraph descriptions in Figure 19.5 on pages 370–373 with the corresponding report in Figure 19.6 on pages 375–393.

3. As you use the working outline to compose the first draft, you can refer to your notes to establish the exact content of each paragraph, or you can write a brief, informal *paragraph outline* for each one. See page 374 for an example. Such "quickie" outlines don't have to be neat; they merely help you write coherent, unified paragraphs quickly. You may prefer to create a paragraph outline for each new paragraph as you come to it, or you may prefer to write outlines for several paragraphs in succession.

Now, let's see this sequence of outlines at work.

The writers are mechanical engineering students engaged in a "real-life" project: the assessment and redesign of an industrial tool. Their reader, Dave Cochrane, owns CT Solutions, which designs and builds equipment used in the maintenance and repair of heavy equipment. He has commissioned this study in hopes of refining his CT83 Pit Jack Adapter so that it may become commercially viable.

A description of the overall process used by the writers appears in page 1 of their report (Figure 19.6, page 380). Figure 19.4 is their planning outline, which they used to guide their research and development process.

After seven weeks of interviews, secondary research, design work, analysis, and computer modelling, the writers produce the detailed working outline presented in Figure 19.5. Such an outline is useful for any writer, but particularly important for a collaborative writing project—this outline keeps the team on track and speeds up the writing process.

ON THE JOB...

Producing a Consultant Report

"A contract lists the expected scope of the project, the objectives, the tasks, the deliverables, the schedule, and the cost. We assign a project manager and a team who gather data, analyze it, write conclusions and recommendations, create supporting visuals, and add appendices. The project manager either writes or coordinates the writing. A completed draft is internally reviewed by people outside the project team, who follow a detailed set of guidelines to ensure quality control. Finally, we print and bind copies and make PDFs. . . ."

—**Dr. Brian Guy, vice-president and general manager, Summit Environmental Consultants**

The title
emphasizes
"recommended"

This statement
essentially poses
the report's
"question"

And here's the
answer to that
question

All headings use
the font type
and font size
that will appear
in the finished
report

Each bullet
represents a
planned
paragraph

The writers will
know if a
transition
paragraph is
necessary when
they write the
first draft

Title: Recommended Changes to the CT83 Pit Jack Adapter

Purpose/Audience
This report will help Dave Cochrane decide whether it is feasible to modify the CT83 Pit Jack Adapter to make it commercially viable, and, if so, what changes are needed.

Thesis (bottom-line) Statement
The CT83 has been transformed into an improved model, the CT04, which has enough promise to warrant making a prototype.

1.0   INTRODUCTION
1.1 Purpose
- purpose of report—improve CT Solutions CT83 Pit Jack Adapter

Figure 1   CT83 Pit Jack Adapter

1.2 Research and Development Method
- seven stages of process—numbered list

2.0   RESEARCH
2.1 Interviews
- lead-in

2.1.1 Automotive Industry
- Bill Sand—no
- transition?

2.1.2 Commercial Transit Industry
- Gerry Hanson—interested
- Hanson's recommendation
- key issue is versatility

2.1.3 Heavy-Duty Industry
- Terry Lockhart—interested
- desires the same design criteria as Hanson

2.2 Examination of Products Available to Industry
- examined: transmission jacks, engineered stands, and heavy-duty engine stands

Figure 19.5 A Working Outline   *(continued)*

2.2.1 Transmission Jacks
- lead-in paragraph
- advantage of the column jack is height, but unstable

Figure 2 Column-type transmission jack
- floor-type jack–larger maximum loading capacity and more stable
- drawback

Figure 3 Floor-type transmission jack

2.2.2 Engineered Stands
- very basic
- the CT83 Pit Jack Adapter falls into this category

Figure 4 Engineered jack stand

2.2.3 Heavy-duty Engine Stands with Rotation Capabilities
- a recent innovation

Figure 5 Heavy-duty rotating engine stand

2.3 Problems and Obstacles
- four main problems: height adjustability, lack of versatility, incorrect material selection and optimization, cost efficiency

3.0   SOLUTIONS

3.1 Redesign Components
- transition paragraph

3.1.1 Casters
- reason for drop forged steel caster with polyurethane tread
- description

Figure 6 Semi-steel 2500 lb. caster

3.1.2 Frame
- reasons for material choice

Figure 7 Frame
- design related to cost

3.1.3 Teleposts
- purpose and specifications (description)

Figure 8 Telepost height adjustment assembly

Each figure is named and inserted where it will appear in the report

This paragraph likely requires a bulleted list

Figure 19.5 A Working Outline   *(continued)*

These specifications will be drafted directly from project notes

3.1.4 Mounting platform
- specifications—result in weight saving and usability (might be two paragraphs)

Figure 9   6061-T6 lightweight mounting platform

3.1.5 Mounting Plates
- specs
- how they will be used

Figure 10 Mounting plate

- introduce Figure 11

Figure 11 Illustration of quick-change mounting posts

3.2 Evolution in Design
- how the design meets the four design needs
- lead-in to Figure 12

Figure 12 CT04 engineered stand improvements

3.3 Room for Growth
- rotation and quick-change posts—why? (connect to Figure 13)

Figure 13 Illustration of concept for engine stand adapter compatible with CT04 frame

3.4 Solution Specifications
- lead-in to Table 1

Table 1 Solution specifications for CT04 Engineered Stand

4.0   COST ANALYSIS

4.1 Redesign
- two main variables
- resolving the cost–design paradox

4.2 Projected Cost Summary
- lead-in to Figure 14's main idea—cost is a factor of scale

Figure 12 occupies a whole page, so its lead-in paragraph will ideally appear at the bottom of the previous page

Figure 19.5 A Working Outline

(continued)

Figure 14 Component costs of manufacturing the CT04 stand

- main point of Figure 14
- check with prototype builders re: cost saving
- introduce Figure 15

Figure 15 Total projected manufacturing cost

- assume a 35% markup
- caution re: figures (list them)

5.0 CONCLUSION AND RECOMMENDATION

- the CT04 addresses all apparent flaws with the CT83
- cannot recommend production because . . .
- should interview craftspeople
- excellent potential

APPENDIX A: Stress calculations

APPENDIX B: CT83 and CT04 engineering drawings and solid models

APPENDIX C: Materials quotations

Figure 19.5 A Working Outline

When you compare this outline with the headings, order of topics, illustrations, and number of paragraphs in the final report, you'll notice that some aspects have changed during the writing process (a perfectly normal occurrence). For example, this outline lists 45 paragraphs, but the final report ended up with 52. Not all possibilities can be anticipated.

As an example of how the writing team works with its working outline, look at the second paragraph in report section 4.1. The working outline's note in Figure 19.5 includes

◆ resolving the cost–design paradox

That note identifies the paragraph's basic idea; then, when it's time to draft the paragraph, the group might have created a simple paragraph outline, such as

◆ cost–design paradox resolved
1. math stress analysis (Appendix A)
2. dimensions/material size affect mfg. cost & vice versa
3. repeat until all criteria met

Now, with the key elements in place, the team drafts a paragraph in the form of a process description (page 10 of the report in Figure 19.6, on page 389):

> This situation created a cost–design paradox that could be resolved only through a process of iteration. Appendix A illustrates the mathematical stress analysis of the CT04's framework to derive the material selection and final sizing of the stand. The cost of manufacturing was then determined according to material size and dimensions, and allowed to change the dimensions and material sizing if a cost reduction was necessary. This procedure was repeated until we established a design that provided solutions to all criteria.

The three-stage series of outlines keeps the group organized and focused. The process also prevents duplicated effort—nearly every word in the working outline ends up in the final report. You might prefer to write your paragraph outlines in pen or pencil, perhaps in paragraph sequences, and then key in the paragraphs as you compose. Find the method that best suits your working style, but use outlines!

A FORMAL ANALYTICAL REPORT

Figure 19.6, on the following pages, presents the report produced from the working outline in Figure 19.5. To save space in this text, the report's 12 appendix pages have not been included. The formatting has been altered somewhat to fit it onto a 6.5″ by 8.5″ sample page, but the essential page design has been retained.

1000 K.L.O. Road, Kelowna, British Columbia, V1Y 4X8

April 19, 2014

Mr. Dave Cochrane
Owner, CT Solutions
6395 Star Road
Vernon, British Columbia
V1B 3J9

Dear Mr. Cochrane:

Re: The CT83 Pit Jack Adaptor

In response to your request, we have completed an analysis of the potential of mass producing the CT83 Pit Jack Adapter. The results of that analysis are provided in the enclosed report, "The CT83/CT04 Pit Jack Adapter and Engineered Stand."

As the "CT04" part of the title suggests, we have modified the device to make it lighter, more versatile, and cheaper to produce. Those modifications have resulted from a process that has included consultation with industry, internet research, engineering analyses of the original design and its modifications, and cost analyses. All the engineering stress calculations point to a strong, versatile, usable piece of equipment. Also, our design modifications have been directed at producing the CT04 as inexpensively as possible. The projected manufacturing cost analysis for the CT04 shows a unit cost of $535.00, based on a mass-production scale. That cost may be further reduced by consulting tradesman and craftsman involved in the production of the first few stands.

At this time, we do not have test data to recommend mass production of the CT04 Modular Engineered Stand. However, we recommend that CT Solutions manufacture a prototype and send it to a suitable testing facility. After the stand has been tested, you will be able to confirm whether the CT04 Modular jack and stand should go into mass production, or undergo further design modifications.

We have enjoyed working on this project and would be pleased to be involved in further development of the CT04 Pit Jack Adapter and Engineered Stand. As a result of our research, we believe the CT04 has strong potential for commercial success.

We would be happy to discuss the report and its recommendations with you, at your convenience. You can contact our group through me, at 250-542-6654, or at vincible@shaw.ca. We look forward to hearing your reactions.

Yours sincerely,

Vince Cummings

Vince Cummings

Figure 19.6 A Formal Analytical Report *(continued)*

THE CT83/CT04 PIT JACK ADAPTER AND ENGINEERED STAND: ASSESSMENT AND PRODUCTION RECOMMENDATION

Prepared for Dave Cochrane
Owner, CT Solutions

Prepared by Vince Cummings, Joe Trainor,
Brandon Vidal, and Greysen Aby

Submitted:
April 19, 2014

Figure 19.6 A Formal Analytical Report

(continued)

Recommendation: CT83 Pit Jack Adapter / CT04 Engineered Stand

SUMMARY

An analysis of the potential of mass-producing the CT83 Pit Jack Adapter has revealed four problems with the CT83 Pit Jack Adapter: a missing height adjustment, lack of versatility, incorrect material choice and usage, and low cost-efficiency. The CT04 Engineered Stand addresses all of these problems.

- Versatility and adjustability concern industry professionals. These problems were addressed by moving to a modular frame incorporating quick-change adapter posts and height-adjustable teleposts. Replacing the rigid mounting pillars of the CT83 Pit Jack Adapter with teleposts allows for a simple but effective height adjustment without the need for hand tools. The quick-change adapter posts featured on the frame increase versatility and will allow future expansion adapters compatible with the CT04 platform.

- The issue of incorrect material specifications has been addressed by a full stress analysis of the structure, followed by selecting the optimum materials and material sizes for the typical working environment of the CT04 Engineered Stand.

- Closely tied to the material choice, a manufacturing cost analysis for the CT04 Engineered Stand has been completed, along with the projected cost of manufacturing one, 50, and 500 units.

The CT04 Engineered Stand has a footprint of 48"H × 36"W × 62"L, and a max loading capacity of 8000 lb. The stand is made out of a combination of AISI 1020 ANN, A-500 Grade "A" structural tubing, and 6061-T6 aluminum, and has a full assembly weight of 344 lb.

The projected manufacturing cost analysis for the CT04 shows that a per-unit cost based on a mass-production scale is $535.00; however, it may be possible to further reduce this cost by consulting tradespeople and craftspeople involved in the production of the first few stands.

At this time, because of incomplete test data, we cannot yet recommend mass-production of the CT04 Engineered Stand, but CT Solutions should manufacture a prototype and send it to a suitable testing facility. After the stand has been tested for an appropriate period of time, it will be much easier to make a sound decision as to whether the CT04 Engineered Stand should go into mass production, or undergo further design modifications.

ii

Figure 19.6 A Formal Analytical Report

(continued)

Recommendation: CT83 Pit Jack Adapter / CT04 Engineered Stand

TABLE OF CONTENTS

iii

Figure 19.6 A Formal Analytical Report   *(continued)*

Recommendation: CT83 Pit Jack Adapter / CT04 Engineered Stand

LIST OF ILLUSTRATIONS

FIGURES

iv

Figure 19.6 A Formal Analytical Report

(continued)

Recommendation: CT83 Pit Jack Adapter / CT04 Engineered Stand

1.0 INTRODUCTION

1.1 Purpose

This report determines how the original CT Solutions CT83 Pit Jack Adapter (Figure 1) can be improved to serve current industry needs. The original design bolted a floor-type transmission jack to the top of its frame, raising the jack off the shop floor by 36 inches. This elevation increase enabled service technicians to work on transport vehicles and heavy-duty equipment in an upright position either in a pit or with a hoist, increasing efficiency by up to 200%.

Figure 1 Model of the Original CT83 Pit Jack Adapter

The CT83 Pit Jack Adapter has generated positive feedback, including several manufacturing requests; however, the current design has some critical engineering defects that must be addressed before it can be manufactured. Therefore, this report investigates the detailed engineering changes necessary to produce the CT83. A recommended redesign of the original CT83 Pit Jack Adapter has resulted from the following process.

1.2 Research and Development Method

1. **Preliminary Research:** Most of the initial information came from a series of interviews and meetings with industry professionals. All interviewees were asked similar questions about their work requirements and where the current design for the jack adapter falls short. Also, we found various components and competing products on the internet.
2. **Analysis of Gathered Data:** After reviewing the data in a series of meetings, our engineering team created a preliminary list of all necessary design changes.
3. **Early Concept for Redesign of the CT83:** All aspects of the redesigned stand had to be combined into a complete, cohesive unit. As a result, the redesigning task was viewed as one large problem to be solved rather than several small problems. Addressing the problems and flaws uncovered by the research was handled collectively by the team, rather than in individually delegated tasks. By methodically attacking all key design faults together, all team members were able to provide input into the redesign of the jack adapter.
4. **Final Research:** Following the redesign stage, the potential prototype design was presented to the previously interviewed professionals for more feedback. This second series of meetings identified a few more minor issues that needed to be addressed.
5. **Final Concept for Redesign of the Adapter:** The second set of industry suggestions and comments helped lead to a final design for the jack adapter.
6. **Mathematical Analysis:** A mathematical analysis of the final adapter concept provided the data to determine material specifications, safety factors, and final dimensions.
7. **Cost Analysis:** Once material sizes and final dimensions were decided, an accurate cost for the new design was determined. The resulting comprehensive cost breakdown for the redesigned adapter for prototype and mass-scale production is provided in Section 4.
8. **Engineering Drawings and Solid Models:** The full set of detailed engineering drawings and production plans, as well as working assembly drawings for the adapter, are found in Appendix B of this report.

1

Figure 19.6 A Formal Analytical Report

(continued)

Recommendation: CT83 Pit Jack Adapter / CT04 Engineered Stand

2.0 RESEARCH

2.1 Interviews

Interviews were conducted within three targeted industries: the automotive industry, the commercial transit industry, and the heavy-duty industry.

2.1.1 Automotive industry

After having been shown models and engineering drawings of the CT83 Pit Jack Adapter, Bill Sand, the shop foreman at Cosworth Specialty Motors, could not see a plausible use for the CT83 in his maintenance environment. He believes that the CT83 Pit Jack Adapter is better suited to the heavy-duty use for which it was originally designed; the CT83 is too large for a general automotive setting: "Most likely, the CT83 would actually hinder maintenance times, and potential benefits we might achieve through extra capacity and stability wouldn't be realized. A regular floor-type or column-type transmission jack is suitable for any situation encountered in our maintenance shop" (personal communication, February 10, 2014).

Because a tool such as the CT83 would not benefit the automotive industry, Sand was unable to provide suggestions for redesigning the CT83 Pit Jack Adapter.

2.1.2 Commercial transit industry

Gerry Hanson, the service manager of Bell Equipment, showed interest in the CT83 and was able to illustrate some scenarios where his facility would benefit from its use. He pointed out several possible design improvements (personal communication, February 19, 2014).

He recommended a height adjustment that would allow the CT83 to be used on a variety of jobs. Because not all maintenance shops have the same equipment, or emulate the exact same working environment, a height adjustment would enable the adapter to conform to the needs of different maintenance facilities.

For Hanson, another key issue is the versatility of the CT83. Having just recently purchased a heavy-duty engine stand capable of 360 degrees of rotation, he provided us with the idea of increasing versatility by enabling different attachments to be placed onto the CT83. For one thing, heavy-duty engine stands capable of rotation are very expensive—he had just paid $5000.00 for a Revolver model. If it could be made possible for the adapter to function both as an engine stand and as a jack adapter, it would not only increase versatility, but also most likely be cheaper for maintenance facilities in the long run. This particular suggestion is explored and realized in further depth later in this report.

This interview illustrated quite clearly that addressing the versatility and ease of use of the CT83 could create serious potential for a modified jack adapter in the commercial transit industry.

2

Figure 19.6 A Formal Analytical Report *(continued)*

2.1.3 Heavy-duty industry

On February 23, 2014, Terry Lockhart, an instructor in the Heavy Duty Trades Department of Okanagan College (OC), showed a great deal of interest in the CT83 Pit Jack Adapter. He provided several possible scenarios where the OC training facility could use such a tool.

Lockhart desires the same design criteria as the professional in the commercial transit industry, especially the height adjustment. An added height adjustment could enable the CT83 to raise heavy equipment transmissions off the facility's concrete floor to a comfortable height, while providing a mobile working platform for the technician. The feedback received from Lockhart at OC was very similar to the feedback received from Hanson at Bell Equipment. Both hope to use the CT83 for much more than just transmission removal.

2.2 Examination of Products Available to Industry

In order to conduct further research, and in an effort to determine whether or not the CT83 is a unique product, we examined three types of tooling currently being used in industry: transmission jacks, bottle jacks, and heavy-duty engine stands (with rotational capabilities). Although engine stands with rotational capabilities are not directly related to this report, or to the CT83 Pit Jack Adapter, several of our research sources mentioned their growing popularity in industry.

2.2.1 Transmission jacks

Two primary types of transmission jacks are used in industry: the column type and the floor type. Both types of jacks have advantages and disadvantages.

As shown in Figure 2, the main advantage of the column jack is its height. The ability to work underneath heavy equipment and commercial transport vehicles in an upright position often increases the technician's efficiency. However, the increased working height makes the jack unstable due to a higher centre of gravity, and the decreased cross-section and inertia of the base result in a decreased maximum loading capacity.

Figure 2 Omega Column-type Transmission Jack (Shinn Fu)

The floor-type jack is shown in Figure 3. Its centre of gravity is much closer to the ground, which, coupled with the increased cross-section and inertia of its base, gives it a much larger maximum loading capacity than the column jack. Its lower operating centre of gravity also makes it much more stable than a column-type jack. The greater stability gives the jack higher maximum loading capacities, and also makes it safer for the technician.

The main drawback of the floor-type jack is the difficulty associated with working underneath heavy equipment so low to the ground. Using a creeper or a similar device becomes necessary, and in most circumstances the technician's mobility becomes severely limited. Its low working height and difficult working conditions compromise the technician's efficiency.

Figure 3 Omega Floor-type Transmission Jack (Shinn Fu)

3

Figure 19.6 A Formal Analytical Report

(continued)

Source: **Figures 2 & 3:** Shinn Fu Company of America

Recommendation: CT83 Pit Jack Adapter / CT04 Engineered Stand

2.2.2 Bottle jacks

Bottle jacks, like the one shown in Figure 4, are usually very basic tools used as support structures, often incorporating a height adjustment. They are rated by their design engineers for a maximum loading capacity and are used to support heavy equipment while it is being serviced.

2.2.3 Heavy-duty engine stands with rotation capabilities

One of the primary redesign elements established in our evaluation of the CT83 was an increase in versatility. Automotive stands with rotational capabilities have been commonplace for some time. However, only recently have engineers deigned heavy-duty engine stands with rotational capabilities to maintain extremely large diesel engines. The type of stand illustrated in Figure 5 is gaining popularity every day in industry. The stable frame and high maximum loading capacity of the CT83 are worth investigating for possible incorporation.

Figure 4 Omega Bottle Jack (Shinn Fu)

The CT83 Pit Jack Adapter falls into the category of engineered stands. A check of several websites for engineered stands similar to the CT83 Adapter revealed nothing similar being currently used in industry.

2.3 Problems and Obstacles

Four main problems with the CT83 were determined through the industrial interviews and an examination of other industrial equipment.

Figure 5 Hein-Werner Rotating Engine Stand (Shinn Fu)

- **Height adjustability:** Due to varying operating environments, and possible varying applications for the CT83, this device really must add height adjustability.
- **Lack of versatility:** An ability to accommodate different types of floor jacks, and possibly even adapters, is currently missing. The redesigned jack adapter must be more versatile than its predecessor. The final design must be able to conform to the existing equipment used at any given shop.
- **Incorrect material selection and optimization:** The CT83 was designed and fabricated with little regard for proper material selection and application. This approach has resulted in a heavy, over-designed adapter that would be expensive to produce. Proper material selection and use will provide the strength, safety, and cost-effectiveness needed in order to move to a mass-producible version.
- **Cost efficiency:** Ultimately, the goal of this report is to determine whether CT Solutions can make money with this new product. A thorough redesign of the adapter must reduce cost while still ensuring that a high level of quality and safety is maintained.

4

Figure 19.6 A Formal Analytical Report

(continued)

Source: **Figures 5 & 6:** Shinn Fu Company of America

3.0 SOLUTIONS

3.1 Redesign Components

Feedback from industry showed that all aspects of the original pit jack adapter would need to be addressed in order to supply potential customers with a useful tool. A detailed explanation of all changes to the CT83 design, moving from the ground up, is illustrated in the following section.

3.1.1 Casters

In light of the environment that the pit jack adapter would be operating in, the following design seems the most suitable. Most maintenance shops will likely have rough, uneven floors with floor dry or dirt on the ground, so a caster with a high loading capacity and smooth manoeuvrability is needed. At first, a solid steel caster seemed to be the best choice due to its high loading capacity; however, the solid steel caster does not provide the pit jack adapter with the required manoeuvrability. Therefore, we chose a drop-forged steel caster with polyurethane tread.

Figure 6 Semi-steel 2500 lb. Caster

The drop-forged steel caster with polyurethane tread has a 6" diameter wheel, a 2" tread width, and a total capacity of 2500 lb. (Figure 6). Because four casters are required, the total capacity that all four casters can take is 10 000 lb. The semi-steel design greatly adds to the capacity of the caster, while the polyurethane tread ensures easy rolling of the pit jack adapter under maximum load. Ease of use is one of the key design considerations for the pit jack adapter, and the steel/poly casters greatly improve the usability while only marginally increasing the price. Suitable casters may be purchased for $21.00 each.

3.1.2 Frame

The frame of the design will end up taking the majority of the load, which will cause bending moments at the base of the load-bearing posts and at the points where the centre channels connect with the side rails. For the frame, our team contemplated changing to a less expensive steel than square tubing. We considered fabricating the frame from I-beam or wide flange; however, suitable sizes were not available and would not have greatly reduced weight or cost. The wall thickness of the tubing was reduced from ⅜" to ¼" while retaining the maximum loading capacity required (Figure 7).

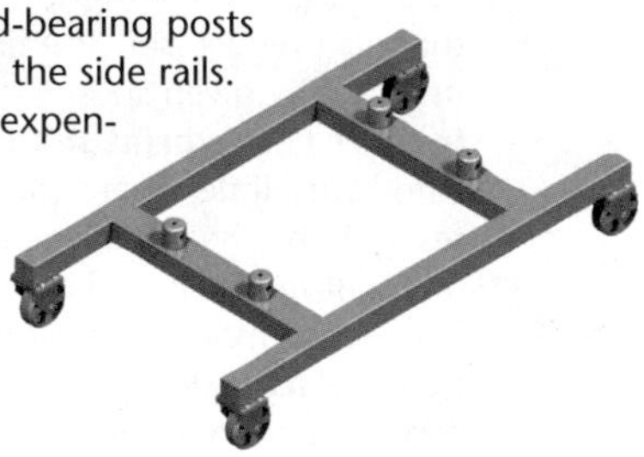

Figure 7 Frame

Figure 19.6 A Formal Analytical Report

(continued)

The frame itself has a maximum loading capacity of 16 000 lb. However, for safety reasons, it should be rated at 8000 lb. The frame is constructed out of A-500 grade "A" structural tubing, the side rails are 3"H × 3"W × ¼"T, and the centre channels are 3"H × 4"W × ¼"T. With a maximum load rating of 8000 lb, the pit jack adapter should have a wide range of applications. For example, the frame should easily be able to support the back end of a city bus (half of the bus's net weight), and theoretically the frame should be able to support the entire weight of the bus without failing.

Cost is decreased by increasing the size of the centre channels. Changing from 3"H × 3"W square tubing to 3"H × 4"W rectangular tubing, we incorporated off-the-shelf teleposts to introduce a simple height adjustment.

3.1.3 Teleposts

With wider frame centre channels, teleposts were incorporated into the design as a simple height adjustment. Each post is rated at 8300 lb. maximum loading capacity; with four posts, the total maximum loading capacity is 33 200 lb. A 33 200 lb. capacity far surpasses what is required for the CT04 design. The posts provide a height range of 18"–33". The posts use a simple design featuring two pieces of round tubing; one piece has a diameter of 3", and the other has a diameter of 2½" (Figure 8). The smaller-diameter piece fits inside the larger-diameter piece and can be secured at the desired height with the use of a shear pin. Each post comes equipped with a fine-tuning height adjuster, which can be used after the pin is secured to acquire the perfect height required for the job.

Since the teleposts arrive ready to use right off the shelf, the only extra costs incurred with this design would be for required bracing materials to be welded between each telepost. This would ensure safety and stability in off-centre loading or side-loading situations. Total costs for the teleposts are $20.50 each and $20.00 for the bracing materials when purchased on a mass-production scale.

**Figure 8
Telepost Height
Adjustment Assembly**

3.1.4 Mounting platform

The mounting platform was created to increase stability to the teleposts at higher adjustments (Figure 9). This design features all-aluminum construction. The weight savings and improved ease of use should more than make up for the extra cost incurred. The mounting platform is made from 6061-T6 aluminum, the side rails are 2"H × 1"W × ⅛"T, the mounting caps are 3" outside diameter × 2.5" inside diameter welded tubing, and the support plates are 3" diameter solid.

If the mounting platform were made from steel, its weight would be around 43 lb. Although this weight is not unmanageable for one person, the length of the platform would make it quite awkward for a single technician to handle. All-aluminum construction reduces the platform weight to about 15 lb. so it can easily be assembled and adjusted by a single operator.

**Figure 9 6061-T6
Lightweight
Mounting Platform**

6

Figure 19.6 A Formal Analytical Report

(continued)

3.1.5 Mounting plates

The mounting plates will be used to bolt any existing floor-type transmission jack to the structure. We chose 4"W × ½"T C1018/1020 flat bar steel for the mounting plates (Figure 10). These plates require quite a bit of machining, and C1018/1020 is easily machined.

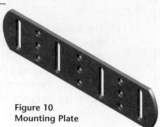

**Figure 10
Mounting Plate**

There is no standard for how a floor-type transmission jack is designed, so each mounting plate will vary from use to use. Several different mounting plates, with different bolt patterns, must be available to accommodate the many existing name-brand floor-type transmission jacks. Providing customized mounting plates will affect the cost due to the machining that will be required for each different plate.

3.2 An Evolved Design

The final design addresses all of the four main pressing design issues: versatility, adjustability, proper material selection/optimization, and cost-efficiency:

- Teleposts provide the design with a height adjustment, cost reduction, and increased usability.
- A modular assembly increases the versatility of the design by featuring the quick-change posts illustrated in Figure 11. The quick-change posts highlighted in the figure can easily adapt to different attachments without requiring hand tools. Adapters fit snugly over the posts on the frame and are secured in place with a locking shear pin. For example, the mounting platform detailed in Figure 9 could be pinned directly to the frame to allow the user to work closer to the ground.

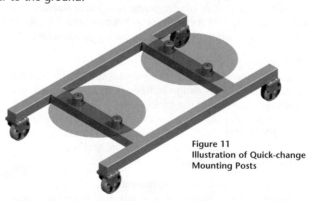

**Figure 11
Illustration of Quick-change
Mounting Posts**

The original pit jack adapter was designed for one specific job with no real engineering in its design. It is now more than just a pit jack adapter. Its increased versatility and adjustability gives this new design a wide range of capabilities. The CT04 Engineered Stand is detailed in Figure 12 on the next page.

7

Figure 19.6 A Formal Analytical Report

(continued)

Figure 12 occupies a whole page, so its lead-in paragraph will ideally appear at the bottom of the previous page

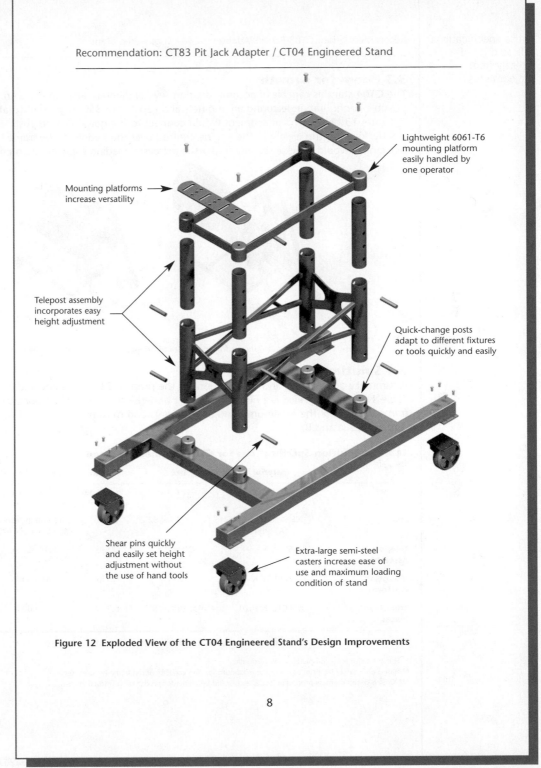

Recommendation: CT83 Pit Jack Adapter / CT04 Engineered Stand

Lightweight 6061-T6 mounting platform easily handled by one operator

Mounting platforms increase versatility

Telepost assembly incorporates easy height adjustment

Quick-change posts adapt to different fixtures or tools quickly and easily

Shear pins quickly and easily set height adjustment without the use of hand tools

Extra-large semi-steel casters increase ease of use and maximum loading condition of stand

Figure 12 Exploded View of the CT04 Engineered Stand's Design Improvements

8

Figure 19.6 A Formal Analytical Report   *(continued)*

These specifications will be drafted directly from project notes

Recommendation: CT83 Pit Jack Adapter / CT04 Engineered Stand

3.3 Room for Growth

The CT04 stand is capable of accommodating several different attachments. An illustration of a possible configuration featuring an engine stand capable of 360 degrees of rotation is provided in Figure 13 below. This attachment would connect to the quick-change posts quickly and efficiently, keeping the weight of the engine centred over the middle of the frame. This load placement would provide stability under the increased loading capacity of an engine.

Figure 13 Illustration of Concept for Engine Stand Adapter Compatible with CT04 Frame

3.4 Solution Specifications

A summary of the sizing and specifications of the proposed CT04 Engineered Stand is set out in Table 1. A more detailed examination of exact dimensions and loading capacities, as well as the frame location of the maximum bending moments and maximum shear, can be found in Appendices A and B.

Table 1 Solution Specifications for CT04 Engineered Stand

	Material	Size	Weight	Loading Capacity	Cost (CAD)
Casters req.	Semi-steel w/polytread	6"W × 2" tread	8 lb. × 4 req.	2500 lb. ea.	$17.70 × 4
Frame	AISI A-500 grade "A" tube	3"H × 36"W × 62"L	204 lb.	16 000 lb. max 8000 lb. with "N"	$150.00
Telepost Assembly	AISI 1020 ANN	3" OD × 2.5" 36"–60" range	73 lb.	8800 lb. ea.	$102.00
Mounting Platform	6061-T6 Al	2.5"H × 17"W × 34"L	15 lb.	> 18 000 lb.	$55.00
Mounting Plates	AISI C1018/1020	4"W × 18"L × ¼"T	11 lb. × 2 req.	> 18 000 lb.	$20.00 avg.
Totals	N/A	48"H × 36"W × 62"L*	344 lb.	8000 lb.**	$397.80***

* Footprint size for a completely assembled unit.
** Based on a safety factor of two, and the maximum loading capacity of the frame (weakest section).
*** Cost is based on mass-production scale, and would be higher for prototype construction.

9

Figure 19.6 A Formal Analytical Report

(continued)

Recommendation: CT83 Pit Jack Adapter / CT04 Engineered Stand

4.0 COST ANALYSIS

4.1 Cost Analysis Procedure and Cost–Design Relationship

Two main variables needed to be eliminated in order to determine an actual manufacturing cost for the CT04:

- The relationship between the specifications and features of the redesigned stand and the stand's final cost is affected by current industry needs and how much industry is willing to pay to satisfy those needs. Therefore, all research and final design content had to be complete before costs could be considered.
- Cost factors such as material choice and final product dimensions could not be properly determined until a full stress analysis had been completed; however, a stress analysis could not be completed until material and dimensions had been chosen.

This situation created a cost–design paradox that could be resolved only through a process of iteration. Appendix A illustrates the mathematical stress analysis of the CT04's framework to derive the material selection and final sizing of the stand. The cost of manufacturing was then determined according to material size and dimensions, and allowed to change the dimensions and material sizing if a cost reduction was necessary. This procedure was repeated until we established a design that provided solutions to all criteria.

4.2 Projected Cost Summary

The projected cost of manufacturing for the CT04 stand varies substantially according to the number of units planned for production. Figure 14 illustrates the projected cost to manufacture the stand for a one-prototype test unit, a 50-unit order, and a mass-production order of 500 units.

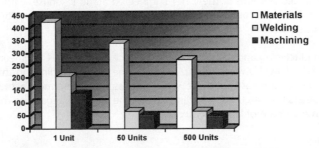

Figure 14 Component Costs of Manufacturing the CT04 Stand

As detailed in Figure 14, the cost of labour to produce the stand is nearly as high as the cost of raw materials and components. Customized jigs and fixtures for the efficient production of the stand should seriously be considered in order to possibly further reduce the cost of the stand.

10

Figure 19.6 A Formal Analytical Report

(continued)

Recommendation: CT83 Pit Jack Adapter / CT04 Engineered Stand

It may also be possible to determine which design factors increase the cost by interviewing tradespeople involved in the actual manufacture of the first prototype. They may well have alternative design suggestions that will reduce the production cost and thus increase the design's profitability.

The complete projected cost of manufacturing with, and without, CT Solutions' sales markup is summarized in Figure 15.

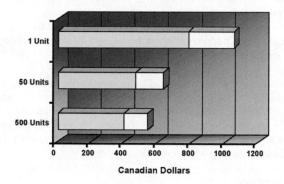

Figure 15 Total Projected Manufacturing Cost

The projected cost of manufacturing and the estimated sale price assume a generic markup of 35% on all labour and components. These projections also assume that CT Solutions will manufacture all fabricated portions of the CT04 Engineered Stand in-house, with no outsourcing of labour. Although this will probably be the case for mass-production, the projected cost and markup may be substantially different.

The cost of manufacturing and the projected retail price for the CT04 Engineered Stand have been summarized below; however, due to several of the reasons outlined in this section, these figures should only be used as a guide. It will be impossible to determine actual pricing until a prototype has been completed and we can evaluate the feedback regarding manufacturing difficulty and the true amount of labour required.

As Figure 15 shows,

*   The cost for one prototype is CDN$777.00; with 35% markup, **CDN$1050.00**.
*   The cost for 50 units is CDN$467.00; with 35% markup, **CDN$631.00**.
*   The cost for 500 units is CDN$397.00; with 35% markup, **CDN$535.00**.

11

Figure 19.6 A Formal Analytical Report

(continued)

Recommendation: CT83 Pit Jack Adapter / CT04 Engineered Stand

5.0 CONCLUSION AND RECOMMENDATION

The CT04 Engineered Stand has started to receive positive industry feedback—the suggested design improvements address all apparent flaws with the CT83 as well as all known consumer requirements. The redesign is quite versatile and has the potential to accommodate virtually any adapter designed for its mounting scheme, as illustrated in Figure 11 on page 7. The future expandability of the CT04 could make this design the foundation of a very large product line for CT Solutions.

At this time, we cannot recommend that the CT04 Engineered Stand enter production because it hasn't been thoroughly tested. Although the design seems sound, modifications to the CT04 may be necessary to ensure customer safety and satisfaction. The next logical step in reaching CT Solutions' ultimate goal of mass production of the stand would be to manufacture a prototype of the CT04 and place it in a suitable testing facility where it would be subject to normal working conditions. After the testing phase, more interviews should be conducted with technicians who have worked with the CT04. Test data should be gathered and analyzed.

In addition, craftspeople involved in the manufacture of the prototype should be interviewed after the first stand has been completed. Their insight about any major difficulties encountered during the manufacturing process may be instrumental in making design changes capable of further reducing the manufacturing cost of the stand, and increasing profitability of the design.

After the test data have been analyzed, it will be much easier to make a sound decision about whether to put the CT04 Engineered Stand into mass production or to further modify the design.

Whatever the outcome of the suggested testing phase may be, the CT04 Engineered Stand has potential for a bright future, and could become one of the cornerstones of the CT Solutions line of high-end tooling.

12

Figure 19.6 A Formal Analytical Report   *(continued)*

Recommendation: CT83 Pit Jack Adapter / CT04 Engineered Stand

REFERENCE CITED

Shinn Fu America. (n.d.) *Professional Lifting Catalog*. Retrieved from http://www.shinnfuamerica
.com/Temp/2354.pdf

13

Figure 19.6 A Formal Analytical Report   *(continued)*

Recommendation: CT83 Pit Jack Adapter / CT04 Engineered Stand

ADDITIONAL SOURCES CONSULTED

Jack-X-Change. (2010). *Catalog.* Retrieved from http://www.jackxchange.com/products/42001

Mott, R.L. (2008). *Applied strength of materials* (5th ed.). New Jersey: Pearson Education, Inc.

OTC Tools. (n.d.). Floor Jack 5078. Retrieved from http://otctools.com/graphics/5078.gif

Snap On Tools. (n.d.). Jack, transmission, high-lift, 1/2 ton. Retrieved from
    http://buy1.snapon.com/catalog/pro_det.asp?P65=&tool=all&item_ID=
    56290&group_ID=1720&store=snapon-store&dir=catalog

Southern-Tool Company. (2010). *Catalog.* Retrieved from
    http://www.southern-tool.com/store/catalog/transmission_jacks.html

14

Figure 19.6 A Formal Analytical Report

for Revising and Editing Analytical Reports

Use this checklist to refine the content, arrangement, and style of your report.

Content

- ◆ Does the report have a clear statement of purpose?
- ◆ Is the report's length adequate and appropriate for the subject?
- ◆ Are all limitations of the analysis clearly acknowledged?
- ◆ Are visuals used whenever possible to aid communication?
- ◆ Are all data accurate?
- ◆ Are all data unbiased?
- ◆ Are all data complete?
- ◆ Are all data fully interpreted?
- ◆ Is the documentation adequate, correct, and consistent?
- ◆ Are the conclusions logically derived from accurate interpretation?
- ◆ Do the recommendations constitute an appropriate response to the question or problem?

Arrangement

- ◆ Is there a distinct introduction, central section, and conclusion?
- ◆ Are headings appropriate and adequate?
- ◆ Are there enough transitions between related ideas?
- ◆ Is the report accompanied by all needed front matter?
- ◆ Is the report accompanied by all needed end matter?

Style and Page Design

- ◆ Is the level of technicality appropriate for the stated audience?
- ◆ Is the writing style throughout clear, concise, and fluent?
- ◆ Is the language convincing and precise?
- ◆ Is the writing in the report grammatical?
- ◆ Is the page design inviting and accessible?

EXERCISES

Prepare an analytical report, using some sequence of these guidelines:

a. Choose a subject for analysis from the list your instructor provides, from your major, or from a subject of interest. Identify the problem or question so that you will know exactly what you are looking for.

b. Restate the main question as a declarative sentence in your statement of purpose.

c. Identify an audience—other than your instructor—who will use your information for a specific purpose.

d. Hold a private brainstorming session to generate major topics and subtopics.

e. Use the topics to make an outline based on the model outline in this chapter. Divide as far as necessary to identify all points of discussion.

f. Make a tentative list of all sources (primary and secondary) that you will investigate. Verify that adequate sources are available.

g. Write your instructor a proposal memo, describing the problem or question and your plan for analysis. Attach a draft bibliography.

h. Use your planning outline as a guide to research and observation. Evaluate sources and evidence, and interpret all evidence fully. Modify your outline as needed.

i. Submit a progress report to your instructor describing work completed, problems encountered, and work remaining. Attach a detailed working outline.

j. Compose an audience/purpose profile.

k. Write the report for your stated audience. Work from a clear statement of purpose, and be sure that your reasoning is shown clearly. Verify that your evidence, conclusions, and recommendations are consistent. Be especially careful that your recommendations observe the critical-thinking guidelines in Figure 19.1 (page 354).

l. After writing your first draft, make any needed changes in the outline and revise your report according to the revision checklist. Include all necessary supplements.

m. Exchange reports with a colleague for further suggestions for revision.

n. Prepare an oral report of your findings for the class as a whole.

COLLABORATIVE PROJECTS

1. Divide into small groups. Choose a subject for group analysis—preferably, a campus issue—and partition

the topic by group brainstorming. Next, select major topics from your list and classify as many items as possible under each major topic. Finally, draw up a working outline that could be used for an analytical report on this subject.

2. Prepare a questionnaire based on your work above, and administer it to members of your campus community. List the findings of your questionnaire and your conclusions in clear and logical form.

MyWritingLab

MyWritingLab: Technical Communication offers the best multimedia resources for technical communication in one, easy-to-use place. You can find a variety of interactive model documents and case studies. There are also extensive guidelines, tutorials, and exercises for Document Design, Writing and the Writing Process, and Research, and a large bank of diagnostics and practice for grammar review.

20

Adding Document Supplements

✳ Explore

Click here in your eText to access a variety of student resources related to this chapter.

LEARNING OBJECTIVES

After reading this chapter, you should be able to

- choose which supplements should be included in a report
- understand how to write clear and accurate titles
- create the supplements required in an assigned report

Supplements help make a long report or proposal more accessible. Readers can refer to any of these supplements or skip them altogether, according to their needs. All supplements, of course, are written only after the document itself has been completed.

Some companies and organizations require a full range of supplements for any long document; others do not. When your audience has not stipulated its requirements, select only those supplements that enhance the informative value of your particular document. Avoid using supplements merely as decoration.

PURPOSE OF SUPPLEMENTS

Document supplements address these workplace realities to reach varied readers:

- ◆ *Confronted by a long document, many readers will try to avoid reading the whole thing.* Instead, they look for the least information they need to complete the task, make the decision, or take some other action.
- ◆ *Different readers often use the same document for different purposes.* Some look for an overview; others want details; others want only conclusions and recommendations, or the "bottom line." Technical personnel might focus on the central section of a highly specialized report and on the appendices for supporting data (maps, formulas, calculations). Executives and managers might read only the transmittal letter and the executive summary. If the latter read any parts of the report proper, they are likely to focus on conclusions and recommendations.

Report supplements can be classified into two groups:

Front matter

1. *Supplements that precede your report* (front matter): transmittal document, cover, title page, summary, table of contents, list of illustrations, acknowledgments

End matter

2. *Supplements that follow your report* (end matter): glossary, appendices, references, endnote pages

COVER

Covers are appropriate only for long documents. Use a sturdy, plain cover with page fasteners. With the cover on, the open pages should lie flat.

Centre the report title and your name 10–13 centimetres (4–5") below the upper edge of your page (many workplace reports include a company name and logo instead of the report author's name).

TRANSMITTAL DOCUMENT

In academic reports, the transmittal memo or letter might follow the title page, bound as part of the report. However, workplace reports usually present the transmittal document separate from the report. These days, the transmittal document is more likely to be an email to which a PDF version of the report is attached. A transmittal document sent with a formal report or proposal addresses a specific reader; it adds a note of courtesy and allows for personal remarks. For instance, your transmittal document might

What to include in a transmittal document

- ◆ acknowledge those who helped with the report
- ◆ refer to sections of special interest: unexpected findings, key visuals, major conclusions, special recommendations, and the like
- ◆ discuss the limitations of your study, or any problems gathering data
- ◆ discuss the need and approaches for follow-up investigations
- ◆ describe any personal (or off-the-record) observations
- ◆ suggest some special uses for the information
- ◆ urge the reader to immediate action

The transmittal document can be tailored to a particular reader, as is Richard Larkin's in Figure 20.1 on the next page. If a report is being sent to a number of people who are variously qualified and bear various relationships to the writer, the information may instead be included within the following basic structure:

INTRODUCTION. Refer to the reader's original request. Briefly review the reasons for your report or briefly describe it. Use a confident, positive tone throughout.

BODY. In the body, include appropriate items from the above list of possibilities.

CONCLUSION. State your willingness to answer questions or discuss findings. End positively, and, where appropriate, suggest follow-up actions.

TITLE PAGE

The title page (see Figure 20.2 on page 399) lists the report title, author's name, name of person(s) to whom the report is addressed or the name of the organization to which it is addressed, and date of submission.

How to prepare a title page

TITLE. The title announces the report's purpose and subject. The title in Figure 20.2 is clear, accurate, comprehensive, and specific. But even slight changes can distort this title's signal.

7409 Trinity Court
Niagara Falls, ON   L2H 3A6

June 2, 2014

Ms. Mimi Lim
Placement Director
Seneca College
1750 Finch Avenue East
North York, ON   M2J 2X5

Dear Ms. Lim:

Here is my analysis to determine the feasibility of a career in technical marketing. In preparing my report, I've learned a great deal about the requirements and modes of access to this career, and I believe my information will help other students as well.

Although committed to their specialities, some technical and science graduates seem interested in careers in which they can apply their technical knowledge to customer and business problems. Technical marketing may be an attractive choice of career for those who know their field, who can relate to different personalities, and who are good communicators.

Technical marketing is competitive and demanding, but highly rewarding. In fact, it is an excellent route to upper-management and executive positions. Specifically, marketing work enables one to develop a sound technical knowledge of a company's products, to understand how these products fit into the marketplace, and to perfect sales techniques and interpersonal skills. This kind of background paves the way to top-level jobs.

If anyone has questions related to the report's information and conclusions, I would be happy to respond.

Sincerely,

Richard B. Larkin

Richard B. Larkin

Figure 20.1 A Letter of Transmittal for a Formal Report

FEASIBILITY OF A CAREER IN TECHNICAL MARKETING

for

Mimi Lim

Placement Director

Seneca College

North York, Ontario

by

Richard B. Larkin,

English 266 Student

June 2, 2014

Figure 20.2 A Title Page for a Formal Report

A TECHNICAL MARKETING CAREER

The version above is unclear about the report's purpose. Is the report *describing* the career, *proposing* the career, *giving instructions* for career preparation, or *telling one person's career story*? Insert descriptive words (*analysis, instructions, proposal, feasibility, description, progress*) that accurately state your purpose.

To be sure that your title forecasts what the report delivers, write its final version *after* completing the report.

PLACEMENT OF TITLE-PAGE ITEMS.   Do not number your title page, but count it as *page i* of your preliminary pages. Centre the title horizontally, 8–10 cm (3–4") below the upper edge. Place other items in the spacing and order shown in Figure 20.2, the title page to a report. Or devise your own system, as long as your page is balanced.

SUMMARY

Chapter 10 defines varieties of summary writing, including the summary (or executive summary) that *accompanies* a formal report or proposal. Many readers who don't have the time or willingness to read your entire report will consider the summary to be the most useful part of the material you present.

Chapter 10 also recommends a step-by-step process that will help you find and condense the elements of a good summary:

- the issue or need that led to the report
- the report's key facts, statistics, findings, and, in some cases, illustrations—this is the material your reader *must* know
- the report's conclusions and recommendations

When you write and edit a summary, make clear connections between the report's data and interpretations. If you find yourself unable to do this in the summary, you'll probably need to revise the original report.

Follow these guidelines for the report summary:

- Make the summary about one-tenth the length of the original, but remember that summaries rarely are shorter than three-quarters of a page or longer than five pages.
- Where appropriate, use a table to summarize key facts and findings.
- Add no new information. Simply give the report's highlights.
- Use the same order of topics in the summary as in the original.
- Adjust the vocabulary to suit the intended reader. An executive summary, for example, includes little technical jargon. When you send report copies to readers with varying levels of expertise, write a different summary for each type of reader.
- Include a graph or other figure in the summary *only* if the illustration is absolutely necessary in understanding the report.

TABLE OF CONTENTS

The table of contents (see Figure 20.3 on page 401) serves as a road map for readers and a checklist for you. If you are using a high-end word-processing program, you can generate a table of contents automatically, provided that you have assigned styles

TABLE OF CONTENTS

Figure 20.3 A Table of Contents for a Formal Report

or codes to all of the headings in your report. If your word-processing program does not have this feature, compose the table of contents by assigning page numbers to headings from your outline. Keep in mind, however, that not all levels of outline headings appear in your table of contents or your report. Excessive headings can fragment the discussion.

Follow these guidelines:

◆ List front matter (summary, list of illustrations), numbering the pages with small Roman numerals. (The title page, though not listed, is counted as page i.) List end matter, such as the glossary, appendices, and endnotes. Number these pages with Arabic numerals, continuing the page sequence of your main report.
◆ Include in the table of contents only headings or subheadings that are in the report; the report may, however, contain subheadings not listed in the table of contents.
◆ Phrase headings in the table of contents exactly as in the report.
◆ List various levels of headings with varying typefaces and indentations.
◆ Use *leader lines* (.........) to connect the heading text to the page number. Align rows of dots vertically, each above the other.

LIST OF FIGURES AND TABLES

Following the table of contents is a list of figures and tables, if needed. When a report has three or more visuals, place the list on a separate page. List the figures first, then the tables. Figure 20.4 shows the list of figures and tables for Larkin's report.

FIGURES AND TABLES

Figure 20.4 A List of Figures and Tables for a Formal Report

DOCUMENTATION

Reference pages are organized according to the format required by your workplace or academic discipline. MLA and APA formats require references to be listed in alphabetical order on a Works Cited page (for MLA) or References page (for APA). CSE and IEEE formats list sources in the same numerical order as they are cited in the report.

Whichever documentation format you use, consider including a separate list of sources you consulted but did not cite under the heading "Additional Sources Consulted." Thus, you will

◆ show your reader that you have indeed covered all the bases,
◆ acknowledge that certain sources influenced your thinking, even if you didn't have reason to cite them specifically, and
◆ serve the reader who wishes to explore the topic further.

GLOSSARY

A glossary alphabetically lists key terms and their definitions, following or preceding your report. Specialized reports often contain glossaries, especially when written for both technical and non-technical readers. A glossary makes key definitions available to non-technical readers without interrupting the flow of the report for technical readers. If fewer than five terms need defining, place them in the report introduction as working definitions, or use footnote definitions. If you use a separate glossary, inform readers of its location: "(see the glossary at the end of this report)." Note, though, that some readers prefer the glossary in the front matter, just before the introduction.

Follow these guidelines for preparing a glossary:

How to prepare a glossary

◆ Define all terms unfamiliar to a general reader (an intelligent layperson).
◆ Define all terms that have a special meaning in your report (e.g., "In this report, a small business is defined as . . .").
◆ Define all terms by giving their class and distinguishing features, unless some terms need expanded definitions.
◆ List your glossary and its first page number in your table of contents.
◆ List all terms in alphabetical order. Boldface each term. You may use a colon to separate the term from its single-spaced definition. This is not necessary if you use a tabular format.
◆ Define only terms that need explanation. In doubtful cases, over-defining is safer than under-defining.

Figure 20.5 on the next page shows part of a non-tabular glossary for a comparative analysis of two techniques of natural childbirth, written by a nurse practitioner for expectant mothers and student nurses. Figure 20.6 (also on the next page) illustrates a partial tabular glossary.

APPENDICES

An appendix follows the text of your report. It expands on items discussed in the report without cluttering the main text. Figure 20.7 on page 405 shows an appendix to a budget proposal. Typical items in an appendix include

What an appendix might include

◆ complex formulas
◆ details of an experiment

- interview questions and responses
- long quotations (one or more pages)
- maps
- photographs
- material more essential to secondary readers than to primary readers
- related correspondence (letters of inquiry, and so on)
- sample questionnaires and tabulated responses
- sample tests and tabulated results
- some visuals occupying more than one full page
- statistical or other measurements
- texts of laws and regulations

GLOSSARY

Analgesic: a medication given to relieve pain during the first stage of labour

Cervix: the neck-shaped anatomical structure that forms the mouth of the uterus

Dilation: cervical expansion occurring during the first stage of labour

Episiotomy: an incision of the outer vaginal tissue, made by the obstetrician just before the delivery, to enlarge the vaginal opening

First stage of labour: the stage in which the cervix dilates and the baby remains in the uterus

Induction: the stimulating of labour by puncturing the membranes around the baby or by giving an oxytoxic drug (uterine contractant), or both

Figure 20.5 A Non-tabular Glossary (Partial)

GLOSSARY

Analgesic   A medication given to relieve pain during the first stage of labour

Cervix   The neck-shaped anatomical structure that forms the mouth of the uterus

Figure 20.6 A Tabular Glossary (Partial)

APPENDIX A

Table 1 Allocations and Performance of Five College Newspapers

	Prairie College	Drake College	Kelsey College	Hollander College	Northern College
Enrollment	1600	1400	3000	3000	5000
Fee paid (per year)	$55.00	$85.00	$35.00	$50.00	$65.00
Total fee budget	$88 000.00	$119 000.00	$105 000.00	$150 000.00	$325 000.00
Newspaper budget	$10 000.00	$6 000.00	$25 300.00	$37 000.00	$21 500.00
					$25 337.14[a]
Yearly cost per student	$6.25	$4.29	$8.43	$12.33	$4.30
					$5.07[a]
Format of paper	Weekly	Every third week	Weekly	Weekly	Weekly
Average no. of pages	8	12	18	12	20
Average total pages	224	120	504	336	560
					672[a]
Yearly cost per page	$44.50	$50.00	$50.50	$110.11	$38.25
					$38.69[a]

[a] These figures are next year's costs/pages for the Northern *Torch*.

Source: Figures were quoted by newspaper business managers in April 2011.

Figure 20.7 An Appendix

The appendix lists items that are important but difficult to integrate within your text.

Do not stuff appendices with needless information, and don't use them unethically for burying bad or embarrassing news that belongs in the report proper. Follow these guidelines:

How to prepare an
appendix

- Include only material that is relevant.
- Use a separate appendix for each major item.
- Title each appendix clearly: "Appendix A: Projected Costs."
- Limit an appendix to a few pages, unless more length is essential.
- Mention your appendix early in your introduction, and refer readers to it at appropriate points in the report: "(see Appendix A)."

Use an appendix for any material that is essential but might harm the unity and coherence of your report.

EXERCISES

1. These titles are intended for assessment, feasibility analysis, causal analysis, or recommendation reports. Revise each inadequate title to make it clear and accurate.
   a. The Effectiveness of the Prison Furlough Program in Our Province
   b. Drug Testing on the Job
   c. The Effects of Nuclear Power Plants
   d. Wood-Burning Stoves
   e. Interviewing
   f. An Analysis of Vegetables (for a report assessing the physiological effects of a vegetarian diet)
   g. Wood as a Fuel Source
   h. Oral Contraceptives
   i. Lie Detectors and Employees

2. Prepare a title page, transmittal document (for a definite reader who can use your information in a definite way), table of contents, and informative abstract for a report you have written earlier.
3. Find a short but effective appendix in one of your textbooks, in a journal article in your field, or in a report from your workplace. In a memo to your instructor and colleagues, explain how the appendix is used, how it relates to the main text, and why it is effective. Attach a copy of the appendix to your memo. Be prepared to discuss your evaluation in class.

COLLABORATIVE PROJECT

Collect samples of various document supplements. As a group, critique the format and content of the supplements.

MyWritingLab

MyWritingLab: Technical Communication offers the best multimedia resources for technical communication in one, easy-to-use place. You can find a variety of interactive model documents and case studies. There are also extensive guidelines, tutorials, and exercises for Document Design, Writing and the Writing Process, and Research, and a large bank of diagnostics and practice for grammar review.

Short Reports

LEARNING OBJECTIVES

After reading this chapter, you should be able to

- appreciate the importance of short reports in business and industry
- compare the structural and page-design elements of semi-formal and formal reports
- adapt an action structure to a wide range of one- to 10-page job-related reports
- choose the format that best suits a given report's content, length, and required degree of formality (letter, memo, email, semi-formal, prepared form)

Short reports form the bulk of the writing done by technologists, technicians, research scientists, engineers, designers, and other technical and business writers. Such reports provide information and analysis that readers use to stay informed and to make practical decisions.

Some reports emphasize *information*:

- ◆ progress reports
- ◆ field trip reports
- ◆ field observations
- ◆ project completion reports
- ◆ periodic activity reports
- ◆ meeting minutes

Other reports focus on *analysis*:

- ◆ feasibility reports
- ◆ causal analyses
- ◆ assessment reports
- ◆ yardstick (comparison) reports
- ◆ justification reports
- ◆ recommendations reports

FORMATS

When a document is quite short (one to three pages) or when its subject matter suits a direct, informal address to the reader, use a *correspondence format*, such as a letter, memo, or email.

When a document is somewhat longer (4 to 10 pages), and when its subject matter is serious enough to warrant a more formal approach, use a *semi-formal report format*. However, use the *formal report format* illustrated in Chapter 19 when you want to influence policy within your organization: the more formal the document, the better chance of its being heeded. So consider "dressing up" even 6- to 10-page reports in formal report clothing if you really want to impress the reader. Table 21.1 summarizes the differences between semi-formal and formal reports.

How to choose among informal, semi-formal, and formal report formats

Table 21.1 Semi-formal and Formal Report Formats

Item	Semi-formal	Formal
Length	4–10 pages	6+ pages
Required Sections		
Transmittal document	optional	yes
Cover	no	optional
Title page	optional—title could be placed on first page of report body	yes
Summary	no	yes
Table of contents	optional	yes
List of illustrations	optional	yes
Glossary	incorporate into text	optional
Introduction	yes	yes
Background	optional	optional
Central analysis	one or more sections	one or more sections
Conclusion	yes	yes
Recommendations	optional	optional
Sources section	optional	yes (if sources were cited)
Sources consulted	optional	optional
Format and Appearance	**Semi-formal**	**Formal**
Headings system	more relaxed; seldom more than two levels of headings; main headings placed where they come on the page	formal; usually three or four levels; each main heading placed at the top of a page
Page numbering	all page numbers placed at same location on page	different for first page of a section than for subsequent pages in that section
Margins	all pages use same margins layout	top and bottom margins for first page of a section are larger than for subsequent pages
Indentation	paragraphs not indented (double space between paragraphs); bulleted and numbered lists may be indented	paragraphs not indented; (double space between paragraphs); bulleted and numbered lists may be indented
Headers	seldom used in short, informal reports	optional, but headers often appear in formal reports

A STRUCTURE FOR ALL PURPOSES

The action structure gives readers what they want

Readers of business reports are generally busy people. They want reports to be as brief as possible. They usually want a report's main point in the first or second paragraph, and they want clear and logical idea patterns. The *action structure* illustrated in Figure 21.1 satisfies those readers' desires.

Action Opening

Using the action structure, you immediately "connect" with your reader by

- referring to an issue that concerns the reader (and you); or
- referring to comments made in a recent meeting; or
- responding to the reader's previous communication (memo, email, letter, telephone call) on the subject.

By making such a connection, you gain the reader's attention. Then, in the same paragraph you summarize the report's main point *or* you preview the approach taken in the report.

Background

Next, you provide any background needed by your reader to understand the report's detailed information and analysis. You may have to

- review the circumstances leading to the issue discussed in the report
- define terms or provide technical background
- review a problem or a proposed solution

Not all reports require background information. Your audience/purpose analysis will help you determine whether your reader needs to be briefed. You may be able to go straight from the opening paragraph to the details section of the report.

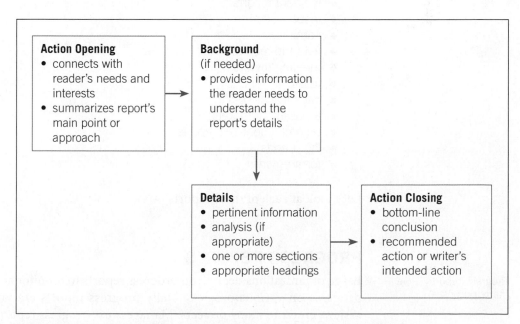

Figure 21.1 An Action Structure for Informal and Semi-formal Reports

Most readers appreciate subject headings for both the background and the details sections, even for a one-page report. Clearly, the number of headings you use depends on the depth and complexity of the report's topic. See this chapter's sample reports for examples of heading usage. As you read the reports, distinguish between standard topical headings ("Project Costs") and talking headings, which speak directly to the reader ("How Much We've Spent").

Details

Essentially, a report's details section answers most, if not all, of the questions posed by any good reporter: *What? Where? When? Who? How? How much? Why?* The order in which you answer such questions depends on the subject matter and your reader's priorities.

Action Closing

Finally, the report provides bottom-line conclusions. Most reports also discuss what should be done next. Some reports recommend action to be taken by the reader or the reader's organization; others state what action the writer intends or proposes. Still other reports simply list possible actions without indicating who might take responsibility.

The action structure can be adapted to develop any type of correspondence or short report. The remainder of this chapter demonstrates the action structure's valuable adaptability to various reports:

The action structure is adaptable

- progress reports
- periodic activity reports
- project completion reports
- incident reports
- inspection reports
- compliance reports
- field trip reports
- meeting minutes
- feasibility reports
- causal analyses
- assessment reports
- recommendations reports
- lab reports
- form reports

Let's look at each of these reports.

PROGRESS REPORTS

Progress reports serve many purposes

Large organizations depend on progress reports to monitor activities, problems, and progress on various projects. Daily progress reports are vital in a business that assigns crews to many projects. Managers use progress reports to evaluate a project and its supervisor, and to decide how to allocate funds. Managers also need to know about delays that could dramatically affect outcomes and project costs.

As well, managers need information to coordinate the efforts of the work groups. For example, a hydro manager responsible for restoring power transmission lines after a severe ice storm will have to coordinate cleanup crews, construction crews, and line crews. A large project, such as a major power line restoration, would require several written *periodic progress reports*.

When work is performed for an external client, the reports explain to the client how time and money have been spent and how difficulties have been overcome. The reports can therefore be used to assure the clients that the work will be completed on schedule and on budget. Many contracts stipulate when progress will be reported. Failure to report on time may invoke contractual penalties.

To inform managers and clients, progress reports should answer these questions:

1. How much has been accomplished since the last report?
2. Is the project on schedule?
3. If not, what went wrong? How was the problem corrected? How long will it take to get back on schedule?
4. What else needs to be done? What is the next step?
5. Have you encountered any unexpected developments?
6. When do you anticipate completion? Or (on a long project) when do you antici- pate completion of the next phase?

Table 21.2 shows how to adapt the action structure to organize the information in progress reports.

Figures 21.2 to 21.4 adapt the action structure to periodic progress reporting. Exactly how you structure a report will depend on the nature of the project and on the aspect you want to stress. You could organize the report by

◆ the different tasks or subcontracts
◆ the amount of work completed versus the amount remaining
◆ phases or time frames identified by the company contact or project manager

Figures 21.2 to 21.4 illustrate outlines for these three possible structures. These out- lines are for the third in a series of reports for a six-month highway reconstruction project. Which structure would you use?

"Problems and Setbacks" should be discussed in an appropriate location

Where needed, any of the three variations could incorporate a special category for a description and possible solutions of "Problems and Setbacks," as shown in Figure 21.2 (page 413). Or this topic could be discussed in another section, if the difficulties do not require special treatment. In Figure 21.3 (page 413), for example, problems and setbacks are discussed under "Work Completed."

Notice that in each of the three sample structures only two main sections are included for reporting actual progress ("road base preparation" and "paving"), but three or more sections could be required, depending on the nature of the work.

Table 21.2 Progress Reports

Section	Content	Comments
Action opening	Identify the project and, in some cases, the project's special requirements. Refer to previous reports, if any. In a sentence or two, summarize progress to date.	The reader connection might be placed in a transmittal letter or memo if a report format is used. A letter or memo report will not need a heading for the opening paragraph or two.
Background	Remind the reader about key background information. Summarize progress that has been previously reported and, where appropriate, summarize previously discussed problems and whether they've been solved. Use appropriate subheadings.	The content depends on ◆ whether progress has been previously reported ◆ whether thorny issues have been previously discussed This section should not exceed two paragraphs. Extra detail can be placed in attachments.
Details of progress	Describe the progress achieved in the most recent reporting period. (If no progress has previously been reported, then start at the beginning of the project.) Use suitable subheadings; organize by ◆ tasks or subcontracts ◆ completed work vs. remaining work ◆ contract time frames or project phases Include, where appropriate, ◆ problems, how they've been handled, and how well ◆ a cost analysis of the work just completed, and the project so far	Detailed test data and other tabulated material should appear in an attachment or appendix. The format (letter, memo, semi-formal report, formal report) will depend on the length of the document and on the writer's assessment of the document's importance. Discussing problems gives the writer an opportunity to justify requests.
Action closing	Provide bottom-line conclusions: ◆ Is the project on time and on budget? ◆ Is the work up to standards? Predict whether the project will be completed on time and on budget. Where appropriate, recommend changes; ask for approval of these recommendations.	All conclusions must be solidly based on material presented in the central section of the report. Subheadings might be appropriate. Where appropriate, the writer should ask for action from the reader.

VARIATION 1: Organized by Task

- Intro: connect with reader/refer to project
- Project description or summary

ROAD BASE PREPARATION
   Progress to date
- in previous reporting period(s)
- work done in period just closing

   Work to be completed
- work planned for next work period
- work planned for periods thereafter

PAVING
   Progress to date
- in previous reporting period(s)
- work done in period just closing

   Work to be completed
- work planned for next work period
- work planned for periods thereafter

PROBLEMS AND SETBACKS
- description
- methods used to overcome the setbacks
- degree of success

COST ANALYSIS
- costs to date
- costs in period just closing

CONCLUSION
- overall appraisal of work to date
- cost appraisal
- conclusions and recommendations re: work

Figure 21.2 A Progress Report Organized by Task

VARIATION 2: Organized by Work Completed

- Intro: connect with reader/refer to project
- Project description or summary

WORK COMPLETED
   Road base preparation
- in previous reporting period(s)
- work done in period just closing
- problems and setbacks
- measures taken to recover from the setbacks

   Paving
- work planned for next work period
- work planned for periods thereafter

WORK REMAINING
   Road base preparation
- work planned for next work period
- work planned for periods thereafter

   Paving
- work planned for next work period
- work planned for periods thereafter

COST ANALYSIS
- costs to date
- costs in period just closing

CONCLUSION
- overall appraisal of work to date
- cost appraisal
- conclusions and recommendations re: work

Figure 21.3 A Progress Report Organized by Degree of Project Completed

Figure 21.4 A Progress Report Organized by Project Time Frame

Two Sample Progress Reports

A workplace progress report

Figure 21.5, on pages 415–420, shows a periodic progress report that organizes its information by distinguishing between "work completed" and "work in progress."

| CASE | Periodic Progress Report |

Art Basran, the writer of the report shown in Figure 21.5, manages the Kelowna, British Columbia, office of MMT Consulting. Art's report is the fourth in a series of oral and written progress reports for his regional manager, Brenda Backstrom, regarding a project that is very important to MMT Consulting. (Chapter 3 features the proposal he wrote to Brenda to get her approval for the project.)

This progress reports the innovative engineering/accounting method (Equivalent Uniform Annual Cost, or EUAC) that was the crux of the proposal memo in Figure 3.4 on pages 49–50. Art also uses the "Recommendations" at the end of the report to argue for one of his preferences: the use of professionally produced graphics in his firm's reports.

For this progress report, Art has chosen a semi-formal report format instead of a memo because of the six-page length and because he is trying to persuade his supervisor to support a new method of analyzing combined engineering and accounting data.

PROGRESS REGARDING MMT'S COST ANALYSIS OF FOUR ALTERNATIVES FOR HAULING COAL FROM JACKSON MINING'S PROPOSED OTHELLO MINE.

For Brenda Backstrom, Regional Manager
MMT Consulting, Calgary Regional Office

Prepared by Art Basran
Kelowna Branch Office Manager
MMT Consulting

Submitted:
March 5, 2015

Figure 21.5 Semi-formal Progress Report Organized by Degree of Completion *(continued)*

INTRODUCTION

The Jackson Mining Company has discovered a high-grade coal deposit in the mountains about 90 kilometres due north of Grand Forks, British Columbia. The company expects the market for this grade of coal to increase dramatically in the near future, and thus anticipates developing the deposit if development and transportation costs are acceptable. Jackson Mining has tentatively named its proposed mine site the Othello mine. MMT Consulting has been authorized to determine the most cost-effective method(s) of transporting the coal from that mine site to the mainline railroad, a distance of approximately 80 kilometres.

Four transport alternatives are being considered:

1. Diesel trucks
2. Diesel trains
3. Electric trains
4. Electric fleets

The criterion used to evaluate each alternative is the Equivalent Uniform Annual Cost (EUAC). The EUAC discussed in this report is in dollars per ton for the haul to the railhead. The four alternatives will be studied over periods ranging from one to 20 years, and at interest rates varying from 6% to 10%.

The EUAC formula provides clear comparisons of the transportation alternatives. Using this method, and capitalizing on hard work by our staff and good surveying weather, the project seems certain to be completed under budget and ahead of schedule. Details are provided in the following sections.

BACKGROUND

The Othello mine under study has anticipated coal sales of 5 million tons annually. Each of the following transportation alternatives would have to handle this volume. Also, the ability to expand in the future would be an asset.

1. **Diesel trucks**
   The CAT 777B truck was chosen for this study because of its load capacity of 95 tons and its output of 920 horsepower. These trucks would be running 24 hours a day.
2. **Diesel trains**
   Each train would be running with six diesel engines and 98 coal cars. It would take 11.2 hours to complete a round trip, and 17.2 hours to produce the amount of coal the train would haul. Thus, the train would haul a complete load and then idle for about six hours until the next load was ready.
3. **Electric trains**
   Each train would run four electric engines and 98 coal cars, with the same haul times as for the diesel train.
4. **Electric fleets**
   Each fleet would consist of three of the above electric trains. The first train would be loaded and then depart. While it is en route, the second train would be loaded, and this load-and-go process would be repeated for the third train. The departures would be timed so that the second train would leave just as the first train crested the highest point of the rail line pass. In this way, the electricity generated by the downgrade trains would be conserved and used to assist in powering the upgrade trains.

1.

Figure 21.5 Semi-formal Progress Report Organized by Degree of Completion   *(continued)*

MMT'S INVOLVEMENT

After the initial negotiations between the respective head offices of Jackson Mining and MMT, our Kelowna office prepared a detailed proposal that Jackson Mining approved on January 6. Since then, two engineers and three technologists from our Kelowna office have researched this project with advice from Brendan Winters, a Penticton chartered accountant who has special expertise in mining projects.

Work Completed

Since January 6, our team has completed the following tasks:
- completed a preliminary survey of the proposed rail line and haul road; the elevations and other data collected from this survey are being used to help calculate tractive and grade resistances for the diesel trucks and for the two types of trains
- researched construction costs for both a rail line and a haul road, and for associated construction costs (such as maintenance buildings)
- researched maintenance costs for both a rail line and a haul road
- researched labour costs for operating the trucks and the trains
- researched the costs of purchasing trucks, diesel trains, and electric trains
- designed and researched the full costs of a complete communication system, which is required to control the proposed mine's transportation system

Work in Progress

We now have all the data required to complete our calculations and analyses. This work will progress in three stages:

1. Using the data collected from our surveys, we will calculate tractive and grade resistances for the diesel trucks and for the two types of trains. A sample of those calculations is provided below in Table 1. Next, to supplement the information regarding resistances, we will calculate the speed, power, and fuel consumption of each transportation type. A sample of those calculations appears in Table 2 on page 3.

Table 1: Tractive and Grade Resistances for Diesel Trucks

Section	Grade (%)	Truck Wt (ton)	Load Wt	Total Wt	Rt (lb.)	Rg	Tr
G	2.0	66.3	95.0	161.3	16 125.0	6450.0	22 575.0
F	1.8	66.3	95.0	161.3	16 125.0	5805.0	21 930.0
E	2.0	66.3	95.0	161.3	16 125.0	6450.0	22 575.0
D	2.1	66.3	95.0	161.3	16 125.0	6772.5	22 897.5

The first column in Table 1 lists the sections into which the road has been divided for analytical purposes. These sections of road have varying grades. Grade resistance (Rg) and tractive resistance (Rt) are found for each section.

Figure 21.5 Semi-formal Progress Report Organized by Degree of Completion   *(continued)*

Table 2:   Speed, Power, and Fuel Consumption for Diesel Trucks

Section	Grade (%)	Tr (lb.)	Power (hp)	Speed (mph)	Dist (mi.)	Time (hr)	BTU (x1000)	Fuel (gal.)	Total Fuel
G	2.0	22 575.0	920.0	12.55	8.4	0.67	1572.5	11.34	42.62
F	1.8	21 930.0	920.0	12.92	7.2	0.56	1309.3	9.44	35.49
E	2.0	22 575.0	920.0	12.55	5.4	0.43	1010.9	7.29	27.40
D	2.1	22 897.5	920.0	12.38	4.8	0.39	911.4	6.57	24.70

The final calculations for resistances and fuel consumption will be added to labour and maintenance costs to determine annual operating costs for each of the alternatives under study. Thus, we will be able to complete the Cost Comparisons table, a draft of which is shown in Draft Table 3 below.

Draft Table 3:   Cost Comparisons

Option	Capital Costs	Annual Operating Costs
Diesel Trucks	$115 539 000	$
Diesel Trains	$149 083 000	$
Electric Trains	$161 783 000	$
Electric Fleets	$201 463 000	$

2. Once the capital and annual costs are found, they can be combined with varying interest rates and study periods to calculate tables of Equivalent Uniform Annual Costs (EUAC). The purpose of converting all costs into an EUAC is to provide a common means of comparing all the alternatives. The capital costs are spread over the length of the study period and added to the annual costs. Thus, a comparison of EUACs will yield a qualitative assessment of the alternatives to go with the quantitative assessment suggested by Draft Table 3.

To provide an idea of how these EUAC tables (and corresponding graphs) will look in the final analysis, we have estimated figures for the four transportation alternatives and placed them in Draft Table 4 and Draft Figure 1 (Appendix).

The final phase of our work for this project will be to produce a comprehensive report of our findings. This report has been outlined, and some of the preliminary material has been drafted. We expect the final version of this report to total about 40 pages, with the bulk of those pages presenting our detailed calculations. We're using the format that we learned from Mykon Communications last March and that proved successful in our final report to the City of Kelowna regarding the proposed multipurpose arena.

3.

Figure 21.5 Semi-formal Progress Report Organized by Degree of Completion   *(continued)*

CONCLUSION

Appraisal of Work to Date
We're satisfied with both the quantity and the quality of the data gathered for the required analysis. Part of this is due to the hard work performed by our Kelowna project team; part is due to our connection to the internet; and part of the credit should go to Brendan Winters, whose experience in gathering and analyzing financial data has been invaluable.

Also, we had some good fortune with the weather during the survey work. Unseasonably warm weather and light snowfalls allowed the surveys to finish one week ahead of our anticipated February 2 completion date and under budget by 15%. (We would have been even more under budget, but we had to rent a helicopter for three days more than anticipated because of very difficult terrain in the Kelso Pass area.)

Overall, it seems that we will be about $14 000 under budget for the project, partly because of the savings in the surveys and partly because Winters's research expertise has cut our estimated research time. Full figures will be provided in our final progress report.

Anticipated Completion Date
We expect to have all calculations completed by March 7. After that, it should take two of our staff (one engineer and one technologist) two working days to produce a final draft of the report for examination at Regional Office. Allowing for examination time and for final editing at the Kelowna office, we should be able to have the finished copy of our report produced by March 14, which is 14 days ahead of the date we had scheduled for delivering our analysis to Jackson Mining.

Recommendations
We have forged a positive working relationship with Jessica Proctor, director of mining development at Jackson Mining. Proctor has indicated that her company would like us to bid on the construction engineering contract for the anticipated road/rail line construction project, which may begin as early as this July. In connection with this possible contract, we recommend two courses of action:

1. Have the Jackson Mining report produced professionally by a graphics firm. We have money to spend because we're under budget for this project. Also, a graphics firm, such as Apex Graphics (which does annual reports and similar documents for major businesses in the Okanagan), could produce a high-quality colour report in one day. My main reason for suggesting this extra expenditure of under $1000 is that Jackson Mining's directors place a high value on professional work. I think it would help project a positive image that will benefit any future proposals we make to this company.
2. Do some preliminary investigation of the parameters and requirements of road construction and rail line construction for Othello Mine's proposed transportation routes. If Jackson Mining presents a Request for Proposal on this construction project in the near future, MMT Consulting will be prepared.

4.

Figure 21.5 Semi-formal Progress Report Organized by Degree of Completion *(continued)*

APPENDIX

Draft Table 4: EUAC for Alternatives at 6% (based on preliminary estimates)

Study Period	Diesel Trains	Electric Trains	Electric Fleets	Diesel Trucks
1	$32.52	$34.91	$43.26	$32.10
2	17.18	18.26	22.53	20.21
3	12.07	12.72	15.63	16.25
4	9.52	9.95	12.18	14.27
5	7.99	8.29	10.12	13.09
6	6.98	7.19	8.75	12.30
7	6.26	6.41	7.77	11.74
8	5.72	5.82	7.04	11.33
9	5.30	5.37	6.48	11.00
10	4.97	5.01	6.03	10.75
11	4.70	4.71	5.66	10.54
12	4.47	4.47	5.36	10.36
13	4.28	4.27	5.10	10.22
14	4.12	4.09	4.89	10.09
15	3.98	3.94	4.70	9.98
16	3.87	3.81	4.54	9.89
17	3.76	3.70	4.40	9.81
18	3.67	3.60	4.27	9.74
19	3.59	3.51	4.16	9.68
20	3.51	3.43	4.07	9.62

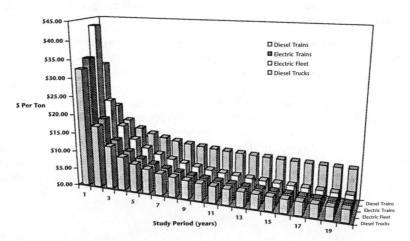

Draft Figure 1: EUAC for Four Alternatives at 6% (based on preliminary estimates)

Figure 21.5 Semi-formal Progress Report Organized by Degree of Completion

A student progress report   The second sample progress report, shown in Figure 21.6, on pages 422–424, comes from Vince Cummings, who is reporting progress on a research project to his project supervisor.

CASE	**Student Project Progress Report**

Cummings is part of the group that produced the recommendations report reproduced in Figure 19.6, starting on page 375. He has taken a leading role in the project; his enthusiasm for that project is reflected in the degree of detail and the tone of phrasing in this progress report. Vince and his collaborators have turned hard work and clear thinking into a successful project, but they have been challenged by the need to create a cost-effective design. Therefore, this progress report includes a section entitled, "Project Problems and Their Solutions," especially because the project supervisor has previously commented on the cost issue.

The heart of this report is the "Project Progress" section, which contains three subsections: "Work completed," "Work in progress," and "Work to be completed." Those components are followed by a project schedule, which the writer knows is a concern for his reader.

Vince attaches his team's working outline (which is reproduced in Figure 19.5, on pages 370–373.)

Occasional progress reports are written for short-term projects that do not have scheduled reporting dates. Either the writer responds to a supervisor's request (or to a client's request), or the writer reports progress in order to elicit reader support.

PERIODIC ACTIVITY REPORTS

The periodic activity report is similar to a progress report in that it summarizes activities over a specified period. But unlike a progress report, which describes activity on a given *project*, a periodic report summarizes the general activities during a given *period*. Manufacturers requiring periodic reports often have prepared forms, because most of the tasks are quantifiable, such as units produced. Still, not all jobs lend themselves to prepared-form reports, and you will have to develop your own format. If so, the action structure can be readily adapted to your purpose.

PROJECT COMPLETION REPORTS

A project completion report is presented as the concluding progress report for a lengthy project, or the only report arising from a short project. Either way, the action structure can form the report's backbone. (Note, though, that in Table 21.3 (page 425) the details section is divided into "Project highlights" and "Exceptions.")

A project completion report by consulting engineer Ken Langedyk can be found in **MyWritingLab: Technical Communication**.

Memorandum

To:	Don Klepp, Professional Communication Department	**Date:** March 7, 2014
From:	Vince Cummings, MET student	
Re:	**Progress Report—PCOM 142 Research Project**	

On February 2, our group proposed an engineering analysis and engineering redesign of a transmission jack adapter for commercial transport settings. Following your approval of our approach, we continued with our research and development. We have encountered relatively few problems, so our team is confident that we'll meet the April 19 deadline for completing the final report. This memo presents details of our progress to date and the direction our group has taken as we near project completion.

Background

As you may recall, our goal is to determine exactly what would be necessary to transform an existing jack adapter into a versatile shop tool capable of meeting current industry needs. As we outlined in our proposal, floor- or column-type transmission jacks are usually used to remove heavy components from the undercarriage of commercial transport vehicles. Each of those jacks has its strengths and limitations; we'd like to blend their individual strengths into one unit.

Our project has evolved substantially since its inception, growing from a simple jack adapter to a modular engineered stand capable of many different configurations. Although our analysis and redesign will still focus on its use as a transmission jack adapter, our research has indicated that end users want the device to adapt to several different working environments.

The following section provides details of the project's progress to date.

Project Progress

Work completed

Our group has worked together to complete all work described in this section.

1. *Preliminary research.* The majority of our research has come from a series of interviews with industry professionals. After talking with Dave Cochrane, our client, we asked all interviewees questions about what their working environments demand of engineered stands (such as the jack adapter) and where the CT83 design falls short. The distilled answers to these questions will be found in sections 2.1.1, 2.1.2, and 2.1.3 of our report. (See the attached outline for details.) We also found internet information about various jack components and about competing products; such products are described in section 2.2.

2. *Analysis of gathered data.* In a series of meetings, our team compiled and reviewed the gathered data, as a basis for design changes.

3. *Early concept for a redesign of the adapter.* We viewed the redesign task as one large problem to be solved, rather than several small problems. As a team, we methodically attacked all key design faults. Building on each other's ideas and knowledge, we were able to efficiently suggest improvements that I believe were more successful than if we had worked separately on the design issues.

4. *Final research.* We presented our potential prototype design to the industry professionals whom we had previously interviewed, to get their responses. This second series of meetings indicated a few more design issues to be addressed.

Figure 21.6 A Student Research Progress Report   *(continued)*

Don Klepp   March 7, 2014   Page 2

5. *Final concept for the redesign.* After once again reviewing the industry feedback, we collectively fine-tuned the prototype design.
6. *Working outline for final analytical report.* By preparing a working outline for our final PCOM 142 report, we have consolidated our thoughts about the project and its results. You'll notice that the attached outline essentially uses a problem–solution pattern, culminating in a cost analysis. Creating the outline has also allowed us to identify and organize our remaining tasks.

Work in progress

Now that we have completed most of the design work and have the report structure in place, we can assign the remaining tasks to individual group members.

1. *Mathematical analysis.* An analysis of the design features of the new jack adapter prototype will provide the information we need to determine material specifications, safety factors, and the adapter's final dimensions. Although these calculations can be time consuming, the proposed design is relatively simple, and will be functioning in static loading conditions. Greysen is leading the team in completing these calculations.
2. *Cost analysis.* Tying in closely with the mathematical analysis is the structure's cost analysis. We will prepare a comprehensive breakdown of the manufacturing costs of the redesigned adapter, based on several production schemes, from small runs to large-scale production. Joe and Brandon are leading this portion of the work.
3. *Engineering drawings and solid computer models.* A full set of detailed engineering drawings and production plans are in the works. Our team started this work very early in the project, and we're well on our way. The mathematical analysis will yield the final dimensions of the adapter, and I will be able to complete the drawings and solid models.

Work to be completed

Our team works well together when collaborating for solid blocks of time. We will complete the majority of the following tasks together, during evenings and weekends.

1. *Draft report.* To gain experience in composing a report of this type, and to edit as we go, we will use the working outline to compose the first draft, which we hope will be close to the final draft.
2. *Oral report preparation.* Once the draft of our written report nears completion, we will be able to work on the oral report.
3. *Completion and submission of the final written report.* As the attached outline shows, the report concludes with a recommendation to build a prototype as the first step in the manufacturing process. I have formatted the report, with all headings, pagination, headers and footers, and spacing ready to go. As you can see in the following project completion schedule, we have left sufficient time to edit, proofread, and polish the document.

Project Completion Schedule

An updated version of the time frame for completing the report follows. We had flexibility early in the project, but now our timelines are rigid because of the looming deadline.

Activity	Time Allotted	Dates	Document Produced
Mathematical analysis & engineering drawings	20 hours	Feb 28–Mar 15	In progress
Write draft of report	20 hours	Mar 6–Mar 22	In progress
Prepare for oral report (due Mar 29)	6 hours	Mar 20	*PowerPoint* show
Revise/edit/refine report	8 hours		Final report due Apr 19

Figure 21.6 A Student Research Progress Report   *(continued)*

Budget to Complete Report
Building a prototype is not part of our project, so our only expenses are travel between Vernon and Kelowna (for interviews and team meetings), photocopying costs, and report printing costs, as follows:

- Transportation costs (total for all members) $80.00
- Photocopying costs $10.50
- Printing (colour printing and paper) $50.00
                                             Total: $140.50

Project Problems and Their Solutions
One thing you impressed upon us, following both the written and oral proposals, was our omission of cost guidelines for the stand. At the time of our proposals, we didn't have a clear idea of costs, so we didn't address the issue. However, we should have clarified that costs are tied closely to design considerations.

Since your feedback regarding this omission, we have kept cost factors in mind through the research and development process. First, we had to learn what industry wants from a tool such as the jack adapter, so that we could modify the design of the original CT83 pit jack adapter. Factors such as material selection, material size, jack dimensions, and fabrication cost will play a large role in the final price of the adapter. Availability of materials and components will affect the manufacturer's cost as will the production scale. We couldn't consider these factors until our R & D was complete.

I'd like to see the stand sell for about $600, because I'm not sure that industry will pay much more than that. I don't know if such a goal is attainable, even if we produce it on a large scale (500 units or more). The "Work in progress" section on the previous page indicates that we're working on the calculations now. When Joe and Brandon have finished that work, we'll be able to plug the results into section 4 of the report.

Conclusion
Our team has no doubts about finishing this project by the April 19 deadline. I hope you agree that the times listed in our completion schedule are realistic.

The size and scope of the project has definitely challenged our abilities and current knowledge base, but I must say that I've learned a lot about engineering in this "communication" project. The project is too much for one student taking a full load of classes, so I'm glad that I'm part of a group whose members share the load. The project's challenge level has kept us motivated, and now we're keen to properly complete what we've started. Another major motivating factor is that we think that our client, Dave Cochrane, may very well make some money with this new design! That's a good thing to take with us to our next job interview.

Thank you for your valuable advice on this project. We will take advantage of your offer to bring our document production questions to you as we swing into the final stages of the project. May we please schedule 15 minutes with you in each of our last three writing labs?

Vince Cummings

Vince Cummings

Attachments: Working outline (3 pages)
                   Sample drawings and solid computer models (4 pages)

Figure 21.6 A Student Research Progress Report   *(continued)*

Table 21.3 Project Completion Report Structure

Section	Content
Action opening	The reader connection depends on the type of reader: ◆ client? ◆ writer's supervisor? (What are the reader's main concerns?) State that the project is complete; briefly describe the outcome.
Background	Review the job's features: purpose, schedule, budget, location, people involved.
Project highlights	Describe the project's main accomplishments (work completed, targets met, results obtained). Discuss problems encountered, the impact of these problems on the project, and how the problems were handled.
Exceptions	Describe the deviations from the contract or project plan (if any)—the work not completed or done differently than planned. Give the reason for each deviation and explain its effect on the final project result.
Action closing	Analyze the reader's main concerns. What type of follow-up is needed?

INCIDENT REPORTS

An incident report resembles news accounts of events. Most of the description is provided as past tense narrative. Note, however, that some incident reports also use present tense to refer to the current situation and future tense to discuss what needs to be done.

INSPECTION REPORTS

Building inspectors, park wardens, gas inspectors, quality-control technicians, and others sometimes use forms to report the results of their inspections. However, often a form is not available or does not suit a particular inspection. In these cases, the adaptation of the action structure shown in Table 21.4 works very well.

The inspection report in Figure 21.7 on the next page deals with a troubling incident: a family, against its will, had been evacuated from its home because of

Table 21.4 Inspection Report Structure

Section	Reader Questions to Answer
Action opening	◆ Why should I read this report? ◆ What is the main result of the inspection?
Background	◆ Why was this inspection conducted? ◆ What was inspected? ◆ Who did the inspection? ◆ When and where did the inspection occur?
Details	What did the inspection reveal? 1. Conditions found: What did the inspectors observe re: the quality of work performed or items provided at the site? In what condition were equipment, facilities, or materials? 2. Deficiencies: What conditions, if any, need to be corrected? Does any work need to be done or redone?
Action closing	◆ Overall, what is the state of the site (facilities, equipment, etc.)? ◆ Does the writer suggest specific actions?

Prairie POWER Corporation

INSPECTION MEMORANDUM

DATE: January 7, 2014
TO: Randall Johnson, Gas & Electrical Inspections Coordinator
FROM: Miranda Ocala, Gas Inspector
RE: **Clogged Masonry Chimney—322 Montcalm Crescent, Saskatoon, SK**

On the evening of January 3, an elderly member of the Smith family, resident at 322 Montcalm Crescent, was rushed to the U of S Hospital Emergency Department. An alert resident suspected CO poisoning and contacted SaskEnergy. Later that day, Melvin Trask of SaskEnergy advised the occupants of the two-storey, single-family residence to vacate because

◆ the chimney was blocked with ice, and
◆ CO concentrations of 0.02% were present, apparently due to spillage of gas combustion products.

On January 4, Keith McLeod and I inspected the gas equipment and found

◆ the masonry chimney was blocked with ice. (We noted a white, lime-like substance on the exterior portion of the chimney that is exposed in the garage.)
◆ the furnace and the water heater were spilling.
◆ the home had evidence of excessive moisture—the windows were frozen shut. A serviceperson from Prairie Heating was present; she opened a small passageway at the top of the chimney's interior. Soon after a draft was established (in 30 minutes), the ice began to melt.
◆ the gas equipment was in good condition. That equipment consists of
    1. a 137 000 BTU standard Lennox furnace with a 6" vent draft hood
    2. a 36 000 BTU John Wood water heater with a draft hood (3" vent).

The furnace and the heater operated satisfactorily as soon as the chimney passage was reasonably clear.

◆ The 1", two-outlet supply pipe was in good condition.
◆ The masonry chimney, constructed of bricks and concrete and lined with tile throughout, seemed in good condition, although our initial inspection was unable to confirm the chimney's interior condition because of the ice buildup.

On January 6, after the ice had thawed, our subsequent inspection revealed damaged tile liner around the vent connector's entry point. This defect, coupled with the exposure of all four sides of the chimney and the recent cold weather, seems to have led to the icing condition. There is evidence that severe icing has occurred before: there is a white substance on the chimney exterior in the garage.

Because the tile liner measures 6¼" by 6¼", and a flexible liner measures 6⅜" OD, I have approved the use of a traditional (shop-made) sectional 6" aluminum liner. This is the most economical method of any acceptable corrections.

Still, the owner, Rod Smith, is annoyed that corrections are necessary to a house built just 14 years ago. He is also very angry that an owner's defect has been issued, precluding occupancy until satisfactory corrections have been made. He has threatened to sue for the costs of housing his family in a hotel until the residence is cleared for occupancy. I suggest that our customer service people speak with Mr. Smith to explain all the ramifications of allowing a family to occupy a home with potential for CO-induced deaths.

Miranda Ocala

Miranda Ocala

Figure 21.7 A Sample Inspection Report

dangerous carbon monoxide levels in the home. The writer, who is relatively new in her position, phrases her observations and opinions very cautiously. In particular, notice that she records details very clearly and that she uses the passive voice wherever possible to emphasize the *results* of the inspection, *not her part in the inspection and subsequent actions.*

COMPLIANCE REPORTS

At every level of government, regulatory agencies require organizations to report the degree to which the organizations have complied with the agencies' regulations. Also, industrial and professional associations require their members to report how the members have complied with codes of business conduct. Some compliance reports can be completed on forms provided by the watchdog agency, but more often the reporting organization has to create its own format and structure for the report. In such cases, the structure outlined in Table 21.5 could form the basis of any particular compliance report. Following Table 21.5, a case study describes a situation that requires a compliance report. **MyWritingLab: Technical Communication** presents the report emanating from that situation.

Table 21.5 Compliance Report Structure

Section	Content	Comments
Action opening	Name the report's purpose. Refer to the act or to the body of the regulations prompting the report. State the degree to which the requirements have been met; list the time frame covered by the report.	The reader connection might be placed in a transmittal letter or memo if a semi-formal format is used. In a letter or memo report, use a subject line that names the specific act and its sections. Often, a compliance report has legal implications, so do not connect with the reader in an informal, friendly manner.
Background	Supply necessary details about regulations for a reader who is not familiar with them. For a report to the regulatory agency, possibly review your organization's compliance record.	Background about the regulations will not be necessary in a report to the regulatory agency.
Details of compliance	Describe specific actions taken to comply with specific regulations, in the order the regulations appear. The report could follow a time pattern— the order in which you've met the reader's expectations. Or organize in the order of importance. Or use an actions–results pattern.	Use bolded headings, bullets, numbered lists, italics, white space, and indents to help make the document easy to read.

(continued)

Table 21.5 *Continued*

Section	Content	Comments
Action closing	Summarize key aspects of the reported activities. State the bottom line, including results, where appropriate. List future methods of complying with regulations or expectations.	In some cases, the report might describe costs of problems of compliance; the conclusion might argue for a relaxed interpretation of the regulations.
Attachments (optional)		Photos, affidavits, test data, shift reports, or other detailed backup data might be appropriate.

CASE | **A Situation Requiring a Compliance Report**

Horizon Environmental Consultants (Horizon) responds to a request from the City of Saskatoon to determine whether landscaping completed at the Goose Hollow housing development complies with regulations set by the city.

Regulations Requiring Compliance

In consultation with the Saskatchewan Environment Ministry, the city had required a 20-metre buffer zone around the wetland, extending outward from the wetland margin, to encompass the following features:

◆ A walking path within the buffer zone may be no nearer than 10 metres from the outer edge of the wetland margin. The 10-metre zone from path to margin shall be a "No Entry" zone.
◆ The path may comprise only natural materials.
◆ Where disturbed by construction, the No Entry zone shall be planted with native grasses. The space between the path and the outer edge of the buffer zone, approximately 6 metres, shall be planted with native prairie shrubs, such as willow, Saskatoon, grey dogwood, and pin cherry.

Observations of Compliance and Non-compliance

During site visits on June 15 and September 7, 2014, biologist Christine Wie and terrestrial ecologist John MacKinnon observed the following project features.

June 15 Site Visit.   The street system had been completed, and construction had begun on several houses north of the wetland. In the wetland area, Cherkewich Landscaping had completed a crushed rock path around the entire wetland perimeter. Wie and MacKinnon observed only one (2.5 metres square) encroachment of the No Entry zone. Some of the Saskatoon bushes that had been planted in the outer buffer zone appeared stressed. At the northeast corner, the observers saw large quantities of silt on the path and well into the wetland margin.

Wie and MacKinnon recommended the following mitigative measures:

◆ Remove as much of the invasive silt as possible.
◆ Construct a silt fence around the perimeter of the wetland for the duration of the construction period.
◆ Space the shrubs at 1 to 1.5 metres and water them with drip lines.
◆ Vary the plantings by adding willow, grey dogwood, and pin cherry.
◆ Plant cattails in the encroached 2.5 metres square No Entry zone.

September 7 Site Visit.　The follow-up visit noted that silt fencing had been placed around the wetland perimeter, except for the entire eastern side of the area, but there was no evidence of silt encroachment on that side of the wetland. The north-side fencing had arrested silt encroachment in several places. Drip lines had been placed to water the planted saskatoon shrubs, although the spacing remained at 3 metres. The landscaping crew had nearly finished planting mixed shrubs in the remainder of the buffer zone. However, some of the shrub spacing well exceeded the recommended 1 to 1.5 metres. The observers noted evidence of cattail seed planting in the 2.5-metre-square No Entry zone.

The observers' compliance report listed the above observations and recommended (1) additional work and (2) a third site visit. **MyWritingLab: Technical Communication** presents that report among its sample documents.

FIELD TRIP REPORTS

Usually, a field assignment is complete only after you have reported on what you observed and what you did. The field trip may have involved a four-hour hike along an abandoned forestry road, or required two weeks of testing pollution levels in salmon-spawning rivers. Regardless of the trip's duration and complexity, you need to perform two main tasks in advance of the inevitable report:

1. Make careful, detailed observations and record them in a notebook or on a voice recorder.
2. Organize those notes to answer your reader's questions in the order your reader would prefer.

Table 21.6 on the next page illustrates a structure that could be used for most field trip reports.

MEETING MINUTES

Many team or project meetings require someone to record the proceedings. Minutes are the records of such meetings. Copies of minutes are distributed to all members and interested parties, to track the proceedings and to remind members of their designated responsibilities. The appointed secretary records the minutes.

When you record minutes, answer these questions:

◆ What group held the meeting? When, where, and why?
◆ Who chaired the meeting? Who else was present?
◆ Were the minutes of the last meeting approved (or not approved)?

◆ Who said what? Was anything resolved?
◆ Who made which motions and what was the vote? What discussion preceded the vote?
◆ Who was given responsibility for which actions?

See Figure 21.8 on page 431 for a sample set of minutes.

Table 21.6 Field Trip Report Structure

Section	Reader Questions to Answer
Action opening	◆ Why are you reporting this? (optional) ◆ In brief, what have you been doing? What did you accomplish?
Background	◆ Who went where? When? Why? ◆ On whose authority? (optional) ◆ How did the writer travel? (optional) ◆ What was the project? (optional)
Details of work accomplished	◆ What did you do? What routine work? Which work specifications were followed? ◆ What work did you perform beyond the routine requirements? ◆ What did you observe? ◆ What meetings, if any, did you have? With whom? What were the results?
Problems encountered	◆ What were the specific problems, if any? Did you identify the causes of these problems? ◆ What specific actions did you take to solve the problems? ◆ Were you successful? If not, why not?
Action closing	◆ What remains to be done? What resources are necessary? Who should perform the work? Have you assigned the work? ◆ Are you requesting support or authorization from me, your reader?

The reports discussed so far in this chapter focus on factual information. Now, the focus shifts to short analytical reports, all of which logically arrive at a conclusion. Readers of analytical reports want more than the facts; they want to know what the facts mean.

This chapter discusses only how to modify them for letter, memo, and semi-formal report versions. Like short informational reports, short analytical reports can profitably use the four-part action structure.

MEETING OF THE CAMPUS RECYCLING COMMITTEE
Room 125,
Student Services Building, March 28, 2014, 4:00 p.m.

Chair: T. Maguire, Jordan College Student Association President

Present: R.W. Siggia, V.P., Administration   T. Singh, 3rd year Arts
J. Klym, Campus Services   J. Cormier-Bauer, 2nd year Engineering
M. O'Connor, Print Services   H. Calvin, 4th year Business
P. Masinkowski, 4th year Phys. Ed.

Guest: John Maravich, Canadian Waste Disposal Ltd.

1. Approval of Agenda
T. Maguire asked to add a presentation by John Maravich and suggested that discussion of bottle recycling as a student association fundraiser be tabled to the next meeting, in order to accommodate the address. T. Maguire called for approval of the amended agenda.
Passed Unanimously.

2. Other Business
John Maravich proposed a business arrangement wherein Canadian Waste Disposal would have exclusive rights to recycle paper products at Jordan College in return for an annual $5000 scholarship to a Jordan College Business student and a commitment to hire Jordan College students on a part-time basis. The projected total volume of business was discussed, along with other details provided in Canadian Waste Disposal's written proposal (see attached). The committee agreed to hold a special meeting in two weeks to discuss the proposal.

Action by: T. Singh and J. Cormier-Bauer will press class presidents to poll students re: their on-campus paper usage.

M. O'Connor and R.W. Siggia will review administrative and academic paper usage.

J. Klym will determine recycling potential for calendars, phone books, and all other campus service publications.

3. Approval of Previous Meeting
After revision of the numbers relating to recycling of library culls, the previous meeting minutes were accepted.
MOVED: R.W. Siggia   SECONDED: J. Klym   **Passed Unanimously.**

4. Recycling Ink Products
H. Calvin and J. Klym presented a report on the types and volume of ink used by on-campus photocopiers and computer printers. Their main findings were that in the last fiscal year:
1. $212 700 was spent on ink cartridges for laser printers, inkjet printers, and photocopiers.
2. Of that amount, $192 654 was used to purchase new cartridges, and the remainder was spent on recycled cartridges. On average, recycled cartridges cost 62% as much as new ones.
3. The latest editions of *Consumer Journal* and *Computer Equipment Monthly* report 93% reliability with recycled cartridges.

MOVED: J. Klym   SECONDED: H. Calvin

That Jordan College adopt a one-year trial period of using recycled cartridges. Discussion centred on the issue of whether this committee has a legitimate right to take this action. R.W. Siggia contended it is Administration's prerogative, but agreed to approach President Monroe with the committee's decision.   **Passed Unanimously.**

Call for adjournment at 5:05 p.m.   **Carried.**

Figure 21.8 A Sample Set of Meeting Minutes

As you examine the structures suggested for short analytical reports, notice a key difference between formal and informal analytical reports. A formal report usually places its main point, its "bottom line," at the end of the report and satisfies a reader's immediate "need to know" by placing a summary in the front matter. An informal report, on the other hand, places the report's main finding in the first or second paragraph, to satisfy the reader's curiosity. While this "front-loading" technique suits most informal reports, it doesn't suit every situation, as you'll see in the discussion of indirect recommendations reports, later in this chapter.

FEASIBILITY REPORTS

The reader of a feasibility analysis will likely want your answer on whether equipment or a project is realistic and practical at the beginning of the report, so the structure in Table 21.7 should work well.

Feasibility reports present a good example of documents that require a background section. That section not only refers to the situation prompting the report; it also lists the all-important assessment criteria. Applying these criteria then provides the backbone of the report.

Figure 21.9 on page 433 provides a sample feasibility report. Notice how it clearly lists the assessment criteria.

Table 21.7 Short Feasibility Report Structure

Section	Content	Comments
Action opening	Refer to the reader's request or the situation requiring analysis. State whether the examined project or equipment is feasible.	The reader connection might be placed in a transmittal document if a semi-formal format is used. A letter or memo does not need a heading for the opening paragraph.
Background	Describe the situation leading to this feasibility study. Explain exactly what kind of feasibility is studied, and list the assessment criteria.	The amount of background depends on the reader's familiarity with the subject. The criteria may have to be justified.
Details of assessment	Apply each assessment criterion, step by step, to the data. Choose suitable criteria—a proposed equipment purchase, for example, could look at cost, warranty, equipment reliability, performance, and compatibility with current equipment.	The title of this section will depend on the kind of feasibility being discussed, and on the reader's priorities.
Action closing	Summarize the results of applying all criteria and state the bottom-line conclusion. If appropriate, recommend approval.	A summary table could be effective. Brochures, test data, financial projections, or other detailed supporting data could be attached.

Ministry of Transportation　　　　　　　　　　**Internal Memo**

DATE: January 10, 2014
TO: Richard Janvier, Engineering Information Systems Coordinator
FROM: Sheri Prasso, Science and Technical Officer
RE: **Sand and Glavine Proposal for Office Networking
(RFP 20021209-EIS)**

As you requested, this report assesses the proposal submitted by Sand and Glavine Systems for an office network for our Vernon Engineering Services office. The proposal has merit, but requires changes before the Ministry of Transportation can accept it.

The RFP placed its focus on making better use of the computer information systems by creating a local area network (LAN). Therefore, I used the following criteria to assess the Sand and Glavine computer network proposal:

1. technical considerations
2. cost
3. training and support
4. efficiency gains

Information for the assessment was collected from current books on the subject, staff at the Vernon Engineering office, and local businesses.

Technical Considerations and Cost
The proposed network will meet Engineering Information Systems' requirements, with minor changes. In particular, the Wang 486 needs to be retained as part of the network (see the attached technical analysis for more detail). These changes put the cost of the network slightly over budget. However, anticipated reductions in cable requirements and installation time should lower the cost. The overall cost of the modified network will be close to the proposal's quoted price of $4000.

Training and Support
Technically, the Sand and Glavine Systems proposed network is simple. Because of this simplicity and the competence of the staff at the Vernon Engineering Office, the proposed training will be sufficient. Unlike training, support was not included in the proposal's quoted price. It was offered at additional cost through monthly service contracts. The Ministry of Transportation has qualified computer support personnel on staff. Purchasing support from Sand and Glavine Systems would duplicate service and add to the direct cost of this network.

Efficiency Gains
Sand and Glavine's proposed computer network will meet Engineering Information Systems' objective of increasing the efficiency of the Vernon office. The network will save time and allow staff to focus their efforts on engineering rather than on file management. Also, Sand and Glavine will install the system on a weekend, saving two days of down time.

Recommendation
If Sand and Glavine Systems resubmits the proposal with the requested changes, it should be adopted.

Sheri Prasso

Attachments: Technical analysis (cable, topology, hardware, and software)
Cost analysis
Task time comparisons

Figure 21.9 A Sample Feasibility Report

CAUSAL ANALYSES

Your reader's concern about the cause of a problem or an incident should direct you to state the report's main point in the lead paragraph, as shown in Table 21.8. Reports about positive issues—an unexpected rise in factory or an improvement in the employee turnover rate—should also name the main cause early in the report.

The causal analysis illustrated in Figure 21.10 on pages 435–437 was commissioned by a homeowner who suspected that a gaping hole under her driveway resulted from her neighbour's faulty driveway design. TerraTech, the consulting firm engaged by the homeowner, presents its findings in detached, carefully measured language. The report writer uses the classic causal analysis technique of identifying and eliminating possible causes to determine the most likely main cause.

A letter format is used for this report because of its relatively short length and straightforward content.

Table 21.8 Informal or Semi-formal Causal Analysis Structure

Section	Content	Comments
Action opening	Refer to the reader's request or to the writer's role in analyzing the identified situation. State whether the cause(s) can be identified and, if so, name the main cause.	The reader connection might be placed in a transmittal letter or memo if a semi-formal format is used. A letter or memo report does not need a heading for the opening paragraph or two.
Background	Describe the situation (or environment) in which the event occurred or in which the problem developed. Provide background about similar problems or situations.	This section should not exceed two paragraphs. If more detail is necessary, it can be added in attachments.
Details of analysis	Describe the step-by-step analytical process and give the results of that process.	Causal analysis usually names possible causes identified from previous experience and based on the relation between an event and prior conditions.
Action closing	Summarize the report's main findings. State the bottom line. If appropriate, recommend remedial or preventative action.	A summary table could be effective. Attached brochures, performance tests, financial projections, or other detailed supporting data could be appropriate.

TerraTech ENGINEERING

1714 Kalamalka Lake Road, Vernon, British Columbia V1G 2N6   Ph. 250-545-0919
Fax 250-545-2020   terra@junction.net

June 20, 2014

Ms. L.P. Garcia
306 Melville Court
Vernon, BC   V1B 2W9

Dear Ms. Garcia:

Re: Causes of the Undercut Driveway at 306 Melville Court

At your request, we have examined the extent of the undercutting of your 17-month-old
driveway and have identified the primary cause of that problem to be water directed from a
neighbouring driveway.

Background

In similar situations that we have analyzed, we've determined that open space can develop
under a concrete pad if the base soil has not been properly packed, or if flowing water has
eroded soil away, or if water has pooled under the pad and thus caused the soil to settle.

Much depends on the soil's composition. In the case of your driveway, the base soil primarily
consists of glacial till, a combination of sand, gravel, larger rocks, and small amounts of organic
material. Typically, when glacial till gets saturated with water, it turns into a slurry that either
flows downhill or collapses into itself, as all loose spaces among the particles are filled.

Investigation and Results

We conducted the investigation in four stages:

1. We determined the nature and extent of damage by examining the concrete deck and by
   probing the empty space beneath it. Our visual inspection of the concrete revealed no cracks or
   sagging of the deck; the driveway has stood up very well. We then interviewed your builder,
   Mark Lambton, who showed us his construction notes. The notes say that the concrete
   subcontractor used twice the normal amount of reinforcing bar within the concrete slab.

   The driveway has not pulled away from the rebar fitted into the garage footings, despite
   23 centimetres of open space under the concrete along the intersection of driveway and
   footings. The drawing on page 2 shows the extent of open space under the driveway, and
   other features of the existing situation. (Numbers show the depth of settling or erosion at
   various points.) The undercutting is quite extensive, as the drawing shows.

2. We ruled out inadequate packing as a cause of the settling. Mark Lambton's notes reveal
   that the driveway base was packed uniformly, that it was watered between packings, and
   that it was packed five times over a four-day period. This exceeds normal practice.

Figure 21.10 A Causal Analysis   *(continued)*

Ms. L.P. Garcia
June 20, 2014
Page 2

3. We ruled out erosion as a major causative factor. There may have been some minor initial erosion along the north edge of the driveway, but erosion could not have caused the irregular pattern of open space beneath the slab. That irregular pattern suggests that pooled water has led to irregular soil settling.

4. We determined the source of the amounts of water required to cause the degree of observed settling. We do not believe that the water has come from your driveway, say in last fall's heavy rains. Your driveway has been designed to channel water down toward a collection drain located two metres from the centre of the garage door. To test our belief, we ran water onto your driveway from two hoses simultaneously. All the water was easily channelled down to the drain.

Then, we investigated water flow from the north-neighbouring driveway that slopes away from the garage, toward the street. That driveway is higher than yours at every point of its length, and the entire driveway slopes down toward the rock-covered depression between the two drives. Moreover, a 1-metre-wide diagonal depression in the neighbouring driveway channels water toward the catchment area immediately adjacent to the spot where your driveway begins to be undercut (see the diagram).

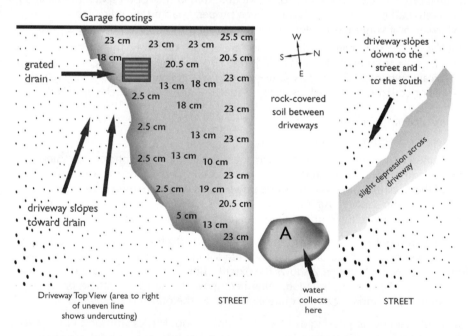

With your neighbour's permission, we ran water onto his driveway for 45 minutes. All of the released water ran down to the catchment area A, from which the water seeped under your driveway and formed pools of settling water. When we re-inspected three hours later, all that water had soaked in.

Figure 21.10 A Causal Analysis   *(continued)*

Ms. L.P. Garcia
June 20, 2014
Page 3

Conclusion
We're confident that the driveway undercutting has resulted from water-induced settling and that the majority of the water has come from your north neighbour's driveway.

Corrective action will require shoring up the space under the driveway with an impermeable base and diverting water away from possible entry points along the north edge of your driveway. We can recommend Majestic Mudjacking for the former task—this company will pump a rapidly hardening slurry of cement, sand, water, clay, and loam under the exposed slab. If you wish, we'll undertake the water diversion.

We believe that you have grounds for requiring your neighbour to pay the costs of all repair work; local bylaws require each homeowner to control the passage of water off their property, so that the water doesn't flow onto a neighbour's property. Please let us know if we can be of further service. Our invoice is enclosed.

Sincerely,
TerraTech Engineering

Dimitri J. Jones, C.E.T.

Figure 21.10 A Causal Analysis

ASSESSMENT REPORTS

Assessment reports essentially use the same approach as feasibility reports, except that assessments (also known as *evaluation reports or investigation reports*) are conducted after a project has been completed, or *after* changes have been made.

Effective assessments describe an issue or a completed process, project, installation, or equipment purchase. A short assessment report provides its main conclusion within the first two paragraphs. Next, it identifies (and in some cases, *justifies*) the criteria used in the evaluation. Applying those criteria, one by one, forms the bulk of the "Details" section that follows. Finally, the "Action Closing" discusses a more detailed version of the opening bottom-line statement and, if appropriate, recommends action.

One common variation type of the "assessment" report is much like a feasibility report because it examines solutions that have not yet been implemented. A *yardstick* or *benchmark* report assesses and compares two or more alternative solutions or equipment proposals. In this type of analysis, appropriate criteria act as yardsticks to compare competing alternatives can be compared. Thus, the analysis is handled consistently and fairly. See Table 21.9's structure for this type of report.

Table 21.9 Yardstick Assessment Report Structure

Section	Content	Comments
Action opening	Define the situation requiring assessment—the problem or challenge. Report the main conclusion.	Only one or two paragraphs are necessary for a letter or memo. Use the heading "Introduction" for a semi-formal report.
Background	Briefly explain possible alternative solutions. Name the assessment criteria and explain how these were developed and selected.	The section could be called "Background" or "Assessment Method." Extensive technical data should be attached to the end of the report, not placed in the background section.
Details of assessment	Evaluate each alternative, according to the assessment criteria. The assessment could be organized by criteria such as cost, durability, product support. Or the alternatives could be examined, one by one: all the criteria would be applied to the first alternative before moving on to the next alternative.	Use a title like "Data and Assessment." Organize by criteria to allow the alternatives to be ranked by each criterion in turn (e.g., the least to the most expensive). Focus on each alternative in turn.
Action closing	Conclude which alternative, or combination of alternatives, best meets the assessment criteria.	An implementation plan could be included, but it shouldn't be the focus of an assessment report.

RECOMMENDATIONS REPORTS

Many recommendations reports respond to reader requests for a solution to a problem; others originate with the writer, who has recognized a problem and developed a solution. This latter type is often called a *justification report*.

The purpose of a justification report is, at heart, similar to that of a proposal—both try to "sell" a solution or influence a decision to be made by the reader. However, these two kinds of documents differ in their approach. A proposal openly attempts to persuade the reader, while the justification report uses an apparently more objective approach: it shows how the facts inexorably lead to the recommended solution.

Whether solicited or unsolicited, recommendations reports could use either a direct pattern or an indirect pattern, depending on the reader's needs and attitudes. The *direct pattern* works best when you can anticipate reader support. Perhaps the reader has accepted similar recommendations in the past. Perhaps your recommended choice is so obvious and clear-cut that there is no other course of action. (See Figure 21.11, on page 440, for such a situation.) Perhaps your recommended action matches company policy and practice. Perhaps the recommendation reflects the reader's own preferred approach or administrative bias.

Table 21.10 Direct Recommendations Report Structure

Section	Content	Comments
Action opening	Refer to the need or problem in a way that your reader will recognize and accept. Use active verbs to recommend action.	Use a heading like "Recommendation" or "Problem and Solution."
Background	Name the alternative solutions you considered and explain the criteria used to assess the alternatives. Briefly explain why you discarded alternatives other than the one you selected.	In some cases, it may not be wise to quickly dismiss potential solutions; your approach may seem arbitrary. In such cases, apply the assessment criteria fully to all alternatives.
Details of assessment	Discuss the benefits, comparative advantages, and drawbacks of your recommended solution. Detail the required resources and costs of your recommended solution.	Use a title like "Features of the Solution" or "Advantages and Requirements" or "How Our Firm Will Benefit."
Action closing	Summarize the main reason for choosing the recommended solution and provide a plan for implementing it. Request authorization for your actions or specify the actions you're asking of the reader.	You could attach detailed supporting information, such as performance tests, brochures, quotes, and financial projections.

Table 21.10 shows the direct pattern for the kind of recommendations that would be received favourably.

The direct recommendations report in Figure 21.11 responds to the following identified need.

CASE | **A Direct Approach for Recommending Action**

At a coal-fuelled electrical power plant, steel outflow pipes carry ash from the coal furnaces to storage lagoons (known as "ash pits" or "lagoons"). Some of these pipes (known as "ash lines") have been eroded and are in danger of failing. The writer knows, as does the reader, that the plant must replace the lines when they are in danger of failing; no other alternative action is available. Still, the recommended action needs to be justified. Also, the report needs to include some background information for the reader, who has recently transferred from a hydroelectric plant.

An *indirect approach* works better in situations where your reader may be unreceptive to your recommendations, or where the recommendations deal with a sensitive issue, such as workplace harassment or strained employer–employee relations. Table 21.11 on page 441 suggests the sequence for an indirect recommendations report.

Direct POWER Corporation   MEMORANDUM

DATE: April 17, 2014
TO: Martin Scherre, Head Engineer, Correl Park Power Station
FROM: Judy Shohat, Plant Engineer, Correl Park Power Station *JS*

RE: Ash Line Replacement

As I mentioned in our plant meeting two weeks ago, some of the ash lines will need replacing. Shimon Barak and I have since examined Production Units 1 through 6 and found potential line failures for the lines leading from Unit 5 and Unit 6.

Recommendation
Correl Park should purchase 915 m of 300 mm (319 mm OD, 9.5 mm WT) commercial-grade black steel pipe for replacement ash line, at an estimated cost of $60 000.

Background
Approximately ⅔ of all ash line failures are detected and repaired with only some welding time required. However, if an ash line fails in the early evening and is not detected until morning, the line downstream of the failure will plug due to reduced flow velocity. When this has happened, we have had to contract a high-pressure washing truck to clean the line at a cost of from $5000 to $10 000. In order to maximize line life, we rotate the lines ⅓ turn every three or four years, to distribute the wear around the inner circumference of the pipe.

Findings
Correl Park Unit 6 has three ash lines—6A and 6B each have one rotation to go, so they will be fine for at least three more years. However, 6C ash line received its third and final turn in May 2009. Last week, this line was examined and rotated in whatever direction exposed the thickest remaining wall to the area of highest wear. A random sampling of thickness readings along this pipe showed an average thickness of 4.47 mm, which is less than the original thickness of 9.5 mm. Experience has shown that an average thickness of equal to or less than ½ the original significantly increases the failure frequency. About 765 metres need replacing.

Currently, No. 5 ash pit is being cleaned out, with the ash being used for road construction. Once the cleaning is complete, it will be necessary to install about 150 metres of ash line to make this pit functional again. (We could then restore Unit 5 to service.) We considered installing the required line with used pipe that we have in stock, but that used pipe is no better than the 6C line that needs to be replaced.

Authorization
I request authorization for the recommended replacement so that bidding for the pipe supply contract can proceed. A Purchase Recommendation and a Technical Specification for the required pipe are attached.

Att.

Figure 21.11 A Direct Recommendations Report

Table 21.11 Indirect Recommendations Report Structure

Section	Content	Comments
Action opening	Refer to the situation in such a way that your reader realizes there's a problem or a need to be addressed. Briefly describe the approach used in this report.	A semi-formal report could have an "Introduction"; a letter or memo would not need a heading for this section. In a letter or memo, "connect" with the reader in the first sentence.
Background	Show the extent of the need or problem by presenting quotes, examples, or supporting statistics. List alternative solutions and explain the criteria used to assess the alternatives.	Either semi-formal or informal reports could use a heading, such as "Problem and Solutions" or "Background".
Details of assessment	Evaluate the alternatives with the identified criteria, starting with the least applicable solution. Present the best alternative last; apply the criteria vigorously.	Use titles like "Assessment of Alternative Solutions" or "Possible Solutions."
Action closing	Summarize your recommendation. Show how it can be implemented. Ask for authorization or specify the actions you're asking your reader to take.	You might attach detailed supporting information such as performance tests, brochures, quotes, and financial projections.

To illustrate the indirect pattern at work, consider the challenge faced by Kim Briere and her colleagues in an engineering technology diploma program.

CASE | **An Indirect Approach for Recommending Action**

Like those of many of her colleagues, Briere's writing skills are not adequate for the Applied English 140 course offered in the first semester of her five-semester program. After lengthy consultations with her colleagues, with a college academic counsellor, and with the chair of the Engineering Technology Program, Briere sends the recommendation illustrated in Figure 21.12 on the next pages to Jake Stroud, the Applied English 140 instructor.

As a result of her audience/purpose analysis, Briere realizes that her needs and those of her colleagues are different from those of Professor Stroud. She's also aware that he is opposed to teaching what he calls "remedial English." He has said that the English 140 class "applies university-level writing skills to real-world situations; there's no time to develop basic skills that university students should bring with them." Briere suspects that Professor Stroud will philosophically oppose her recommendations and that he'll resist her recommended action because it means more work for him. Therefore, she places her recommendations at the end of the report, after leading her reader through an analysis that shows that the majority of the Applied English 140 students really do require some writing instruction.

MEMORANDUM

TO: Professor Stroud, English Department   **DATE:**   October 16, 2014

FROM: Kim Briere, Applied English 140

RE: **Improving Performance in English 140**

As you have stated in class, we have a severe problem in English 140: the overwhelming majority of the 38 students have failed at least two of the first three assignments so far. Students like me, who are committed to success, are very concerned. So we've consulted an academic counsellor and the engineering dean and we've identified the solution that is presented in this memo.

Thirty-one of us have met three times to discuss the issue. We need to solve the problem of low grades and we need to pass this class to move on to English 150 next semester. Many of us have also discussed the issue with our department chair, who warns us that we need a solid grounding in English to do well in our engineering program.

We see a problem for you, too. It must be difficult to have to correct so many things in our memos, letters, and reports. Also, we ask so many questions about grammar and sentence basics in class that you don't have time to present your full lecture.

The cause of the problem seems to be our "inadequate grasp of the English language" (your comment in the October 12 class). We agree with your assessment. In our meeting yesterday, 29 of us found that we lost an average 21 marks for mechanical errors and poor paragraphing on the last assignment!

So what can we do? We must satisfy three criteria. We need to

1. improve our English grades to succeed in our program of study
2. build our writing skills for future careers
3. find a practical, immediate solution

Following the advice of an academic counsellor, who showed us the Harvard Case Study model, we've applied the above criteria to four options:

1. work harder and spend more time on our assignments
2. lobby the college to reinstate the drop-in Writing Centre, which disappeared after last year's budget cuts
3. drop the English 140 course now
4. arrange for special tutorial sessions

The table on page 2 summarizes our thoughts about the four options.

Figure 21.12 An Indirect Recommendations Report   *(continued)*

Professor Stroud   **October 16, 2014**   **Page 2**

Options	Improve Grades	Build for Future?	Practical and Immediate?
Work harder and longer	Not likely—we still need basic skills	No—we're held back by lack of basic skills	No—we carry seven classes each and spend too much time on English now!
Lobby for writing centre	Yes—individual examples and instruction could help us develop basics	Yes—could build the base we need	No—the college is still in a deficit situation and the bureaucracy moves too slowly
Drop the course	Perhaps if we take the course later, we'll succeed	We don't know how to build writing skills on our own	No—we need a solution this semester
Arrange for tutor(s)	Yes—we can develop the skills, individually and collectively	Yes—we need to build skills to get to the next level	This is the only practical possibility of the four options, if we can get the needed assistance

As you can see, we have only one viable option. In order to make that option work, two things have to happen:

1. We need times and a place to meet. Dean Cartwright has arranged a classroom for 4–7 p.m. on Mondays, Wednesdays, and Fridays. She has also found $1000 to pay a tutor or tutors.

2. We need your help to direct a tutor (or tutors); you have the best idea of our needs. Also, can you help us find one or two tutors? Perhaps you know capable retired professors or graduate students.

Please support our recommended action. We'd like to start no later than next Monday, so may we have your response in Thursday's class?

Kim Briere

Kim Briere

Figure 21.12 An Indirect Recommendations Report

The Harvard Case Study model can be used to plan indirect recommendations

A variation of the indirect recommendations report is known as the Harvard Case Study model, which many business schools use as their primary learning and teaching tool. Listed below are the basics of that logical method of solving problems and viewing issues:

1. *Identify the critical success factors* in a given situation. These factors are used later in the process to evaluate proposed solutions. (In the scenario in Figure 21.12, the critical factors of passing English 140 and succeeding in future careers depend on English writing skills.)
2. *Define the problem* (e.g., "the overwhelming majority of . . . students have failed at least two of the first three assignments").
3. *Identify the causes* (e.g., "inadequate grasp of the English language").
4. *Develop possible solutions* that address the root causes of the problem (e.g., "work harder and longer"; "lobby for writing centre"; "drop the course"; "arrange for tutors").
5. *Choose evaluation criteria* that emanate from the critical success factors (e.g., "improve grades"; "build for future"; "practical and immediate"). Apply these criteria equally to the potential solutions. (See the evaluation grid in Figure 21.12.)
6. *Choose the solution* that comes closest to satisfying the combined evaluation criteria (hire tutors); rarely does a solution completely meet all criteria, so pick the best of the lot.
7. *Recommend a detailed plan of implementation*—who should do what, when, how? (See the closing eight lines in Figure 21.12.)

The Case Study Model has many practical applications

This method is useful in all sorts of situations: deciding among competing bids, solving technical problems, making career choices, choosing courses of study, dealing with difficult people, and so on.

LAB REPORTS

The academic lab reports you write during your studies are different from the reports you'll write in industry. First, the names are different; at university or college, you submit *lab reports*; at work, you'll produce *laboratory reports* or *test reports*. The purpose also differs: your university or college lab reports help you learn material or prove a theory, while "real-world" laboratory tests have practical applications, such as the following:

◆ Water samples are tested to determine if a water treatment plant is working properly.
◆ Car seat child-restraint systems are tested to see if they are safe and effective.
◆ Soil core samples are tested to determine if a PCB-contaminated site has been decontaminated.

Various formats and requirements exist for academic and industrial settings. You will have to adapt to the specific requirements at your workplace. However, all laboratory and test reports use the general pattern shown in Table 21.12 on page 445. A sample lab report and links to online lab reports are found at **MyWritingLab: Technical Communication**.

Table 21.12 Lab Report Structure

Section	Content	Comments
Introduction	Name and define the subject; review the subject's significance. Or indicate how this test fits into a project or routine procedure.	This section could also include the scope of the research. It might discuss the rationale for the research or the objective of the research. Sometimes, this is expressed as a question to be answered, sometimes as a hypothesis to be proved or disproved.
Equipment and procedure	Where appropriate, describe the design of the investigation.	This section might also be called "Materials and Methods."
	List materials, instruments, and equipment.	
	Describe, step by step, how the test, experiment, or study was done.	Use the passive voice, third-person narrative, in the past tense.
	Describe methods for observing, recording, and interpreting results.	Use the passive voice, third-person narrative, in the past tense.
Results	List recorded observations.	Relate these results to the methods used to achieve them.
	Provide detailed relevant calculations.	
Conclusion	Analyze the percentage of error and the possible causes of error.	This section is often called "Discussion."
	Answer:	You might also answer:
	◆ Do the results answer the questions?	◆ Was the hypothesis proven?
	◆ Was the research objective met?	◆ Are these results consistent with other research?
	◆ Do you have doubts about the results? Why?	◆ Are there implications for further research?

FORM REPORTS

Many reporting situations can be handled with pre-printed forms (or electronic templates). Daily and weekly progress reports, for example, often use forms to keep clients and supervisors informed about a project. In many jobs, the best time to learn how to use job-specific forms is during the orientation period: the first two to three weeks on the job. Ask questions about the purpose of each form and the expected standard of completion. Also ask to see completed sample forms. (Some supervisors will prefer to show you how to use a given form only when that form is needed, and not before.)

Employers have passed on the following hints for successfully completing form reports:

◆ Read headings or questions on the form carefully. If necessary, ask directions, or look for models (precedents). Do not assume that you've guessed correctly.

◆ Before writing or keying the form, read the entire form and make some quick notes of what to include.

◆ Choose *exact* words and phrases, not approximate descriptive language.
◆ Use jargon only if necessary; perhaps a non-technical person will read your report.
◆ Analyze the audience and the purpose for the report, and provide *all* necessary detail.
◆ Write or print neatly. On multiple-copy forms, press firmly!
◆ Check for errors in facts and figures, spelling, or logic.
◆ Know deadlines and stick to them. Remember that form reports are not designed for the writer's convenience; they're used to help you provide information quickly, while the information is still useful to the reader.

CHECKLIST

for Revising and Editing Short Reports

Use this checklist as a guide to revising and refining your short reports.

◆ Have you chosen the best report format for your purpose and audience?
◆ Does the letter or memo use proper format?
◆ Does the subject line forecast the contents of the letter or memo?
◆ Does the semi-formal report format contain the appropriate elements?
◆ Are readers given enough information for an informed decision?
◆ Are the conclusions and recommendations clear?
◆ Did you make the right choice between the direct and indirect patterns of presenting the report's bottom line?
◆ Are paragraphs single-spaced within and double-spaced between?

◆ Do headings, charts, or tables appear whenever needed?
◆ If more than one reader is receiving copies, does the letter or memo include a distribution notation to identify other readers?
◆ Does the semi-formal report's title page name other readers?
◆ Is the writing style clear, concise, exact, fluent, appropriate, and direct?
◆ Does the document's appearance create a favourable impression?
◆ Have you included useful details, such as supplementary attachments, enclosures, or appendices?

EXERCISES

1. Identify a dangerous or inconvenient area or situation on campus or in your community (endless cafeteria lines, a dimly lit intersection, slippery stairs, a poorly adjusted traffic light). Observe the problem for several hours during a peak-use period. Write a justification report to a *specifically identified* decision maker, describing the problem, listing your observations, making recommendations, and encouraging reader support or action.

2. Assume that you have received a $10 000 scholarship, $2500 yearly. The only stipulation for receiving instalments is that you send the scholarship committee a yearly progress report on your education, including courses, grades, school activities, and cumulative average. Write this year's report.

3. In a memo to your instructor, outline your progress on your term project. Describe your accomplishments,

plans for further work, and any problems or setbacks. Conclude your memo with a specific completion date.

4. Keep accurate minutes for one class session (preferably one with debate or discussion). Submit the minutes in memo form to your instructor.

5. Write a recommendations report (choose one):
   a. You are a consulting engineer to an island community of 200 families suffering a severe shortage of fresh water. Some islanders have raised the possibility of producing drinking water from salt water (desalination). Write a report for the Island Trust, summarizing the process and describing instances in which desalination has been used successfully or unsuccessfully. Would desalination be economically feasible for a community of this size? Recommend a course of action.
   b. You are a health officer in a town less than 1 kilometre from a massive radar installation.

Citizens are disturbed about the effects of micro-wave radiation. Do they need to worry? Should any precautions be taken? Find the facts and write your report.

c. You are an investment broker for a major firm. A long-time client calls to ask your opinion. She is thinking of investing in a company that is fast becoming a leader in fibre optics communication links. "Should I invest in this technology?" your client wants to know. Find out, and give her your recommendations in a short report.

d. The "coffee generation" wants to know about the properties of caffeine and the chemicals used on coffee beans. What are the effects of these substances on the body? Write your report, making specific recommendations about precautions that coffee drinkers can take.

e. As a consulting dietitian to the school cafeteria in Belleville, you've been asked by the school board to report on the most dangerous chemical additives in foods. Parents want to be sure that foods containing these additives are eliminated from school menus, insofar as possible. Write your report, making general recommendations about modifying school menus.

f. Dream up a scenario of your own in which information and recommendations would make a real difference. (Perhaps the question could be one you've always wanted answered.)

COLLABORATIVE PROJECT

Organize into groups of four or five and choose a topic upon which all group members can take the same position. Here are some possibilities:

◆ Should your college or university abolish core requirements?
◆ Should every student in your school pass a writing proficiency exam before graduating?
◆ Should courses outside one's major be graded pass/fail at the student's request?
◆ Should your school drop or institute student evaluation of teachers?
◆ Should all students be required to be computer literate before graduating?
◆ Should campus police carry guns?
◆ Should dorm security be improved?
◆ Should students with meal tickets be charged according to the type and amount of food they eat, instead of paying a flat fee?

As a group, decide your position on the issue. Brainstorm collectively to justify your recommendation to a stipulated primary audience in addition to your colleagues and instructor. Complete an audience/purpose profile, and compose a justification report. Appoint one member to present the report in class.

MyWritingLab

MyWritingLab: Technical Communication offers the best multimedia resources for technical communication in one, easy-to-use place. You can find a variety of interactive model documents and case studies. There are also extensive guidelines, tutorials, and exercises for Document Design, Writing and the Writing Process, and Research, and a large bank of diagnostics and practice for grammar review.

22

Workplace Correspondence: Letters, Memos, and Email

Explore

Click here in your eText to access a variety of student resources related to this chapter.

LEARNING OBJECTIVES

After reading this chapter, you should be able to

- recognize and use accepted letter and memo formats
- adapt the format of each email to suit the content, purpose, and audience
- follow guidelines for appropriate, effective, readable emails
- consider the interpersonal aspects affecting correspondence and adapt the action structure to reflect the document's purpose and audience

---

Many reports are completed by a team of writers for multiple readers, but letters, memos, and email are usually written by a single person for one or more definite readers. Also, workplace correspondence is more direct and personal than reports and often has a persuasive purpose, so proper tone is essential. You want your reader to be on your side. Because successful relationships depend on two-way transactions in which both participants meet their needs, you must use a "you" attitude. You must look at the situation from the reader's viewpoint.

Various correspondence media are used for a wide variety of technical and business communications. This chapter introduces factors common to all these media, discusses issues that concern each separate medium, and presents current formats.

Letters appeared on the scene first, followed in the mid-20th century by memos. Then, as that century drew to a close, new technologies brought facsimile transmission (faxes), email messaging, computer-based instant messaging, and phone-based text messaging. Now, social-networking media are being used more and more in business contexts.

Recent developments have confirmed these certainties:

1. Rapid technological change will bring new varieties of electronic correspondence, *but*
2. We will want relatively permanent records of that correspondence, *and*
3. The basic principles of successful correspondence will work in any correspondence medium, paper-based or electronic.

Although letters and memos are being replaced by email, text messaging, and other electronic formats, this chapter introduces letter and memo format because

Why letter and memo formats are discussed in this chapter

1. Letters and memos still have uses, particularly for messages that go beyond the 200–300 words that can comfortably fit on an email screen;

2. The boldface, italics, or bullets of formatted messages may not be properly "read" by the receiver's email software, so writers often attach a formatted letter or memo to a brief introductory email; and

3. Knowing the history and format of letters and memos helps email writers understand and adapt aspects of letter-style and memo-style email messages.

LETTER FORMAT

Canadian and American business correspondence blends ordered, elaborate formatting with streamlined, direct phrasing. (By contrast, European letters often *look* less formal but are phrased more formally and elaborately.)

Letter formats have evolved

The traditional *semi-block format* (Figure 22.1) and *block format* (Figure 22.2) are now out of fashion. They have largely been replaced by *full-block format* (see Figure 22.3 on the next page) because its left-justified set-up saves typing time. The full-block format is starting to give way to an even more streamlined layout, which was introduced as the "simplified letter" by the U.S. National Office Management Association in the early 1960s. Figure 22.4 shows a version of the *simplified format.*

Today's simplified format responds to a growing discomfort with the formal greeting ("Dear …") and complimentary close ("Yours …"), which better suit an earlier era. Notice that a subject line replaces the salutation and that the complimentary close is eliminated. In other respects, a simplified letter copies a full-block letter, although the simplified format's overall appearance resembles contemporary memos. In other words, letter and memo formats are becoming similar.

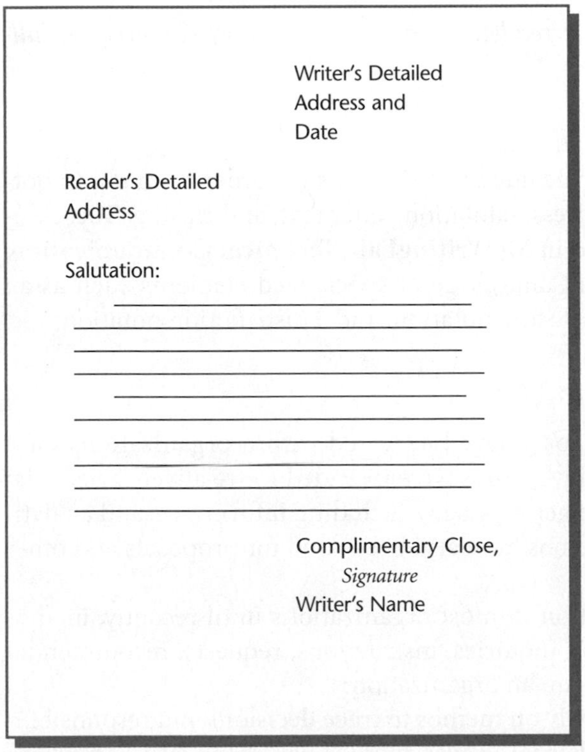

Figure 22.1 A Semi-block Format

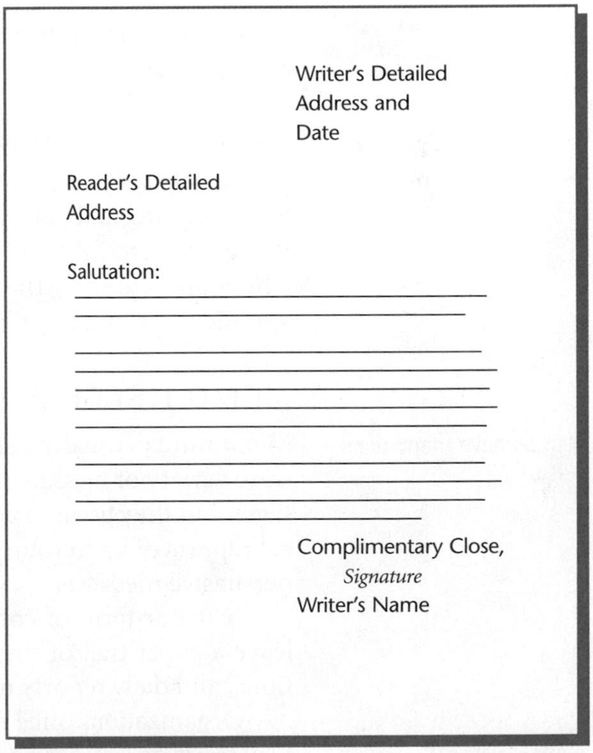

Figure 22.2 A Block Format

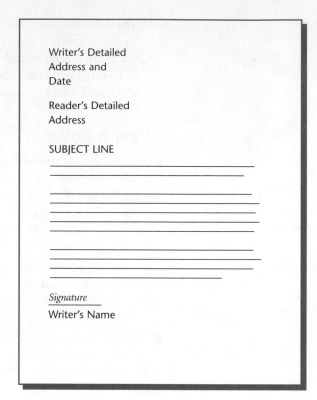

Figure 22.3 A Full-block Format

Figure 22.4 A Simplified Format

If you're not sure of the reader's preferred letter format, use the more conservative full-block format.

Elements of Letters

Business letters have traditionally included five elements (in order from top to bottom): heading and date, inside address, salutation, letter text, and closing. These elements are described and illustrated in **MyWritingLab: Technical Communication**, which also explains the placement and usage of specialized elements such as an attention line, a subject line, an enclosure notation, and a distribution notation.

MEMO USAGE AND FORMAT

Memos have many uses

Memoranda (usually called "memos") have been used within organizations for a wide variety of messages. Originally, memos were intended for relatively brief messages, but they became used for longer messages, including informative and analytical reports of up to four pages. Memos have also been used for proposals and other persuasive messages.

A major form of communication in most organizations until recently, memos leave a paper trail of the directives, inquiries, instructions, requests, recommendations, and daily reports needed to run an organization.

Organizations once relied heavily on memos to trace decisions and responsibilities, track progress, and recheck data. However, memos have been largely replaced by email or by wikis, blogs, and social networking media.

Regardless of the medium used, writers need to remember that their messages can have far-reaching ethical and legal implications. Information must be specific, unambiguous, and accurate. Electronic messages can be retrieved, and many are printed so that the recipient has a record.

NOTE *These days, short memos have largely been replaced by emails or text messages, or even tweets. Memos longer than one page are often replaced by email attachments. When the attachment is intended for a reader **within** the organization, the attachment is most often formatted as a memo, as a briefing note, or as a short informal report.*

Basically, there are two types of memos: intra-office (within an office) and inter-office (between offices of the same firm). Both use the same type of format, with minor variations. That format is described in detail in **MyWritingLab: Technical Communication**.

ELECTRONIC MAIL

The most widely used application on the internet is electronic mail, known simply as email. The Radicati Group, which tracks email and instant messaging, estimates that around the world 192 billion emails will be sent *every day* by 2016.

Email Format

In effect, email combines written correspondence and one-on-one conversation.

> **ON THE JOB...**
>
> **The Ascendancy of Email**
>
> *"A lot of my communication now is email. Initially, there's a consultant meeting with the engineers, the architect, and the developer. From there, my early communication with the client is through preliminary engineering drawings. But the whole process now is moved forward through a steady stream of emails, most of which go to all the interested parties. This collaborative process helps us catch potential problems, if everyone reads their emails!"*
>
> —**Tom Guenther, consulting structural engineer**

Many email messages are relatively casual in tone and format, reflecting the democratic, "free" nature of the medium. However, as uses for email increase and as software becomes more sophisticated, email messages can look more formal, and writers are taking more care with their phrasing, especially in business settings.

A survey of 45 Canadian organizations by Mykon Communications has revealed that email format has evolved into three main levels of formality:

Short, casual emails have largely been replaced by texting and by social media

♦ *Personal, brief notes.* Similar to the unstructured, casual messages sent in the early stages of email development, these notes are often carelessly written and contain spelling errors. They're fine for chatting but not for business and professional messages. (In any event, many people now send text messages or tweet when they have short messages to send.)

Memo-style emails reflect their memo heritage

♦ *Memo style.* Email messages sent within an organization tend to look like standard, informal memos, with carefully constructed block-format paragraphs. Shorter internal emails tend to look less formal than the ones sent externally, especially if the message is only one or two paragraphs. However, longer internal emails sometimes take on characteristics of formal memos and letters: bulleted or numbered lists, boldface and italic text, spell-checked text, and even boldface headings.

Some internal emails even use salutations ("Hello," "Good afternoon," "Greetings") and complimentary closes ("Regards," "Best regards," "Cheers,"

"Sincerely"). Still, 84 percent of respondents to the Mykon survey said that they and their fellow workers tend to send less formal-looking internal emails.

◆ *Letter style.* Formal business emails sent to clients, customers, or suppliers often have the characteristics of formal business letters—bulleted or numbered lists, headings, boldface, and italics help clarify meaning and impart a professional look.

Respondents to the Mykon survey said they use salutations: 40 percent use formal greetings such as "Dear Ms.," while 60 percent said that they use less formal salutations. Formal address is used most often in making an initial business contact. Letter-style emails also use complimentary closes—70 percent of the respondents regularly use closes like "Regards," "Best regards," and "Sincerely."

Alternatively, external emails are used to introduce the reader to attached formal documents (e.g., in Word, WordPerfect, or PDF formats). PDF is particularly favoured by engineering firms and others who want to prevent readers from altering an attached document's findings and recommendations.

Emails often attach technical or business documents

Figure 22.5 illustrates a note-style email, and Figure 22.6 shows a letter-style email (see page 453). An email inquiry is illustrated later in this chapter.

GUIDELINES   **FOR EMAIL FORMAT**

1. *Use a clear subject line to identify your topic* ("Subject: Request for Beta Test Data for Project #16"). This line helps recipients decide whether to read the message immediately and helps the recipient file and retrieve the message.

2. *Refer clearly to the message to which you are responding* ("Here are the Project #16 Beta test data you requested on October 10").

3. *Choose the degree of formality that reflects your reader and your purpose.* The more important the message, the more formal the format. Short, abrupt messages, such as the following, do not require salutations or closings: "The attached file summarizes the tender process you requested. If you'd like to discuss the process, please call."

4. *Use block format.* Don't indent paragraphs. Keep paragraphs and sentences short.

5. *Where appropriate in formal emails, use graphic highlighting* (headings, bullets, numbered lists, boldface, and italics) to improve readability and impart professionalism. However, keep in mind that a little goes a long way.

6. *Avoid wildly varying line lengths on your receiver's screen by limiting line length.* Set Word Wrap to create line lengths of 64 characters or less.

7. *Where appropriate, use formal salutations and closings for letter-style emails* ("Dear Sir," "Sincerely"), but use less formal greetings and closings for most emails ("Hello," "Regards").

8. *Close with a signature section* that names you, your company or department, your telephone and fax number, and any other information the recipient might consider relevant. If you have the technology, include your electronic signature.

9. *Do not write in FULL CAPS,* unless you want to SCREAM at the recipient!

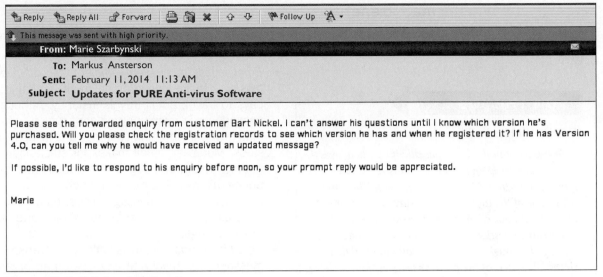

Figure 22.5 A Note-style Email

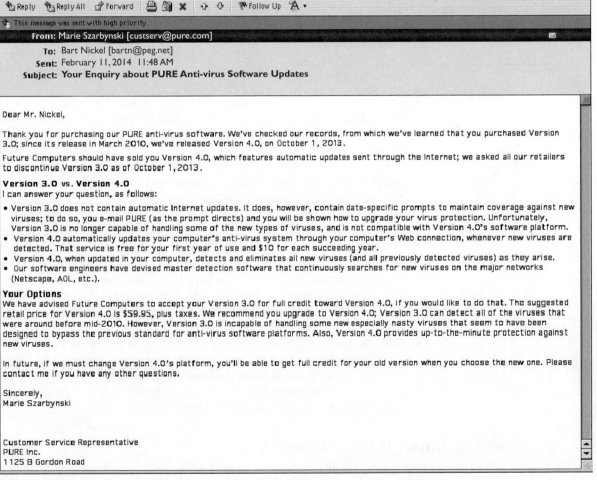

Figure 22.6 A Letter-style Email

Email Benefits

Email facilitates communication and collaboration

Compared with phone, fax, or conventional mail (or even face-to-face conversation, in some cases), email offers benefits:

- *Email is fast, convenient, efficient, and relatively unintrusive.* Unlike conventional mail, which can take days to travel, email travels instantly. Although a fax network can transmit printed copy rapidly, email eliminates paper shuffling, dialing, and a host of other steps. Moreover, email makes for efficiency by eliminating "telephone tag." It intrudes less than the phone, allowing the recipient to choose when to read and respond to a message. Thus, email and texting have almost entirely replaced memos and letters in many settings.

- *Email is democratic.* Email allows for transmission of messages by anyone at any level in an organization to anyone at any other level. For instance, the mail clerk conceivably could email the company president directly, whereas a conventional memo or phone call would be routed through the chain of management or screened by administrative assistants (Goodman 33–35). In addition, people who are ordinarily shy in face-to-face encounters may be more willing to express their views in an email conversation.

- *Email is excellent for collaborative work and research.* Collaborative teams keep in touch via email, and researchers contact people who have the answers they need. Especially useful for collaborative work is the email function that enables documents or electronic files of any length to be attached and sent for downloading by the receiver. To further enhance this advantage, many organizations merge their email systems with collaboration software. The Radicati Group predicts a $1 billion increase in the "on-premises email and collaboration market" from 2009 to 2013.

- *Email provides a record of agreements.* Among other benefits, this record makes it easier to hold people to their commitments. Despite this relative permanence many emails are printed and stored as paper documents. A 2007 *Backbone* article by Gail Balfour says that "Canadian firms are printing more than ever" (20). It seems that people like paper's tactility, portability, and readability (24). Also, technological advances have made printers and printing more affordable.

- *Email is good for sending a single message to multiple recipients at the same time.* Even though the recipients will not likely open the message at the same time, there's a record of when the message was sent.

- *Current email software allows for well-designed "pages,"* either on the email screen itself or as a formal attached document.

Email Privacy Issues

Gossip, personal messages, or complaints about the boss or a colleague—all might be read by unintended receivers. Employers often claim legal right to monitor *any* of their company's information, and some of these claims can be legitimate:

Monitoring of email by an employer may be legal

> In some instances, it may be proper for an employer to monitor email, if it [the employer] has evidence of safety violations, illegal activity, racial discrimination, or sexual improprieties, for instance. Companies may also need access to business information, whether it is kept in an employee's drawer, file cabinet, or computer email. (Bjerklie 15)

Email privacy can be compromised in other ways as well:

Email offers no privacy

◆ Everyone on a group mailing list—intended reader or not—automatically receives a copy of the message.
◆ Even when deleted from the system, messages often live on for years, saved in a backup file.
◆ Anyone who gains access to your network and your private password can read your document, alter it, use parts of it out of context, pretend to be its author, forward it to whomever, plagiarize your ideas, or even author a document or conduct illegal activity in your name.

Email privacy concerns have led to encryption technology and to "self-destructing" emails that exist for a limited time only. Encrypted messages are scrambled so that they can be opened only by someone with a password or with the answer to a predetermined question. Encryption can also prevent forwarding or altering a message. After they're first opened, self-destructing emails expire within a range of 10 seconds to two weeks, as chosen by the sender. (Or the sender can choose to not have the message expire.) Meanwhile, the sender keeps a copy of the sent message.

For a fee of less than a dollar, emails can even be sent as "registered mail," with all the features that registered paper mail provides. Registered email users, such as lawyers and technology professionals, wish to guard against email-related liability—email is increasingly being used as evidence in court cases.

Email Quality Issues

The ease of communicating via email has added to the quantity of information shared, but not always to its quality. As the following examples show, information quality can be compromised by email communication:

Email does not always promote quality in communication

◆ The ease of sending and exchanging messages can generate overload and junk mail—a party announcement sent to 300 employees on a group mailing list, or the indiscriminate mailing of a political statement to dozens of newsgroups (*spamming*).
◆ Some electronic messages may be poorly edited and long-winded.
◆ Off-the-cuff messages or responses might offend certain recipients. Email users often seem less restrained about making rude remarks (*flaming*) than they would be in a face-to-face encounter.
◆ Recipients might misinterpret the tone. *Emoticons* or *smileys*, punctuation cues that signify pleasure :-), displeasure :-(, sarcasm ;-), anger :-<, and other emotional states, offer some assistance but are not always an adequate or appropriate substitute for the subtle cues in spoken conversation. Also, common email abbreviations (FYI, BTW, HAND—which mean "for your information," "by the way," and "have a nice day") might strike some readers as too informal.

Email Offshoots

Internet forums

Internet email forums allow various groups to discuss topics of common interest. One such forum, Power Globe, brings together persons interested in electrical power engineering at **www.ece.mtu.edu/faculty/ljbohman/peec/globe**. Here, subscribers discuss technological trends, announce conferences and new publications, post job openings, and request information.

E-newsletters

Newsletters delivered by email provide news stories, background information, member updates, and human-interest articles to such diverse groups as cross-country ski clubs, Manchester United soccer fans, and fashion design aficionados. In-house electronic newsletters cover company news, industry developments, and stories about employees. These company e-newsletters can provide a forum for employee feedback that is missing from printed newsletters.

Converged messaging (also known as *unified messaging* or *UM*) is rapidly gaining popularity—it had 1 million users worldwide in 2000; by 2005, this number had jumped to 40 million. The technology uses Voice over Internet Protocol (VoIP) and converging software to send phone messages, faxes, and text messages as email attachments to a receiver's email inbox.

Converged messaging offers several advantages:

Advantages of converged messaging

- ◆ The technology saves time and effort by directing all messages to one device; therefore, the receiver is more likely to make the effort to look at all incoming phone calls and faxes.
- ◆ When the receiver sees who has phoned, he or she can tell at a glance which messages need to be heard and will likely need a quick response. Using traditional voice mail, receivers are less likely to plough through many messages to discover which ones are important enough to warrant an immediate response.
- ◆ The technology is compatible with smartphones and PDAs, so receivers can pick up all their messages when they're out of the office.
- ◆ As a consequence of all this, converged messaging facilitates timely communication, an important consideration in today's business climate.

Complex versions of converged messaging systems for large companies require expensive software, VoIP hardware, and a subscription to a telephone system vendor. The cost of the total package for a company can run "well into five or six figures," according to Info-Tech Research's Carmi Levy (Blackwell). However, less expensive alternatives are available for small businesses, and Levy believes that converged messaging could be particularly useful for small businesses that rely on quick responses to customer demands.

Recent smartphone developments have extended converged messaging to mobile users who can view their friends' presence status and who can send instant messages to everyone in a designated address book. A survey released in October 2012 by Infonetics Research said that "smartphones and tablets will be the two most widely used devices for UC in 2013, passing traditional computers and deskphones" (Myers).

Email's Future

Internet forums

Some observers have predicted that the growing use of social-networking sites, wikis, internal blogs, and texting, especially among teens and young adults, will lead to email's demise. A strong challenge comes from innovative in-house social networks that allow employees to share information and request assistance as they collaborate or work on parallel projects.

GUIDELINES   FOR USING EMAIL EFFECTIVELY

Recipients who consider an email message poorly written, irrelevant, offensive, or inappropriate will only end up resenting the sender. These guidelines offer suggestions for effective email use.

1. *Check and answer your email daily.* Like an unreturned phone call, unanswered mail annoys people. If you're really busy, at least acknowledge receipt and respond later.

2. *Don't use email when a more personal medium would be better.* Sometimes, an issue is best resolved with a phone call or personal visit.

3. *At work, consider email to be correspondence, not conversation.*

4. *Assume that your email correspondence is permanent and could be read by anyone at any time.* **Electronic mail is not a private medium.** Increasingly, employers monitor their email networks—Forrester Research's 2006 survey of 294 large U.S. companies revealed that 38 percent of the companies hire staff to read outbound employee email and 46.9 percent regularly audit outbound email ("Virtual Teamwork").

5. *Don't use email to send confidential information.* Avoid complaining, evaluating, criticizing, or saying anything private. And don't use the company email network for personal correspondence or for anything not work-related. **Violating company email policy can result in severe penalties.** The Forrester Research survey found that 31.6 percent of companies had fired employees for violations and 52.4 percent had disciplined employees during the previous 12 months ("Virtual Teamwork").

6. *Check your distribution list before each mailing,* to be sure the message reaches only the intended primary and secondary readers.

7. *Before you forward an incoming message to other recipients, obtain permission from the sender.* Assume that everything you receive is the sender's private property.

8. *Write subject lines that catch the reader's attention and that provide a context for the message.* Good subject lines are like good news headlines. Use action-oriented phrasing in the subject line and the opening paragraph to suggest how the reader can use the information contained in the email.

9. *Limit your message to a single topic.* Remain focused and concise. (Yours may be just one of many messages confronting the recipient.) Don't ramble.

10. *Limit your topic to a single screen, if possible.* Don't force recipients to scroll needlessly.

11. *Don't use email for a detailed discussion.* Send a letter or memo or report instead. In cases where time is of the essence, especially for in-house recipients and others who can easily receive and print your form of word-processed documents, attach your paper document to an introductory email. By doing so, you include the best features of both media.

12. *Carefully check your spelling, grammar, and tone.* Especially, spell recipients' names correctly! Even short messages reflect your image.

13. *Pause before hitting the "send" button.* Catch errors and omissions. Make sure that attachments are attached.

14. *Clean out your mailboxes.* Mail that piles up in employee inboxes and deleted folders can eventually clog up a company's email database.

Such "Enterprise 2.0" tools have advantages over email. "Employees become more engaged, especially those working in off-site locations. Such networks can also streamline operations and put an end to annoying, time-wasting and hard-to-follow e-mail chains between large groups of employees. Unlike e-mail, enterprise social networks can store and keep detailed records of public exchanges between employees and track how ideas evolve" (Weeks).

An example of an in-house messaging and project management system

One example of an in-house communication system is found at Toronto-based Klick Health, a healthcare marketing agency. Its own system, Genome, has replaced email within the organization, although Klick Health still uses email to communicate with clients. Klick Health's CEO, Leerom Segal, says that Genome, a messaging and project management platform, provides a much more open and collaborative system than standard email.

> Segal says that Genome indicates when an employee is going to miss a deadline; it helps a staff member who's trying something new to connect with a co-worker who's recently completed a similar task; and it allows workers to quickly ask others for help with a problem (Borzykowski).

Will the use of custom-designed social-networking and collaborative tools "spell the end of email altogether"? Not so fast, says marketing consultant Marti Trewe: "email still rules the roost and the number of email users is expected to be at 3.8 billion by 2014, up from 2.9 billion in 2010."

Email will survive

A 2012 poll by Ipsos Global Public Affairs supports the belief that email will be around for a long time. The worldwide poll found that 85 percent of internet subscribers use email to communicate. *Digital Trends* writer Geoff Duncan explains that email's continuing popularity rests on its "ubiquity and very low fundamental requirements." By contrast says Duncan, "Social networking services including blogs, Twitter, Facebook, and others all have relatively high technological bars. Web-based services are constantly dropping older browsers, or introducing new features that only work on particular platforms."

The last word on this subject belongs to Ray Tomlinson, the engineer who is often referred to as the "father of email." Tomlinson points out that there is still nothing quite like email, despite the efforts of Facebook and others to create a replacement. He points to email's "specific audience"—as opposed to Facebook's billboard approach—and he notes that email doesn't require a receiver to be at the ready, as instant messaging requires (Hachman).

INTERPERSONAL ELEMENTS OF WORKPLACE CORRESPONDENCE

Besides presenting the reader with an accessible and inviting design, effective correspondence enhances the relationship between writer and reader. Various interpersonal elements forge a *human* connection, as the following guidelines explain.

Focus on Your Reader's Interests: The "You" Perspective

In face-to-face conversation, you unconsciously modify your statements and expression as you read the listener's signals: a smile, a frown, a raised eyebrow, a nod. In a telephone conversation, a voice provides cues that signal approval, dismay, anger, or confusion. When you write a letter, memo, or email, however, you face a major disadvantage; you can easily forget that a flesh-and-blood person will react to what you say or seem to say.

Correspondence that displays a "you" perspective subordinates the writer's interests to those of the reader. Besides focusing on what is important to the reader, the "you" perspective conveys respect for the reader's feelings and attitudes.

To achieve a "you" perspective, put yourself in the place of the person who will read your correspondence; ask yourself how the reader will react to what you have written. Even a single word, carelessly chosen, can offend. In writing to correct a billing error, for example, you might feel tempted to say this:

A needlessly offensive tone

Our record keeping is very efficient and so this obviously is your error.

Such an accusatory tone might be appropriate after numerous failed attempts to achieve satisfaction on your part, but in your initial correspondence it will alienate the reader. The following version is more considerate. Instead of indicting the reader, it conveys respect for the reader's viewpoint.

A tone that conveys the "you" perspective

If my paperwork is wrong, please let me know and I will send you a corrected version immediately.

Use Plain English

Workplace correspondence too often suffers from *letterese*, those tired, stuffy, and overblown phrases some writers think they need to make their communication seem important. Here is a typically overwritten closing sentence:

Letterese

Humbly thanking you in anticipation of your kind cooperation, I remain

Faithfully yours,

Although no one *speaks* this way, some writers lean on such heavy prose instead of simply writing this:

Clear phrasing

I will appreciate your cooperation.

Here are a few of the many old standards that would be better presented in plain English:

Letterese	Plain English
As per your request	As you requested
Contingent upon receipt of	As soon as we receive
I am desirous of	I want
Please be advised that I	I
This writer	I
In the immediate future	Soon
In accordance with your request	As you requested
Due to the fact that	Because
I wish to express my gratitude	Thank you

Be natural. Write as you would speak in a classroom or office.

Anticipate Your Reader's Reaction

Like any effective writing, good correspondence does not just happen. It is the product of a deliberate process. As you plan, write, and revise, answer these questions:

1. *What do I want the reader to do, think, or feel after reading this correspondence?* Do you want him or her to offer you a job, give advice or information, answer an inquiry, follow instructions, grant a favour, enjoy good news, accept bad news?
2. *What facts will my reader need?* Do you need to give measurements, dates, costs, model numbers, enclosures, other details?
3. *To whom am I writing?* Do you know the reader's name? When possible, write to a person, not a title.
4. *What is my relationship to my reader?* Is the reader a potential employer, an employee, a person doing a favour, a person whose products are disappointing, an acquaintance, an associate, a stranger?

Answer those four questions *before* drafting correspondence. After you have a draft, answer the following three questions, which pertain to the *effect* of your correspondence. Will readers be inclined to respond favourably?

1. *How will my reader react to what I've written?* With anger, hostility, pleasure, confusion, fear, guilt, resistance, satisfaction?
2. *What impression of me will my reader get from this correspondence?* That you are intelligent, courteous, friendly, articulate, pretentious, illiterate, or confident?
3. *Am I ready to sign my correspondence with confidence?* Think about it.

Send correspondence only when you've answered each question to your satisfaction.

Decide on a Direct or Indirect Plan

The reaction you anticipate from your reader should determine the organizational plan of your correspondence: either *direct* or *indirect*.

- ◆ Will the reader feel pleased or neutral about the message?
- ◆ Will the message cause resistance, resentment, or disappointment?

Each reaction calls for a different organizational plan. The direct plan puts the main point right in the first paragraph, followed by the explanation. Use the direct plan when you expect the reader to react with approval or when you want the reader to know immediately the point of your letter (e.g., in good-news, inquiry, or application letters—or other routine correspondence).

If you expect the reader to resist or to need persuading, consider an indirect plan. Give the explanation *before* the main point (as in refusing a request, admitting a mistake, or requesting a pay raise). An indirect plan might make readers more tolerant of bad news or more receptive to your argument.

Whenever you consider using an indirect plan, think carefully about its ethical implications. Never try to deceive the reader—and never create an impression that you have hidden something.

STRUCTURES FOR WORKPLACE CORRESPONDENCE

Trial and error, along with perceptive analysis of readers and purposes for letters, memos, and emails, will allow you to find the best ways to organize your workplace correspondence. The best structure for each message will depend on the circumstances. You can start by using the audience/purpose analysis form to choose the content of the message. Then you can adapt the action structure described in Chapter 21 for virtually any letter, memo, or email you write.

Let's review that structure:

1. *Action opening.* Connect with the reader's interest and, where appropriate, summarize the full message in one or two sentences.
2. *Background.* Provide any information the reader needs to understand the main message that follows. Many letters and memos don't need this section; in others, you'll be able to fit a one-sentence background overview into the introductory paragraph.
3. *Details.* Present the main message. (Answer the reader's questions; describe your idea; present your information and analysis; argue your case; explain your point of view.)
4. *Action closing.* Request the desired action from your reader and/or describe the action you intend to take. Provide information that the reader needs to perform the requested action.

Examples of direct and indirect approaches appear in this chapter

Now let's see some examples of how that structure can work for workplace messages. The scope of this text does not allow coverage of every kind of letter, memo, or email you might eventually write in your working life, but a careful study of the following examples will help you learn how to adapt the action structure to many other kinds of messages.

Notice that some messages require a *direct approach:* the main point or bottom line is provided in the opening paragraph because you've realized that's what the reader wants, and you've concluded that your own purpose will not be jeopardized by doing so. Other messages require an *indirect approach*, where the opening paragraph previews the document's structure and the main point comes at the end of the document.

The inquiry (letter or email) structure described in Table 22.1 (and then illustrated in Figure 22.7; see page 462) employs a direct pattern; the reader will want to immediately know the message's main purpose. Then, seeing the potential for business, the recipient will be motivated to read carefully.

Figure 22.7 illustrates the phrasing that might be used for the message that Table 22.1 outlines. The situation outlined in Table 22.2 on page 463 also uses a direct pattern: stating the requested action and its contribution to the *reader's* goals will get the reader's attention and help hold that attention throughout the memo.

Figure 22.8 on page 464 illustrates the phrasing that might be used for the message that Table 22.2 outlines. The writer feels comfortable using Rich Text Format for this email, including boldface headings and quotation marks, because both the sender and receiver use Outlook Express as their email software.

Table 22.1 Inquiry (Letter or Email) Structure

Section	Situation: Inquiry about updates for PURE anti-virus software Reader: PURE Inc. Customer Services Department
Action opening	State purpose for writing/general nature of inquiry (paragraph 1)
Background	Say how long you've had the PURE software installed; say whether you purchased a CD-ROM version or downloaded it from the internet (paragraph 2)
Details	Ask detailed questions about the updates: ◆ range of virus types detected ◆ methods of receiving updates/frequency of updates ◆ company's main strategy for discovering viruses ◆ costs of updates/types of payment accepted by PURE (paragraph 3 uses numbered or bulleted list)
Action closing	Provide email address and phone number; ask PURE rep to respond; give deadline for response, if that's an issue (paragraph 4)

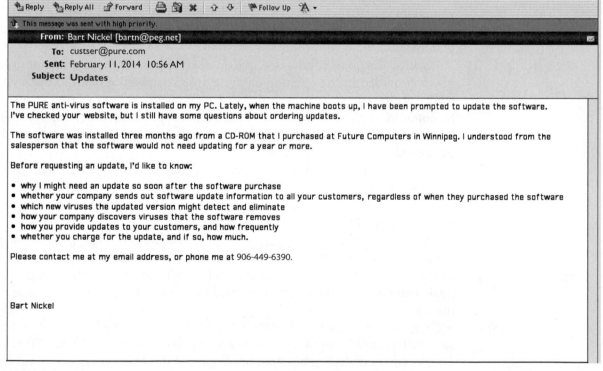

Figure 22.7 A Sample Email Inquiry

Table 22.2 Request Memo Structure

Section	Situation: Request for funds to design a company website Reader: General Sales Manager Writer: Sales Representative
Action opening	State the main action that needs to be taken and the primary reason for that action; connect with reader's interests relative to this action (paragraph 1)
Background	Describe the circumstances leading to this situation or events affecting the situation (paragraph 2)
Details	Provide a detailed description of what is being requested or needs to be done: ◆ staffing required ◆ equipment and software required and whether the company currently has these requirements in inventory ◆ timelines and schedule ◆ costs (one or more paragraphs)
Action closing	Request the specific desired action or reader's approval for the writer's proposed action (final paragraph)

Next, the structure for a claim letter (also known as an adjustment letter or a complaint letter) in Table 22.3 also uses a direct pattern. The writer immediately gets to the point.

Table 22.3 Claim Letter (or Email) Structure

Section	Situation: Claim for damages Reader: Manager, Shipping Department, Genesis Computer Systems Co. (an Edmonton wholesaler) Writer: Home-based computer consultant, Fort McMurray, Alberta
Action opening	State the problem or the action requested (in this case, claim for a system damaged in shipment because of poor packing) (paragraph 1)
Background	Name the original order number, the packing slip number, the invoice number, and the method and date of shipment (paragraph 2)
Details	Provide the details of the claim: ◆ Describe the nature and extent of the equipment damage and estimate the damage. ◆ Refer to how the damaged equipment is being returned to the supplier. ◆ Describe the loss of business (the customer went to another dealer). ◆ Refer to enclosed documents. (paragraph 3)
Action closing	Close politely, but firmly—say what you expect to be done, and confirm that you desire continued business relations with the reader (paragraph 4)

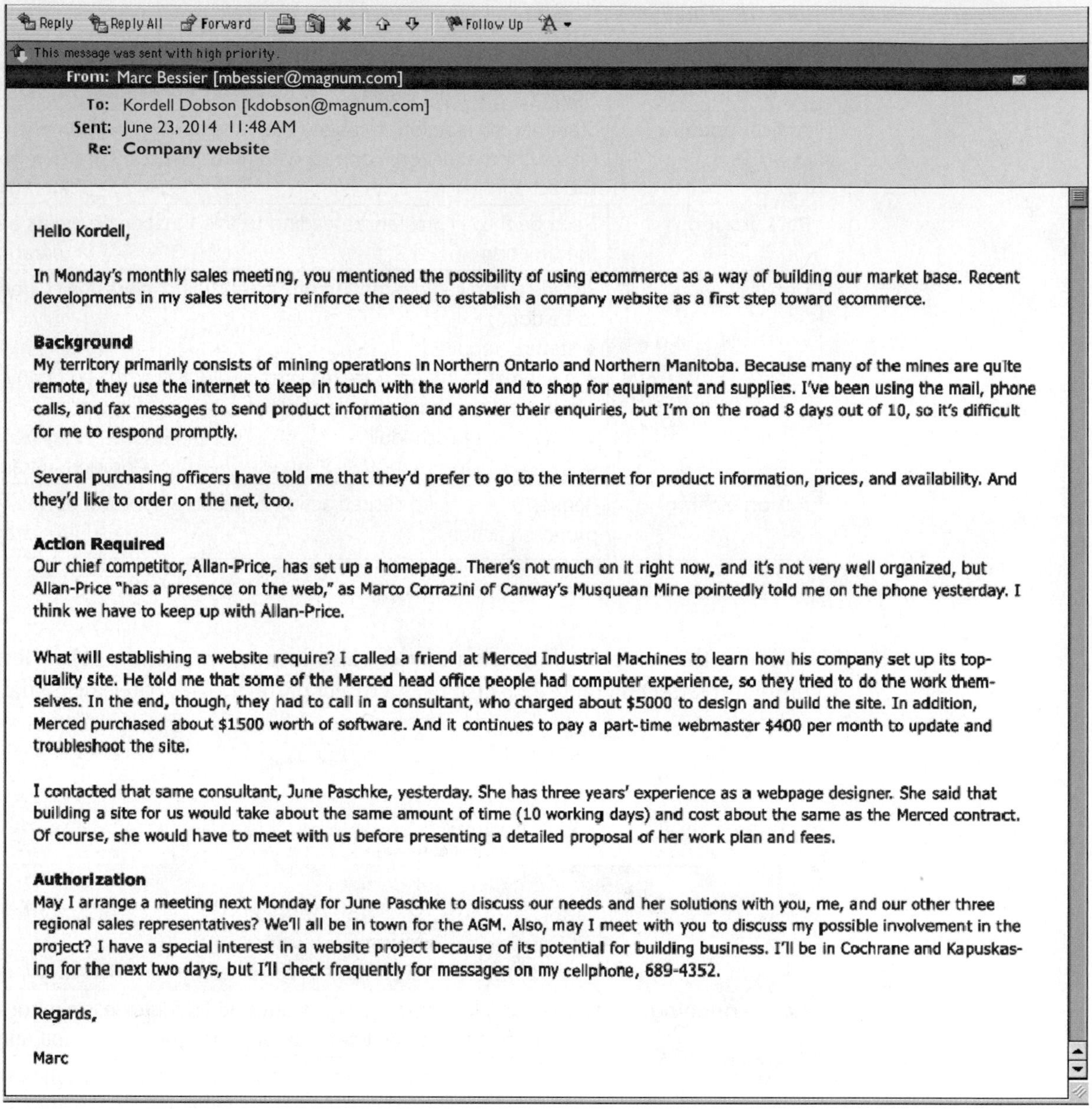

Reply Reply All Forward Follow Up

This message was sent with high priority.

From: Marc Bessier [mbessier@magnum.com]

To: Kordell Dobson [kdobson@magnum.com]

Sent: June 23, 2014 11:48 AM

Re: **Company website**

Hello Kordell,

In Monday's monthly sales meeting, you mentioned the possibility of using ecommerce as a way of building our market base. Recent developments in my sales territory reinforce the need to establish a company website as a first step toward ecommerce.

Background

My territory primarily consists of mining operations in Northern Ontario and Northern Manitoba. Because many of the mines are quite remote, they use the internet to keep in touch with the world and to shop for equipment and supplies. I've been using the mail, phone calls, and fax messages to send product information and answer their enquiries, but I'm on the road 8 days out of 10, so it's difficult for me to respond promptly.

Several purchasing officers have told me that they'd prefer to go to the internet for product information, prices, and availability. And they'd like to order on the net, too.

Action Required

Our chief competitor, Allan-Price, has set up a homepage. There's not much on it right now, and it's not very well organized, but Allan-Price "has a presence on the web," as Marco Corrazini of Canway's Musquean Mine pointedly told me on the phone yesterday. I think we have to keep up with Allan-Price.

What will establishing a website require? I called a friend at Merced Industrial Machines to learn how his company set up its top-quality site. He told me that some of the Merced head office people had computer experience, so they tried to do the work them-selves. In the end, though, they had to call in a consultant, who charged about $5000 to design and build the site. In addition, Merced purchased about $1500 worth of software. And it continues to pay a part-time webmaster $400 per month to update and troubleshoot the site.

I contacted that same consultant, June Paschke, yesterday. She has three years' experience as a webpage designer. She said that building a site for us would take about the same amount of time (10 working days) and cost about the same as the Merced contract. Of course, she would have to meet with us before presenting a detailed proposal of her work plan and fees.

Authorization

May I arrange a meeting next Monday for June Paschke to discuss our needs and her solutions with you, me, and our other three regional sales representatives? We'll all be in town for the AGM. Also, may I meet with you to discuss my possible involvement in the project? I have a special interest in a website project because of its potential for building business. I'll be in Cochrane and Kapuskasing for the next two days, but I'll check frequently for messages on my cellphone, 689-4352.

Regards,

Marc

Figure 22.8 A Request

As the preceding tables show, the direct, four-part action structure can be adapted for a variety of situations. Some of those situations, such as a refusal of a request, will require an indirect pattern. Table 22.4 presents a suggested structure for refusing a request.

When Kordell Dobson writes to refuse Marc Bessier's request (see Figure 22.9 on page 466), note the phrasing Kordell uses.

Table 22.4 Refusal Message Structure

Section	Situation: Inquiry about updates for PURE anti-virus software Reader: PURE Inc. Customer Services Department
Action opening	Express appreciation for the sales rep's initiative; acknowledge that there has recently been discussion of creating a company website (paragraph 1)
Background	Without actually saying "no" just yet, explain the circumstances that preclude creating a website at this time (perhaps the firm is in the process of contracting a web consultant or joining forces with another firm, or perhaps the company is analyzing its entire marketing strategy) (paragraph 2)
Details	Soften the bad news by ◆ placing it in the middle of the paragraph ◆ using the passive voice (not "We cannot grant your request at this time," but "Funds are therefore not available for the project at this time") ◆ focusing on the reasons for the refusal, not on the refusal itself (part of paragraph 2, or perhaps a new paragraph)
Action closing	Thank the reader for his or her commitment and ideas; encourage the reader to pursue an interest in website design; assure the reader that he or she will be consulted when the project is next discussed (final paragraph)

LENGTH OF WORKPLACE CORRESPONDENCE

All letters and memos should be as short as you can make them. Restrict them to one page, if you can. (Emails should fit on one screen.) Although one-page messages are not always possible, you can restrict some messages to one page by placing the main message in a cover letter and the remaining details in enclosures. For instance, a complex order for supplies might be organized in order sheets enclosed with a cover letter that

- ◆ authorizes the purchase, designates the delivery method, and names the source of your information about the supplies;
- ◆ quickly summarizes the nature of the order and refers to the enclosed order "forms" (which provide details of items, quantities, order numbers, prices, taxes, shipping, and overall costs); and
- ◆ closes by saying how you will pay and when you expect delivery.

WRITE CORRESPONDENCE EFFICIENTLY

Clear thinking leads to efficient writing

When you write workplace correspondence, write efficiently. Very few one-page messages should take more than 45 minutes to plan, compose, and polish. You will, however, take longer if you have to rewrite major sections of letters or memos. That kind of rewriting can be avoided by clear thinking in the early stages of the process.

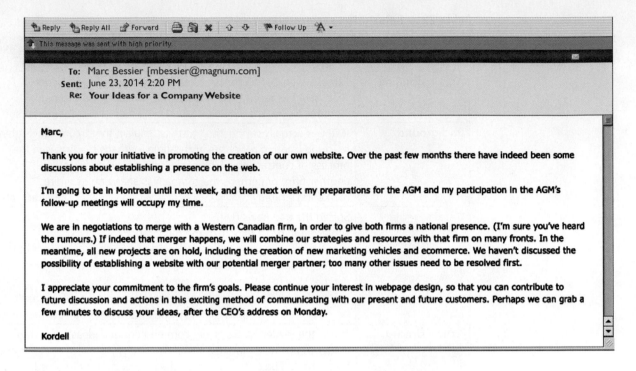

Figure 22.9 A Refusal of a Request

So remember:

- ◆ consider your audience's needs, interests, and priorities as well as the purpose of the document
- ◆ use your audience/purpose analysis to choose the content and arrange that content
- ◆ work from an outline that identifies the content and/or purpose of each paragraph
- ◆ revise the content and structure before writing sentences and paragraphs

You'll be amazed at how efficiently you'll write.

Also remember that how you use workplace correspondence will affect your credibility within your organization. Your colleagues' and supervisors' perceptions of your worth will be affected by the degree of care and skill you put into your letters, memos, and emails.

TEXT AND INSTANT MESSAGING

SMS and IM are fine for short messages, but too limited for detailed messages

In the late 1990s, *text messaging* (known primarily outside Canada and the U.S. as SMS, for "short messaging services") and *instant messaging* (IM) started to become popular with young people who saw the computer-based and cellphone-based service as replacements for phone calls and email. Text messages travel via asynchronous cellphone transmission: a sender types and sends a message, and signs off; a response message, if any, will come on a separate transmission. By contrast, instant messaging operates synchronously—writers correspond online together, via computer or text-enabled cellphone. IM also allows people to indicate whether they're online or offline.

Immediacy is the strength of these media, not carefully crafted messages. Screen size and real-time messaging limit the medium to brief bursts of communication (about 160 characters, maximum). So these media are fine for "I've sent the package today. Should get there soon," but totally inadequate for explaining why the package wasn't sent earlier.

Electronic messaging is growing, especially in Asia, the Pacific region, and Europe. Worldwide, text messages sent to and from portable phones were expected to reach 9.6 trillion in 2012, according to a survey conducted by Portia Research (Mobi). As of October 2012, according to an estimate by the Canadian Wireless Telecommunications Association, Canadians send "10 million text messages per hour" (LaSalle). These figures are expected to continue to grow dramatically.

Although text and instant messaging are used primarily for social connections, they do offer advantages in work settings:

◆ Receivers are more likely to respond to the "ping" of an IM or text message than to an email.

◆ Text and IM are better than phone calls when privacy is an issue, when silent conversation is required, or when excessive background noise makes phone usage difficult.

◆ IM allows for real-time, inexpensive collaboration among two or more workers when specific details are being discussed.

◆ Special subscription services such as ipipi.com offer text messaging between computers and cellphones anywhere in the world, which can dramatically reduce the costs of long-distance communicating.

◆ Smartphones and PDAs allow users to direct their correspondents to stored images, websites, email messages, and even to audio and video.

◆ Large companies can subscribe to "message-brokering" software that prioritizes thousands of text messages coming from sources anywhere in the world. Then the software can place the messages in queues of decreasing relevance and send the messages to appropriate employees within the company. For example, an engineer working on a project in Africa might have a specialized question that a senior engineer at the company's North American head office could answer.

Many workers prefer to be contacted via text message, rather than by phone or by email. A survey of 2277 Americans by The Pew Research Center's Internet & American Life Project learned that 31 percent of respondents prefer text message contact over other methods (Smith). ThoughtWorks Managing Director Craig Gorsline confirms that preference in his field: "A high percentage of the programmers I know are very reluctant to check their email, but they do respond to texts."

EXERCISES

1. Collect samples of workplace writing—letters, memos, and email messages. In those samples, look for instances of letterese, clichés, and wordy and unclear phrasing. Improve the phrasing where necessary.

2. Check the tone of emails, memos, and letters that you examine. Do you see phrasing that might antagonize the reader? Do you see phrasing that places the writer's needs and interests ahead of the reader's? If so, rephrase.

3. Examine workplace emails, letters, and memos to see if they use a variation of the four-part action structure. (See pages 461–65.) If the action structure has not been used for a given message, would the message be more effective if it were restructured?

4. Monitor your next composition of a one-page letter, memo, or email. Record how long you take to complete each stage of the writing process.

5. Evaluate whether certain emails could have been boiled down to text message size. What information and what interpersonal messages would have been lost in translation?

6. Describe a situation where a text message could be used to get someone's attention and to direct that reader to an email or to a document posted to a blog or file sharing site.

7. Write a letter to a supplier ordering parts or supplies for a piece of equipment. Reformat (and perhaps edit) that letter to be sent as an email. Evaluate whether a text message would be able to accommodate the content of the order.

8. Decide whether a letter or an email or a text would be best for informing a supplier about damaged goods that you've received from that supplier.

9. A customer claims that you have not completed an order (or work that was promised). This customer has a reputation for finding excuses for not paying the full amount of provided goods or services, so you have carefully fulfilled all aspects of the contracted transaction and you have carefully documented these aspects. Write a letter (or email) that provides these details and requests full payment.

COLLABORATIVE PROJECT

In conjunction with other members of your class, survey correspondence practices of businesses in your part of Canada. Ask your survey respondents about the following topics:

- which types of correspondence (letter, memo, email, fax, text message) get used most often
- what standards the business requires for formatting letters and letter-style emails sent to external receivers
- what guidelines or restrictions the company has for email and text messaging

Also, ask for a sample letter, a sample internal memo, and a sample formal email.

Members of your group could conduct the survey in person. If so, send a copy of the survey questions with an accompanying letter or email first, and then telephone for an appointment. Emphasize that information found in the sample letters and memos will be held confidential. Also, offer to make a copy of the survey's statistical results available to each participating business.

Alternatively, the entire survey could be conducted by mail or by email.

NOTE *All correspondence sent during the survey should be approved by a class-appointed editing team.*

MyWritingLab: Technical Communication offers the best multimedia resources for technical communication in one, easy-to-use place. You can find a variety of interactive model documents and case studies. There are also extensive guidelines, tutorials, and exercises for Document Design, Writing and the Writing Process, and Research, and a large bank of diagnostics and practice for grammar review.

23

Job-Search Communications

Explore
Click here in your eText to access a variety of student resources related to this chapter.

LEARNING OBJECTIVES

After reading this chapter, you should be able to

- Use a variety of research and self-assessment tools to understand a job market and to assess your competitive position within that market.
- Use appropriate communication media to introduce and sell yourself to decision makers within a specific job market.
- Plan and conduct an organized campaign for each new job search.
- Sell your unique package of skills and personal attributes; relate relevant components to the job you're currently pursuing.
- Differentiate yourself from your competition in your sales material (résumé, letters, website, etc.) and in interviews.

Technology has given job seekers many new tools, but today's job searches require the same basic approach and communication skills that have been used for some time. This chapter discusses seven steps in a successful job search:

1. Assessing what you have to offer
2. Identifying which kinds of jobs to pursue
3. Researching the job market
4. Producing résumés and other job-search materials
5. Devising a campaign strategy for contacting employers
6. Performing well in interviews
7. Responding to job offers

Being informed about careers and specific positions in your field, preparing to meet professional requirements, and then following a systematic job-search strategy will increase your future job satisfaction.

SELF-ASSESSMENT INVENTORY

Assessing what you have to offer employers is not easy to do, especially if you have not received detailed, accurate job-performance appraisals from previous employers. Self-assessment is even more difficult if you have had little job experience to use as a basis for evaluating your skills and aptitudes. Assessing your package of employable characteristics requires objectivity and an organized approach. Fortunately, tools, such as the self-inventory shown in Figure 23.1 on page 470, help provide that structured objectivity.

An objective, organized self-appraisal

Completing a Self-Inventory

The self-inventory chart's four sections (quadrants) can be completed in any sequence. The chart is very useful if you fill each quadrant with your transferable skills (see the next section), work habits, attitudes, and/or knowledge, the full package you bring to an employer. You can use this chart to choose content for your résumés, to help write an application letter, or to prepare for a job interview.

Strengths readily apparent to an employer

QUADRANT 1. Quadrant 1, "apparent strengths," features those qualities that an employer would discover by reading your résumé, by interviewing your references, or by interviewing you. For example, your work experience may have developed skills required by the position you're applying for. Or your university major may have developed skills required by the position. By highlighting such strengths in your résumé and cover letter, you can create the image you want to project. Reviewing quadrant 1 will also help you prepare your strategy for job interviews.

Strengths not necessarily apparent

QUADRANT 2. The "subtle strengths" in quadrant 2 will be more difficult to project to employers. These strengths may include such qualities as your loyalty, capacity for long hours and hard work, or desire to succeed in your career. Such qualities are difficult to build into traditional (reverse chronological) résumés, but you could use some of the following techniques to highlight your subtle strengths:

- Start your résumé with a profile section that summarizes the main positive features you bring to a particular position.
- In your résumé, include sections of relevant skills and attitudes.
- Attach reference letters that mention your less obvious strengths.
- Cite those subtle strengths in your application letter.

Limitations readily apparent will concern an employer

QUADRANT 3. Quadrant 3 includes those characteristics that you and the employer will readily identify as potential limitations on your ability to do the job. The most common limitation for applicants trying to enter a given field is lack of experience in

Apparent to Audience	Subtle or Not Obvious
1. Strengths	2. Strengths
3. Limitations	4. Limitations

Figure 23.1 A Self-inventory

that area. These limitations will certainly be considered when the employer chooses which applicants to interview, and the interviewer will either ask about these limitations or will wait to see what you have to say about them.

So what can you do? Trying to avoid discussing your obvious limitations does not work. You need to address the issue directly by showing that you have indeed developed the skills and understanding needed for the job, through university or college work placements, class projects, volunteer work, and/or leisure activities. (If you don't meet the requirements for the position, you should not apply!)

Limitations not likely apparent

QUADRANT 4. Quadrant 4 contains those limitations that the employer will not discover unless you blurt them out. You may have an aversion to writing the daily reports required by your job, for example. Or you may be a "night person" who has difficulty getting to work on time in the morning. Or you may not like the customer service component of a position that otherwise appeals very much to you. Or perhaps your perfectionist tendencies bring you into conflict with others.

There are two main issues here, an ethical issue and a practical one. Ethically, you should not present yourself for a position where you are not prepared to work wholeheartedly. You should also not apply for a position if you have attitudes that you are not prepared to change and that you know will hamper your performance in the job. On the other hand, if you are aware of a potential limitation that you are in the process of overcoming, there may be no need to reveal the problem when the interviewer asks you to comment about your limitations.

ON THE JOB...

Attitude

"Without question, attitude is the number one factor in choosing whom to hire. The knowledge, experience, and skills listed in a résumé help a candidate get an interview. At that point, I always tried to gauge attitude and personal aptitude for the job. Does the candidate go that extra mile? Has he or she had previous experience, even in other work fields or in volunteer work, that would have developed transferable skills? Can we see enthusiasm? A desire for self-improvement? An ability to work effectively with others? A sense of responsibility? All those are part of how I define 'attitude'. . . ."

—Jan Bath, CADD Manager, MMM Group

Your Transferable Skills

Employers look for these skills

Another very useful tool in determining your employable characteristics is a list of the transferable skills that you have previously developed. As Figure 23.2 on page 472 demonstrates, such a list includes proof ("validating experience") as well as the skills themselves.

So how will you identify the skills you have that will be

ON THE JOB...

Finding Your Transferable Skills

"Transferable skills are often the key to opening new doors . . . [and] these abilities will be even more highly sought by employers in the future. As technological advances require workers to communicate more quickly and frequently, a premium will be placed on professionals with elevated communication, organization, and critical-thinking skills."

—Robert Half International, "Finding Your Transferable Skills"

Skill	Above Average	Average	Below Average	Validating Experience
Programming in Java	✓			• Programming project in CoSci 356—mark: 92% • Co-op work experience at Fundy Industries—good evaluation
Supervising/coaching/ leading	✓			• Coached jr. lacrosse 3 yrs • Bell captain at Lakeshore Lodge for 2 summers
Report writing		✓		• Completed 5 major reports at college—avg grade: 70% • Collaborated on documentation report at Fundy Industries— involved in all phases of 95-page report
Keying	✓			• Timed at 64 words per minute
Problem solving	✓			• Won software troubleshooting competition 2 yrs at college

Figure 23.2 A Segment of a Transferable Skills Inventory

useful in a new job? You could start by listing all the skills you've developed and used in previous jobs and volunteer experience. Recognize the potential value of any skill that you might apply in a new position. (Figure 23.2 includes examples of skills honed through school, work, and/or volunteer experience.)

Many of the transferable skills valued by employers are communication skills: writing, speaking, listening, reading, and interpersonal. Such skills are featured in the report "Employability Skills 2000+" by the Corporate Council on Education, a program sponsored by the Conference Board of Canada. (See the In Brief box on pages 473–74 for more details.)

PERSONAL JOB ASSESSMENT

The long-term goal for your job search should be job satisfaction, so after you have thoroughly examined what makes you employable, you should choose what kinds of positions to pursue. Before drawing up a list of positions for which to apply, you will benefit from thinking about your life goals and reasons for working.

Your Career Orientation

Why do you work?

What motivates you to work? What benefits and satisfactions most attract you to certain types of positions? Knowing the answer to these questions may help you choose your career path and help you in your job-search activities.

Here is a set of categories that reflect workers' job expectations:

1. *Getting ahead.* This worker looks for advancement within an organization and starts looking elsewhere if the organization doesn't have such opportunities.
2. *Getting rich.* This worker is primarily interested in a position's financial benefits. Even though he or she may find the work stimulating, this person will take less interesting positions if the pay is better.

Employability Skills Profile: Employability Skills 2000+

FUNDAMENTAL SKILLS

The skills needed as a base for further development.

You will be better prepared to progress in the world of work when you can:

Communicate

◆ read and understand information presented in a variety of forms (e.g., words, graphs, charts, diagrams)
◆ write and speak so others pay attention and understand
◆ listen and ask questions to understand and appreciate the points of view of others
◆ share information using a range of information and communications technologies (e.g., voice, email, computers)
◆ use relevant scientific, technological, and mathematical knowledge and skills to explain or clarify ideas

Manage Information

◆ locate, gather, and organize information using appropriate technology and information systems
◆ access, analyze, and apply knowledge and skills from various disciplines (e.g., the arts, languages, science, technology, mathematics, social sciences, and the humanities)

Use Numbers

◆ decide what needs to be measured or calculated
◆ observe and record data using appropriate methods, tools, and technology
◆ make estimates and verify calculations

PERSONAL MANAGEMENT SKILLS

The personal skills, attitudes, and behaviours that drive one's potential for growth.

You will be able to offer yourself greater possibilities for achievement when you can:

Demonstrate Positive Attitudes and Behaviours

◆ feel good about yourself and be confident
◆ deal with people, problems, and situations with honesty, integrity, and personal ethics
◆ recognize your own and other people's good efforts
◆ take care of your personal health
◆ show interest, initiative, and effort

Be Responsible

◆ set goals and priorities balancing work and personal life
◆ plan and manage time, money, and other resources to achieve goals
◆ assess, weigh, and manage risk
◆ be accountable for your actions and the actions of your group
◆ be socially responsible and contribute to your community

Be Adaptable

◆ work independently or as a part of a team
◆ carry out multiple tasks or projects
◆ be innovative and resourceful: identify and suggest alternative ways to achieve goals and get the job done

TEAMWORK SKILLS

The skills and attributes needed to contribute productively.

You will be better prepared to add value to the outcomes of a task, project, or team when you can:

Work with Others

◆ understand and work within the dynamics of a group
◆ ensure that a team's purpose and objectives are clear
◆ be flexible: respect, be open to and supportive of the thoughts, opinions, and contributions of others in a group
◆ recognize and respect people's diversity, individual differences, and perspectives
◆ accept and provide feedback in a constructive and considerate manner
◆ contribute to a team by sharing information and expertise
◆ lead or support when appropriate, motivating a group for high performance
◆ understand the role of conflict in a group to reach solutions
◆ manage and resolve conflict when appropriate

Participate in Projects and Tasks

◆ plan, design, or carry out a project or task from start to finish with well-defined objectives and outcomes
◆ develop a plan, seek feedback, test, revise, and implement
◆ work to agreed quality standards and specifications
◆ select and use appropriate tools and technology for a task or project
◆ adapt to changing requirements and information

(continued)

FUNDAMENTAL SKILLS
Think and Solve Problems
- assess situations and identify problems
- seek different points of view and evaluate them based on facts
- recognize the human, interpersonal, technical, scientific, and mathematical dimensions of a problem
- identify the root cause of a problem
- be creative and innovative in exploring possible solutions
- readily use science, technology, and mathematics as ways to think, gain, and share knowledge, solve problems, and make decisions
- evaluate solutions to make recommendations or decisions
- implement solutions
- check to see if a solution works, and act on opportunities for improvement

PERSONAL MANAGEMENT SKILLS
- be open and respond constructively to change
- learn from your mistakes and accept feedback
- cope with uncertainty

Learn Continuously
- be willing to continuously learn and grow
- assess personal strengths and areas for development
- set your own learning goals
- identify and access learning sources and opportunities
- plan for and achieve your learning goals

Work Safely
- be aware of personal and group health and safety practices and procedures, and act in accordance with these

TEAMWORK SKILLS
- continuously monitor the success of a project or task and identify ways to improve

Source: This document was developed by members of the Conference Board of Canada's Employability Skills Forum and the Business and Education Forum on Science, Technology, and Mathematics.

3. *Getting secure.* Some workers will sacrifice organizational status and high salaries for long-term security.
4. *Getting control.* This employee wants to control his or her workday to reduce stress and to be more productive. Sometimes, this person equates "control" with controlling others.
5. *Getting high on work.* This person loves the work he or she does and would likely do it for no pay if a paycheque were not necessary.
6. *Getting balanced.* Many employees stay in jobs that no longer challenge them because they see their work as just one part of their lives, along with family life, recreation, and community involvements.

Workers want to live balanced lives

Workplace analyst Barbara Moses reports that "recruiters say that people at all levels, all ages, and of both sexes are putting [work–life balance] on the negotiating table as a 'must' in their work" ("What Makes" C2). John Izzo, author of *Values Shift*, adds that meaningful work, feeling good about one's organization and one's colleagues, and feeling involved in the organization are all important motivators, especially for younger workers (qtd. in Immen, "Shifting Values").

Work–life balance has recently been more difficult to achieve

In late 2012, Linda Duxbury and Christopher Higgins released the results of their third major study of work–life balance in Canada. They looked at 25 000 Canadians employed full-time in 71 public, private and not-for-profit organizations across Canada. They found that work demands and stress levels have risen, flexible work arrangements are rare, and life satisfaction has gone down. The report summary is at **http://newsroom.carleton.ca/wp-content/files/2012-National-Work-Long-Summary.pdf**.

Keep in mind, though, that many people **are** able to balance their work and the other components of their lives.

Choosing Positions to Pursue

Choosing what types of work to do involves a simple four-step process:

What type of work suits you best?

- ◆ *Step 1.* List the positions you would like to have and the jobs you would like to do. Don't worry about your qualifications for these positions just yet.
- ◆ *Step 2.* List the positions that you could currently handle, based on your self-inventory of employable characteristics and transferable skills. If you're close to completing an educational or training program, you could also list those positions you'll soon be able to handle successfully.
- ◆ *Step 3.* Check which positions are on both lists, and organize them on the basis of two criteria: (1) which positions appeal to you the most, and (2) which positions best suit your qualifications. It's important to be realistic at this point, but it's also important to try to identify what you consider ideal employment. Also list positions for which you'll be qualified after you get more work experience.
- ◆ *Step 4.* Confirm that you have included the full range of positions open to people with your interests and qualifications. For example, a person with a degree in technical writing might consider working as a webpage designer and webmaster, or a B.Sc. graduate with a major in botany might look at developing new varieties of plants for a commercial greenhouse, or a civil engineering technologist might explore the opportunities for a consultant contracting for small towns in a rural region.

The internet has many sites that help people choose career paths, such as

- ◆ The Keirsey Temperament Sorter, at **http://keirsey.com**
- ◆ Canada Job Futures, at **http://jobfutures.ca/en/interface-be.shtml**

So now you have a list of what types of positions to pursue. The next task in your job search is to learn what's available.

JOB-MARKET RESEARCH

Again, you need an organized approach:

- ◆ Identify information needs.
- ◆ Identify information sources.
- ◆ Use a variety of research tools to ensure that you find all available positions, including those in the "hidden job market."

Identify Information Needs

Unfortunately, wanting to hold a particular position does not guarantee that such a position is available. You'll need to research the job market to learn several key things:

Know the job market

- ◆ which of your desired positions are available now, and which might be available in the near future

- which companies are expanding and thus requiring additional workers
- which companies have internship programs or will accept "volunteer employees"
- what qualifications are required for specific positions
- when and how companies recruit for seasonal employees and/or permanent employees
- which firms contract some or all of their work, and how you can monitor these firms' ongoing contract requests
- which business trends and expansion of government or commercial operations can be anticipated for the foreseeable future

Your first research step, then, should be to decide the scope of your information search. In your current job search, do you need to know only which companies are hiring, or will soon be hiring, people like you? Or are you interested in the full range of information in the list above? Your information requirements will determine which of the following job market information sources you use.

Identify Information Sources

Know how to research the market

Literally hundreds of printed and electronic information banks can help you find employers and jobs. Some of these will list available positions; others will help you direct your spoken or written inquiries.

The following information sources should provide valuable leads:

1. *Campus placement office.* Most universities and colleges have placement offices. Those institutions that offer co-op work programs provide information about co-op employers; also, evaluation reports written by former co-op students are on file. Most institutions have online career resources, so go to your school's homepage and look for "jobs," "employment," "career centre," or "career resources." For an excellent example, visit Seneca College's "Guide to Internet Job Search Sites" at **www.senecac.on.ca/student/careerservices/students/ search-a-job/job-search-sites.html**.
2. *Canada Employment Centres and Job Futures publications.*
3. *Career Planning Annual.* Published by the University and College Placement Association, it provides information on employer-members who recruit at the college and university level.
4. *Job Outlook (Annual).* An annual, national forecast of the hiring intentions of employers published by the U.S. National Association of Colleges and Employers (NACE), it examines issues related to the employment of new university and college graduates. For an overview, go to **www.naceweb.org**.
5. *Canadian Trade Index.* It provides information about Canadian manufacturers.
6. *D&B Canada's Canadian Key Business Directory.* It has information about businesses in Canada.
7. *Standard & Poor's Register of Corporations, Directors, and Executives.* This register alphabetically lists the products and services, officers, and telephone numbers for more than 35 000 corporations in Canada and the United States.
8. *Newspaper feature articles* focus on companies and executives, expansion plans, new products, and key appointments.
9. *Trade or professional journals and associations.* These list opportunities, trade shows, seminars, and meetings that provide opportunities for networking.

Employers and job seekers research job boards and other online information sources

Most of the above publications have an internet presence. Increasingly, job seekers and employers are going to the internet to glean information about each other. Job seekers, especially those about to graduate from university or college, will find the following sites useful:

www.monster.ca and www.monster.com
www.careerbuilder.ca
http://employment.gov.bc.ca/index.php?rLoad=1
www.workopolis.com
www.cacee.com
www.workinfonet.bc.ca, **www.onwin.ca**, or **http://mb.workinfonet.ca**

Sites such as Monster.ca and Workopolis.com include an occupational profile section, massive résumé databases, and posts of currently available positions. They also feature a résumé builder that guides users step by step through the process of creating or updating a résumé.

Searching the web can frustrate and overwhelm job seekers because of the sheer volume of available information. So use the focus function found on most job boards to zero in on certain jobs and eliminate others from your search.

Visit company websites

You can learn a great deal about some companies by visiting their homepages in addition to gleaning information from websites, such as the ones listed above. Also, many companies list job openings on their webpages.

Join professional associations and visit their online forums and job boards

Additional information can be discovered in blogs, special-interest forums, and professional association newsletters. Power Globe, an IEEE (Institute of Electrical and Electronic Engineers) offshoot, is an internet email forum for people interested in electric power engineering. Among other topics, subscribers regularly post positions coming open in their companies. Another typical example, Canadian-Universities.net, posts jobs at **www.canadian-universities.net/Employment/Jobs/index.html**. No matter what field you're in, whether it's medical research or fashion design, you'll be able to find a relevant online professional forum or job board exclusive to your field. Getting this kind of inside information provides another incentive to join the professional association in your field.

Internet career research is increasing

Some industry insiders, such as Katherine Spencer Lee, executive director of Robert Half Technology, believe that many job seekers "place too much faith in the internet [which] has made it easier for those on a job hunt to identify open positions but . . . not to actually land jobs." However, business writer Virginia Galt reports that the number of Canadians visiting online career sites increased from 2 million in 2002 to 7.6 million in 2006 ("Surfing"). Contrary to Spencer Lee's pessimism about hiring success, a U.S. Conference Board survey of 5000 respondents found that 38.2 percent of those applicants receiving job offers felt that their internet searches had been instrumental in their getting hired. That figure was ahead of offers through newspaper ads (23.9 percent) or networking contacts (27.8 percent; Galt "Surfing").

Social-networking sites can provide valuable information about potential employees

EMPLOYERS PROSPECT FOR POTENTIAL EMPLOYEES. Employers also use the internet to find and recruit potential workers. Of course, they go to sites where job seekers have posted résumés, but recruiters also search discussion forums, blogs, conference proceedings, online newsletters, college and university award postings, social-networking sites, and many other online sources. In the United States, 68 percent of firms "use social media or social networking sites for recruiting purposes." A June 2009 survey of Canadian hiring managers and recruiters found that more than two-thirds use social media, with 69 percent using LinkedIn and 44 percent using Facebook (Grant).

Sites such as Facebook (about one billion active members) and LinkedIn (over 150 million) provide opportunities to network in social and professional ways with people who share interests, backgrounds, and experience (Dobrota C1). Social-networking sites, such as Facebook, also allow members to log on with work-related addresses, not just university addresses. Coming from the opposite direction, career site Jobster.com now enables social networking (Dobrota C7).

Going a step further, Facebook has combined with CareerBuilder to help recruiters find job candidates. According to CareerBuilder, its featured Facebook position gives employers access to 67 million potential employees (Bloomberg C2).

Job seekers should be careful in their online posting

These sites provide useful social- and career-networking tools, but they reveal much more than names—unless protected with privacy passwords, they function as a public display, and recruiters can find indications of a potential candidate's personality, attitudes, and judgment. "Seventy-five percent of recruiters use search engines to check out job candidates . . . and 26 percent said they'd eliminated candidates from consideration because of information found online," according to a 2005 survey of executive recruiters (Marron, "Employers"). Salacious photos or unpleasant text can damage a job candidate's chances when an employer learns about questionable posts. A survey from the National Association of Colleges and Employers revealed that over 60 percent of employers who look at self-disclosing internet sites use their gleaned information to help decide whom to hire (Marron, "Employers").

To avoid negative reactions from employers who find you on the web, follow these guidelines (Dobrota C7; Marron "Employers"):

Guidelines for an online presence that won't come back to haunt you

◆ Keep your website or blog updated.
◆ Make sure your site is well designed and that all the links work.
◆ Use privacy settings to restrict access to your site only to those who have a password or special authorization.
◆ Don't post questionable videos, pictures, or information online.
◆ Check to see whether someone else has posted material that puts you in a bad light. If so, try to get the negative material removed or corrected.
◆ Post positive information about your work experience, leisure activities, and background. Include sensible, insightful comments about your industry or current happenings.
◆ Avoid placing silly, tasteless, or opinionated material online.
◆ Don't post critical comments about your boss or your company.
◆ Use your online connections to keep in touch with former colleagues or classmates.

Employers use social media sites to not only find talent, but also to **screen potential employees**

Targeted searches will find you if you have a LinkedIn account

Phenix Management International reports that special software can now quickly scan a host of sites to identify and evaluate potential candidates, and then deliver a short list of candidates to a hiring department or search firm. This efficient technology benefits companies and those candidates who have a web presence. Also, software such as that used by RecruitIn (**www.recruitin.net**) allows employers to quickly find potential candidates who have LinkedIn accounts. The user chooses a country, keys in a job title, names a location or other keyword, and clicks "Open in Google." For example, a February 7, 2013, search using the keywords "technical writer" and "engineering" immediately identified 4340 Canadian technical writers who have engineering experience.

Inquiries

Inquiries can elicit valuable information

One way to learn about unadvertised openings is to inquire in person, by mail, or by email.

WHY INQUIRE? You should inquire about potential employment if you don't know whether a company will have a position for someone like you and if you're not sure whether the company's available position would suit your interests and qualifications. Also, a well-crafted inquiry will tell you which additional skills you will have to develop.

The following advice applies particularly to letter inquiries, but much of the advice could also be applied to telephone and in-person inquiries.

Your main challenge in sending unsolicited inquiries is getting the reader to respond. To meet that challenge, you must do two main things:

How to improve your chances of getting a reply

1. You must show your reader that you have something to offer the company. The best way to do this is to review your qualifications quickly and refer your reader to an enclosed résumé for more details. Remember, though, that you're not applying for a named position in this letter, so do not write a sales pitch. Save that for any application letter that you later send to this same reader.
2. You must make it easy for the reader to respond. Even the kindest, best-intentioned reader will place your inquiry at the bottom of a priority list if a response to your letter requires a lot of time and effort.

TIPS. Here are some tips for reducing your reader's effort and thus increasing the chances of getting a response to your inquiry:

- Use a structure that the reader can easily follow.
- Place your questions in a numbered or bulleted word list.
- Make sure those questions follow a logical order. (The first logical question for most job inquiries will be whether the employer has, or anticipates having, openings in your field.)
- Make each question absolutely clear. (Show your letter to others before sending it.)
- Ask a reasonable number of questions; five seems to be the limit.
- Do not ask questions that you could easily answer by reading the company's annual report, visiting its website, or phoning the company's personnel office.
- Use appropriate tone and phrasing: positive, assertive, polite, concise, business-like, fresh (no clichés), and energetic (active verbs).
- Make it easy for the reader to respond: where feasible, offer to phone or visit your reader to get the information.
- Display the "you" attitude by focusing on the nature of the work involved, rather than on how you might benefit.

The inquiry illustrated in Figure 23.3 on page 480 was sent by a civil engineering technology student who was responding to an invitation he had received while on a co-op job placement. (His accompanying résumé appears in Figure 23.8, pages 491 and 492.

453B Gordon Drive
Kelowna, BC   V1W 2T1

January 27, 2014

Mr. Miro Betts
International Survey Systems
1199 – 11th S.W.
Calgary, AB   T2R 1K7

Dear Mr. Betts:

In late October, at Greenwood Consultants in Nelson, your demonstration of real-time kinematics further kindled an interest I've developed in GPS surveying techniques. Your invitation to contact ISS this winter has led to this inquiry.

As you may remember, I'm studying civil engineering technology at Okanagan College. Also, my interest in GPS has prompted me to read widely on the subject. That's partly why I was so enthusiastic about the accuracy of the equipment you demonstrated in Nelson. Now, I'm eager to get a chance to work with a company such as yours.

Consequently, I'm wondering

1. whether ISS will have an opening for someone like me, this May to September,
2. what the work would entail and what equipment might be used,
3. what qualifications ISS would require, and
4. whether I should upgrade my qualifications to be ready to work for ISS this summer. (Please see the attached résumé for my current qualifications.)

Would it be possible for me to meet with you or another representative to discuss what I might contribute to ISS? I am willing to travel to meet anywhere in Alberta or British Columbia. You can contact me at my email address, chris@silk.net, or you can leave a phone message at 250-876-3422, and I'll respond to your call as soon as I return from class.

Sincerely,

Chris Hendsbee

Chris Hendsbee

Enclosure: Résumé

Figure 23.3 A Sample Job Inquiry

Another form of "prospecting" for jobs is an introduction letter. You might write this letter yourself, or someone else might "introduce" you in a letter or in the course of a conversation with an employer. For advice on writing self-introductions, see page 501. Usually, self-introductions are sent to firms whose activities appeal to that job seeker, but which have not advertised any openings. The ideal outcome would be an informal interview with someone in that organization.

Job prospecting can also be managed in informational interviews in which a small-business owner or a department manager consents to visit with a job seeker. Ostensibly, this type of interview is designed to help the job seeker get industry and career information, but often the conversation turns toward the job seeker's employment skills and suitability for a position with that particular business. Many employers readily assent to such interviews because they are on the lookout for new talent and because they feel good about contributing to someone's career development.

Networking

Finding employees through existing networks has advantages for employers and for job seekers

Many positions are not advertised; advertising is expensive and time-consuming, so many companies rely on informal methods of finding employees. In a survey of 343 employers conducted by ERE Media and Classified Intelligence in 2006, 75 percent of respondents said that employee-referral programs are very effective in recruiting employees. Some firms encourage referrals by paying employees who refer prospective employees. ERE's recruiting guru, John Sullivan says that an organized program of soliciting employee referrals produces high-quality hires from outside the company and inside the company.

"Prospecting" for jobs can take many different forms

Flowork International asserts that "80% of job opportunities are found through our contacts" ("Top 5 Ways"). Members of your contact network can alert you about impending retirements, company expansion, or new programs that require staff. Those same contacts can tell employers about your imminent appearance on the job market, or your desire to join that company. Here's one scenario that illustrates how the two-way process can work.

Jenny Roy recently graduated from the environmental sciences program at the University of Waterloo. After graduation, she landed a four-month contract with Cambrian Consultants, a private firm investigating the link between water quality and fish stocks in Georgian Bay. But that contract ran its course, and Jenny had no luck finding other employment through newspaper ads, internet sites, environmental journals, or her inquiries to a variety of private and government operations.

ON THE JOB...

Active Networking

"I think it's very important to 'cultivate' your contact network. You have to keep in touch with people, for possible job leads, but also to exchange professional information. And when you're actively pursuing a particular position, it helps to be persistent. Follow up after submitting an application package; follow up again after having been interviewed. Make sure they recognize and remember your name. . . ."

—**Jan Bath, CADD Manager, MMM Group**

In desperation, she turned to several people who had been influential in her life:

◆ her mother, Adrienne Lavois, who worked as an administrative secretary in the Department of Indian Affairs in Ottawa
◆ her uncle, Pierre Riley, a retired geologist in Montreal
◆ her mentor, Dr. Howard Cash, who taught freshwater biology at the University of Waterloo
◆ a former classmate, Sandy Travers, who worked as a lab assistant at the University of Guelph
◆ her high school volleyball coach, Nancy St. Jean, who ran a fishing lodge on Lake Temagami

Initially, these five people agreed to watch for possible positions for Jenny. They offered to distribute résumés for her and to talk to their friends, relatives, business acquaintances, and colleagues. At that point, Jenny had five people "searching" for job leads. Figure 23.4 shows how she might have diagrammed her network:

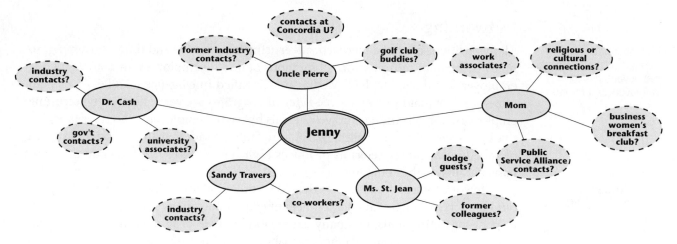

Figure 23.4 A Basic Network

Jenny originally thought only in terms of her five front-line searchers. She didn't anticipate how far the network could develop. Indeed, she didn't know that dozens of strangers would be contributing to her job search. Let's see how just one of the contact lines developed (Figure 23.5):

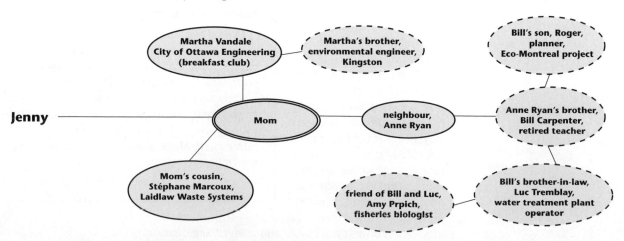

Figure 23.5 An Expanded Network

In addition to Jenny's mother, eight other people became aware of a bright, eager young person looking for work. For various reasons, all of them were on alert for suitable employment opportunities and were prepared to pass their information along. The amazing thing is that only two of them, Jenny's mother's neighbour and Jenny's mother's cousin, have ever met Jenny!

If you think that the basic assumption behind the networking concept is naive, and if you say that there's no motivation for people to actively seek employment for someone whom they've never met, remember two things:

Why networking works

1. The person farther down the network (Bill Carpenter's brother-in-law, for example) is not really doing a favour for Jenny; he's doing it for his friend (or brother-in-law, in this case).
2. Each person in the network is just relaying information, which is not a time-consuming or onerous task. Most often, this type of information is passed along during casual, friendly conversation.

Most jobs are not formally advertised

Jenny's situation, though fictional, is common enough. She can get a tip about an opportunity from any one of the many "researchers" in her network. Thus, she has a strong chance of tapping into the huge hidden job market—the vast majority of available positions are not formally advertised. However, she doesn't have to wait for things to happen; she can regularly contact persons in her network to update her situation or just to stay in touch. Internet social-networking sites and text messaging facilitate such contact maintenance.

Networks can be established in many different venues

Useful networks are carefully maintained

Other forms of networking opportunities can be found at trade shows, public meetings, job fairs, sports clubs, and social events. Keith Ferrazi, who calls himself "the ultimate networker," has written a book on the subject, *Never Eat Alone*. He claims to have 5000 contact names stored on his smartphone, people who will answer when he phones (qtd. in Timson).

EXTENDING YOUR NETWORK. Joann S. Lublin reports that "surveys show that most candidates find work through networking." However, says Lublin, many job seekers "feel frustrated with standard strategies such as tapping friends for referrals" ("Tricks of the Trade"). The *Wall Street Journal*'s Sarah Needleman advises going to the "perimeter of your social circle," to such persons as "fellow parishioners, members of your industry trade group, and alumni of your college" ("Widen Social Circle"). Also, an online social site such as Twitter allows job seekers to make contacts and help employers learn about you.

CIO.com's C.G. Lynch has consulted career and social media experts to get advice on using Twitter effectively. Here's a summary of what he has learned:

- ◆ Identify people in your industry and get in their radar by following their tweets. You can also use Twitter's search tool to locate keywords that will help you find tweets and the people who posted them.
- ◆ Thoughtfully respond to tweets that particularly interest you, taking care to put "@" in front of the person's Twitter name. By doing so, you increase the chance of being followed back.
- ◆ Build a follower list on Twitter long before you need to contact people about possible job openings.
- ◆ Keep your Twitter posts professional in tone and focused on work-related issues; don't talk about highly personal issues or trivial daily occurrences.

◆ In your Twitter profile, link to a blog if you have one, or establish a LinkedIn presence and link to that site.
◆ In order to have your tweets noticed by recruiters, endear yourself to them by referring candidates to them or by passing information to them.

Why personal contact is important

USING YOUR NETWORK FOR REFERRALS.   Donald Asher, an expert on converting higher education to career opportunities, says that "one of the biggest mistakes job seekers make is to spend all of their time looking for work online." He says "there's a critical missing part, that last little link in the chain, when one human being connects with another, in real life." So Asher recommends that you contact employers directly through emails, pings, tweets, IMs, or phone calls. Or better yet, he says, get someone to refer you to the employer. Asher quotes global staffing guru Gerry Crispin: "never ever apply without having an employee refer you." Crispin and Mark Mehler conducted a 2011 survey that found that "46% of all hires at top performing firms are referrals, while for all firms, they range between 28% and 39.9% of all hires" (Sullivan, "10 Compelling Numbers").

PRODUCING JOB-SEARCH MATERIALS

The most common sales material produced by job searchers is a *résumé*. It provides an objective, organized record of key facts about the applicant and summarizes the applicant's skills.

Uses for Résumés

Résumés accompany letters of introduction, application letters, and inquiries. You can send them by fax or email; you can post e-résumés on the internet; you can send them as follow-ups to self-introduction telephone calls; and you can leave them with each employer in a campaign of visiting business offices and other places of employment. You can also use a résumé to accompany a business proposal.

Preparing Your Résumé

Create more than one résumé, or create one master resume from which you select segments for particular job applications

You should modify your basic résumé to match each position for which you apply. Different positions emphasize certain requirements more than others, so you might place your education first in one résumé and your work experience first in another. A third résumé might expand the amount of detail regarding the relevant skills you've developed through your volunteer experiences.

BASIC CONSIDERATIONS.   Employers generally spend 15 to 45 seconds initially scanning a résumé. They look for an obvious and persuasive answer to the question, "What can you do for us?" Wallace Immen reports that the "new reality" is that "a résumé must scream your value or it will hit the trash." One way to get attention, recommends career coach Sharon Graham, is to draw attention to and validate your accomplishments by using "dollars, percentages, and other values to show measurable results" (Graham).

How to get your résumé noticed

Employers are impressed by a résumé that

1. looks good (conservative, tasteful, uncluttered, on quality paper);
2. reads easily (headings, typeface, spacing, and punctuation provide clear orientation); and
3. provides information the employer needs to decide whether to interview the applicant.

Employers generally discard résumés that are mechanically flawed, cluttered, sketchy, or difficult to follow. Don't leave readers guessing or annoyed; make the résumé perfect.

Most résumés include

- name and contact information
- educational background
- work experience
- transferable skills
- interests, leisure pursuits, volunteer activities, awards, and skills
- references

In the past, some résumés open with an "Objective" section that identifies the applicant's immediate career goal, or with a "Profile" section that highlights the applicant's qualifications and relevant attributes.

Select and organize material to emphasize what you can offer. Don't just list *everything*; be selective. (We're talking about *communicating*, not just delivering information.) Don't abbreviate, because some readers might not know that abbreviation and because other readers may infer that you're lazy and take shortcuts. Use punctuation to clarify and emphasize, not to be "artsy."

Tailor the résumé to the job requirements

Begin your résumé well before you plan to use the document as part of your job search. You'll need time to produce a first-class résumé. Most job seekers find it useful to create two or three different résumés that have different emphases and different lengths, and quickly modify them to suit positions that suddenly appear.

CAUTION

Never invent or misrepresent credentials. Your résumé should make you look as good as the facts allow. Distorting the facts is unethical and counterproductive. Companies routinely investigate claims made in résumés, and people who have lied are fired or dismissed as a viable candidate.

"Seven out of every 10 backgrounds we screen has some sort of embellishment on their résumé," says Marco Confuorto of United Risk Partners, a company that performs background checks on job candidates (qtd. in Morsch). Hiring managers are

---

> **ON THE JOB...** ▶
>
> **Omit Career Objectives**
>
> *"The days of including a career objective and/or professional summary are over. It's a waste of valuable space. Instead, just address this with a sentence in your cover letter about how the position you're applying for fits into your overall career plan. Get to business by starting with accomplishments and facts that are relevant to the job posting."*
>
> **—Justin Thompson, CareerBuilder researcher and writer**

not amused: a CareerBuilder survey found that 43 percent of managers "said they'd automatically dismiss a candidate who fibbed" ("Lying about Lying on Résumés").

If there are gaps in your résumé, periods when you seem to have done nothing, don't fill them in with bogus information. Instead, explain them in your application letter, or use a functional résumé that lists your skills and accomplishments under headings, not a chronological listing of your employment.

NAME AND ADDRESS. In a heading, include your full name, mailing address, phone number, and email address. Make your name stand out by using a larger type size and boldface. If you anticipate an address change after a certain date, include both your current and future addresses and the date of the change. Make it easy for an employer to contact you!

EDUCATIONAL BACKGROUND. If your education is more substantial than your work experience, place it first for emphasis. Begin with your most recent school and work backwards, listing degrees, diplomas, and schools attended beyond high school (unless the high school's prestige, its program, or your achievements warrant its inclusion). List the courses that have directly prepared you for the job you seek. Where applicable, name co-op work placements or special projects that have helped you develop relevant skills and knowledge.

WORK EXPERIENCE. If you have solid experience related to the job you seek, list it before your education. Start with your most recent experience and work backwards, listing each significant job. Describe your exact duties in each job, indicating promotions. Where applicable, describe skills developed in certain positions.

The standard heading "Work Experience" might not adequately describe your situation. If that's the case, here are some alternative ways of highlighting your valuable experience:

Use effective, appropriate headings

◆ Provide two categories, "Related Experience" and "Other Work Experience." In the first category, include related volunteer experience, as long as it is substantial enough to merit inclusion. In the second category, list those positions not related to the position for which you're currently applying. Include details of non-related work experience to show that you've had significant experience and to highlight such qualities as loyalty, leadership ability, or ability to learn new skills quickly.
◆ Feature a special category called "Volunteer Experience." A person's experience in a service club or other organization will develop skills needed for a particular position, sometimes at a very high level.
◆ Create a separate category called either "Practicum Experience" or "Co-op Experience" if you've had significant work experience through a university or college work placement.

PERSONAL DATA. An employer cannot legally discriminate on the basis of gender, age, religion, race, national origin, disability, or marital status. Therefore, you aren't required to provide this information.

ON THE JOB...

Evidence of Transferable Skills

"When we hire engineers or technologists, we look for a cross-section of experience that would have built related skills and knowledge, even if that experience wasn't in our specific field. For example, someone with a background in building houses would develop the habit of thinking in three dimensions, which is very useful in civil engineering. Another example is the engineer we hired to be a project manager, partly because she had had extensive experience setting up a chain of retail stores across Canada. . . ."

—**Jan Bath, CADD Manager, MMM Group**

PERSONAL INTERESTS, ACTIVITIES, AWARDS, AND SKILLS. List leisure activities, sports, and other pastimes. Employers use this information to learn whether you will easily fit into an existing work team. Also list memberships and offices held in teams and organizations. Employers value team skills and leadership potential. If your volunteer experience is not directly related to the position for which you're applying, include that volunteer work in this personal activity section; employers know that people with well-rounded lifestyles are likely to take an active interest in their jobs. But be selective—list only those items that reflect qualities employers seek.

Extracurricular activities can be particularly valuable to recent university and college graduates who have limited work experience in their field. Community volunteer work, academic competitions, debating societies, student government, theatrical productions, student newspapers, and athletics build communication skills and qualities of leadership, commitment, and team building, all of which employers value (Bloom).

Highlight skills that match the position you're pursuing

You should group lists of skills in broad categories, such as communication skills, administrative abilities, computer and technical skills, and team work skills. Also, you should not necessarily include all your transferable skills in a given résumé. Instead, highlight those skills that match the position that you're applying for.

REFERENCES. List three to five people *who have agreed to provide* strong, positive assessments of your qualifications and personal qualities. Some of their support will come in the form of requested letters or email responses. Others will be asked to complete reference forms. Or you could request reference letters from employers, teachers, or others who have closely observed your work habits and skills.

Some employers use references to narrow the list of applicants to a shortlist. Others will contact references after interviews have been conducted. Usually, the motivation for these follow-up calls is to confirm interviewers' perceptions or to investigate aspects that troubled the interviewers during the interview.

Often, then, your references' comments can make or break your application. So choose those references carefully:

How to choose references

◆ Select references who can speak with authority about your ability and character.

◆ Choose among previous employers, professors, and respected community figures who know you well enough to *comment concretely* on your behalf. (Opinions without detailed, concrete observations will not usually convince alert employers.)

◆ Do not choose members of your family or friends not in your field.

◆ When asking someone to be a reference, don't simply ask, "Could you please act as one of my references?" This question leaves the person little chance to refuse, but this person might not know you well or might be unimpressed by your work and therefore might provide a watery reference that does you more harm than good. Instead, ask, "Do you know me and my work well enough to provide a strong reference?" This second approach gives your respondent the option of declining gracefully or it elicits a firm commitment to a positive recommendation. Heidi Allison, who manages a firm that checks references, recommends that job seekers should carefully evaluate how their references might react when they're contacted, for a variety of reasons (Allison "Seven Deadly Myths of Job References").

Your listing of references should include each person's name, position title or occupation, full address, phone number, email address, or other method of contacting the person. You might also indicate the nature of the person's reference (work, academic, character, supervisory) if that's not obvious from your listing of the person's occupation.

> **NOTE**
>
> *There is some controversy about whether to include references in a résumé. Some job-search counsellors suggest a line saying, "References available upon request." These advisors argue that the employer will then call you to get those references and you'll have an opportunity to make a positive impression. The problem is that if other qualified applicants have listed references in their résumés, the employer can save time and effort by interviewing those applicants instead of you.*

Counsellors at universities and colleges that have job-placement offices will sometimes advise graduates to list the address of the placement office where employers can get your list of references. The problem with that arrangement is that you lose control of which references are provided for certain applications.

Organizing Your Résumé

Three main organizational patterns

Organize your résumé in the order that conveys the strongest impression of your qualifications, skills, and experience. Depending on your background, you can arrange your material in reverse chronological order, functional order, or a combination of both.

REVERSE CHRONOLOGICAL ORGANIZATION.   In a reverse chronological résumé you list your most recent experience first, moving backwards through your earlier experiences. Use this arrangement to show a pattern of job experiences or progress along a specific career path. Reverse chronology suits applicants who have a well-established record of relevant experience and education. Many employers like this traditional organization because they're used to it and they find it easy to read. Marcel Dionne's résumé in Figure 23.6 on page 489 shows an example of the traditional reverse chronological order.

FUNCTIONAL ORGANIZATION.   In a functional résumé, you emphasize skills, abilities, and achievements that relate specifically to the job for which you are applying. Use this arrangement if you have limited job experience, gaps in your employment record, or are changing careers. See Carol Hampton's résumé in Figure 23.7 on page 490 for an example.

COMBINED ORGANIZATION.   Most employers prefer chronologically ordered résumés because they are easier to scan. However, electronic scanning of résumés calls for a more functional pattern. One alternative is a modified-functional résumé, which preserves the logical progression that employers prefer, but which also highlights your abilities and job skills, as in Christopher Hendsbee's résumé, shown in Figure 23.8 on pages 491–92.

Marcel Dionne

144 Avenue
Champlain Quebec, PQ   G7E 1R6
Cell 418-451-3565   Email: mdionne@avoir.ca

EDUCATION
2008–2012

Software Engineering (Bachelor's degree)
University of Waterloo
Learned to apply a full range of software engineering principles
through scenarios based on concurrent object-oriented software
system designs.

Projects: 2nd year—helped Prof. Warsaw develop tracking software
for Kitchener Trucking's "smart tire" program
3rd year—developed software reliability model and
automatic detection of software failures for Waterloo
Region District School Board
4th year—redesigned maintenance scheduling software for
the University of Waterloo's boiler heating system

EXPERIENCE
Nov. 2012
to present

Software Consultant (contract position)
City of Quebec
Worked with City engineers to redesign Quebec City's transit
management software. Co-responsible for designs and responsible
for implementing the program, including training 12 city employees
to work with the software. The project will be completed by
June 10, 2013.

Summer Relief Worker

Summers
2007–2010,
and May to
Oct. 2012

Canada Courier, Montreal and Hamilton
Worked in all aspects of the parcel delivery business: truck driver,
dispatch worker, delivery tracking troubleshooter, order desk clerk,
packing department clerk.
A highlight of this experience was the opportunity to troubleshoot
problems with Canada Courier's parcel tracking software. Rewrote
the software to make it more reliable.

PROFILE

Interests include distance running, computers, and computer-
generated music. Single. 24 years old. Capable of working long,
productive hours.

REFERENCES

Professor John Marks
Computer Engineering Dept.
University of Waterloo
Ph. 519-888-4532, ex.5430
Email: marks@coulomb.uwaterloo.ca

Bill Spender
Operations Manager
Canada Courier
1780 St. Laurent Blvd.
Montreal, QC   G9T 2W2
Ph. 514-986-2987

Professor Viktor Warsaw
Computer Engineering
University of Waterloo
Ph. 519-888-4532, ex.5444
Email: warsaw@coulomb.uwaterloo.ca

Lise Tremblay
Engineering Contract Manager
City of Quebec
Place Centrale
Quebec, PQ   G7E 4R4
Ph. 418-655-2517

Figure 23.6 A Reverse Chronological Format Résumé (compressed to fit on one page)

Carol Hampton

196-4068 Minster Avenue
Toronto, ON   M6W 2R5
Phone: 416-445-5333
Email: comphamp@collect.ca

OBJECTIVE
To apply computer installation, configuration, and troubleshooting skills to a networked environment

SKILLS AND ABILITIES
Analytical
- Have shown an aptitude for determining client network requirements
- Proficient at solving math-based problems (top marks in Mathematics in first two semesters of college program)
- Proficient at discovering programming errors

Design
- Won city-wide competition for high school webpage design, May 2010
- Proficient in desktop publishing operations—currently design advertising for George Brown College's student newspaper

Research
- Researched comparative study of firewall software packages for Professional Communications II class
- For a 30-minute class oral presentation, researched the San Francisco Project's development of Java applications
- Industry experience with Farnham Industries, Mississauga—found ready-made solutions for problems with a networked configuration. Saved the firm $6000, the cost of a custom-designed solution (summer experience, 2011)

Interpersonal
- Worked closely with the staff at Farnham Industries to implement network solutions
- Gained the trust and cooperation of staff who were at first reluctant to try new methods, especially those introduced by a "student"
- Engaged in peer counselling in high school and at George Brown College

EDUCATION
Currently completing the last semester of the two-year Computer Systems Technician diploma program at George Brown College, Toronto. Qualified to function as a user support specialist, network administrator, or business network trainer.

REFERENCES

Terry MacArthur
Systems Analyst Consultant
Toronto
416-442-3454
tersystem@toronto.ca

Marjorie Prystai-Alvarez
Personnel Coordinator
Farnham Industries
Mississauga
905-657-7439
staff@farnham.on.ca

Jens Husted
Computer Networking Instructor
Information Technology Department
George Brown College
416-415-2010
husted@gbcscitech.ca

Figure 23.7 A Functional Format Résumé

Christopher Hendsbee
453B Gordon Drive
Kelowna, BC   V1W 2T1
Phone: 250-876-3422
Email: chris@silk.net

Education

Civil Engineering Technology
Okanagan College, Kelowna, BC   2012–present
The program stresses practical applications of civil engineering theory.

Experience

Surveyor
Deep Woods Consultants, Nelson, BC   May–Nov. 2013
Surveyed forestry roads, property lines, bridge sites, and topographics.
Became familiar with many types of total stations and data collectors.
Led and supervised a surveying crew. Responsible for all the notes.

Equipment Operator
Lucky Logging, Keremeos, BC   July–Aug. 2012
Operated a *966* Cat loader and stacked processed logs. Operated the
excavator with a Styer processing head, and worked on the grapple skidder.
Built landings and fire guards. Greased and repaired equipment.

Release Driver
Sterile Insect Release Program, Osoyoos, BC   Aug.–Sept. 2011
Followed a marked route on an ATV. Every morning, the manager would drop
the sterile moths and I would release them over my route. Recorded field
notes and addressed questions and problems raised by farmers.

Labourer
Worked as a ranch hand, fencing company assistant, and assistant orchardist.

Special Skills

Proficient with Windows XP and Vista. Also proficient with software programs:
WordPerfect; Microsoft Works (spreadsheets, word processing, databases,
charts/graphs, communication); Microsoft Office 2010 (Word, Excel, Access,
PowerPoint); Harvard Graphics presentation package; Microsoft Publisher;
AutoCAD version 18.0, AutoCAD Light; and MiniCAD
Proficient with compasses
Very familiar with GPS
Proficient at operating heavy equipment
Proficient at operating chainsaws and a large variety of hand and power tools

. . . 2

Figure 23.8 A Combination Format Résumé: Chronological and Functional   *(continued)*

Part **V** Applications

Christopher Hendsbee/Page 2

Interests and Activities	Archery (bronze medal in the BC Summer Games), hunting, hiking, swimming, weightlifting, trap shooting
Accreditations, Licences, and Memberships	S100 Wild Fire Suppression Certificate Hunter's licence Avalanche course Driver's licence Level one first aid Firearms acquisition certificate Student member, Applied Science Technologists and Technicians
References	**Dave Muster** Owner, Lucky Logging, Keremeos, BC RR#1 Newhouse Road Keremeos, BC V0X 1D0 Ph. 250-499-5609 **Jean McElroy** Survey Manager Deep Woods Consultants Nelson, BC V1L 5T9 Ph. 250-825-4565 **Bill Waterburn** SIR Supervisor RR#1 Barcello Road Cawston, BC V8X 5Y2 Ph. 250-499-2366 **Bob Bradley** Department Chair Civil Engineering Technology Okanagan College 1000 K.L.O. Road Kelowna, BC V1Y 4X8 Ph. 250-762-5445 (ext. 4590)

Figure 23.8 A Combination Format Résumé: Chronological and Functional

Image Projection

When you plan, write, edit, and periodically adjust your résumés, evaluate them according to how well they project your desired image. If, for example, you want to be seen as energetic, active, and enthusiastic, be sure to include information about your volunteer activities and outdoor leisure pursuits. Further project the desired image by choosing active verbs, such as *led*, *organized*, *built*, *maintained*, and *completed*.

Also, have your résumé project an image consistent with the position's requirements. For example, if you're applying for a position where you'd spend half your time in the field collecting samples and the other half assessing those samples and reporting the results, use language and provide facts that reveal your writing skills and the meticulous side of your nature, as well as your physical fitness.

VISUAL RÉSUMÉS.   In a highly competitive environment, some job seekers are using visual résumés to stand out from the crowd. Online tools like VisualCV (**http:// www.visualcv.com**) and Vizualize.me (**http://vizualize.me**) assist in creating colourful, well-structured résumés. Vizualize.me has a feature that converts a LinkedIn profile into a résumé with one mouse click. The idea is to create documents that are more attractive than text-heavy résumés. However, if most people start producing visual résumés, they will no longer stand out.

QUICK RESPONSE CODES.   Another way to stand out is to place a Quick Response (QR) code on your business card. These small, square barcodes appear "on everything from advertisements to company home pages to cereal boxes. All someone needs to do is scan the code with a smartphone and the information that's contained within is pulled up" (Alexander). So business cards that you distribute at a job fair or networking at a social event will allow employers to access your LinkedIn profile, your blog, or examples of your work, anything in fact that you would want a prospective employer to see.

Electronic Résumés

Increasingly, employers are favouring electronic résumés over paper documents; that trend is illustrated by Hamilton steelmaker ArcelorMittal Dofasco, which does almost all of its recruiting online (Galt "Net Ties Job Seekers"). This explosion of online recruiting has been abetted by electronic headhunters, such as Recruitsoft, but many companies now post job positions on their corporate websites.

Companies see many advantages:

◆ They can save up to 75 percent of the average cost of US$6000 of recruiting and hiring employees, according to Recruitsoft President Louis Tetu (Ray).
◆ The hiring cycle is dramatically shortened; hundreds of applicants can be screened, candidates can be interviewed, and an employee hired, all in two or three days, instead of in two to six weeks.
◆ Sophisticated software screening helps companies find the best available talent, even among employed workers not actively seeking a new job.

In addition to standard online résumés, some job seekers produce hypertext documents with links to the applicant's placement dossier, publications, and support material. Certain links (such as references) can be password-protected so that only employers can access these links. Multimedia résumés incorporate sound, video, and animation. (See the advice on preparing electronic and printed *portfolios*, starting on the next page.)

Know how to adapt your résumé so that it doesn't get rejected by automatic scanners

ELECTRONIC PRESENTATION OF RÉSUMÉS. Hard-copy résumés still work for many situations, but to tap into the total job market, you'll need to adapt your résumé to be scanned by an electronic scanner or to be placed online. Electronic storage of scanned or online résumés offers employers an efficient way to screen countless applicants, to compile a database of applicants for later openings, and in theory at least, to evaluate all applicants fairly.

AUTOMATED SELECTION OF SCANNED RÉSUMÉS. Using special software, a computer captures and searches the printed or online image for keywords (nouns instead of traditional "action verbs"). Those résumés containing the most keywords ("hits") make the final cut (Pender). Such software has limitations that sometimes fail to accurately rank highly qualified applicants, but large companies still use this screening technique for large batches of applications. (For example, Starbucks Corp. received 7.6 million applications in 2011.)

GUIDELINES FOR SENDING E-RÉSUMÉS

Katharine Hansen, a researcher and writer for Quintessential Careers, says that the best format for submitting electronic résumés is a *text résumé*, also known as *plain-text* or *ASCII text*. This format carries the *.txt* file extension. Plain text strips nearly all of a document's formatting, so it lacks visual appeal. However, it resists viruses and it's versatile. It can be

- ◆ posted on job boards,
- ◆ pasted into email messages,
- ◆ converted into HTML for web use, or
- ◆ easily scanned into a database.

Printed scannable résumés that are converted to electronic form with optical character recognition software "are being used less and less frequently by employers" because it's simpler to have applicants email their résumés directly.

Katharine Hansen also recommends trying Rich Text Format (*.rtf* file extension) for certain applications because much of, say, Word's formatting is retained in Rich Text. And, like ASCII plain text, Rich Text is much less vulnerable to viruses than a Word document. For more advice about choosing a format for your e-résumé, see Hansen's article, "Your E-résumé's File Format Aligns with Its Delivery Method" (**www.quintcareers.com/e-resume_format.html**).

HOW TO PREPARE AN E-RÉSUMÉ.   Using nouns as keywords, list all your skills, credentials, and job titles. (Help-wanted ads are a useful source for keywords.)

- *List specialized skills.* Marketing, C++ programming, database management, user documentation, internet collaboration, software development, graphic design, hydraulics, fluid mechanics, editing, surveying, soil testing, water-quality monitoring, job-site management
- *List general skills.* Teamwork coordination, conflict management. oral communication, report and proposal writing, troubleshooting, bilingual
- *List credentials.* Student member, Institute of Electrical and Electronic Engineers, certified PADI Open Water diver, B.Sc. electrical engineering, top 5 percent of class
- *List job titles.* Manager, director, supervisor, intern, coordinator, project leader, technician, trainer

If you lack skills or experience, emphasize your personal qualifications: analytical skills, energy, efficiency, flexibility, imagination, motivation.

Use keywords that you find in job postings

Richard H. Beatty, author of *The Ultimate Job Search*, suggests that you search several job sites for your desired position. Then in the various job postings you've found, highlight all the nouns and noun phrases that reflect key qualifications for that kind of position. The words and phrases that repeatedly crop up in those postings will form the bulk of the keywords an employer will consider important in evaluating your résumé.

When preparing a printed and scannable résumé, use a plain typeface, 10-to-14-point type, and white paper. Use boldface or full caps for emphasis. Avoid fancy fonts, underlining, bullets, slashes, dashes, parentheses, or ruled lines (Pender). Left-justify all information within the résumé: many searchable résumé systems do not understand indenting or columns. Avoid a two-column format, which often is jumbled by scanners that read across the page. Figure 23.9 illustrates a computer-scannable résumé.

Practical advice about how to prepare and post e-résumés may be found at Rebecca Smith's website, **www.eresumes.com**, and on pages 42–46 of Centennial College's Co-op Education Employment Guide, which is available at **www.centennialcollege.ca/adx/aspx/adxGetMedia.aspx?DocID=5245.**

Remember, an electronic résumé does not have to be as attractive as your "normal" résumé. Personnel departments that use computer databases are looking for content, not appearance.

Employment Portfolios

A good portfolio may provide a competitive edge

In a competitive environment, candidates striving for the best jobs need to provide something special. Professional portfolios (or career portfolios) are a major step beyond résumés. Portfolios allow job seekers not only to list their skills and personal abilities, but also to demonstrate and validate them.

This book does not have the space to fully discuss portfolios, but such coverage is available in Helene Martucci Lamarre's *Career Focus* and at websites sponsored by wikiHow (**www.wikihow.com/Create-a-Career-Portfolio**), by Ball State University's Career Center (**http://cms.bsu.edu/About/AdministrativeOffices/CareerCenter/Prep/Resumes/Portfolios.aspx**), and by Quintessential Careers (**www.quintcareers.com/job_search_portfolio.html**).

Carol Hampton
196–4068 Minster Avenue
Toronto, ON   M6W 2R5
Phone: 416-445-5333
Email: comphamp@collect.ca

OBJECTIVE
To apply computer installation, configuration, and troubleshooting skills in collaboration with others.

QUALIFICATIONS
Programming in Java, HTML, Logo, Pascal, C++. Network design and troubleshooting. Determining client network requirements. Advanced computer mathematics. Webpage design. Newspaper advertising design. Internet research. Usability testing. Desktop publishing of installation and maintenance manual. Writing, designing, and testing hardware upgrade manuals. Research on comparative study of firewall software packages.

EXPERIENCE
FARNHAM INDUSTRIES, 7800 CONCEPT COURT, MISSISSAUGA. Solutions for problems with a networked configuration. Network solution implementation. Liaison with staff. Positive relations. Development of trust and cooperation.

EDUCATION
GEORGE BROWN COLLEGE, TORONTO. Computer systems technician diploma. Qualified as a user support specialist, network administrator, and business network trainer.

REFERENCES
Terry MacArthur
Systems Analyst Consultant
Toronto
416-442-3454
tersystem@toronto.ca

Marjorie Prystai-Alvarez
Personnel Coordinator
Farnham Industries
Mississauga
905-657-7439
staff@farnham.on.ca

Jens Husted
Computer Networking Instructor
Information Technology Department
George Brown College
416-415-2010
husted@gbcscitech.ca

Figure 23.9 A Computer-Scannable Résumé

The content may vary with each application

CONTENT.   Portfolio content varies with the candidate's field and the nature of the job being pursued, but most portfolios will contain some or all of the following:

- a targeted résumé
- supporting documentation—performance appraisals, mark transcripts, internship or co-op reports, reference letters, copies of certificates, diplomas, or degrees, references to candidate achievement in newspaper or newsletter articles
- lists of workshops or conferences attended and other professional-development activities, including membership in associations
- records of community service and team involvement
- artifacts (samples of your work or visual record of your work)—reports and other writing samples, photos, video or audio recordings, project summaries, brochures, *PowerPoint* presentations, representative components of software developed or debugged
- a list of accomplishments
- detailed categorical listing of skills and abilities—with each skill and level of demonstrated competence, and the background or specific experiences that validate your assessment of your skill level. Skills and personal attributes might include the following categories: writing and speaking, problem solving, time and resource management, technical, teamwork, conflict management, self-confidence, initiative, drive, imagination, commitment to quality, flexibility, and goal achievement

Some portfolios also include a career summary and goals and perhaps a professional philosophy that guides and drives you.

Portfolios must be well organized and professional-looking

ORGANIZATION AND FORMAT.   Professional portfolios are usually bound in a professional three-ring binder with a table of contents and plastic section dividers. Helene Martucci Lamarre recommends the following structure:

- cover page and table of contents
- résumé
- skills and abilities
- academic record and/or professional record
- certifications, diplomas, degrees, awards
- community service

Other sections may be included as desired.

Chapters 12, 13, and 17 provide advice on graphics and page design that can be applied to portfolios. Remember to build a unifying visual identity in the portfolio, by using logos, fonts, boldface, colours, lines, and other graphics in a consistent way.

Electronic portfolios can include dynamic features and hyperlinked depth

E-PORTFOLIOS.   Portfolios can be presented on CDs or on websites. These electronic versions can be quite dynamic because they can link to online sources, and they can feature video records of your work and achievements or statements from people who support your candidacy. An online portfolio requires that you have the skills to produce one, or the money to hire a site developer. Also, you need a space on the web. This latter challenge can again be met by using free or inexpensive website hosting by such outfits as Yahoo Web Hosting, Béhance, Dotster, or FreeYellow.

Their main limitation is their restriction on web space, with usually no more than 25 MB for each site. A good alternative is Viewbook.com, especially if your portfolio is full of visual images.

LinkedIn provides another option for an online portfolio. This business-networking site allows you to import and tweak your résumé to create a customized profile. At the site, you can post photos and link to your blog, your Twitter account, your website, or a YouTube video. You can also recommend business contacts in your network and request recommendations from others.

Manage access to your online portfolio

USING PORTFOLIOS.   You should probably password-protect your online portfolio, and provide the password to employers to whom you're applying for work. That warning explains why you'll not likely be able to gain access to portfolios currently on the web. However, a sample of one style of online portfolio can be viewed at **http://larrylabelle1947.wix.com/lindasmithportfolio?goback=.gde_1164347_member_72415089**.

Your résumé and/or covering letters should refer to the availability of an online portfolio. You should take your printed portfolio with you to job interviews, but you shouldn't thrust it upon interviewers; you should ask whether they would like to peruse it, in order to get a more complete idea of your potential worth to their organization. You'll not want to leave that portfolio with them for a prolonged time because you may need it for other interviews. By the way, don't be disappointed if the panel isn't interested in viewing the portfolio during an initial interview—when they've narrowed down the field to one to three candidates, that's when they'll want more in-depth evidence of your abilities.

CASE   # E-portfolio Analysis and Planning

An employment portfolio appears in **MyWritingLab: Technical Communication**. At that site, click on "Resources" and then click on both "TechComm." Under "TechComm" click on "ePortfolio" and then open the "Sample ePortfolio."

The person in the sample portfolio wants to use it to serve more than one kind of career path. His primary career has been in radio broadcasting, with an emphasis on hockey play-by-play. However, he has expanded his role with the Junior "A" hockey team that currently employs him. His experience and skill sets now include sales, marketing, and business management. He could, therefore, branch out in other business fields.

Consequently, his portfolio needs a range of categories that indicate his varied background and range of employment capabilities. Also, his portfolio would not be complete without samples of his radio work, so it must include audio clips of his performance behind the microphone.

Analysis
Study the portfolio and answer these questions:

1. Do the portfolio's 10 categories adequately cover the candidate's range of interests, skills, and career potentials?

2. Is "Career Development" a suitable and accurate title for his work history?
3. The "Broadcasting Skill Set" is the first of the three lists of transferable skills. Should it come first? Should "Sports PR" or "Business Skill Set" be moved forward if the portfolio is being used to apply for those kinds of positions?
4. Would the supporting written and video statements and the audio portfolio help you strongly consider this candidate for a broadcasting or sports business position?

CONTACTING EMPLOYERS

It's important to have a plan for contacting employers. Much time and money can be wasted on pursuing the wrong positions or pursuing them at the wrong time. You need to plan whom you contact, how you contact them, in what order, and when. You also need to follow up initial contacts. Pick suitable times to contact employers by phone or in person; for example, most managers don't want to get job inquiries at the end of their work week. Don't drop off a résumé at a restaurant during its lunch or dinner rush. Try to pick a time when you know that the recipient will be free to speak with you. If possible, call a receptionist to learn the best time to call or appear.

Campaign Strategies

The job-search campaign you wage will vary with your circumstances. If you search for a new job while employed, your search will be more selective and will need to be more discreet than if you are unemployed or at university.

Searching for a satisfying job involves a carefully planned campaign of contacting employers. The excerpt from a campaign log in Figure 23.10 illustrates the required planning and detailed record keeping. Such a log becomes especially important if your campaign involves many contacts over a prolonged period.

Employer	Contact	Response	Interview	Follow-up	Status	Comments
Trendmark Electronics, Kanata	Inquiry letter/ Mar. 17 (see file)	Mar. 28; phone call Jim Lund, R & D encouraged me to apply	Not yet	Mar. 29—sent thank-you letter; asked to tour Ottawa plant	Hopeful— J.L. said my Qualifications "looked fine"	Phone by Apr. 10 if no further news
Ames Research, Hull	Applied—letter Mar. 21 for Electronics Researcher	None so far	Not yet	Not yet	Highly desirable but doubtful; many experienced people will apply	Will send positive follow-up letter if no response by Apr. 4— prepare letter!

Figure 23.10 An Excerpt from a Campaign Log

| CASE | **Three-Part Campaign Strategy** |

Many employment advisors recommend tiered campaigns. For example, Tom Smith, a graduate of the Industrial Electronics program at Saskatoon's SIAST Kelsey Institute, used the following strategy to win a position that would give him job satisfaction after his May graduation.

Tier 1

Because of his military supervisory service, Tom wanted to start his electronics career as a quality-control supervisor or production-shift supervisor. He found four firms that had such positions—three in Saskatoon and one in Calgary. Using the library and the internet, he set out to learn all he could about the four companies, their business prospects, and their range of positions. He used his contact network, which included friends (and friends of friends) who worked at three of the companies, to try to learn about possible openings.

He devised a detailed campaign of introduction letters, follow-ups, and requests for informational interviews, to exhaust every possibility of being employed by one of the four companies. He placed a time limit of six weeks on Tier 1 activities before moving on to Tier 2.

Tier 2

Tom's secondary interest was in electronics research, so he identified several firms in Halifax, Moncton, Ottawa, Saskatoon, Calgary, Montreal, and Vancouver engaged in such research. He did some preliminary research on these firms but delayed further action until Tier 1 activities were completed.

Tier 3

In case the first three months of job searching at his Tier 1 and Tier 2 levels proved fruitless, Tom was prepared to go to his backup plan, Tier 3. Here, Tom would look for a position as a glider flight instructor (he developed these skills in the military and out of personal interest) or as an electronics research assistant.

Result

Tom did not have to go to Tier 2 or Tier 3 of his plan. In week 2 of his campaign, one of the Saskatoon firms advertised a quality-control position; also, early in week 3, Tom learned that the other three firms were sufficiently interested in Tom's qualifications to invite him for an exploratory meeting. The Calgary firm brought him back for a second meeting and then a formal interview for a production manager position.

He was also formally interviewed for the quality control position in Saskatoon. One week later, he was offered both positions for which he had been formally interviewed. He accepted the production manager job in Calgary, though the pay was lower, because that company would give Tom more opportunities to pursue a management career.

Perhaps your next campaign will not need to be as elaborate as Tom's. But his approach had one major advantage—he gave himself every chance to land the kind of position that would give him *maximum job satisfaction* at that stage.

Contact Methods

An inquiry letter (or in-person inquiry) offers a good method of getting noticed by an employer. Indeed, it's not uncommon for the authors of well-written inquiries to be invited for exploratory interviews; sometimes, an employer will offer a position to an impressive interviewee, even if the employer hadn't intended to hire just yet.

Still, the inquiry is primarily a research tool, not a method of selling your qualifications. (See the advice about inquiries earlier in this chapter.) So, if you already know that you would like to work for a particular employer, but it doesn't seem that this employer has openings, you may want to introduce yourself.

A letter of introduction is quite straightforward. You introduce your qualifications and your interest in the company. You name the position(s) that would appeal to you and refer the reader to the pertinent experience in your enclosed résumé. The action closing should indicate when you're available and how the reader can easily contact you. Perhaps you'll also request a meeting to discuss future employment opportunities.

An introduction letter does not apply for a specific available position, so you can't present a full sales pitch showing your suitability for a given position. But you can say enough about yourself and your career goals to engage the reader's interest. The introduction letter should be short, though; you have to know when to stop and let your résumé speak for you.

An application letter is more formal and persuasive in tone. You will write an application when you know that a position is available and when you're certain that you have what it takes to fulfill the position's requirements. Each application must be tailored to the specific position for which you're applying, so you have to think clearly about the letter's content and phrasing.

First, you need to remember that *unsolicited applications* require a different approach from *solicited applications*.

UNSOLICITED APPLICATIONS. If your contact network has informed you of an opening, and if that position has not been formally advertised, your main challenge is to deal with a reader who is not anticipating your application. It's possible that the reader doesn't even want applications; the reader may be using his or her own contacts to find candidates. If that's the case, you must get and hold the reader's interest by immediately appealing to the reader's needs and preferences.

Finding the right appeal is not easy. For starters, you'll have to research the position, to learn what type of work is done, which qualifications are essential, and, if possible, what characteristics the employer would like the new employee to have. If the employer has not yet drawn up a set of position requirements or employee characteristics, that employer may be receptive to a wider range of candidates than if the position had been advertised. The following is an unsolicited application scenario.

CASE	**Unsolicited Application**

Royale Ltée, a Montreal-based courier company, is expanding from metropolitan Montreal to all of southern Quebec and Southern Ontario. It has recently been purchased by a trucking consortium, Bouchard Transport, which has injected cash into the operation.

Margin notes:

A self-introduction can be an effective prospecting tool

Appeal to the reader of an unsolicited application letter

The writer of the letter in Figure 23.11 on page 503 is Marcel Dionne, a graduate of McGill University. His cousin, Michelle Legaré, who works as an accountant for Bouchard Transport, has given Marcel the background details and said that she's heard that Royale Ltée will need someone to help develop the logistics of expanding the courier operation. Marcel worked four summers for Canada Courier when he was at university, so he has a good idea of what Michelle means. His recent degree in software engineering has led to an eight-month contract to help redesign Quebec City's transit management software, but that contract will finish in three weeks. By acting quickly, he has found out what he can about Royale Ltée. Among other things, he has learned that Jean-Guy Ryan is the general manager and that Ryan and the other Royale Ltée managers have been successful running an informal, democratic style of operation that encourages employees to suggest ways of improving service and controlling operating costs.

Figure 23.11 shows how Marcel's unsolicited application letter tries to appeal to Jean-Guy Ryan's needs and priorities. Marcel's accompanying résumé appears in Figure 23.6 on page 489.

SOLICITED APPLICATIONS.   Solicited applications, such as the one illustrated in Figure 23.12 on page 504, require a similar structure to the unsolicited letter. However, the opening paragraph will be more straightforward—you simply apply for the position by name, you refer to the advertisement, and you preview your sales pitch.

Show that you understand the position's requirements

The application letter's *sales pitch* depends on your accurate analysis of the advertised position's requirements. The employer has carefully chosen those requirements and will expect your application to show how you can fulfill them. Therefore, if your reading of the position description leaves you with questions, write or phone the employer to get answers. Your main challenge is to prove that you have what the employer needs, so you *must* understand those needs.

When you construct your sales argument, look at your application from the employer's point of view. As in all business letters, answer the reader's question: *What's in it for me?* Two hints for answering that question are as follows:

Show that you can meet the position's requirements

1. Sell your qualifications and personal qualities as a *package*. The whole is greater than the sum of its parts, but only if you show how one set of skills complements and strengthens another set. For example, a former McDonald's manager who has earned a computer science degree will be better able to apply that computer knowledge because of previous business experience.
2. Highlight that package with a phrase or fact that identifies you and you only. Marketers refer to this technique as "positioning your product." You "position" yourself in a special place in the reader's mind. That positioning idea should reflect the strongest connection between your package and the employer's set of requirements.

Create your own brand and sell it to employers

The term *branding* can also be used to describe the special kind of image that each job seeker needs to cultivate and project. "Personal branding . . . is about understanding what is unique about you—your accomplishments, experience, attitude—and using that to differentiate yourself from other job hunters" (Levinson and Perry).

144 Avenue Champlain
Quebec, PQ   G7E 1R6
May 18, 2014

M. Jean-Guy Ryan, General Manager
Royale Ltée
2408, boulevard Laurentian
Montreal, PQ   G6H 5T5

Dear Monsieur Ryan:

Michelle Legaré, an accountant with Royale's parent company, Bouchard Transport, has told me about your expansion plans. Please consider my offer to help your courier firm develop software and logistics to allow a rapid, smooth transition from an urban courier to a regional operation.

My background in software engineering, my experience with Canada Courier, and my current work for the City of Quebec place me in a unique position to contribute to your firm. The enclosed résumé provides details of my software engineering degree; you'll notice that most of my class projects focused on developing solutions for logistical problems. That focus resulted from my experience at Canada Courier in Montreal and Hamilton; as a summer relief worker, I became familiar with all aspects of the delivery business. In my last stint with Canada Courier, I was asked to troubleshoot problems with its parcel tracking software. That experience has helped me manage several very difficult logistical issues in Quebec City's transit management software.

A growing company needs energetic, innovative people who wish to grow with the organization. The enclosed reference letters show that I match that description. Two of the references listed in my résumé, Professor Marks and Canada Courier's Bill Spender, will be able to comment on my problem-solving abilities and work ethic.

I'll be available for employment in three weeks, when my project contract in Quebec City finishes. However, my supervisor has given his approval for me to go to Montreal to discuss my application for a position with your firm, at any time you set. I really hope you do invite me for an interview because I believe that my skills and interests match Royale's upcoming needs. You can contact me at my email address (on the résumé) or at my cellphone, 418-451-3565.

Sincerely,

Marcel Dionne

Marcel Dionne

Enclosure: Résumé

Figure 23.11 An Unsolicited Application

108-3118D 33rd Street West
Saskatoon, SK   S7L 6K3
May 7, 2014

Mr. Dirk Benefeld, General Manager
Altitude Electronics
2310 Hanselman Avenue
Saskatoon, SK   S7L 6A4

Dear Mr. Benefeld:

Re: Quality Control Supervisor

Please consider my application for the quality control supervisor position that you advertised on page D18 of yesterday's *StarPhoenix.* This position requires what I have to offer.

As the enclosed résumé details, I'm about to graduate from Kelsey's Industrial Electronics program, which provides a comprehensive and practical overview of current electronics applications. The course also teaches us how to think and how to keep abreast of the rapidly changing electronics field. My marks have consistently been in the top three of the class in each of the six trimesters of the program. This reflects my work ethic and my attention to detail. It also reflects the solid theoretical grounding I received from two years of engineering studies at the University of Saskatchewan.

Instructor Al Schlatter, who's listed as a reference in my résumé, has agreed to tell you about my fabrication, assembly, troubleshooting, and CAD skills.

From my inquiries, I've learned that your quality control position requires strong leadership communication skills. The attached reference letters comment on my performance as president of the Kelsey Students' Association this year. As president, I've been able to hone skills that I've developed as a cadet leader and glider pilot instructor for the Department of National Defence.

I'd appreciate the opportunity to discuss the quality control supervisor position with you. If you agree that my qualifications match the position's requirements, you can contact me, Monday through Saturday, 7:30 a.m. to 8:00 p.m., by calling the Students' Association office, 306-652-0980. I can be available for an interview at any time you arrange. I'll be available for work after May 29, the final day of exams.

Yours sincerely,

Tom Smith

Tom Smith

Enclosures: Résumé
                Industrial Electronics Course Overview and Grades
                Reference letters (3)

Figure 23.12 A Solicited Application

Business guru Tom Peters has taken the branding idea a step further. He believes that workers need to convey their uniqueness and the "special value" that they each bring to their workplace. Peters introduces his Brand You™ concept at **www.tompeters.co.uk/pages/toolbox_brand.htm**. For a more complete discussion of this powerful idea, read his book *The Brand You 50*.

Three letters in this chapter illustrate successful positioning (or "branding"):

1. The inquiry letter in Figure 23.3 on page 480 positions the writer's enthusiasm for GPS surveying.
2. The unsolicited application in Figure 23.11 on page 503 presents three faces of the writer's special collection of relevant skills. (The opening sentence in paragraph 2 even uses the phrase "unique position.")
3. The solicited application in Figure 23.12 on page 504 positions the applicant's unusual and special combination of electronics training, leadership skills, and communication skills.

CAUTION　　*Use a positioning statement like the one in Figure 23.12 only if you can prove that you know exactly what the position requires.*

In phrasing your sales pitch, project a *you* attitude by

- discussing the position's requirements *before* matching your qualifications to those requirements
- using objective, third-person phrasing wherever possible (e.g., "the enclosed résumé lists that related experience") or "you" phrasing (e.g., "you can reach me at . . ." rather than "I can be reached at . . .")

Close the sale!

Application letters require a *positive, businesslike closing*. Say when you're available for work or for an interview, and indicate the easiest way for the reader to reach you. Request an interview or express your desire for an interview, because gaining an interview invitation is the reason you wrote this letter! Finally, you might close by stating your desire for the position or by reiterating your key argument for considering your application.

Figure 23.12 shows Tom Smith's response to an opening for a quality-control supervisor. Figure 23.13 is the advertisement to which Tom replied.

ON THE JOB...

Hiring Criteria

"Our website job descriptions signal what we seek in a candidate, whose combination of academic background and experience must match a given position. Then, a candidate must be able to accept new challenges, to be flexible, to work well within a team, and to have a professional attitude. People in leadership positions have to manage others and to communicate very well. Entry-level positions require people who have a positive attitude, who are willing to work hard and learn, and who are able to work with minimal supervision. And all of our staff must be able to write well because of the report writing component of the job—some of our people spend up to 80% of their time writing and editing reports and proposals. Our people are also involved in generating business through direct contact with clients, so their oral communication skills are important, too. . . ."

—Dr. Brian Guy, vice-president and general manager, Summit Environmental Consultants

Quality Control Supervisor

The company: Altitude Electronics specializes in electronic instrumentation and controls for the aerospace industry—satellites, launch rockets, and high-altitude weather balloons. Its award-winning products result from superior design and meticulous attention to detail in the various stages of the manufacturing process.

The position: You will be one of two quality control supervisors. You will test products, report deficiencies, supervise production-worker methods, and work in conjunction with the production manager to train production line assemblers.

The successful candidate will have . . . : A two-year electronics diploma from an accredited Canadian college, or equivalent. Demonstrated leadership skills and above-average communication skills (interpersonal, oral, written, listening). Familiarity and some skill in fabrication, component assembly, soldering, and product finishing. Demonstrated skills in electronics troubleshooting and CAD/CAM.

Apply to Dirk Benefeld, General Manager, Altitude Electronics
2310 Hanselman Avenue
Saskatoon, SK S7L 6A4

Figure 23.13 Job Advertisement

Tom's first thought was, "It's as if they wrote this ad for me!" Then he analyzed the advertisement as follows:

- The position requires a combination of technical skills and communication skills. That's exactly what I have to offer, so I should be able to make a good case.
- The ad expresses pride in the company's accomplishments ("award-winning," "superior design"), so I should emphasize my own drive to succeed.
- Mr. Benefeld is on the Industrial Electronics industry advisory board, so I shouldn't have to sell the program to him, just remind him of how practical and comprehensive it is.
- Probably, my competitive advantage is my background of leadership positions and communication experience, so that should be stressed in the résumé as well as in the letter. And I'll choose my references and reference letters to emphasize the leadership communication.
- The ad mentions "meticulous attention to detail," so I'd better show that side of my character.

ONLINE APPLICATIONS. Each year, more and more job seekers use sites, such as *LinkedIn* and *Indeed* (**http://ca.indeed.com/jobs?q=Apply+Canada**), to find open positions and even to apply online for positions. For example, on December 10, 2012, Indeed listed 16 673 jobs available in Canada.

Still, job seekers and employers have difficulty finding specifically what they want to see on a site like LinkedIn. In response, software developers have developed search engines just for LinkedIn. For example, RecruitIn allows an employer to go to

LinkedIn with search parameters that will help the employer identify matches from among people who have posted profiles on LinkedIn.

A Montreal-based company called *interJob* has identified a slightly different niche to serve in the online job market. According to its website (**http://www.interjob.ca**), "Only 20 percent of all job offers are posted in newspapers and on the Internet," largely because advertising is expensive for companies and because most medium-sized and smaller companies don't have the resources to continually update job postings on their own websites. Social networks bridge the gap to some extent but according to interJob, social networks are "effective only for those with a solid background of professional experience or for those with a well-established contact network. Much time and energy must be spent on them."

So, in order to bring together employers and potential employees, interJob offers a fee-for-service model that posts job seekers' résumés and lists positions as they become available. Job seekers who post on interJob have an advantage, says interJob, because employers believe that if a candidate is paying to send their résumé (even if the cost is purely symbolic), they're showing more seriousness and interest in the job search process."

EMPLOYMENT INTERVIEWS

How interviews work

A successful application will lead to an interview, sometimes to a series of interviews. Perhaps just one person will interview you, but more often you can expect to be quizzed by a panel. At a minimum, expect to be interviewed by your potential immediate supervisor and by someone from the company's human resources division. Smaller firms might be represented by the firm's manager, a project supervisor, and a member of the work team that has the open position.

Candidates may be subjected to psychological testing or role-played job simulations

Companies realize that choosing the wrong candidate can be very expensive. Therefore, many firms go beyond the standard interview. They might start by interviewing several candidates to narrow the field to two or three, who will then be invited to return for more in-depth interviews. (An extreme form of the elimination interview process comes in the form of "speed interviewing," in which candidates at job fairs are questioned for 10 to 15 minutes by each firm attending the job fair.) In addition to conversational interviews, many firms now use a battery of psychological tests to decide whether a candidate suits the job and will fit into the work group and company culture. Employers may also place candidates in simulated or actual work situations to gauge performance under pressure; sometimes, such simulations will last a whole day (E. White).

Psychologist Dr. Jordan Peterson has lectured on psychological testing for the Rotman School of Management at the University of Toronto. He believes that "traditional job interviews are proving to be poor predictors of success . . . because the [interview] questions are not standardized." Instead, he says, well-designed testing identifies the most capable candidates for a particular job and weeds out dishonest or insincere candidates (Immen, "Prospective Hires" C1). Richard Wajs, the president of an executive search firm, is another advocate for interview testing. He points out that "psychological tests can also identify issues that need to be followed up in interviews" (qtd. in Immen, "Prospective Hires" C2). Candidates may be subjected to psychological testing or role-played job simulations.

Companies, such as Weyerhaeuser, take up to two full days to evaluate a candidate, through various interviews, tests, job simulations, and plant tours. This thorough approach helps an employer learn much about a candidate's character and emotional intelligence as well as the person's specific skills and verbal intelligence.

Interview Preparation

Prepare thoroughly

Whether you engage in a single traditional interview or in a gruelling round of tests and interviews, careful preparation is the key to a productive experience. Prepare by learning about the company in trade journals and industrial indexes. (Learn about the company's products or services, its history, its prospects, its branch locations.) Go the company website, request company literature, such as its annual report, and speak with people who know about the company.

If you're applying to a government or municipal agency, try to learn what will affect the growth or downsizing of that agency, and try to learn if there's room for personal growth or career advancement within the agency. Also, learn about the agency's range of services. If you can't find information of this sort before the interview, ask questions during the interview.

Fascinating "insider information" may be found at websites that present the results of surveys and personal interviews with employees (**www.vault.com**) or with recruiters and human resources managers (**http://career-advice.monster.com**). Some of this information is free, but sites like Vault.com charge a subscription fee.

Prepare also by thinking about your reasons for wanting the job and what you have to offer the employer. Your self-inventory and your list of transferable skills will help you prepare. Also, look at your résumé from the interviewer's point of view—think of the questions you would ask if you were the interviewer. If you have difficulty thinking of such questions, have friends go through your résumé to look for issues that will elicit questions.

As you prepare, you will find the following "game plan" helpful:

1. List your strengths in order of priority.
2. Think of the subtle strengths you need to bring out.
3. Consider the apparent limitations you should be ready to counteract.
4. Think of subtle limitations you need not introduce.
5. Devise an overall strategy for projecting your desired image.
6. List things you want to learn about the job and the company.

An Interview Timeline

Standard interviews follow a fairly predictable timeline:

◆ *In the first 10 to 20 seconds*, the panel reacts to the candidate's non-verbal behaviour—his or her appearance, way of moving, degree of eye contact, vocal tones, type of handshake, and so on. According to communications trainer Patti Wood, "most hiring decisions are made within the first 10 seconds of a meeting, before you even sit and talk" (Nazareno). That assertion may be debatable, but first impressions can be very important.

◆ *In the next 5 to 10 minutes*, the meeting's chair establishes the pattern for the interview and tries to put the candidate at ease, by making small talk or by asking "information questions" that encourage the candidate to perform naturally.

- ◆ *During the bulk of the interview*, the candidate should expect to discuss a number of challenging topics and, if the panel has adequately prepared its strategy, to face some of the same questions in different forms—the successful candidate will consistently and honestly answer such questions, not change his or her story according to what seems expedient at the moment.
- ◆ *At the end of most interviews*, the candidate gets a chance to ask questions. Also, the lead interviewer might introduce such business aspects as salary and benefits, union affiliations, selection timelines, and the procedure for informing candidates of the decision.

Questions

It's possible to anticipate many of the questions you'll be asked, such as

- ◆ Why do you wish to work here?
- ◆ What do you know about our company?
- ◆ What do see as your biggest weakness? Biggest strength?
- ◆ What kind of supervisor do you prefer?
- ◆ Do you prefer to work under close supervision in a structured environment, or more independently within broadly stated guidelines?
- ◆ What are your short-term and long-time career goals?

You can anticipate three main categories of questions: information questions, high-risk questions, and opportunity questions.

Categories of interview questions

INFORMATION QUESTIONS. These easy questions come early in the interview and enable you to provide factual background information about your education and experience. The interviewer will assess how you present yourself, what you know about the firm, and how you "fit" the culture of the organization. (*Why did you apply for this job? Tell me about your program at university.*)

HIGH-RISK QUESTIONS. These questions can destroy your chances for a job if handled poorly. They probe what the interviewer believes to be your potential weaknesses related to the job requirements. If you can successfully deflect these questions in the middle of the interview, then you will have an opportunity to sell yourself in the next stage. (*Why did you leave your previous job? What difficulties have you faced in previous jobs? What are your limitations? What kinds of things would bother you sufficiently to leave our organization? Why were you laid off?*)

OPPORTUNITY QUESTIONS. These questions usually appear at the end of the interview and provide the opportunity for you to convince the interviewer that you can make a contribution to the organization. Your answers should be bold (not arrogant) and confident. (*What are your strengths? How can you contribute to our product or service? How do you see yourself fitting into our organization? Tell us about yourself.*)

Here are additional potential interview questions.

- ◆ What has been your greatest work achievement in the past year?
- ◆ What made you choose the career you're preparing for?
- ◆ What made you choose the college or university you're attending? Did you choose well?

- ◆ What studies did you like most in university? Least?
- ◆ How do you feel about relocating?
- ◆ In your opinion, do your grades accurately indicate what you've achieved academically?
- ◆ How would you solve the following problem?
- ◆ If you could go back five years, what would you do differently?
- ◆ How do you feel about overtime work? Travelling on the job?
- ◆ Describe your least favourite boss or teacher.
- ◆ What kind of references would you receive from former employers?
- ◆ What do you feel this company can do for you?
- ◆ Describe yourself as a fellow student or employee would.
- ◆ What current world event concerns you?
- ◆ How would you move Mount Robson? If you were ice cream, what flavour would you be? If you were the leader of this organization for a week, what major initiative would you try to push forward?

The last three groups of questions might seem bizarre, but they fit into a category of question that attempts to catch the candidate off guard and to see how he or she responds to being startled.

YOUR QUESTIONS.   Also, you will have a chance to ask your own questions. You should have a list of written questions with you, in an unobtrusive notebook. Refer to this notebook when you're asked if you have questions. Focus on the work you would be doing and the conditions of employment—ratio of office work and fieldwork, travel involved, level of responsibility, opportunities for further training and skill development. If your questions have already been answered in the interview's give-and-take, mention them anyway, to show that your primary interest is in the work.

If at all possible, do not ask about salary and benefits. After you receive an offer of employment, you can then discuss employment benefits. Or, if you must know, call the company's human resources office. Usually, however, these issues will not pose a problem because a company representative will review salary and benefits in a closing stage of the interview.

Good answers present appropriate levels of detail

Answers

Answer questions directly and fully. Some closed-ended questions like, "Are you familiar with Microsoft *Office Pro*?" probably merit only a short response of one to three sentences. However, the majority of interview questions are open-ended, as in, "What are your main strengths?" Such questions need to be answered fully enough to satisfy the interviewer's interest, though not so fully that you totally exhaust any interest. If you don't provide enough detail to support your responses, the interviewer will doubt your assertions. So give examples of your strengths and skills and achievements; back up your statements with reasons and specific details; recount incidents that establish that you've used certain techniques or completed certain tasks in the past.

Interviews can place you under pressure, so two kinds of preparation are essential:

1. You need to anticipate questions and rehearse ways to answer them.
2. You should practise using impromptu speaking techniques. Chapter 24 describes impromptu organizing methods that will allow you to control nervousness and to perform under pressure.

Here are some other hints for successful job interviews:

Job interview hints

- If possible, research the range of candidates who might apply for the same position. Know your competition, and how you compare.
- Dress appropriately for the interview, as if this occasion were your most important day at work. One approach is to research the firm's dress policy, and dress one notch above that standard.
- Arrive early for the interview. Occasionally, the interview team will give you a task to perform or a case study to analyze before you're called into the interview room.
- Remember that you'll be the "guest" and the lead interviewer will be the "host," so don't be too socially aggressive—wait for the host to initiate a handshake, wait to be asked to take a chair, and allow the interviewers to guide the conversation.
- Maintain eye contact much of the time, but don't stare.
- Maintain a relaxed but alert posture. Don't slouch. Don't fidget. Show your interest by leaning forward slightly. Smile when appropriate. Be yourself, your best self.
- Answer questions truthfully; skilled interviewers will ask the same controversial question or probe the same issue in different ways to see if your answers are consistent.
- When you don't know what to answer to a question, say so. In some cases, you could explain what process you might use to determine the answer to that question.
- Don't blurt things you don't need to mention. (See quadrant 4 of the self-inventory on page 470.)
- Don't be afraid to allow silence. An interviewer may deliberately stop talking, to observe your reaction and to induce you into revealing your inner thoughts.
- Remember to smile, and show other non-verbal responsiveness to your interviewers. You'll improve your chances of getting the job—no one likes to work with a grouch!
- When the interviewer says the interview is ending, or hints that it's drawing to a close (closes the question folder or puts away his or her pen), don't wear out your welcome. Restate your interest, ask when you might expect further word, thank the interviewer, and leave.
- In general, treat the interview as potentially one of the most important conversations in your life!

Follow-ups to Interviews

Within a few days after the interview, refresh the employer's memory with a letter or email restating your interests. "A thoughtful follow-up carries as much clout as a cover letter, many experts say" (Lublin, "Thoughtful"). Here's an example:

Thank you for the opportunity to discuss your technologist position on Wednesday.
Learning about your planned plant expansion has strengthened my desire to work for Kraftsteel Industries. Also, our conversation has confirmed my belief that I could contribute to your company's continuing growth, through my CAD skills and design

capabilities. I'm especially interested in working with your design projects that use new lightweight steel materials.

   If you require further information, please call me on my cellphone at 519-767-9901.

Opinions vary about the format to use in follow-up notes. While many experts prefer email, some such as Lori Black of Warren Recruiting and Mark Hordes of the Sinclair group prefer handwritten letters (Sixel).

RESPONDING TO JOB OFFERS

If all goes well, your strategy will result in an offer for a position in your first tier of choices. If so, you'll likely have no difficulty accepting with enthusiasm. If you receive an offer by phone, ask when you'll receive a written offer. Respond to that letter with a formal *letter of acceptance*. Your response may serve as part of your contract, so spell out the terms you are accepting. Here's an acceptance letter, written by the same person as the above follow-up letter.

   I am delighted to accept your offer of the mechanical engineering technologist position at your Hampstead plant, with a starting salary of $32 400. I understand that there will be a six-month probationary period and that I will be eligible for raises after one year.

   As you requested, I will phone Pat Larsen in your Human Resources office for instructions on a reporting date, physical exam, and employee orientation.

   I look forward to a satisfying career with Kraftsteel Industries.

You may have to refuse offers, perhaps because you learned things during the job interview that gave you reason to believe that the position was not for you, or perhaps because you have accepted another position that better meets your current job objectives. So, even if you refuse by telephone, write a prompt, cordial letter of refusal, explaining your reasons and allowing for future possibilities. The writer of the above acceptance letter phrased a refusal this way:

   Although I was impressed with your new hydro-forming technology and the efficiency of your auto parts plant, I am unable to accept your offer of a position as an assistant shift foreman at your plant.

   I have accepted a position with Kraftsteel Industries because Kraftsteel has offered me the chance to join its design team. CAD design, as I mentioned in our recent interview, is one of my main interests.

   In the future, if opportunities arise with your hydro-forming design group, I would appreciate the chance to be considered for a position with that group.

   Thank you for your interest in me and for your courtesy.

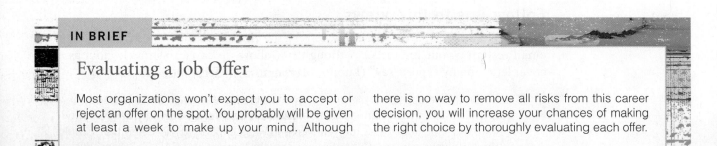

IN BRIEF

Evaluating a Job Offer

Most organizations won't expect you to accept or reject an offer on the spot. You probably will be given at least a week to make up your mind. Although there is no way to remove all risks from this career decision, you will increase your chances of making the right choice by thoroughly evaluating each offer.

The Organization

Background information on the organization can help you decide whether it is a good place for you to work.

Is the organization's business or activity in keeping with your own interests and beliefs? It will be easier to apply yourself to the work if you are enthusiastic about what the organization does.

How will the size of the organization affect you? Large firms generally offer a greater variety of training programs and career paths, more managerial levels for advancement, and better employee benefits than do small firms. Large employers also have more advanced technologies in their laboratories, offices, and factories. However, jobs in large firms tend to be highly specialized, with relatively narrow responsibilities. By contrast, jobs in small firms may offer broader authority and responsibility, a closer working relationship with top management, and a chance to clearly see your contribution to the success of the organization.

Should you work for a fledgling organization or one that is well established? New businesses have a high failure rate, but for many people, the excitement of helping create a company and the potential for sharing in its success more than offset the risk of job loss.

Is the organization in an industry with favourable long-term prospects? The most successful firms tend to be in rapidly growing industries.

Where is the job located? You need to consider the cost of living, the availability of housing and transportation, and the quality of educational and recreational facilities in a new location. Consider the time and expense of commuting and whether it can be done by public transportation.

Where are the firm's headquarters and branches located? Although a move may not be required now, future opportunities could depend on your willingness to move to these places.

The Nature of the Work

Even if everything else about the job is good, you will be unhappy if you dislike the daily work. Determining in advance whether you will like the work may be difficult. However, the more you find out about it before accepting or rejecting the job offer, the more likely you are to make the right choice. Ask questions like the following.

Does the work match your interests and make good use of your skills? The duties and responsibilities of the job should be explained in enough detail to answer this question.

How important is the job in this company? An explanation of where you fit in the organization and how you are supposed to contribute to its overall objectives should give an idea of the job's importance.

Are you comfortable with the supervisor? Do employees seem friendly and cooperative? Does the work require travel? Does the job call for irregular hours? How long do most people who enter this job stay with the company? High turnover can mean dissatisfaction with the nature of the work or something else about the job.

The Opportunities

A good job offers you opportunities to grow and move up, learn new skills, increase your earnings, and rise to positions of greater authority, responsibility, and prestige. The company should have a training plan for you. You know what your abilities are now. What valuable new skills does the company plan to teach you?

The Salary and Benefits

Wait for the employer to introduce these subjects. To know if an offer is reasonable, you need to estimate what the job should pay. Several sources could help provide this information: friends who recently were hired in similar jobs; teachers and the staff in the college placement office; the internet, the library, or your school's career centre. Allow for cost-of-living differences between a large city and a smaller city, town, or rural area.

Consider benefits as well as salary. Benefits can add a lot to your base pay, especially extended health insurance and pension plans. Other benefits include life insurance, paid vacations and holidays, and sick leave. Benefits vary widely among smaller and larger firms, among full-time and part-time workers, and between the public and private sectors. Learn what the benefit package includes and how much of the costs you must bear.

Source: Adapted excerpts from U.S. Department of Labor.

SPECIAL CONSIDERATIONS FOR CONTRACTORS

Hire yourself!

If you're unable to land a desired position in the part of Canada where you want to live, you may have to hire yourself! Or perhaps you have an entrepreneurial spirit and you wish to run your own operation. If so, you might provide healthcare services; you might offer your engineering or project-management skills and knowledge in short-term contracts; you might provide consulting services; you might offer testing or maintenance or quality-control services to firms that aren't large enough to hire full-time staff; you might provide seasonal recreation management. Every field has opportunities for contract work.

Advantages for employers

After all, contract "employees" offer real advantages for employers who get on-demand workers who aren't eligible for employee benefits or severance pay. Employers get qualified professionals who are motivated to do a really good job because they want repeat business and referrals to other employers. Moreover, the level of job performance can be specified in the contract.

Advantages for contractors

As a contractor, you also benefit: you get a variety of work with different clients; you gain tax benefits from being able to deduct your expenses; you are your own boss, and you decide how much work to take on; and, although you may have to work long, hard hours, you have some control of which days and hours you work. On the negative side, you must deal with a varying income flow, and you must spend time to generate business.

Business networks provide valuable leads

Much of this chapter's advice aimed at those seeking salaried positions applies to an entrepreneur's search for work. In particular, contractors need to maintain networks of personal and business contacts. The following groups can generate useful leads: membership in professional associations; placement on contractors' lists; memberships in service clubs, the chamber of commerce, or informal breakfast clubs. Once contacts have been established, they must be maintained through "pinging."

How to "ping"

Jim Gray defines *pinging* as "the targeted, upbeat contact by telephone, email, and handwritten note with clients, prospects, colleagues, friends, and supporters." He adds that "those who ping effectively know that the world of work is, at its core, about relationships" and that successful relationships of any kind depend on mutual benefits.

> **NOTE** *In certain economic climates, self-employment statistics can dramatically increase. In April 2009, for example, 37 000 Canadians turned to self-employment, thus accounting for over half of the 61 800 self-employed positions created from April 2008 to April 2009 (Scoffield and Galt B1).*

Gray suggests that one's pinging list should be restricted to people who can provide good business leads, and that one should contact those people about once a quarter. He also recommends brief, positive pinging messages that clearly indicate one's reason for the contact.

Blogs can help you obtain work

Blogs can be used to position a contractor as an expert worker in his or her field. In "Musings That Matter: Blog to a Job," reporter Mary Gooderham cites several examples of people who have leveraged their blog for job-hunting purposes (C1). The article also advises what to include in one's blog (C5).

for Revising and Editing Job-Search Correspondence

Use this checklist to refine the content, arrangement, and style of your letters and résumés.

Content

- Is the letter addressed to a specifically named person?
- Does the letter contain all of the standard parts?
- Does the letter have all needed specialized parts?
- Have you given the reader all necessary information?
- Have you identified the name and position of your reader?

Arrangement

- Does the introduction immediately engage the reader and lead naturally to the central section?
- Are transitions between letter parts clear and logical?
- Does the conclusion encourage the reader to act?
- Is the format correct?
- Is the design acceptable?

Style

- Is the letter in conversational language (free of "letterese")?
- Does the letter reflect a "you" perspective throughout?

- Does the tone reflect your relationship with your reader?
- Is the reader likely to react favourably to this letter?
- Is the style throughout clear, concise, and fluent?
- Is the letter grammatical?
- Does the letter's appearance enhance your image?

Résumés

- Have you adapted the content, structure, and format of your résumé to suit the specific position you're pursuing?
- Have you made sure that your résumé is absolutely free of errors in content, spelling, and grammar?
- Have you used section headings that accurately identify your package of employable characteristics?
- Have you made it easy for your reader to find key pieces of information?
- Does your résumé's appearance enhance your image?

EXERCISES

1. Complete a self-inventory and a list of transferable skills. Use these self-assessment tools to
   - identify the employable characteristics that form the core package you'll "sell" for any position you may pursue
   - identify which skills, qualifications, and personal qualities you will emphasize for particular positions
   - prepare for questions job interviewers will ask you
2. Develop a strategy for your next job search.
   List the types of positions you should pursue.
   - Determine which positions are currently available and which may become available soon.
   - Decide which position(s) to pursue in the first tier of your campaign and the methods you'll use to contact employers.
3. Design and write two or more résumés that you can adapt for the range of positions you expect will interest you. Adapt those résumés for placement in an online résumé bank.
4. Analyze a job posting that describes a position that interests you. Which qualifications does the employer's

advertisement stress? How? Do you have the qualifications and personal attributes required by the position? If you have relevant skills that have not been developed (and demonstrated) through directly relevant work experience, how will you show that you can indeed perform the tasks required in the position?
5. Write a letter applying for an advertised position. Show how your combined qualifications, skills, and personal qualities match the employer's stated needs.
6. Write a letter applying for a non-advertised position. Show that you understand the position's requirements and that you can meet them. Place your primary appeal in the opening paragraph and elaborate in subsequent paragraphs.
7. Prepare a strategy sheet for a real or imagined job interview. As you do so, plan answers for questions likely to be asked.
8. Plan two versions of your own employment portfolio, one print version and one online version.
9. Establish a presence on LinkedIn. Start by creating an account at **www.linkedin.com**. Then, create what is essentially an online portfolio by following LinkedIn's prompts.

COLLABORATIVE PROJECTS

1. In conjunction with other members of your class, prepare an inventory of positions that may be available to graduates of your program of study. Alternatively, prepare an inventory of summer part-time jobs. Among other information, provide the following:
    - descriptions of the positions, their related duties, and required qualifications
    - expected dates of availability and whether the positions will be advertised externally
    - background information about the companies offering the positions
    - salary ranges and benefit packages
    - names of contact persons, their telephone or fax numbers, and mailing or email addresses

2. Form groups of four to six persons who will practise interviewing skills.

- In advance of the practice interviews, have each person distribute a job description, an application for that position, and a résumé to each of the other group members.
- For each person to be interviewed, assign two interviewers who will review the applicant's letter and résumé while preparing questions to be asked during the interview.
- Each interview will involve one interviewee, two interviewers, and one or more observers who will take notes and report their observations during the post-interview debriefing.
- Each interview should last 15 to 30 minutes.
- Immediately following each interview, the observers will lead a discussion of how well the interviewee handled the questions and to what extent the interviewee's non-verbal behaviour contributed to his or her desired image.

MyWritingLab

MyWritingLab: Technical Communication offers the best multimedia resources for technical communication in one, easy-to-use place. You can find a variety of interactive model documents and case studies. There are also extensive guidelines, tutorials, and exercises for Document Design, Writing and the Writing Process, and Research, and a large bank of diagnostics and practice for grammar review.

Oral Presentations

LEARNING OBJECTIVES

After reading this chapter, you should be able to

- recognize the value of being able to speak well at work
- develop and polish a public speaking style that is rooted in your personality and in the best aspects of your private speaking style
- plan and structure your talks but not over-rehearse or read your talks
- pay attention to non-verbal aspects of delivering oral presentations

---

Think about the kinds of situations where your speaking skills will lead to success at work. Include informal situations (your firm's general manager stops you in the hallway to ask how a project is progressing) as well as formal situations (you make a half-hour sales presentation to an important client).

Think about opportunities you'll have to

- persuade people to consider, perhaps even accept, your ideas
- provide listeners with the information they need
- foster the image you want and need to project (knowledgeable? skillful? dynamic? competent? supportive?)
- help listeners solve problems, find answers, or resolve issues
- maintain or improve working relationships with your colleagues
- show your clients that you value their business
- explain how things work
- instruct others how to perform tasks
- maintain or build your status within the organization

ON THE JOB...

Leaders Need to Speak Well

"Anybody in any level of leadership must be able to influence, and influence comes from the ability to use oral communication to build relationships. And so our company does a lot of speaker training."

— **Craig Gorsline, President and Chief Operating Officer, ThoughtWorks, a custom software developer**

Now, take your reflections a step further. Think about your reactions to a speaker you've recently heard in a classroom or meeting room. What judgments did you make about that speaker's competence based on the way the speaker spoke? To what degree did the speaker's presentation style affect your willingness to believe the speaker's information and agree with the speaker's ideas? Did you "buy" what she or he had to say? As a result of these sets of reflections, are you now more aware of how effective presentation skills could advance or hinder your career?

Speaking on the Job

"When I was a civil engineering co-op student, I conducted meetings and acted as a liaison between local municipal council members, contractors, and property owners, especially when we were doing projects for the Village of Alert Bay. I found that it was important to prepare for both the formal and informal meetings. I learned that you have to be persistent in getting decisions and information from those responsible. Most of these people were busy, so I had to impress on them that their answers to my questions were vital to the success of the project. Many times, I had to remind them that timely decisions had a direct bearing on project cost. 'We can change the direction of this line now at a cost of $100, or we can change it later at a cost of $2000!'

In my current role as project manager, when I am talking to our clients, I try to be concise about I'm asking for. It's important to make sure that you know what is going on with the project, but don't be afraid to admit if you don't have all the details. People respect you more for honesty than they do for knowing it all. And if you don't have an answer to their questions, get back to them with the answer as soon as possible. . . ."

—**Mark Hall, Project Manager, Urban Systems**

SPEAKING SITUATIONS FOR TECHNICAL PEOPLE

Many engineers, technologists, technicians, scientists, and other "technical workers" prefer to design, test, build, or improve things rather than talk or write about the details of their work. But often these same people can succeed only by communicating their information, instructions, explanations, and analyses. Also, they must communicate effectively to get the facilities, staff, or equipment they need.

Here is a random selection of situations where a technical person's speaking skills could be very important:

- answering key questions in a job interview
- presenting an informal proposal at a department meeting
- introducing a sewer-line plan to residents who don't want the line extended into their rural subdivision
- announcing a layoff to affected employees
- selling a formal proposal to a client
- presenting a technical paper or a successful project at a convention
- explaining a medical condition and treatment plan to a patient

Your degree of success in such situations depends upon your speaking skills and the preparation you put into each presentation.

FACTORS OF SUCCESSFUL SPEAKING

Ask almost anyone about the number-one factor of speaking successfully and that person will reply, "confidence." Or, possibly, the response might reflect a negative view: "You have to overcome fear and anxiety to speak well."

Fear, anxiety, nervousness—these feelings are understandable. After all, others judge us by the way we speak, and our work effectiveness frequently depends on our ability to explain and persuade. We need to focus on factors that make us successful.

You already know the secret to effective speaking—confidence! However, if you currently don't have much confidence, you might find that statement rather hollow. If that's the case, the following hints will probably interest you.

1. *Have something to say.* Usually, you'll be asked to speak if your experience or role means that you ought to have information and insights that the audience will be glad to hear. But even if you're not an expert on a subject, you can research that subject so that you have plenty to offer. The net result is that your confidence level rises because you know that you have a worthwhile message.

How to build confidence

2. *Know your purpose.* Being clear about the purpose of your presentation helps you focus on the key ideas and information to stress. When you have that focus, when you know exactly what you're trying to accomplish, and when you know how to achieve your purpose, you'll feel a surge of confidence.

Here are some common purposes for talks: to inform, explain, persuade, entertain, teach, involve emotions, or inspire. Usually, you will combine two or more purposes.

3. *Know your audience.* Knowing your listeners' needs and interests should help to further boost your confidence because you can shape your talk to meet your audience's expectations and priorities. Some presenters also prefer to speak to people whom they know because they feel supported by a familiar group of faces. However, the main advantage of a known audience is that you can anticipate audience reactions and you can plan accordingly.

4. *Know the speaking environment and situation.* You can reduce pre-talk jitters and the chance of being flustered during a presentation if you know what to expect. Will the audience have heard other speakers before you get up? Will you be introduced as the highlight speaker? If you're making a sales presentation or oral proposal, how many people will be at the meeting, and who will be the prime decision makers? Also, will your competitors have already presented their sales pitches, or will they present later?

You'll also find it useful to *scout the room in advance.* Learn the location of the electrical outlets, the lectern, the viewing screen, and the light switches. Plan your placement of DVD players, laptop and data projection units, or overhead projectors. Check whether sightlines are obscured by pillars or by your audio-visual (AV) placement. Determine where you can move around. Check that the seating arrangements suit your planned presentation. Above all, get a "feel" for the room. You'll feel more confident.

5. *Know how to perform.* This last factor is the most difficult to manage because it takes practice to develop one's performing style. However, here's a good starting point: *Be yourself!* Whatever style you develop, it should grow out of your normal conversational patterns. Depending on your personality and on the style required for the situation, you will eventually be able to function at any of the following four levels.

Four levels of speaking

a. *Conversational level.* At its best, the conversational style is intimate, relaxed, and natural, but not particularly forceful. It may include bursts of liveliness but generally stays rather low-key, with just enough volume and projection to allow the audience to hear.

b. *Heightened conversational level.* Speaking at an augmented conversational level may be the best level for beginning speakers. At this level, the speaker pumps up the volume. He or she increases the energy level and emphasizes keywords more forcefully. The speaker is somewhat animated and involved with the audience.

c. *Performance level.* Now, the speaker is more dramatic. He or she uses vocal emphasis, pacing, pauses, varying volume, and varying vocal tones to really engage listener interest and to signal the full meaning. Both speaker and listener are aware that this is a performance, but the speaker retains a natural spontaneity.

d. *Oratorical level.* This is suitable for large audiences only. In every way possible, the speaker uses dramatic techniques and theatrical gestures to stimulate audience interest and to amplify the speaker's message. This is definitely not a normal way of speaking, but the oratorical style works for political rallies, exhortations by sports team coaches, and motivational speakers.

So where do you fit on the above scale? Perhaps you can operate on any one of the levels, depending on the situation.

FOUR PRESENTATION STYLES

There are four main ways to deliver oral presentations:

1. You can read a prepared script.
2. You can memorize the whole talk.
3. You can speak impromptu, with little or no prior preparation.
4. You can speak extemporaneously, with a rehearsed set of notecards.

Only highly trained professionals are proficient at *reading* presentations. Untrained readers lose eye contact with their audience, so the audience feels excluded. Also, most people do not read well—their delivery sounds artificial and the audience quickly becomes bored. Further, a scripted presentation allows little chance for revision in mid-delivery.

For several reasons, then, you should not read a speech, not even a complex technical presentation. Your audience for such a presentation would prefer a handout along with your comments about how to read and interpret that document. Still, your job might require you to read a prepared statement. If so, use a double-spaced, large-type script with extra space between paragraphs. Rehearse until you are able to glance up from the script periodically without losing your place.

Like scripted deliveries, most *memorized* speeches don't sound natural. They sound, well, "memorized." Also, you cannot change the content or tone of the speech, even if audience reactions make it obvious that you need to take a different approach. And, of course, the pressure of giving a speech easily plays havoc with your memory. You only have to forget one phrase for whole sections of the memorized talk to temporarily disappear from your memory.

The *impromptu* style brings what the previous two styles lack—spontaneity and real contact with the audience. However, "speaking off the cuff" has its own dangers, most notably the lack of a clear line of development. Impromptu speakers can ramble and thus make very little sense to their listeners. Also, the impromptu approach can put so much pressure on you that you can't retrieve from memory what you know and believe. Further, you run the risk of phrasing things rather awkwardly and ineffectively.

The *extemporaneous* style makes the most sense for nearly all presentations because it provides the benefits of careful preparation (research, structured outline, strong introduction and conclusion, planned transitions, and rehearsed AV aids), as well as the direct contact and spontaneity that this method allows.

DELIVERING EXTEMPORANEOUS TALKS

The extemporaneous process

Reduce nervousness by focusing on your message and your audience's reactions

Briefly, here's how the extemporaneous technique works. The speaker

1. chooses appropriate material and organizes that material in an outline
2. carefully prepares an introduction and conclusion
3. transfers the whole talk to a series of brief notes on notecards
4. rehearses with the notes and with the planned AV materials
5. refers to notes while delivering the talk

The rehearsals are critical to success. In the rehearsals, the speaker can

- find the most effective ways of expressing points
- develop clear bridges between various parts of the talk
- learn the best way to integrate AV aids and supporting examples or stories
- determine how long the planned talk takes to deliver and modify the content accordingly
- modify the notecards where necessary

NOTE *Rehearsals should be spoken aloud, not read silently or simply thought over. It's important to know how it **feels** to make your points and transitions.*

How to rehearse

You should rehearse thoroughly, but not to the point where you memorize the talk. At that point, you can get trapped into saying things one way only, and thus lose your ability to respond to audience feedback. One of the major advantages of the properly rehearsed talk is that you know what you are about to say and have the freedom to use the best way to phrase ideas to reach a given audience at a given moment. You talk directly to the audience with occasional glances at your notecards.

In the end, speaking directly to the audience is the best way of overcoming stage fright. If you concentrate on getting your message to your audience, and if you heed your audience's reactions, you soon forget about yourself and your natural stage fright. You focus on communicating!

IMPROMPTU SPEAKING

Although impromptu speaking is not recommended for any situation where you have adequate time to prepare, certain situations occur where impromptu skills can help you achieve your goals. Job interviews come immediately to mind, for example. Also, you'll frequently be asked questions after your presentations or you'll be asked for your opinion during meetings.

There's Nothing to Fear but Fear Itself

Quite simply, many people don't perform well as impromptu speakers because of panic. Suddenly, the pressure is on and your brain freezes. Later, perhaps just a few seconds later, you know exactly what you should have said, but it's too late. You have spouted some gibberish and feel like a fool.

Here's What to Do

Flustered speakers need something that will allay the panic long enough to get their brain moving again. At the same time, they need a device to comb through their brain to retrieve what they know about the subject. And finally, that magic "something" must be used to make sense of what the speakers have found in their brain.

Sound impossible? Actually, the solution is really quite simple and effective—the speaker needs an *instant organizing device*, such as a *statement-elaboration pattern*. If you're asked in a job interview why you chose to study at a certain university, you might pause for a second and say,

> I chose this university mainly because the graduates from my program are qualified for a wide range of positions, and most of those graduates get employed in their field.

The above statement buys you a little time to think, but more important you have an idea and you have an organizing method! The statement-elaboration pattern allows you to expand upon each of the main parts of the statement:

- when and how you chose your university and which other universities you investigated
- the "wide range of positions"
- some of the "employed" graduates you know (or statistics that you may happen to know)

Knowing that you have a "method" allows the panic to subside long enough for you to retrieve and organize what you really do have to say. You're not stuck! Now, this method requires practice before you can get really proficient with it, but you'll be amazed at how well you can teach yourself to use this pattern and the following impromptu organizing patterns.

1. *Chronological.* Time patterns are built into our language and thinking. For example, we readily understand the *past/present/future* pattern. Or we can organize our thoughts according to a series of dates that lead up to the present.

   This technique could work for a 28-year-old who has completed an engineering program. In a job interview, she is asked why she chose engineering. She might reply that starting when she was 10 years old, she used to help her father repair equipment at the family-run amusement arcade. Then, when she was 14, she won a science contest with a design for a gravity pump for the family home's fish pond. When she was 19 . . .

   And so it goes. Chronological patterns are easy to use because we organize our brains with certain milestones and accompanying memories.

2. *Narrative.* Similarly, we can use story patterns to retrieve key bits of information from our brains and then use the same story (or "narrative") to organize our thoughts on a subject. For example, if the engineering grad in the job interview mentioned above were asked to explain whether she could effectively supervise

a work crew on a forest road deactivation project, she could tell a story about a parallel experience she had during a work placement last year. At various points, she could interrupt her narrative to highlight the skills she demonstrated during that earlier experience.

3. *Topical arrangement.* In a job interview for a technologist position, a 37-year-old graduate of an engineering technology program might be asked about his strengths relative to the position being discussed. He might say,

> I can answer that by discussing three main areas: my previous factory experience, my talent for computer drafting and design, and the valuable experience I gained during my four-month work placement at your firm last year.

Then, he would discuss each topic in turn.

4. *Cause–effect.* This two-part structure is described by its name. It can start with causative factors and work toward the results of those factors, or it can start with a description of a situation and work back to the causes of that situation. For example, the head of a highway maintenance team was recently asked why a certain highway had so many potholes during the previous winter. He began by briefly describing the physical features of some of the more common types of potholes and then recounted the special factors that result in those types of potholes, factors that were particularly prevalent that winter.

Now, before we get lost in a consideration of potholes, let's remember the main benefit of that highway maintenance spokesperson's use of the cause–effect pattern—long after listeners forgot the details of his explanation, they remembered that he seemed sensible and trustworthy!

PREPARING YOUR PRESENTATION

Plan the presentation systematically, to stay in control and build confidence. For our limited purposes here, we will assume that your presentation is extemporaneous.

Research Your Topic

Do your homework. Be prepared to support each assertion, opinion, conclusion, and recommendation with evidence and reason. Check your facts for accuracy. Your listeners expect to hear a knowledgeable speaker; don't disappoint them.

Begin gathering material well ahead of time. Use summarizing techniques from Chapter 10 to identify and organize major points. If your presentation is a spoken version of a written report, most of your research has been done, but you may need to introduce material not found in the written version in order to appeal to listeners rather than readers.

Aim for Simplicity and Conciseness

Keep your presentation short and simple. Boil the material down to a few main points. Listeners' normal attention span is about 20 minutes. After that, they begin tuning out. Time yourself in practice sessions and trim as needed. If the material requires a lengthy presentation, plan a short break, with refreshments if possible, about halfway.

Anticipate Audience Questions

Consider those parts of your presentation that listeners might question or challenge. You might need to clarify or justify information that is new, controversial, disappointing, or surprising.

Outline Your Presentation

When you're preparing a talk, especially a lengthy, complex presentation, you can use the same kinds of organizing techniques as in a written document. However, the patterns in a written report might not suit an oral presentation.

CASE	An Oral Version of a Written Report

The report outline on pages 370–73 in Chapter 19 suits a written technical report, but not an oral version of the report. If the report's authors give their report orally to their client, Dave Cochrane, they might revise their working outline as follows:

INTRODUCTION

Connect with listener
Background

- express appreciation for opportunity to work on this project
- review the requirements for a redesign: four main "problems"
- show **Slide 1:** seven stages of R & D process

CONCLUSION AND RECOMMENDATION

Provide bottom line early
in talk—listener will be
receptive

- the CT04 addresses all apparent flaws with the CT83
- cannot recommend production because . . .
- should interview craftspeople
- excellent potential
- **Transition:** now, let's look at our rationale for that recommendation

RESEARCH SUMMARY

Factual basis for ideas
proposed later

- interviews with Bill Sand, Gerry Hanson, Terry Lockhart
- what they said
- examined: transmission jacks, engineered stands, and heavy-duty engine stands

DESIGN SOLUTIONS
Components

An overview of the new
design, mostly through
visuals

- casters—sample of actual caster
- frame—**Slide 2:** computer model (discuss reasons for material choice)
- teleposts—purpose and specs- **Slide 3:** telepost height-adjustment assembly
- mounting platform—**Slide 4:** computer model
- mounting plates—**Slide 5:** mounting plate

CONCLUSIONS

Arguments that support
the recommended design

- design meets the four design needs
- room to add additional features

COST ANALYSIS

Finally, the crucial cost analysis

◆ resolving the cost–design paradox
◆ projected cost summary
◆ **Slide 6:** component costs of manufacturing the CTO4 stand
◆ **Slide 7:** total projected manufacturing cost
◆ assume a 35 percent markup
◆ caution re: figures (list them)

ACTION CLOSING

Maintain business connection with listener

◆ wish Dave Cochrane success with the new CTO4
◆ offer to participate further in the project
◆ invite questions

Carefully Prepare Your Introduction and Conclusion

Intriguing introductions

Speech *introductions* are very important. You must gain audience attention, establish your credibility, and lead in to the content of your talk. In the process, establish the mood of your presentation. Three groups of related techniques for immediately engaging your listeners are listed in Table 24.1.

Table 24.1 Effective Methods for Engaging Listeners

Establish the Topic	Grab Audience Attention	Connect with Your Audience
◆ Refer directly to subject.	◆ Refer to recent events.	◆ Establish common ground.
◆ Provide a pre-summary.	◆ State a startling fact or idea.	◆ Promise a reward.
◆ Delimit the topic.	◆ Give an illustration or a story.	◆ Present your credentials.
◆ Provide background info.	◆ Ask a rhetorical question.	◆ Justify the topic.
◆ Define terms.		

Strong conclusions

The *conclusion* to your talk provides your last chance to confirm your main point. Sometimes, you may also want to finish on an emotional high or persuade your listeners to act in a certain way. And, no matter what the purpose of your talk has been, your closing phrasing should signal a strong sense of conclusion; you should never have to tell your audience, "That's the end of my talk."

Here are some techniques for closing strongly:

◆ Summarize your main points.
◆ Request a specific action from your audience.
◆ Challenge the audience to reach certain goals.
◆ Declare *your* intended action.
◆ Quote a memorable phrase or piece of poetry.
◆ Finish with a powerful example or story.
◆ Repeat a key phrase from the introduction.
◆ Show how you've fulfilled a promise made in the introduction.

Some combination of two or more of the above techniques will suit any presentation.

Build Bridges for Your Listeners

Most audience members don't listen attentively all the time; they tune in and out. When they do tune back in, you need to help them understand what direction the talk has taken while they were "away." Not all audience members wander, but even the best listeners find it difficult to follow speakers who do not provide bridges from one section of a talk to the next.

The following bridging techniques can help the best and worst of listeners stay on your idea track:

Keep listeners on track

1. *Build bridges at the beginning.* Preview the presentation's structure so that your listeners know where you're headed and can anticipate the relationships among the topics you'll be discussing. If your presentation is lengthy and complex, an agenda in poster form or a handout would help the audience.
2. *Use the organization structure that best suits the topic.* A presentation about overcoming speaker nervousness could use a problem–solution pattern. A presentation about the development process for a residential subdivision could employ a chronological pattern, perhaps with key deadline dates assigned to each stage of the process.
3. *Build word bridges during the presentation.* Transitional phrases can take several forms:

   ◆ Confirm the direction of the presentation. (*Now let's consider some other methods of reducing waste.*)
   ◆ Change the direction of the presentation. (*We've identified the problem; now it's time to find a solution.*)
   ◆ Lead into an example or a detailed list of evidence. (*If you want assurance that this program actually works, here's a case for you to consider . . .*)
   ◆ Develop an idea further. (*In addition to their contribution to the local ecology, these marsh lands help local tourism . . .*)

4. *Use repetition to confirm key ideas and to reorient wayward listeners.* You could repeat a key phrase (as Martin Luther King did with "I have a dream"). Or you could repeat an idea without making the repetition obvious (*Nothing to fear but fear itself; You're your own worst critic; Believe in your abilities as much as we do*). You could summarize the points you've made in a preceding section—these *stage summaries* help those who've not listened as attentively as they might have.

Plan Your Visuals

Visuals increase listeners' interest, focus, understanding, and memory. Select visuals that will clarify and enhance your talk—without making you fade into the background.

DECIDE WHEN VISUALS WILL WORK BEST.   Use visuals to emphasize a point and enhance your presentation whenever showing would be more effective than merely telling.

DECIDE WHICH VISUALS WILL WORK BEST.   Will you need numerical or prose tables, graphs, charts, illustrations, computer graphics? How fancy should these visuals be? Should they impress or merely inform?

LIST YOUR VISUALS IN YOUR NOTES.   You can list your visuals as in the working outline on pages 524–25. Or you can create two columns, with the content notes in the left-hand column and the corresponding visuals in the right-hand column.

DECIDE HOW MANY VISUALS ARE APPROPRIATE.　Use an array of lean, simple visuals that present material in digestible amounts instead of one or two overstuffed visuals that people end up staring at endlessly.

DECIDE WHICH VISUALS CAN REALISTICALLY BE CREATED.　Fit each visual to the situation. The visuals you select will depend on the room, the equipment, and the production resources available.

Fit each visual to the situation

How large is the room, and how is it arranged? Some visuals work well in small rooms but not large ones, and vice versa. How well can the room be darkened? Which lights can be left on? Can the lighting be adjusted selectively? What size should visuals be to be seen clearly by the whole room? (A smaller, intimate room usually is better than a room that is too big and cavernous.)

What hardware is available (slide projector, opaque projector, overhead projector, film projector, video player, data-projection unit, projection screen)? How far in advance does this equipment have to be requested? What graphics programs are available? Which is best for your purpose and listeners?

What resources are available for producing the visuals? Can drawings, charts, graphs, or maps be created as needed? Can transparencies (for overhead projection) be made or slides collected? Can handouts be typed and reproduced? Can multimedia displays be created?

SELECT YOUR MEDIA.　Fit the medium to the situation. Which medium or combination is best for the topic, setting, and listeners? How fancy do listeners expect this to be? Which media are appropriate for this occasion?

Fit the medium to the situation

- ◆ For a weekly meeting with colleagues in your department, jotting notes on a blank transparency, chalkboard, or whiteboard might suffice.
- ◆ For interacting with listeners, you might use a whiteboard or chalkboard to record audience responses to your questions.
- ◆ For immediate orientation, you might begin with a poster or flip chart sheet listing key visuals/ideas/themes to which you will refer repeatedly.
- ◆ For helping listeners take notes, absorb technical data, or remember complex material, you might distribute a presentation outline as a preview or provide other handouts.
- ◆ For a presentation to investors, clients, or upper management, you might require polished and professionally prepared visuals, including computer graphics, multimedia, and state-of-the-art technology.

Figure 24.1 on pages 528–529 presents various common media in approximate order of availability and ease of preparation.

Prepare Your Visuals

As you prepare visuals, focus on economy, clarity, and simplicity.

BE SELECTIVE.　Use a visual only when it truly serves a purpose. Use restraint in choosing what to highlight with visuals. Try not to begin or end the presentation with a visual. At those times, listeners' attention should be focused on the presenter instead of the visual.

Whiteboard/Chalkboard

Uses • simple, on-the-spot visuals
• recording audience responses
• very informal settings
• small, well-lit rooms

_____

Tips • copy long material in advance
• write legibly
• make it visible to everyone
• use washable markers on whiteboard
• speak to the listeners—not to the board

Poster

Uses • overviews, previews, emphasis
• recurring themes
• formal or informal settings
• small, well-lit rooms

_____

Tips • use 51 cm × 76.5 cm (20" × 30")
posterboard or larger
• use intense, washable colours
• keep each poster simple and uncrowded
• arrange/display posters in advance
• make each visible to the whole room
• point to what you are discussing

Flip Chart

Uses • a sequence of visuals
• back-and-forth movement
• formal or informal settings
• small, well-lit rooms

_____

Tips • use an easel pad and easel
• use intense, washable colours
• work from a storyboard
• check your sequence beforehand
• point to what you are discussing

Handouts

Uses • presentation of complex material
• help listeners follow along
• help listeners take notes
• help listeners remember

_____

Tips • staple or bind the packet
• number the pages
• try saving distribution for the end
• if you must distribute up front, ask
listeners to await instructions
before reading the material

Figure 24.1 Selecting Media for Visual Presentations

(continued)

Film and Video

Uses
- necessary display of moving images
- coordinated sound and visual images aid understanding and help persuade
- added credibility and impact

Tips
- carefully introduce segment to be shown; tell viewers what to see or expect
- show only those segments needed to make your points
- if the segment is complex, or happens very quickly, play it more than once
- where appropriate, use video slow-motion replay
- always practise with the segment beforehand

Overhead Projection

Uses
- on-the-spot or prepared visuals
- overlaid visuals
- formal or informal settings
- small or medium-sized rooms
- rooms needing to remain lit

Tips
- use cardboard mounting frames for your acetate transparencies
- write discussion notes on each frame
- check your sequence beforehand
- turn off projector when not using it
- face the audience—not the screen
- point directly on the transparency
- use text slides to follow and confirm your spoken points. DO NOT READ THE TEXT IN THE PROJECTED SLIDES!

Networked Computers

Uses
- training sessions
- meetings—to display some combination of text, audio, slides, and/or video found online
- collaborative work on a document, a design, or a website

Tips
- explain the purpose of a networked session in advance
- use a networked session for small groups only, either in a room or in an online "virtual room"
- use a networked session for interactive discussion, not a lecture

Computer Projection

Uses
- sophisticated charts, graphs, maps
- multimedia presentations
- formal settings
- small, dark rooms

Tips
- take lots of time to prepare/practise
- work from a storyboard
- check the whole system beforehand
- have a default plan in case something goes wrong
- face the audience—not the screen
- use text slides to follow and confirm your spoken points. DO NOT READ THE TEXT IN THE PROJECTED SLIDES!

Figure 24.1 Selecting Media for Visual Presentations

MAKE VISUALS EASY TO READ AND UNDERSTAND.   Think of each visual as an image that flashes before your listeners. They will not have the luxury of studying the visual at leisure. Listeners need to know at a glance what they are looking at and what it means. Here are some guidelines for achieving readability.

GUIDELINES   **FOR READABLE VISUALS**

- ◆ Make visuals large enough to be read anywhere in the room.
- ◆ Don't cram too many words, ideas, designs, or type styles onto a visual.
- ◆ Keep wording and images simple.
- ◆ Boil your message down to the fewest words.
- ◆ Break things into small sections.
- ◆ Summarize with keywords, phrases, or short sentences.
- ◆ Use 18- to 24-point type size and sans serif typeface (J. White, *Great Pages* 80).

In addition to being able to *read* the visual, listeners need to *understand* it. Guidelines for achieving clarity follow.

GUIDELINES   **FOR UNDERSTANDABLE VISUALS**

- ◆ Display only one point per visual—unless previewing or reviewing (J. White, *Great Pages* 79).
- ◆ Give each visual a title that announces the topic.
- ◆ Use colour sparingly to highlight keywords, facts, or the bottom line.
- ◆ Use the brightest colour for what is most important (J. White, *Great Pages* 78–79).
- ◆ Label each part of a diagram or illustration.
- ◆ Proofread each visual carefully.

Audience members pay more attention to visuals than to words on slides

LOOK FOR ALTERNATIVES TO WORD-FILLED VISUALS.   Instead of presenting mere overhead versions of printed pages, explore the full visual possibilities of your media. For example, a writer might find it difficult to use words to describe the subtle nuances of light and colour displayed by the image in Figure 24.2. Whenever possible, use drawings, graphs, charts, photographs, and other visual tools discussed in Chapter 12.

Figure 24.2 Images More Powerful Than Words

Source: Don Klepp

USE ALL AVAILABLE TECHNOLOGY. Using desktop publishing systems and presentation software, such as *PowerPoint*, you can create professional-quality visuals and display technical concepts. Using hypertext and multimedia systems, you can create dynamic presentations that appeal to the listener's multiple senses. Using automated graphics and a laser pointer, you can deliver a smooth, dynamic presentation. These are just a few of the possibilities inherent in the technology.

CAUTION   *Do not use the "whiz-bang" features of presentation software for their own sake. Ensure that each feature you use has a legitimate purpose, and is not being used just to impress your audience with your command of current technology! Also, save* PowerPoint *slides in the older .ppt format as well as the most recent format you're using.*

CHECK THE ROOM AND SETTING BEFOREHAND. Make sure that you have enough space, electrical outlets, and tables for your equipment. If you will be addressing a large audience by microphone and plan to point to features on your visuals, be sure the microphone is moveable. Pay careful attention to lighting, especially for whiteboards, chalkboards, flip charts, and posters. Bring a pointer if you need one.

Prepare Notecards

Notecards build your confidence—you know that you won't be stuck for something to say. Think of notecards as insurance against going blank in front of your audience. More important, notecards help keep you from wandering. When you work from a structured set of notes, there's less chance of going off on tangents.

GUIDELINES   **FOR CREATING NOTECARDS**

1. *Use card stock or heavy paper* (32 lb. or heavier), so the notes don't shake or droop.
2. *Choose a size that suits you.* Probably the smallest effective size is 7.5 cm by 12.5 cm (3″ × 5″). Remember that the larger the card, the more you can place on it.
3. *Use no more than 10 cards.* Four to six cards will suffice for most talks.
4. *Write on one side of each card.* Number the cards sequentially in the upper right-hand corner.
5. *Use point form to remind yourself of the things you want to say.* Write large, legible points. Do not print your speech out in full; you'll be tempted to read the material, and if you succumb to the temptation, you'll lose your audience.
6. *Write sentences for the few critically important statements that you want to get exactly right* (no more than three to five per talk).
7. *Include*
   - headings or names of sections of the talk
   - your main points, in point form
   - key statistics, facts, or quotes that you may not remember
   - reminders to yourself—*Breathe! Don't pace! Pause here. Slow down! Look at them!*
   - AV usage cues—*Overhead: Table 1; draw on whiteboard; dim lights*
   - time limits for certain parts of the presentation

However, you do need to discipline yourself. Notes will help only if you regularly check them to make sure that you're still on course and that you haven't omitted key points. And you do need to rehearse with the cards so that you're sure that you can read them and that you're comfortable with them. Figure 24.3 illustrates two of the notecards that the transmission jack redesign team might use for its presentation to Dave Cochrane.

Rehearse Your Talk

Reread the advice on page 521 to refresh your memory of how and what to rehearse. Pay particular attention to the introduction and conclusion, both of which must be presented smoothly and confidently. If possible, record one of your rehearsed deliveries to determine what might be improved. Then, use the evaluation checklist in Figure 24.5 on page 542 to guide your improvements.

NOTE   *Most speakers find that their rehearsal time is about 80 percent of actual performance time, unless nervousness causes them to forget material or to speak too quickly.*

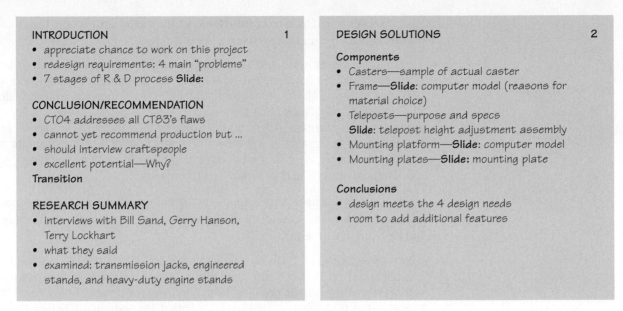

INTRODUCTION 1
- appreciate chance to work on this project
- redesign requirements: 4 main "problems"
- 7 stages of R & D process **Slide:**

CONCLUSION/RECOMMENDATION
- CTO4 addresses all CT83's flaws
- cannot yet recommend production but ...
- should interview craftspeople
- excellent potential—Why?

Transition

RESEARCH SUMMARY
- interviews with Bill Sand, Gerry Hanson, Terry Lockhart
- what they said
- examined: transmission jacks, engineered stands, and heavy-duty engine stands

DESIGN SOLUTIONS 2

Components
- Casters—sample of actual caster
- Frame—**Slide:** computer model (reasons for material choice)
- Teleposts—purpose and specs
  Slide: telepost height adjustment assembly
- Mounting platform—**Slide:** computer model
- Mounting plates—**Slide:** mounting plate

Conclusions
- design meets the 4 design needs
- room to add additional features

Figure 24.3 Sample Notecards

DELIVERING YOUR PRESENTATION

You have planned and prepared carefully. Now, consider the following simple steps to make your actual presentation enjoyable instead of terrifying.

1. **Connect with Your Audience**

   A successful presentation involves relationship building with the audience.

   - *Get to know your audience.* Try to meet some audience members before your presentation. We all feel more comfortable with people we know. Don't be afraid to smile.
   - *Display enthusiasm and confidence.* Audience members prefer lively, engaging speakers. So, overcome your shyness; research indicates that shy people are seen as less credible, trustworthy, likeable, attractive, and knowledgeable. Also, use downward inflections at the end of sentences, to project self-confidence.
   - *Be reasonable.* Don't make your point at someone else's expense. If your topic is controversial (layoffs, policy changes, downsizing), decide how to speak candidly and persuasively with the least chance of offending anyone. Avoid personal attacks. For example, a groundwater consultant discussing his or her evaluation of the town's water supply should assess the system's strengths and limitations, and not criticize those who manage that system.
   - *Don't preach.* Speak like a person talking—not like someone giving a sermon. Use *we*, *you*, *your*, and *our* to establish commonality with the audience. Avoid jokes or wisecracks.

2. **Adjust Your Presentation Style**

   On pages 519–520, we discussed four levels of speaking; try to operate at least at a heightened conversational level and move to a performance level or even an oratorical level when you're comfortable at that level and when the occasion demands a more dramatic approach.

- *Use natural movements and reasonable postures.* Move and gesture as you normally would in conversation (except for more dramatic moments), and maintain reasonable postures. Avoid foot shuffling, pencil tapping, swaying, slumping, fidgeting, or nervous pacing.
- *Adjust volume and rate.* Adjust your volume to the size of the audience and to the situation—for example, you'll need to speak louder when a projector fan is running. When you use an amplified microphone for a large audience, speak more intimately, as if to one person or to a small group.

Nervousness causes some speakers to gallop along and mispronounce things. Slow down and enunciate clearly. Vary your rate to reflect the changing mood or content of the talk.

- *Maintain eye contact.* Look directly into listeners' eyes. With a small audience, eye contact is one of your best connectors. As you speak, establish eye contact with as many listeners as possible. With a large group, maintain eye contact with those in the first rows. Establish eye contact immediately—before you even begin to speak—by looking around.
- *Avoid vocal fillers.* We use fillers such as *uhh, eh, like,* and *um* in normal conversation out of habit, often to signal non-verbally that we're not finished speaking and that we don't want to be interrupted. However, frequent use of such fillers can annoy audiences.
- *Vary your tone.* Good speakers skillfully use the right tones to signal the emotional content of their messages. A variety of vocal tones helps listeners stay awake!

3. **Manage the Communication Flow**
Do everything you can to keep things running smoothly.

- *Be responsive to listener feedback.* Assess listener feedback continually and adjust the talk as needed. If you are labouring through a long list of facts or figures and people begin to doze or fidget, you might summarize. Likewise, if frowns, raised eyebrows, or questioning looks indicate confusion, skepticism, or indignation, you can backtrack with a specific example or explanation. You might also prepare more than one example to illustrate a given point, so that you can select the best example to match the audience's current mood.
- *Stick to your plan.* Say what you came to say, then summarize and close—politely and on time. Don't punctuate your speech with digressions that pop into your head. Unless a specific anecdote was part of your original plan to clarify a point or increase interest, avoid excursions. We often tend to be more interested in what we have to say than our listeners are!
- *Leave listeners with something to remember.* Before ending, take a moment to summarize the major points and re-emphasize anything of special importance. Are listeners supposed to remember something, or have a different attitude, or take a specific action? Let them know!
- *Allow time for questions and answers.* If questions suit the format of the event, at the very beginning, tell your listeners that a question-and-answer period will follow. Observe the following guidelines for managing listener questions diplomatically and efficiently.

GUIDELINES   **FOR MANAGING LISTENER QUESTIONS**

- Announce a specific time limit for the question period to avoid prolonged debates.
- Listen carefully to each question.
- If you can't understand a question, ask that it be rephrased.
- Repeat every question, to ensure that everyone hears it.
- Be brief in your answers.
- If you need extra time to answer a question, arrange for it after the presentation.
- If anyone attempts lengthy debate, offer to continue after the presentation.
- If you can't answer a question, say so and move on.
- End the session with "We have time for one more question," or some such signal.

CONDUCTING WEBINARS

Web-based seminars, which are known as *webinars*, are growing in popularity both for in-house training and for industry special interest groups. "They use your computer's Website browser to display presentation materials and other applications important to the webinar topic, with the audio portion of the presentation provided either through your computer speakers or over the phone" (IFEA).

Jim Gray, a Toronto-based speaker and communication skills coach, says that "successful webinars happen when you take the focus off yourself and place it directly on the members of your audience."

Gray offers the following advice for hosting successful webinars:

1. Choose someone to handle the technical components of the webinar, so that you can focus on hosting.
2. Place a time limit of one hour (or less) on the session.
3. Use a clean, straightforward structure that participants will easily follow.
4. Engage the audience with survey questions before the session begins and once or twice during the session.
5. Have visual materials and or short digital clips available to spice things up and to confirm main points.
6. Don't lecture—a conversational approach that welcomes online feedback and lively exchanges will be more enjoyable and memorable for all.

Keep the session lively and invite audience reactions

SPEAKING AT MEETINGS

A recent internet search for sites discussing "effective meetings" led to 397 *million* results! Clearly, there's a market for advice on how to conduct effective meetings. Here, we focus on preparing for meetings and on the skills required to speak successfully at meetings.

The Meeting Manager's Role

In most cases, a leader is responsible for calling meetings, setting the goals for a meeting, and distributing agendas. That leader might run meetings as part of his or her job duties, or a task group might elect one of its members to organize and chair its meetings.

SETTING THE AGENDA.   Group members often contribute agenda items, but the chair must organize those items into a smooth-flowing agenda. That agenda should do more than simply list topics; it should help participants prepare for their roles in the meeting. Figure 24.4 below shows such an agenda, which involves employees of Mountain Environmental Consultants. Present at the meeting will be the firm's managing partner (René Aubois), two members of the surveying team (Orvald Tomsen and Jas Dhaliwal), and three members of an environmental assessment team (Joe Silveira, engineer, and Rita Cherneski and Chris Dohrmann, technologists).

CHAIRING THE MEETING.   Usually, the designated meeting manager chairs the meeting. This person introduces the meeting's goals, provides necessary background, and establishes the maximum meeting length. The chair also keeps the discussion on

| Meeting purpose: | Establish a procedure for an environmental assessment of the proposed Chetwekoo Beach reclamation project |
| Place and time: | Mountain Environmental board room. 7:30 a.m., 27 March 2014 |

Item	Responsible	Supporting information	Discussion time
1. Announcements and project overview	René	Attached Appendix A	3 to 5 min.
2. Geological overview	Joe	Provided by Joe at meeting	3 to 5 min.
3. Potential surveying times and problems	Orvald and Jas	Provided by Orvald & Jas, if necessary	3 to 5 min.
4. History of two similar assessments we've done	Joe	n.a.	3 to 5 min.
5. **Discussion:** procedure for surveying and for assessing the impact of reclaiming the beach	All—review all required resources and M.O.E. regulations	All: bring pertinent documents	15 to 25 min.
6. **Action plan:** for the environmental assessment project	All		15 to 25 min.
Follow-up: write business proposal for submission to the Chetwekoo Regional District	René, with input from all	Material generated during meeting	**Deadline for proposal:** April 7

Figure 24.4 Sample Meeting Agenda

track and gives everyone a chance to speak. The chair should summarize consensus decisions and guide debates, from the time that motions are made to the final decision. In situations where participants regulate themselves, the chair guides the discussion; in other situations, the chair must wield his or her authority to control undisciplined participants.

Notice that the agenda in Figure 24.4 reflects an inclusive, democratic leadership style that encourages participation. Some leaders, however, favour a more autocratic style in which meeting participants have less influence than the leader.

The Meeting Participant's Role

Many meeting participants simply show up at meetings, with no prior preparation. The participants in the Mountain Environmental meeting (see Figure 24.4) apparently will not have such an option. Still, participants should not need a totally structured agenda to spur them to prepare.

You can get ready to contribute to meetings by

- making notes about points you'd like to make about certain agenda items
- gathering relevant information and bringing pertinent documents
- preparing visuals to help support your points

In other words, you should prepare to do more than simply occupy a chair at meetings. When you do speak, the following guidelines will help you contribute to a productive meeting.

GUIDELINES   **FOR SPEAKING EFFECTIVELY AT MEETINGS**

1. *Plan and rehearse your comments* when you're responsible for introducing the discussion or for reporting to the group.

2. *Prepare visuals to illustrate and support your points.* Effective visuals can reduce meeting time and improve the chances of reaching consensus. Designated presenters should particularly consider using visuals, but other participants might bring their own visuals.

3. *Speak only if you can contribute meaningfully to the discussion.* Do not talk just to hear your voice or to maintain your status within the group.

4. *Don't waste time during discussions.* Do not repeat what someone else has already said. Don't repeat what you said earlier. Don't ramble while you discover what you have to say—know the basic structure of your point before you speak.

5. *Listen carefully to others' comments.* Also, if you're not certain of another's intended meaning, ask for clarification before responding.

6. *Express your views tactfully.* Focus on the facts and how the facts relate to the item being discussed. Don't attack others, even though it may be necessary to disagree with others' conclusions or opinions.

7. *In sales meetings, focus on providing value to your listeners.* Speak to your listeners' needs, interests, and priorities.

Electronically Conducted Meetings

Occasionally, not all those who want to or should attend a meeting can do so in person. Under those circumstances, technology can save the day. Teleconferencing and videoconferencing allow all key persons to participate. Often, the majority of participants will be situated in the "host" location with one or more others in another location. Such meetings save travel time and costs, which is why many internal business meetings and job interviews are now conducted in this manner.

The market for videoconferencing equipment and services is growing rapidly. Thus, Cisco (Unified MeetingPlace), Microsoft (Live Meeting), Citrix (GoToMeeting), and others are working hard to earn a share of that market. The core offerings from these web collaboration providers include phone audio, VoIP, or videoconferencing integrated with desktop file sharing, text chat, and shared whiteboard (Skeen).

Clearly, conference phone calls are easier to set up than videoconferences, and the equipment is much cheaper, but the presence of visual non-verbal cues in videoconferencing is often worth the trouble. Technical glitches can mar both events, but cost and convenience are leading to increased usage, especially with new computer technology.

Web-based videoconferences require each participant to have a computer with built-in camera, microphone, and speakers, and the participants communicate in a "virtual" room as if they were sitting next to each other. The chief advantage of internet conferences is that the equipment and software are inexpensive compared to the equipment required for conference rooms. However, companies, such as Canada's ADCOM Videoconferencing, offer a full range of installations from USB-based personal desktop systems to large conference room set-ups with multipoint capabilities.

In addition to reducing travel time and costs, videoconferencing offers other benefits:

- Internal communication can be improved, especially for companies with far-flung operations.
- Fewer iterations of a product need to be developed, because instant feedback can be provided.
- Teams can be managed more efficiently by frequent "personal" contact.
- Out-of-town attendees can be present at meetings using videoconferencing and live webcasts. Companies such as TELUS have been using these methods to allow shareholders to attend their quarterly and annual meetings.
- All meetings can be recorded and archived, so that those who miss a session can view the event, complete with the accompanying visuals presented at the meeting.

Skill and care are required to maximize the benefits of any communication technology. The following guidelines for effective videoconferencing are loosely modelled on detailed guidelines posted by the University of Warwick at its website, **www2. warwick.ac.uk/services/ldc/resource/eguides/cmcguidelines**.

1. *Prepare well for a videoconference.* Make a list of all contact numbers and addresses. Ensure that the equipment is working at all locations and that participants know how to use it. Decide who's calling whom. Establish the time and the difference in time zones. Check that you can access each other's machines through the firewalls, if you are using the internet. Distribute large paper or electronic files before the meeting.

2. *Start with a clear introduction.* The meeting's objectives should be laid out. Also, participants should be reminded of materials they may need at various points later in the meeting. If the meeting is following Robert's Rules of Order or other regulations, a previously approved agenda will be distributed.

3. *Speak very clearly in all electronically mediated meetings.*

4. *Integrate the brief delays that occur because of the nature of the technology.* Longer delays can also result when microphones have to be turned on and off. These pauses may be awkward, so the session leader and all participants must

   ◆ leave longer pauses than in face-to-face conversation before passing the conversation to the next person
   ◆ wait until the other person has clearly finished before continuing
   ◆ always give a verbal response, not just a non-verbal cue

5. *Fix image and sound problems of desktop-to-desktop meetings.* Improve the camera sightlines as much as possible and, using picture-in-picture, ensure that you're in the camera image at all times. Audio echo can be solved by using headphones or disabling the duplex function. A mute button can be used to cut out extraneous noises such as coughing and paper shuffling.

6. *Encourage everyone to respond and contribute.* Room-to-room sessions often inhibit participation, so ask the moderator or meeting chair to carefully structure the proceedings so that participants in the multiple locations know when they will have opportunities to communicate.

7. *Pre-record physical demonstrations of procedures or equipment* before the meeting.

8. *Encourage follow-up discussions* by setting up chat rooms or discussion boards.

NOTE *New technologies are making audio and videoconferencing easier than ever. For example, Skype software provides internet audio calls and video calls that can morph into group calls. For examples of how business can use Skype technology, see the case studies at* **www.skype.com/intl/en/business/case-studies/maxim.**

IN BRIEF

Cross-Cultural Audiences May Have Specific Expectations

Imagine that you've been assigned to represent your company at an international conference or before international clients (e.g., of passenger aircraft or mainframe computers). As you plan and prepare your presentation, remain sensitive to various cultural expectations.

For example, some cultures might be offended by a presentation that gets right to the point without first observing formalities of politeness, well wishes, and the like.

Certain communication styles are welcomed in some cultures, but considered offensive in others. In southern Europe and the Middle East, people expect direct and prolonged eye contact as a way of showing honesty and respect. In Southeast Asia, this may be taken as a sign of aggression or disrespect (Gesteland 24). A sampling of the questions to consider:

◆ Should I smile a lot or look serious (Hulbert, "Overcoming" 42)?
◆ Should I rely on expressive gestures and facial expressions?
◆ How loudly or softly, rapidly or slowly should I speak?

(continued)

- Should I come out from behind the lectern and approach the audience or keep my distance?
- Should I get right to the point or take plenty of time to lead into and discuss the matter thoroughly?
- Should I focus only on the key facts or on all the details and various interpretations?
- Should I be assertive in offering interpretations and conclusions, or should I allow listeners to reach their own conclusions?

- Which types of visuals and which media might or might not work?
- Should I invite questions from this audience, or would this be offensive?

To account for language differences, prepare a handout of your entire script for distribution after the presentation, along with a copy of your visuals. This way, your audience will be able to study your material at their leisure.

IN BRIEF

Great Speeches, Great Speakers

What makes a speech great? Does it depend on the contents of the speech and the circumstances in which it is delivered, or are great speeches made only by great speakers?

An analysis of great speeches, in their historical contexts, reveals certain commonalities that will help you in your own speeches:

- In all cases, the speakers know what is important to their audiences, and they choose language and examples that speak to the listeners' dreams and fears.
- Great speakers "connect" with their audiences through personal disclosure and direct emotional appeals to listeners' needs and concerns. Great speakers don't talk down to their audiences, but they don't pander to them, either. They treat audiences with respect.
- Successful speakers know when to use humour. When used appropriately, humour can create a bond between speaker and listener. But inappropriate use of humour can alienate audiences.
- Most great speakers know how to impart natural emotion in their voices, gestures, facial expressions, and body movements. Their non-verbal messages never seem forced or contrived. As their speeches

develop, they ride a rhythm and cadence of memorable phrases.
- Great speeches are eloquently phrased, but not necessarily elaborately phrased. Often, eloquence comes from using the exactly right, simple phrase at exactly the right moment—"I have a dream" (Martin Luther King); "If they fool you once, shame on them; if they fool you twice, shame on you!" (Tommy Douglas).
- Most of all, great speakers have something really important to say at a time when it needs to be said.

Here are some websites that have collected speech transcripts or audio clips of memorable speeches and speakers:

- CBC Radio Archives—Tommy Douglas, the father of Canadian Medicare, **http://archives. cbc.ca/IDC-1-74-851-4958/people/tommy_ douglas/clip4**
- Collections Canada—transcripts of speeches by Canadian leaders, **www.collectionscan-ada.gc.ca/primeministers/h4-4000-e.html**
- Carmine Gallo reveals the approach and the techniques that made Apple's evangelist, Steve Jobs, such an effective public speaker, **www.slideshare.net/cvgallo/the-presenta-tion-secrets-of-steve-jobs-2609477**

EXERCISES

1. In a memo to your instructor, identify and discuss the kinds of oral reporting duties you expect to encounter in your career.
2. Design an oral presentation for your class. (Base it on a written report.) Make a sentence outline, and a storyboard that includes at least three visuals. Practise your presentation with a digital audio recorder, a video recorder, or a friend. Use the checklist in Figure 24.5 (page 542) to evaluate your delivery.
3. Observe a lecture or speech, and evaluate it according to the evaluation checklist. Write a memo to your instructor (without naming the speaker), identifying strong and weak areas and suggesting improvements.
4. In an oral presentation to the class, present your findings, conclusions, and recommendations from the analytical report assignment in Chapter 19.
5. Evaluate the speaker in the video at **http://g4tv.com/videos/44277/dice-2010-design-outside-the-box-presentation** or at **http://fury.com/2010/02/jesse-shells-mindblowing-talk-on-the-future-of-games-dice-2010**. Do you find his presentation entertaining? If so, why? Do you believe his comments? If so, what elements of his presentation contribute to his credibility and persuasiveness?

COLLABORATIVE PROJECT

Exercises 2, 3, and 4 may be done as collaborative projects.

MyWritingLab

MyWritingLab: Technical Communication offers the best multimedia resources for technical communication in one, easy-to-use place. You can find a variety of interactive model documents and case studies. There are also extensive guidelines, tutorials, and exercises for Document Design, Writing and the Writing Process, and Research, and a large bank of diagnostics and practice for grammar review.

Checklist for Oral Presentations

Presentation Evaluation for (name/topic) _____

Comments

Content

☐ Began with a clear purpose.   _____

☐ Showed command of the material.   _____

☐ Supported assertions with evidence.   _____

☐ Used adequate and appropriate visuals.   _____

☐ Used material suited to this audience's
   needs, knowledge, concerns, and interests.   _____

☐ Acknowledged opposing views.   _____

☐ Gave the right amount of information.   _____

Organization

☐ Presented a clear line of reasoning.   _____

☐ Used transitions effectively.   _____

☐ Avoided needless digressions.   _____

☐ Summarized before concluding.   _____

☐ Was clear about what the listeners
   should think or do.   _____

Style

☐ Seemed confident, relaxed, and likeable.   _____

☐ Seemed in control of the speaking situation.   _____

☐ Showed appropriate enthusiasm.   _____

☐ Pronounced, enunciated, and spoke well.   _____

☐ Used appropriate gestures, tone,
   volume, and delivery rate.   _____

☐ Had good posture and eye contact.   _____

☐ Answered questions concisely and convincingly. _____

Overall professionalism: Superior_____ **Acceptable** _____ **Needs work** _____

Evaluator's signature:_____

Figure 24.5 An Evaluation Checklist for Presentations

Using Electronic Media

LEARNING OBJECTIVES

After reading this chapter, you should be able to

- understand what readers expect from webpages
- write appropriate web-based content
- understand how to write for blogs and wikis
- recognize different types of social networks and their uses

We are now well into the third decade of the internet, which has assumed an increasingly significant role in our personal, professional, and business lives. While printed materials remain important in business and industry, digital media linked to the internet are quickly assuming a dominant role in presenting information.

This chapter first looks at the basics of writing and designing webpages and then moves to more recent digital media formats: blogs, wikis, and social media.

WEBPAGES

The internet has over two billion users, with six new connections every second (BBC News). The highest percentage of users is in North America, where four out of every five persons use the internet. However, the greatest growth has recently been in Africa and Asia, according to **internetworldstats.com (Minimarts)**, and in May 2012 Cisco projected global internet traffic to increase threefold by 2016.

> Mobile traffic will rise steadily and explosively over the next four years, Cisco projects.... The Internet, increasingly, will be portable. (Garber)

There are many reasons for this phenomenal growth. However, the factor that most concerns this text is its ability to present information. Web-based material provides two main advantages over print material:

- Information placed on the internet can be updated quickly, at minimal cost, and without the expense of shipping a new document. For example, if an organization discovers a mistake in some key information presented on its website, this information can be revised immediately, and readers can be informed by a simple email message or notice on the site.
- The web also offers a level of interactivity that printed texts do not. Readers can search the information, forward material to others, ask questions, download files, and click on links that lead to other sources of information. Various types and levels of information, suitable for different audiences, are easy to access from the same page.

Web-based material has advantages compared to print material

Audience and Purpose

The web can be accessed by countless readers, so we need to think carefully about our intended audience and purpose for a webpage. Who will be the primary readers? Are they potential customers seeking product or support information? Are they people with questions about a medical condition? Because web-based information is easy to forward to others, we should consider the many potential secondary audiences. For instance, the site in Figure 25.1 is written and designed for general readers who do not need a specialized science degree to gain general information about safety in the Canadian food chain. However, links from that page can take specialized readers to scientific studies.

As writers, we need to learn the purpose of every webpage. What do people want to do with the information? The writers of the content in Figure 25.1 avoid technical terms, use questions to address readers' main concerns, and provide links to other sites. Readers expect to access exactly what they are looking for, and they will leave a page very quickly if they don't find what they want.

Who creates websites

Website development is a complex process that changes as the related technologies evolve. Most commercial and organizational websites are developed and maintained by a team trained in graphic design, information architecture, marketing, computer programming, technical writing, and other areas. In smaller organizations (a small non-profit, for example), the team may be smaller, but rarely does a single person create and maintain a professional website. Large or small, any website development process benefits greatly from having a technical communicator on the team.

Figure 25.1 Example of a Webpage Designed Primarily for Non-technical Readers
Source: Courtesy of the Government of Canada's Food Inspection Agency.

Technical communicators bring special attention to issues such as audience and purpose, clear writing style and tone, and use of visuals.

How People Read Webpages

In general, when people read a webpage, they want information quickly, and they typically share the following expectations.

- ◆ *Accessibility.* Webpages should be easy to enter, navigate, and exit. Instead of reading word for word, readers tend to skim, looking for key material without having to scroll through pages of text. They look for chances to interact (e.g., links to click on), and they want to download material quickly.
- ◆ *Worthwhile content.* Webpages should contain all the information readers need and want. Content (such as product and price updates) should be accurate and up to date. Readers look for links to other high-quality sites as indicators of credibility. They look for a "search" tool and for accessible contact information.
- ◆ *Sensible arrangement.* Readers want to know where they are and where they are going. They expect a reasonable design and layout, with links easily navigated forward or backward. They look for navigation features to be labelled ("Company Information," "Ordering," "Job Openings," and so on).
- ◆ *Clean, crisp page design.* Readers want a page design that is easy to navigate quickly, with plenty of white space and a balance of text, visuals, and colour.
- ◆ *Good use of visuals and special effects.* Readers expect high-quality visuals (photographs, charts and graphs, company logos) used in a balanced manner (not too many on the page). Special effects, such as fonts that blink, can be annoying.

One person rarely tackles all the areas described above. There are two areas—writing the content and designing the pages—where technical communicators can make contributions as part of a web design team.

Writing for the Web

When writing content for webpages, remember that readers will be busy. They will lack the patience to wade through long passages of prose. The following guidelines are offered for writing web content for busy readers.

GUIDELINES **FOR WRITING WEBPAGES**

1. *Chunk the information.* Break long paragraphs into shorter passages that are easy to access and quick to read. Chunking is also used in paper documents, but it is especially important for webpages. Chunking can also be used to break information into sections that address different audiences.

2. *Write in a readable style.* Write clear, concise, and fluent sentences. Use a friendly but professional tone. Avoid abbreviations and technical terms that some audience members might not understand.

3. *Keep sentences short.* Long sentences not only bog down the reader, but also may display poorly on a webpage.

(continued)

4. *Keep paragraphs short.* Long paragraphs can make a webpage look prose-heavy. Make your online text at least 50 percent shorter than what you would write in a print document.

5. *Catch reader attention in the first two paragraphs.* Jakob Nielsen suggests that "the first two paragraphs must state the most important information."

6. *Write in a factual, neutral tone.* Even on overtly political websites (such as a site for a political candidate), readers prefer writing that is fact-based and maintains a neutral tone.

7. *Choose words that are meaningful.* Start headings, subheads, and bulleted items with "information-carrying words" (Nielsen) that have immediate meaning for readers. Instead of using a word like "Introduction," you might say "Strategic thinking that achieves ambitious goals" as in Figure 25.2.

8. *Write with interactive features in mind.* Use hyperlinks to provide more information about a technical term or a concept; think about when to link to other webpages (within and outside your site). When you do create outside links, consider the ethical and legal implications.

9. *Remember that most webpages are globally accessible.* Avoid confusing readers for whom English is not a first language. Avoid violating cultural expectations. For more on writing for global audiences, see page 36–37.

Designing Webpages

An effective webpage that people will want to read and explore has a clean, attractive, uncluttered design. The screen strikes a good balance between text and visuals, offers inviting and complementary colours, and uses white space effectively. Ample margins, consistent use of fonts, and clear headings all contribute to the design. The following guidelines offer design suggestions.

GUIDELINES   FOR DESIGNING WEBPAGES

1. *Use plenty of white space.* Cluttered webpages are frustrating. White space gives the page an open feel and allows the eye to skim the page quickly.

2. *Provide ample margins.* Margins keep your text from blurring at the edges of the computer screen.

3. *Use an unjustified right margin, which makes for easier reading.*

4. *Use hyperlinks to direct readers to other information but don't overuse hyperlinks.* More than 10 in a column is excessive.

5. *Use a consistent font style and size.* Don't mix and match typefaces randomly.

6. *Don't use underlining for emphasis.* Use underlining only for hyperlinks.

(continued)

7. *Use ample headings.* People skimming a webpage look for headings as guides to the content areas they seek.

8. *Use visuals (charts, graphs, photographs) effectively.* Excessive visuals confuse readers. Inadequate visuals cause readers to avoid the page. Therefore, arrange visuals in an F-shaped pattern: a horizontal visual across the top, and small square visuals along the left vertical margin.

9. *Use a balanced colour palette.* Colour can make a webpage attractive and easier to navigate. Colours of text and visuals need to reflect the theme of the webpage. For more on colour, see Chapter 12.

Paul Heron, the owner of a consulting firm, says "people need to feel engaged in less than 10 seconds when they land on your website in order to stay there" (Brent). He recommends enabling Google Analytics to determine the "bounce rate"—the number of people who leave a site within a few seconds of landing there. A bounce rate of over 30 percent indicates that changes need to be made.

His firm's site, which appears in Figure 25.2, illustrates the advice provided in the above guidelines:

Figure 25.2 A Well-Written and -Designed Webpage
Source: Courtesy of Complex2Clear and Conrad Ferguson and Associates.

- Key blocks of information stand out because white space is used effectively. The page features four horizontal "white areas," each of which presents four blocks of information.
- The wide margins allow this homepage to fit newer wide screens or narrower older computer monitors.
- Hyperlinks in each section provide quick access to more detail.
- The same clean font style is used throughout the page, with different sizes for body text and headings, with one exception, the firm's logo: **complex2clear**
- The four horizontally placed headings at the page's midpoint correspond to the four hyperlinked videos immediately below those headings.
- The photos and verbal messages that scroll at the top of the page also correspond to the midpoint headings. The attractive photos draw the viewer to the "MORE" invitation. The photographs are nicely balanced with the text and complement powerful messages such as "Strategic thinking that achieves ambitious goals" and "Bid proposals that aim higher than the short list."
- Primary headings are red; secondary headings are black. These dramatic colours reflect the dynamic image Complex2Clear wishes to project.

Global issues that concern webpages

Most webpages originating in North America are written in English, but not every reader speaks English as a first language. Therefore, write in clear, simple English, in a way that makes translation easy. Avoid cultural references and humour. These references are not only hard to translate but can also be offensive to some cultures. Offer different language options. For a global audience, websites should include links to information in different languages. Use colours and visuals appropriately. If your website is aimed at specific cultural groups, find out whether certain colours or images might offend.

BLOGS

Technorati (**http://technorati.com**) monitors web developments. It tracks and provides links to over 1.3 million *weblogs*, which are a form of online journal. These journals, more commonly known as *blogs*, provide more than information, ideas, and visuals; they allow visitors to engage in online forums. Topics include the full range of human interests, experience, and knowledge. Basic blogs, most of which are hosted for free, provide text, photo placement, and limited graphics. Premium-level blogs allow for advanced formatting, links, and even video.

Most blogs are written by individual bloggers

According to Technorati, about 92 percent of blogs are created by individual "bloggers," attracted by the opportunity to express themselves and to engage in conversation with people who share their interests. Blogs are particularly good at serving discourse communities, ranging from people interested in technical writing to hang-gliding enthusiasts.

Corporate blogs can reach various types of readers and viewers

Also, all sorts of businesses use blogs to provide general information and product updates, to interact with their clients, and to market their products and services. Good examples include Google (**http://googleblog.blogspot.com**) and Microsoft. Companies also host staff blogs that invite posts from "expert" employees and that are aimed at employees.

Some companies host several blogs. For example, "Dell has a number of corporate blogs, focusing on many specific topics, including technology, investor relations, products, education, and health information technology. Dell organizes all of its blogs under one landing page, which is used to curate featured content from across the blogs, while also including a feed of the latest blog posts, a directory of blogs, and a search bar that scours content from all the blogs" (Swallow).

How to maximize a blog's usefulness

According to *Backbone*'s Andrew Rideout, companies can get maximum benefit from their corporate blogs if they hire a dedicated blogger and use the blog to "establish the company as a trusted source of information and opinion about its industry or market" (41). Moreover, says Rideout, "it's important to emphasize the soft sell" because "overtly self-promotional messaging reeks of desperation and can undermine a company's online presence" (41). He also recommends combining the blog with an RSS syndication service or "blog aggregation system to collect relevant posts" (42).

Blogs can also benefit an individual's career. "Especially for someone just out of school, a blog is an excellent supplement to a résumé," Adrienne Waldo says. "It serves as a sort of enhanced writing sample because it allows employers a unique look at your personality in addition to seeing that you can, in fact, write. It also shows that you're tech-savvy and motivated—both extremely important qualities to have in today's job market" (qtd. in Balderrama).

Varieties of blogs

Technology has spawned several blog varieties: *vlogs* (composed of videos with complementary text), *linklogs* (focusing on links to other blogs and websites), *sketchlogs* (portfolios of sketches and other artwork), and *photoblogs*. A *moblog* is written on a mobile phone or PDA. Blogs that feature streaming audio are called *podcasts*, an amalgam of "iPod" and "broadcast" that was coined by the *Guardian*'s Ben Hammersley in 2005. Several internet sites offer menus of free podcasts—iTunes.com, PodcastAlley.com, and Podcast.com are among the better sites.

Podcasts can serve a wide variety of purposes

Podcasting was initially meant for people to distribute their own "radio shows," but it is now used in many other ways: lectures and recorded discussions, museum audio tours, motivational talks, safety messages, foreign-language study, and podcasted versions of public radio shows. Subscription services such as iTunes will even send podcasts auto-

> ### ON THE JOB...
>
> **Podcasting and Receiver Control**
>
> *"In the digital age, users expect that they should be able to access information on their time, their way. People want to be able to place material on their iPod or other MP3 device and listen to it in the car, on the bus, or wherever they want...."*
>
> —**Marc Arellano, new media producer**

matically to subscribers' PCs so that subscribers can listen to the podcasts at their leisure. The majority of podcasts are produced by individuals, but companies also podcast. Three of the earliest companies to jump on the podcasting bandwagon were *TV Guide*, Purina Pet Care, and BMC Software.

For some time, Microsoft has encouraged its employees to share their expertise through the company's Academy Mobile, which hosts thousands of podcasts each month. Microsoft supplies the tools and rewards podcasting employees with points

Benefits of Podcasting

"Podcasting has great potential for training applications. Also, it allows employees to tell their 'stories' and to contribute to the shared wealth of information about how their industry or company works and about how things could be done better. It's very satisfying for employees to have their opinions and experiences recorded and distributed to a wider audience. At the same time, employers can talk about the things they like about their company and its products or services. So, within an organization, podcasts and vlogs help build a sense of community and permit a space for sharing best practices, and externally these genuine employee expressions help build the company's brand and make it easier to recruit new employees. People are influenced by this 'credible buzz'...."

—**Marc Arellano, new media producer**

that can be redeemed for gifts. The podcasts can be either audio or a combination of audio and video. Also, they are available for various platforms, including mobile devices (Desprez).

As with email, bloggers and podcasters should be careful about what they post. Recent media reports have cited cases where employees have been fired for indiscreet blog posts (Gray, Jeff). Also, tech-savvy employers are searching through the blog universe for blogs, vlogs, and podcasts involving prospective employees. An applicant's star may rise because of what the employer finds, or the applicant's chances of employment may be erased completely.

USING WIKIS FOR TECHNICAL COMMUNICATION

A wiki (from the Hawaiian phrase *wiki wiki*, meaning "quick") is a type of website used to collect and keep information updated. Most people are familiar with Wikipedia, the online encyclopedia that allows anyone to write and revise its content. Companies and other organizations also use wikis to update related content. As with blogs, organizations maintain both internal and external wikis, depending on the project and need.

Internal Wikis

Wikipedia is just one of many available wikis

Internal wikis provide a one-stop site for employees seeking information about a project or topic. Employees can add to and update the content; the wiki keeps track of the date of entries and revisions. Members of an engineering team working on an airline hydraulic system, for example, might create a wiki to track the latest information about fluid pressure, landing gear, wings and flaps, and so forth. As the project develops, the wiki could be updated by the engineers and other team members. To ensure confidentiality, internal wikis can only be accessed by authorized users. Usually a login and password are required.

External Wikis

External wikis allow people to locate and update content that is specific to their area of expertise. Because external wikis are open to countless people with all sorts of knowledge, these sites can be excellent sources of information. But the lack of gatekeeping by a central editor (as in peer-reviewed journal articles or books) also creates the potential for inaccuracy. A well-managed external wiki, one that balances open access with some level of editorial control, can be a valuable resource.

Take a look at **www.plastics.inwiki.org/Main_Page**, a wiki designed for chemists and technicians who work with plastics. These specialists can access the information and also contribute, revise, and update entries.

Wikis and New Ways of Working

Blogs and wikis allow groups to communicate, share knowledge, and work without commuting to the office. These tools also keep a record of the communication. And unlike email, which can get confusing when an inbox begins to overflow with messages, blogs and wikis keep information orderly, with postings dated in reverse chronological order and changes noted by date and time (Shinkle 49).

GUIDELINES   **FOR WRITING AND USING BLOGS AND WIKIS**

- *Use standard software.* Readers expect blogs and wikis to have a certain look and feel. Adapt tools such as Blogger (**www.blogger.com**) and Wikispaces (**www. wikispaces.com**) for confidential workplace environments.

- *Keep web writing guidelines in mind.* Like webpages, blogs and wikis work best when information is written in chunks, using short sentences and a clear, readable style.

- *Write differently for internal versus external audiences.* Internal audiences do not require as much background. External audiences may require more explanation.

- *For an internal blog, think carefully before posting comments.* An internal blog is considered workplace communication. Keep your ideas factual and tone professional.

- *For an external blog, focus on the customer's priorities and needs.* A friendly, encouraging tone goes a long way in good customer relations.

- *Check blog and wiki entries for credibility.* Check the content to see when it was last updated (on a wiki, this information is usually under the "History" tab).

- *Before editing or adding to a wiki, get your facts and your tone straight.* If you spot some type of error, be diplomatic in your correction—and be sure you aren't creating some other type of error.

USING SOCIAL NETWORKS FOR TECHNICAL COMMUNICATION

Much recent buzz has attended the development of *social-networking* sites that can be accessed through computers or smartphones. Sites such as Facebook, MySpace, Flickr, Twitter, and LinkedIn meet the twin human needs to socialize and to share. Claiming hundreds of millions of users, sites such as Facebook are also used to facilitate marketing and public-relations efforts. Although there are many more sites from which to choose, the following discussion focuses on three that dominate the Canadian scene: Facebook, Twitter, and LinkedIn.

The Value of Social-Networking Media

As Flowork International puts it, "it's not just who you know in business, but also what you know about them and what they know about you" (Flowork International, "Why You Should Conduct"). Flowork's Dr. J. P. Hatala counsels *strategic social networking* that takes advantage of online "friendships" for business purposes. He says that all employees should be trained "to think strategically about all their business relationships" and that public relations and sales are now part of the roles for all employees. He sees a double benefit: the company maintains or improves its market share and employees develop their own circles of contact while increasing their visibility within the company (Flowork International, "Top Five").

Extracting value from social media

In his book *Six Pixels of Separation* and on the blog of the same name, Twist Image's Mitch Joel provides a number of ideas for extracting value from social-networking media (also called *social media*). Start with a strategy, he says, and choose content channels and online social networks that match that strategy. He also suggests reading other persons' blogs and Facebook posts, listening to others' podcasts, and checking out YouTube videos before adding one's voice to the mix.

J.P. Hatala suggests conducting a social network audit that examines all the contact persons in your network. For each person, suggests Hatala, you should record what he or she does, what resources and information that person has, and how each of you might help the other. Then, "attach personal goals to specific contacts" and "measure the frequency of contact you have with each network member" (Flowork International, "Why You Should Conduct a Social Network Audit").

FACEBOOK.   Estimates of Facebook's community of users vary from 500 million to 1 billion, depending on how one measures that online network. (On October 4, 2012 Facebook CEO Mark Zuckerberg updated his status to announce that Facebook had more than 1 billion active users every month.)

The site's global popularity can be traced to its ability to "tap into the genetically wired human need to socialize and share while at the same time giving users a set of tools customized to their needs" (Harvey 28).

It's also become a portal and navigation tool for socializers—people who spend a lot of time on social-networking sites and who feel overwhelmed by the sheer amount of information available online. Such people "trust what their friends have to say and [so] social media act as information filtration tools.... Socializers gravitate towards and believe what is shared with friends and family" (Gibs).

How to maximize Facebook's business potential

It's not easy to take advantage of all the possibilities offered by a site such as Facebook, so Ean Jackson and others in the International Internet Marketing Association have compiled a list of suggested techniques. That list of 61 ways to leverage Facebook for business appears at **www.backbonemag.com/facebook.asp**.

Still, the process can be time-consuming and bewildering, so some small business owners are hiring social-networking consultants to get the ball rolling (Needleman, "Facebook Status").

TWITTER.   According to the Pew Research Centre's survey of U.S. Twitter usage, (**http://pewinternet.org/Reports/2012/Twitter-Use-2012.aspx**), "As of February 2012,

> **ON THE JOB...**
>
> *"Social networking sites are good for business. They allow companies to easily establish an online community and build stronger relationships with customers, partners, and industry experts. When I worked for TalkSwitch (**www.talkswitch.com**), we created a group on Facebook in early 2008. We launched the group with a promotion that saw several hundred people join. Now that I represent Nike Golf, I find that Twitter helps people to 'follow' our company and our products. It's all about building an online community where people can find out more about your company any time, from anywhere—especially on mobile phones. Social networking sites are important marketing tools that build and maintain relationships and that get information to interested parties."*
>
> —**Michael Doyon, Nike Golf territory manager, Ottawa**

some 15% of online adults use Twitter, and 8% do so on a typical day." However, "10 percent of the service's users account for 90 percent of tweets" (Swansbury and Singer-Vine), and the vast majority of Twitter users have very few followers.

Most Twitterers who do have a following work hard at providing a stream of information and links to other data sources. The primary advantage of Twitter microblogging—tweets are restricted to 140 characters—is that frequent updates can easily be posted and followed. All this tweeting takes time, of course. It also requires someone who has an outgoing and engaging personality. Thus, some companies are hiring professional Twitterers to post a series of tweets about the company, the company's products, and the writer's reactions to those products. Tweets can be used to extend and augment customer service and to provide brief technical updates or link to more detailed technical specifications, updates, or descriptions. Also, Twitter posts can help to involve future users in product development (Holson).

Advantages and drawbacks of using Twitter

LINKEDIN. This site was designed for business purposes, such as making and maintaining business contacts, recruiting employees, conducting market research, connecting with researchers in a given field, and providing or soliciting recommendations from colleagues. The site is free, it's relatively simple to use, and it provides its members a Google presence (Yoakum).

LinkedIn is suited for business and professional networking

The first step in taking advantage of LinkedIn's networking advantages is to build a profile. Whitefield Consulting Worldwide columnist James Yoakum recommends a profile that is complete and has substance. The profile's "Headline" introduces your personal brand to readers, so it must be accurate and it must reflect the way you want people to see you. Then your "Summary" should pique reader interest so that users will want to read your entire profile. Finally, the "Work Experience" section should be detailed and descriptive, says Yoakum.

Yoakum recommends making "numerous" connections with other LinkedIn members: "people with more than 25 connections are four times more likely to be approached by others" (Yoakum). He also suggests joining alumni groups, business associations and interest groups. Then, you need to be active in those groups, joining discussion forums, asking and answering questions, and posting your knowledge activity descriptions.

The Future of Social-Networking Media

Clearly, these media are more than just a fad. Their popularity builds as new tools and applications are added weekly. Each medium has its strengths:

- *Facebook*'s social portal model provides a single alternative to email, YouTube, Flickr, instant messaging, and MySpace, which explains its large following. Although most people use it for social purposes, it can be effective as a marketing and professional connection tool (Thornton).
- *Twitter* particularly appeals to the technically adept, who use it to provoke discussion, to promote business, and to follow lines of inquiry. Its ability to provide instantaneous responses allows it to "tap into the collective consciousness of others on the internet, which makes it a powerful search engine" (Thornton).
- *LinkedIn* is designed for technical and business people who wish to connect with people in their industry or profession for a variety of business and professional reasons.

Two other factors will help decide whether the above social media last very long in a rapidly changing universe of electronic communication media. First, these media have to find a self-sustaining model of making money from their free services. Also, will social-media users succumb to "account fatigue"? After all, it takes many hours a week to follow others on Twitter, to post to blogs and Facebook, to contribute to group discussions, and to generally maintain one's "responsibilities" in the connected world. However, Carmi Levy, who left the Info-Tech Research group to become a technology analyst at AR Communications, says that advances in technology and larger online communities will soon make it easy for users to integrate and control the features of online sites (El Akkad).

SOCIAL NETWORKS AND ACCESS TO MEDICAL RESEARCH.   Social networks can help organizations disseminate cutting-edge technical and scientific research—and get feedback from people who may not be experts but who have first-hand experience. In one case, the Mayo Clinic used Twitter to announce a new research study about celiac disease (an autoimmune disease that prevents the digestion of wheat gluten). The Clinic was able to track how the Twitter messages spread and to send a prepublication copy of the article to select readers from the Twitter stream. Mayo gave these readers permission to blog about the study (adapted from Ruiz).

SOCIAL NETWORKS AND NEW WAYS OF LEARNING.   Blogs, wikis, and social networks are changing the face of online learning. According to one source, "[a] growing number of online courses are requiring students to participate in blogs, wikis, or gamelike simulations" (Clark 46). These same tools are also being used for corporate and government training.

SOCIAL NETWORKS LISTENING POSTS.   Media such as Twitter and Facebook are very useful for helping companies learn what their customers are saying about products and services provided by those companies. For example, Dell Computers has established social media listening centres and "has trained thousands of employees in the art of social media listening and engagement" (Mansfield 42). ING Direct Canada, a specialist in online banking, has extensively used Facebook to "listen" to its customers and to actively respond to complaints and inquiries.

Ethical and Legal Considerations

Social media can be an effective way to reach customers, stay connected, and do research. But they also harbour potential for information abuse. One such example is "stealth marketing," in which bloggers who publish supposedly objective product reviews fail to disclose the free merchandise or cash payments they received for their flattering portrayals.

People need to be able to trust the information they receive. Furthermore, in workplace settings, co-workers need to have confidence in the discretion of their colleagues. As with email, any employee who discusses proprietary matters (see Chapter 22, page 457) or who posts defamatory comments can face serious legal consequences—not to mention job termination. Also like email, blogs, wikis, and social networks carry the potential for violations of copyright or privacy (see Chapter 22, page 455). To avoid such problems, corporate blogs often have a moderator who screens any comment before it is posted.

To reinforce ethical, legal, and privacy standards, companies such as IBM have established explicit policies that govern social media. See, for example, "IBM Social Computing Guidelines," at **www.ibm.com/blogs/zz/en/guidelines.html**.

Many companies understand that social media can reach thousands of potential customers, but that social media usage entails risk. Employees assigned to work on social marketing may waste incredible amounts of time surfing around in Facebook. Also, if companies make a mistake (say an inaccurate claim about a product) on a social network, those potential customers can turn the mistake into a public relations nightmare (Stephen Baker 48–50).

CHECKLIST

for Blogs, Wikis, and Social Networks

◆ Is the information you post on a blog, wiki, or social network written in a way that is appropriate for the audience and purpose?

◆ Is the blog or wiki organized, accessible, and current?

◆ If the blog is internal, is the information appropriate for the organization?

◆ If the blog is external, does it reflect positively on the organization?

◆ If the wiki is internal, is the information current, accurate, and confidential?

◆ If the wiki is external, is there any editorial oversight?

◆ Is the social network the most appropriate one for the intended readers (Facebook for social networking,

LinkedIn for job and career building, YouTube for video)?

◆ Does the site respect the privacy of its users?

◆ Is the tone friendly, yet professional?

◆ If video, photographs, or other copyrighted information are included on a site, has permission been granted to use this material?

◆ For personal sites (such as Facebook), does the information you've posted portray you in a way you would want a potential employer to see?

◆ Does the posting satisfy ethical and legal standards?

EXERCISES

Corporate Blog

1. Examine a corporate blog maintained by a company, such as General Motors (at **www.gmblogs.com**), Dell (at **http://direct2dell.com**), or F-Secure Security Labs (**www.f-secure.com/weblog**). Assess the style of writing (formal, informal, technical, lively, serious, etc.). Evaluate the apparent objectivity and degree of "sales hype" in the blog posts—do you trust that the comments in the articles are honestly expressed, or do you suspect that the company's marketing department has manipulated the posts' content?

Digital

2. Wikis are powerful tools for collecting information on a topic and for keeping this information updated. Pick a topic or issue that is relevant to your community or your major. For instance, a local highway construction project

may be confusing to commuters; a wiki could keep the construction schedule, long-term plans, and technical details organized in one location. Use the internet to research the technical aspects of how to set up a wiki. Then do an audience and use analysis to determine what level of technicality the wiki will require.

Global

3. Blogs, wikis, and social networks have global reach—you can use these tools to collaborate with team members from other countries. Locate a wiki related to your major. See if the wiki is available in languages other than English. Write an email to your instructor explaining what you found.

COLLABORATIVE PROJECT

In teams of 3 to 4 students, pick an on-campus organization (a club, an athletic group, a professional society for students). What are some ways a social-networking page, such as Facebook, might help the organization network with current members and reach out to potential new members? Discuss what information you would place on the site and what you would omit.

MyWritingLab

MyWritingLab: Technical Communication offers the best multimedia resources for technical communication in one, easy-to-use place. You can find a variety of interactive model documents and case studies. There are also extensive guidelines, tutorials, and exercises for Document Design, Writing and the Writing Process, and Research, and a large bank of diagnostics and practice for grammar review.

WORKS CITED

Adams, Gerald R., and Jay D. Schvaneveldt. *Understanding Research Methods*. New York: Longman, 1985. Print.

The Aldus Guide to Basic Design. Seattle, WA: Aldus, 1988. Print.

Alexander, James. "What's a QR Code & Do I Need One for My Job Search?" *Business Insider* 18 Jul. 2011. Web. 8 Feb. 2013. <http://www.businessinsider.com/whats-a-qr-code-and-do-i-need-one-for-my-job-search-2011-7>.

Allison, Heidi. "Seven Deadly Myths of Job References." *Allison and Taylor*. n.d. Web. 3 Dec. 2013.

American Psychological Association. *Publication Manual of the American Psychological Association*. 6th ed. Washington, DC: APA, 2010. Print.

Archee, Raymond K. "Online Intercultural Communication." *Intercom* [Newsletter of the Society for Technical Communication] Sept./Oct. 2003: 40–41. Print.

"Are We in the Middle of a Cancer Epidemic?" *University of California at Berkeley Wellness Letter* 10.9 (1994): 4–5. Print.

Arellano, Marc. Message to the authors. 2 June 2004. Email.

———. "Single Sourcing." Presentation at Okanagan University College. 12 Mar. 2004.

Armstrong, William H. "Learning to Listen." *American Educator* Winter 1997–98: 24+. Print.

Asher, Donald. "Why College Grads Still Can't Get a Job." *MSN Encarta* n.d. Web. 17 Aug. 2009. <http://encarta.degreesandtraining.com/articles.jsp?article=featured_why_grads_cant_still_get_job&GT1=27001>.

Baker, Russ. "Surfer's Paradise." *Inc.* Nov. 1997: 57+. Print.

Baker, Stephen. "Beware Social Media Snake Oil." *Bloomberg Businessweek* 14 Dec. 2009: 48–50. Print.

Balderrama, Anthony. "Why Blogging Is Good for Your Career." *CareerBuilder* 3 Aug. 2009. Web. 4 Jan. 2010. 18 Sept. 2013. <http://www.careerbuilder.com/Article/CB-1293-Getting-Hired-Why-Blogging-Is-Good-For-Your-Career>.

Balfour, Gail. "Want a Paperless Office? Forget It." *Backbone Magazine* Jan./Feb. 2007: 20–24. Print.

Barbour, Ian. *Ethics in an Age of Technology*. New York: Harper, 1993. Print.

Barker, Larry J., et al. *Groups in Process*. 3rd ed. Englewood Cliffs, NJ: Prentice-Hall, 1987. Print.

Barnett, Arnold. "How Numbers Can Trick You." *Technology Review* Oct. 1994: 38–45. Print.

Barnum, Carol, and Robert Fisher. "Engineering Technologists as Writers: Results of a Survey." *Technical Communication* 31.2 (1984): 9–11. Print.

Bashein, Barbara, and Lynne Markus. "A Credibility Equation for IT Specialists." *Sloan Management Review* 38.4 (Summer 1997): 35–44. Print.

Baumann, K.E., et al. "Three Mass Media Campaigns to Prevent Adolescent Cigarette Smoking." *Preventive Medicine* 17 (1988): 510–30. Print.

BBC News. "Super Power: Visualising the Internet." 11 Oct. 2012. Web. <http://news.bbc.co.uk/2/hi/technology/8552415.stm>.

Beamer, Linda. "Learning Intercultural Communication Competence." *Journal of Business Communication* 29.3 (1992): 285–303. Print.

Beatty, Richard H. *The Ultimate Job Search*. Indianapolis: JIST Publishing, 2006. Print.

Beaumont, Malcolm. "Focus on a Freelance Project." *Communicator* Autumn 2005. Web. 16 Dec. 2007.

Bedford, Marilyn S., and F. Cole Stearns. "The Technical Writer's Responsibility for Safety." *IEEE Transactions on Professional Communication* 30.3 (1987): 127–32. Print.

Begley, Sharon. "Is Science Censored?" *Newsweek* 14 Sept. 1992. Print.

Benson, Phillipa J. "Visual Design Considerations in Technical Publications." *Technical Communication* 32.4 (1985): 35–39. Print.

Bertolucci, Jeff. "What's So Great about Google Voice?" *itBusiness.ca*. 30 June 2009. Web. 7 June 2009. <http://www.itbusiness.ca/it/client/en/home/DetailNewsPrint.asp?id=53712>.

Bjerklie, David. "E-mail: The Boss Is Watching." *Technology Review* 14 Apr. 1993: 14–15. Print.

Blackwell, Gerry. "From Mixed Messages to One Single E-mail." *Globe and Mail* 31 Jan. 2007, natl. ed.: B8. Print.

Blackwell, Tom. "'You Are Not God,' Walkerton Health Watchdog Told." *Ottawa Citizen* 11 Jan. 2001, final ed.: A3. Print.

Blatchford, Christie. "Koebel Knew It Killed, but Drank It Anyway." *National Post* 21 Dec. 2000, natl. ed.: A1. Print.

Bloom, Richard. "'Extracurriculars' Vital on Curriculum Vitae." *Globe and Mail* 19 Jan. 2007: C1. Print.

Bloomberg.com News Service. "Facebook Forges Link with CareerBuilder." *Globe and Mail* 4 Apr. 2008: C2. Print.

Blum, Deborah. "Investigative Science Journalism." *A Field Guide for Science Writers*. Ed. Deborah Blum and Mary Knudson. New York: Oxford, 1997. 86–93. Print.

Boiarsky, Carolyn. "Using Usability Testing to Teach Reader Response." *Technical Communication* 39.1 (1992): 100–02. Print.

Borzykowski, Bryan. "Want Better Intraoffice Communication? Get Rid of Email." *Globe and Mail* 8 Mar. 2013: B9. Print.

Bradbury, Danny. "Video Conferencing Finally Takes Wing." *Backbone Magazine* 17 Nov. 2008. Print.

Branscum, Deborah. "bigbrother@the.office.com." *Newsweek* 27 Apr. 1998: 78. Print.

Brierley, Sean. "Beyond the Buzzword: Single Sourcing." *Intercom* [Newsletter of the Society for Technical Communication] Jan. 2002: 15–17. Print.

Brody, Herb. "Great Expectations: Why Technology Predictions Sometimes Go Awry." *Technology Review* July 1991: 38–44. Print.

Brownell, Judi, and Michael Fitzgerald. "Teaching Ethics in Business Communication: The Effective/Ethical Balancing Scale." *Bulletin of the Association for Business Communication* 55.3 (1992): 15–18. Print.

Bryan, John. "Down the Slippery Slope: Ethics and the Technical Writer as Marketer." *Technical Communication Quarterly* 1.1 (1992): 73–88. Print.

Buckler, Grant. "The Virtual Reality of Getting Good Help." *Globe and Mail* 18 Oct. 2006: E5. Print.

Burghardt, M. David. *Introduction to the Engineering Profession*. New York: Harper, 1991. Print.

Busiel, Christopher, and Tom Maeglin. *Researching Online*. New York: Addison, 1998. Print.

Byrd, Patricia, and Joy M. Reid. *Grammar in the Composition Classroom*. Boston: Heinle, 1998. Print.

CAIB. *The CAIB Report* Vol. 1. 20 Aug. 2003. *Columbia Accident Investigation Board*. 16 May 2004. Web. 1 June 2010. <http://caib.nasa.gov>.

Callahan, Sean. "Eye Tech." *Forbes ASAP* 7 June 1993. Print.

Canada. Copyright Act. R.S., 1985, c. C-42. Web. 1 June 2010. <http://www.canlii.org/en/ca/laws/stat/rsc-1985-c-c-42/latest/rsc-1985-c-c-42.html>.

"Canada's Wireless Industry: A Global Success Story Continues." *Canadian Wireless Tele-communications Association*. 17 Aug. 2007. Web. 1 June 2010. <http://www.cwta.ca/CWTASite/english/index.html>.

Canadian Press. "Doctor Faces Rough Ride at Walkerton Inquiry." *Ottawa Citizen* 8 Jan. 2001: A5. Print.

———. "Notifying Government about Tainted Walkerton Water Standard Practice for Lab." 20 Oct. 2000. Online database: Canadian MAS FullTEXT Elite. 22 June 2001.

———. "Ontario Ministry of Environment Wins Inaugural Code of Silence Award." 26 May 2001. Online database: Canadian MAS FullTEXT Elite. 22 June 2001.

Canalys. "North American Smart Phone Shipments to Exceed 65 Million Units in 2010." 17 Mar. 2010. Web. 26 May 2010. <http://www.canalys.com/pr/2010/r2010033.html>.

Candib, Saul. "Saving Time, Money, and Expertise with a Self-Help Support System." *Microsoft TechNet* 5.5 (May 1997). Print.

Caswell-Coward, Nancy. "Cross-Cultural Communication: Is It Greek to You?" *Technical Communication* 39.2 (1992): 264–66. Print.

Chauncey, Caroline. "The Art of Typography in the Information Age." *Technology Review* Feb./Mar. 1986: 26+. Print.

Christians, Clifford G., et al. *Media Ethics: Cases and Moral Reasoning*. 2nd ed. White Plains, NY: Longman, 1978. Print.

Chung, Stan. Message to the authors. 16 June 2004. Email.

Clark, Gregory. "Ethics in Technical Communication: A Rhetorical Perspective." *IEEE Transactions on Professional Communication* 30.3 (1987): 190–95. Print.

Clark, Kim. "New Answers for E-learning." *US News and World Report* 21 Jan. 2008: 46. Print.

Clark, Thomas. "Teaching Students to Enhance the Ecology of Small Group Meetings." *Business Communication Quarterly* 61.4 (Dec. 1998): 40–52. Print.

Clement, David E. "Human Factors, Instructions, and Warnings, and Product Liability." *IEEE Transactions on Professional Communication* 30.3 (1987): 149–56. Print.

Cochran, Jeffrey K., et al. "Guidelines for Evaluating Graphical Designs." *Technical Communication* 36.1 (1989): 25–32. Print.

Coe, Marlana. *Human Factors for Technical Communicators*. New York: Wiley, 1996. Print.

Cole-Gomolski, Barb. "Users Loathe to Share Their Know-How." *Computerworld* 17 Nov. 1997: 6. Print.

Coletta, W. John. "The Ideologically Biased Use of Language in Scientific and Technical Writing." *Technical Communication Quarterly* 1.1 (1992): 59–70. Print.

Communication Concepts, Inc. "Electronic Media Poses New Copyright Issues." *Writing Concepts*. Reprinted in *Intercom* [Newsletter of the Society for Technical Communication] Nov. 1995: 13+. Print.

Constine, Josh. "Quora Launches Blogging Platform with Mobile Text Editor to Give Every Author a Built-In Audience." *Tech Crunch* 23 Jan. 2013. Web. 28 Jan. 2013. <http://techcrunch.com/2013/01/23/quora-launches-blogging-platform-with-mobile-text-editor-to-give-every-author-a-built-in-audience>.

Consumer Product Safety Commission. *Fact Sheet No. 65*. Washington, DC: GPO, 1989. Print.

Cooper, Lyn. "Listening Competency in the Workplace: A Model for Training." *Business Communication Quarterly* 60.4 (Dec. 1997): 75–84. Print.

Corbett, Robert. Personal interview. 28 Sept. 2009.

Cotton, Robert, ed. *The New Guide to Graphic Design*. Secaucus, NJ: Chartwell, 1990. Print.

Coulthard, Glen. Telephone interview. 3 June 2004.

Council of Science Editors. *Scientific Style and Format: The CSE Manual for Authors, Editors, and Publishers*. 7th ed. New York: Cambridge UP, 2006. Print.

Cross, Mary. "Aristotle and Business Writing: Why We Need to Teach Persuasion." *Bulletin of the Association for Business Communication* 54.1 (1991): 3–6. Print.

Crossen, Cynthia. *Tainted Truth: The Manipulation of Fact in America*. New York: Simon, 1994. Print.

Dana-Farber Cancer Institute. *Facts and Figures about Cancer*. Boston: Dana-Farber, 1995. Print.

Daniel, Diann. "7 Technologies That Changed the World." *PCWorld* 19 Feb. 2009. Web. 1 June 2010. <http://www.pcworld.com/article/159892/7_technologies_that_changed_the_world.html>.

Daugherty, Shannon. "The Usability Evaluation: A Discount Approach to Usability Testing." *Intercom* [Newsletter of the Society for Technical Communication] Dec. 1997: 16–20. Print.

Davidovic, Milan. "Getting Started in Single Sourcing." Toronto Chapter STC *Communication Times* Mar. 2003. Web. 28 Feb. 2004.

Debs, Mary Beth. "Collaborative Writing in Industry." *Technical Writing: Theory and Practice*. Ed. Bertie E. Fearing and W. Keats Sparrow. New York: Modern Language Association, 1989: 33–42. Print.

Devlin, Keith. *Infosense: Turning Information into Knowledge*. New York: W.H. Freeman, 1999. Print.

Dobrota, Alex. "When Previous Lives Can Come Back to Haunt." *Globe and Mail* 11 Oct. 2006, natl. ed.: C1, C7. Print.

Dombrowski, Paul M. "*Challenger* and the Social Contingency of Meaning: Two Lessons for the Technical Communication Classroom." *Technical Communication Quarterly* 1.3 (1992): 73–86. Print.

Dowd, Charles. "Conducting an Effective Journalistic Interview." *Intercom* [Newsletter of the Society for Technical Communication] May 1996: 12–14. Print.

Dragga, Sam, and Gwendolyn Gong. *Editing: The Design of Rhetoric*. Amityville, NY: Baywood, 1989. Print.

Dube, Rebecca. "No Emails Please. I'm Trying to Work." *Globe and Mail* 18 Feb. 2008: L1–L2. Print.

Dumont, R.A., and J.M. Lannon. *Business Communications*. 3rd ed. Glenview, IL: Scott, 1990. Print.

"Earthquake Hazard Analysis for Nuclear Power Plants." *Energy and Technology Review* June 1984: 8. Print.

Ebden, Theresa. "Lessons from the Slippery Slope." *Globe and Mail* 26 Mar. 2009, natl. ed.: E1. Print.

El Akkad, Omar. "The Medium Is No Longer the Message." *Globe and Mail* 10 Mar. 2009: A3. Print.

El Akkad, Omar and Suzanne Bowness. "Telework or Teamwork? The Office Evolves." *Globe and Mail* 27 Feb. 2013: B1, B6. Print.

Evenson, Brad. "Bacterial Strain One of World's Most Dangerous." *National Post* 26 May 2000, natl. ed.: A9. Print.

Farkas, David, and Jean B. Farkas. *Principles of Web Design*. Toronto: Longman, 2002. Print.

Felker, Daniel B., et al. *Guidelines for Document Designers*. Washington, DC: American Institutes for Research, 1981. Print.

Fineman, Howard, "The Power of Talk." *Newsweek* 8 Feb. 1993: 24–28. Print.

Finger, Hedley. "The Joy of Single Sourcing." *Soltys* n.d. Web. 28 Feb. 2004.

Finkelstein, Leo, Jr. "The Social Implications of Computer Technology for the Technical Writer." *Technical Communication* 38.4 (1991): 466–73. Print.

Fisher, Anne. "My Team Leader Is a Plagiarist." *Fortune* 27 Oct. 1997: 291–92. Print.

Flood, Sally. "All Play and More Work." *Computing*. 23 Mar. 2006. Web. 17 Feb. 2010.
  <http://www.computing.co.uk/computing/analysis/2152597/play-work>.

Flowork International. "Top 5 Ways to NOT Get a Job in Today's Economy." *Flowork Social
  Capital Development*. 28 July 2009. Web. 14 Aug. 2009. <http://www.prlog.
  org/10294134-top-5-ways-to-not-get-job-in-todays-economy.html>.

———. "Why You Should Conduct a Social Network Audit!" *Flowork Social Capital Development* n.d. Web. 5 Jan. 2010. <http://www.socialnetworkaudit.com/about_whysna.php>.

FOX News. "How Many Teens Play Video Games? All of Them." *FOX News.com*. 17 Sept.
  2008. Web. 16 Feb. 2010. <http://www.foxnews.com/story/0,2933,423402,00.html>.

Friedland, Andrew J., and Carol L. Folt. *Writing Successful Science Proposals*. New Haven,
  CT: Yale UP, 2000. Print.

Fugate, Alice E. "Mastering Search Tools for the Internet." *Intercom* [Newsletter of the Society for Technical Communication] Jan. 1998: 40–41. Print.

Gabor, T. Shane et al. *Beyond the Pipe*. Mar. 2001. Web. 30 May 2004. <https://ozone.
  scholarsportal.info/bitstream/1873/6762/1/10294156.pdf>.

Galt, Virginia. "Drive-by Interruptions Taking a Toll on Production." *Globe and Mail*
  21 Apr. 2007: B9. Print.

———. "Net Ties Job Seekers to Employers." *Globe and Mail* 8 July 2003: B5. Print.

———. "Surfing Smoothes Job Hunt's Ups and Downs." *Globe and Mail* 11 Nov. 2006: B12. Print.

Garber, Judith Ellison. "The Future Growth of the Internet, in One Chart (and One
  Graph)." *The Atlantic*. 30 May 2012. Web. 11 Oct. 2012.

Gartiganis, Arthur. "Lasers." *Occupational Outlook Quarterly* (Winter 1984): 22–26. Print.

Gaudiosi, John. "Video Games Take Bigger Role in Education." *Reuters.com*. 10 Dec. 2009.
  Web. 10 Feb. 2010. <http://www.reuters.com/article/idUSTRE5B92DW20091210>.

Gesteland, Richard R. "Cross-Cultural Compromises." *Sky* May 1993: 20+. Print.

Gibaldi, Joseph, and Walter S. Achtert. *MLA Handbook for Writers of Research Papers*.
  5th ed. New York: Modern Language Association of America, 1999. Print.

Gibs, Jon. "Social Media: The Next Great Gateway for Content Discovery?" The Nielsen
  Wire blog. 5 Oct. 2009. Web. 5 Jan. 2010. <http://blog.nielsen.com/nielsenwire/online_
  mobile/social-media-the-next-great-gateway-for-content-discovery>.

Gilbert, Nick. "1-800-ETHIC." *Financial World* 16 Aug. 1994: 20+. Print.

Gilsdorf, Jeanette W. "Write Me Your Best Case for…." *Bulletin of the Association for Business Communication* 54.1 (1991): 7–12. Print.

Girill, T.R. "Technical Communication and Art." *Technical Communication* 31.2 (1984): 35.
  Print.

———. "Technical Communication and Ethics." *Technical Communication* 34.3 (1987):
  178–79. Print.

———. "Technical Communication and Law." *Technical Communication* 32.3 (1985): 37. Print.

Glidden, H.K. *Reports, Technical Writing, and Specifications*. New York: McGraw, 1964. Print.

Golen, Steven, et al. "How to Teach Ethics in a Basic Business Communications Class." *Journal of Business Communication* 22.1 (1985): 75–84. Print.

Goodall, H. Lloyd, Jr., and Christopher L. Waagen. *The Persuasive Presentation*. New York:
  Harper, 1986. Print.

Gooderham, Mary. "It Was a Light Bulb for Me." *Globe and Mail* 18 Oct. 2006: E1, E3. Print.

———. "Musings That Matter: Blog to a Job." *Globe and Mail* 16 May 2007: C1, C5. Print.

Goodman, Danny. *Living at Light Speed*. New York: Random, 1994. Print.

Gorsline, Craig. Personal interview. 4 Oct. 2012.

Graham, Sharon. "Use Proven Strategies To Create SMART Accomplishment Statements."
  Sharon *Graham, Canada's Career Strategist*. © 2013. Web. 3 Dec. 2013.

Grant, Tavia. "Weekend Workout: Social Networking Sites." *Globe and Mail* 8 Aug. 2009:
  B14. Print.

Gravelle, Christian. "Maximizing the Value Provided by a Big Data Platform." Branham Group Inc. June 2012. Web. 19 Sept. 2012. <http://www.branhamgroup.com/research-reports-1/maximizing-the-value-provided-by-a-big-data-platform>.

———. "Top 10 Trends Shaping the Technical Communication Industry." *Branham Group Inc.* Jan 2012. Web. 1 Oct. 2012. <http://www.adobe.com/content/dam/Adobe/en/products/technicalcommunicationsuite/pdf/top-10-trends-shaping-tomorrow-tci-whitepaper-jan2012.pdf>.

Graves, Kelly. "Managing Workplace Conflict." Conflict Resolution Series. *Project Mechanics.* 19 Oct. 2005. Web. 31 Oct. 2006.

Gray, Jim. "Reach Out and 'Ping' Someone." *Globe and Mail* 9 Feb. 2007: C2. Print.

Gribbons, William M. "Organization by Design: Some Implications for Structuring Information." *Journal of Technical Writing and Communication* 22.1 (1992): 57–74. Print.

Grice, Roger A. "Focus on Usability: Shazam!" *Technical Communication* 42.1 (1995): 131–33. Print.

Gudykunst, William B. *Bridging Differences: Effective Intergroup Communication.* 3rd ed. London: Sage, 1998. Print.

Gunning, Robert. *The Technique of Clear Writing.* Rev. ed. Toronto: McGraw-Hill Ryerson, 1968. Print.

Gurak, Laura, and John M. Lannon. *A Concise Guide to Technical Communication.* New York: Longman, 2001. Print.

Hall, Edward T. *Beyond Culture.* Garden City, NY: Anchor, 1989. Print.

Halpern, Jean W. "An Electronic Odyssey." *Writing in Nonacademic Settings.* Ed. Dixie Goswami and Lee Odell. New York: Guilford, 1985. 157–201. Print.

Hansen, Katharine. "Your E-resume's File Format Aligns with Its Delivery Method." *QuintCareers.com* n.d. Web. 1 June 2010. <http://www.quintcareers.com/e-resume_format.html>.

Harcourt, Jules. "Teaching the Legal Aspects of Business Communication." *Bulletin of the Association for Business Communication* 53.3 (1990): 63–64. Print.

Hart, Geoff. "Accentuate the Negative: Obtaining Effective Reviews through Focused Questions." *Technical Communication* 44.1 (1997): 52–57. Print.

Hartley, James. *Designing Instructional Text.* 2nd ed. London: Kogan Page, 1985. Print.

Harvey, Ian. "Facebook Your Business." *Backbone Magazine* Sept./Oct. 2008: 24–28. Print.

Haskin, David. "Meetings without Walls." *Internet World* Oct. 1997: 53–60. Print.

Hauser, Gerald. *Introduction to Rhetorical Theory.* New York: Harper, 1986. Print.

Hayakawa, S.I. *Language in Thought and Action.* 3rd ed. New York: Harcourt, 1972. Print.

Hays, Robert. "Political Realities in Reader/Situation Analysis." *Technical Communication* 31.1 (1984): 16–20. Print.

Hein, Robert G. "Culture and Communication." *Technical Communication* 38.1 (1991): 125–26. Print.

Hill-Duin, Ann. "Terms and Tools: A Theory and Research-Based Approach to Collaborative Writing." *Bulletin of the Association for Business Communication* 53.2 (1990): 45–50. Print.

Hilligoss, Susan. *Visual Communication: A Writer's Guide.* New York: Longman, 1999. Print.

Holler, Paul F. "The Challenge of Writing for Multimedia." *Intercom* [Newsletter of the Society for Technical Communication] July/Aug. 1995: 25. Print.

Holson, Laura M. "Wanted: Professional Twit." *Globe and Mail* 25 May 2009: L57. Print.

Hopkins-Tanne, Janice. "Writing Science for Magazines." *A Field Guide for Science Writers.* Ed. Deborah Blum and Mary Knudson. New York: Oxford, 1997. 17–29. Print.

Horton, William. "Is Hypertext the Best Way to Document Your Product?" *Technical Communication* 38.1 (1991): 20–30. Print.

———. "Mix Media, Not Metaphors." *Technical Communication* 41.4 (1994): 781–83. Print.

Howard, Tharon. "Property Issue in E-mail Research." *Bulletin of the Association for Business Communication* 56.2 (1993): 40–41. Print.

Huff, Darrell. *How to Lie with Statistics.* New York: Norton, 1954. Print.

Hughes, Michael. "Rigor in Usability Testing." *Technical Communication* 46.4 (1999): 488–95. Print.

Hulbert, Jack E. "Developing Collaborative Insights and Skills." *Bulletin of the Association for Business Communication* 57.2 (1994): 53–56. Print.

———. "Overcoming Intercultural Communication Barriers." *Bulletin of the Association for Business Communication* 57.2 (1994): 41–44. Print.

Hume, Mark. "Fired Ex–Public Servant Wins Whistle Blower Award." *Globe and Mail* 12 Dec. 2007, natl. ed.: S1. Print.

Humphreys, Donald S. "Making Your Hypertext Interface Usable." *Technical Communication* 40.4 (1993): 754–61. Print.

Immen, Wallace. "Job Hunting 101." *Globe and Mail* 18 Feb. 2009: C1–C2. Print.

———. "The Next Great Curse: Self-Inflicted ADD at Work." *Globe and Mail* 7 July 2006: C1, C5. Print.

———. "Prospective Hires Put to the Test." *Globe and Mail* 26 Jan. 2005: C1. Print.

———. "Shifting Values: More Than a Paycheque." *Globe and Mail* 29 Apr. 2005: C1. Print.

———. "Working Remotely Gains Popularity." *Globe and Mail* 20 March 2012: B17. Print. 23 Oct. 2012.

Ingram, Mathew. "News Flash: Wikipedia Has Errors in It." *Globe and Mail.* 9 Apr. 2009. Web. 25 Aug. 2009. <http://www.theglobeandmail.com/blogs/ingram-2_0/news-flash-wikipedia-has-errors-in-it/article776327>.

"International Scientific Review Calls for Communities to Think *Beyond the Pipe* in the Protection of Ontario's Water Resources." Stonewall, MB: Ducks Unlimited Canada, 2 Apr. 2004. Web. 30 May 2004.

Jameson, Daphne A. "Using a Simulation to Teach Intercultural Communication in Business Communication Courses." *Bulletin of the Association for Business Communication* 56.1 (1993): 3–11. Print.

Jang, Brent, and Patrick Brethour. "Whistle Blower Has Borne 'A Lot of Pain and Agony.'" *Globe and Mail* 18 Oct. 2006: A1, A12. Print.

Janis, Irving L. *Victims of Groupthink: A Psychological Study of Foreign Policy Decisions and Fiascos.* Boston: Houghton, 1972. Print.

Johannesen, Richard L. *Ethics in Human Communication.* 2nd ed. Prospect Heights, IL: Waveland, 1983. Print.

Johne, Marjo. "Yearning for the Sounds of Silence in Workplace." *Globe and Mail* 25 Oct. 2006: C5.

Journet, Debra. Unpublished review of *Technical Writing*, 3rd ed., by John M. Lannon.

Karakowsky, Len. "Corruption-Proofing Your Career." *Globe and Mail* 14 Dec. 2007, natl. ed.: C1. Print.

Kavur, Jennifer. "Soft Skills Are Sexy: 10 Soft Skills Techies Need and Five Ways to Get Them." *Computerworld Canada.* 19 June 2009. Web. 6 July 2009. <http://www.itworldcanada.com/Pages/Docbase/ViewArticle.aspx?id=idgml-44ae9ce6-7a58-458a&sub=485038>.

Kawasaki, Guy. "Get Your Facts Here." *Forbes* 23 Mar. 1998: 156. Print.

———. "The Rules of E-mail." *Macworld* Oct. 1995: 286. Print.

Kelley-Reardon, Kathleen. *They Don't Get It Do They?: Communication in the Workplace—Closing the Gap between Women and Men.* Boston: Little, Brown, 1995. Print.

Kelman, Herbert C. "Compliance, Identification, and Internalization: Three Processes of Attitude Change." *Journal of Conflict Resolution* 2 (1958): 51–60. Print.

King, Ralph T. "Medical Journals Rarely Disclose Researchers' Ties." *Wall Street Journal* 2 Feb. 1999: B1+. Print.

Kinneavy, James L. *A Theory of Discourse.* Englewood Cliffs, NJ: Prentice-Hall, 1971. Print.

Kirsh, Lawrence. "Take It from the Top." *Macworld* Apr. 1986: 112–15. Print.

Kohl, John R., et al. "The Impact of Language and Culture on Technical Communication in Japan." *Technical Communication* 40.1 (1993): 62–72. Print.

Kotulak, Ronald. "Reporting on Biology of Behavior." *A Field Guide for Science Writers.* Ed. Deborah Blum and Mary Knudson. New York: Oxford, 1997. 142–51. Print.

Koudsi, Suzanne. "Actually, It Is Like Brain Surgery." *Fortune* 20 Mar. 2000: 233–34. Print.

Krauss, Clifford and John Schwartz. "BP Will Plead Guilty and Pay Over $4 Billion." *The New York Times* (online version) 15 Nov. 2012. Web. 8 Jan. 2013. <http://www.nytimes.com/2012/11/16/business/global/16iht-bp16.html?pagewanted=all&r=1&>.

Kremers, Marshall. "Teaching Ethical Thinking in a Technical Writing Course." *IEEE Transactions on Professional Communication* 32.2 (1989): 58–61. Print.

Lambert, Steve. *Presentation Graphics on the Apple Macintosh.* Bellevue, WA: Microsoft Corporation, 1984. Print.

Larson, Charles U. *Persuasion: Perception and Responsibility.* 7th ed. Belmont, CA: Wadsworth, 1995. Print.

Lavin, Michael R. *Business Information: How to Find It, How to Use It.* 2nd ed. Phoenix, AZ: Oryx, 1992. Print.

Learn2 Corporation. "Learn2 Online Tutorials." *Learn2.com* n.d. Web. 1 June 2010.

Leggett, Hadley. "Wikipedia to Color Code Untrustworthy Text." *Wired Science.* 30 Aug. 2009. Web. 31 Aug. 2009. <http//www.wired.com/wiredscience/2009/08/wikitrust>.

Leki, Ilona. "The Technical Editor and the Non-native Speaker of English." *Technical Communication* 37.2 (1990): 148–52. Print.

Levinson, Jay C., and David E. Perry. "Your Brand Is Your Edge to Stand Out from the Crowd." *Globe and Mail* 10 July 2009: B12. Print.

Lewis, Philip L., and N.L. Reinsch. "The Ethics of Business Communication." Proceedings of the American Business Communication Conference. Champaign, IL, 1981. In *Technical Communication and Ethics.* Ed. John R. Brockman and Fern Rook. Washington, DC: Society for Technical Communication, 1989. 29–44. Print.

Lima, Paul. "Calling All Workstations." *Technology Quarterly* 1.3 (2006): 20–25. Print.

——. "Small Firms Far from Leading Edge on IT." *Globe and Mail* 18 Oct. 2006: E2. Print.

Littlejohn, Stephen W. *Theories of Human Communication.* 2nd ed. Belmont, CA: Wadsworth, 1983. Print.

Littlejohn, Stephen W., and David M. Jabusch. *Persuasive Transactions.* Glenview, IL:. Scott, Foresman, 1987. Print.

Lublin, Joann. "Thoughtful Thank-Yous Carry Clout." *Globe and Mail* 22 Feb. 2008: C2. Print.

——. "Tricks of the Job Hunting Trade." *Globe and Mail* 15 Nov. 2008: B18. Print.

"Lying about Lying on Resumes." *Globe and Mail* 25 Oct. 2006: C2. Print.

Lynch, C.G. "How to Use Twitter to Job Hunt." *CIO.* 25 Feb. 2009. Web. 11 Aug. 2009. <http://www.cio.com/article/482324/Twitter_Tips_How_to_Use_Twitter_to_Job_Hunt>.

MacKenzie, Nancy. Unpublished review of *Technical Writing,* 5th ed., by John M. Lannon. Print.

Mackin, John. "Surmounting the Barrier between Japanese and English Technical Documents." *Technical Communication* 36.4 (1989): 346–51. Print.

Maeglin, Thomas. Unpublished review of *Technical Writing,* 7th ed., by John M. Lannon. Print.

Makarenko, Jay. "Judicial System and Legal Issues." *Copyright Law in Canada: An Introduction to the Canadian Copyright Act.* 13 Mar. 2009. Web. 1 June 2010. <http://www.mapleleafweb.com/features/copyright-law-canada-introduction-canadian-copyright-act#history>.

Marketwire. "Canadian Study Reveals Lack of Team Collaboration Despite Its Positive Business Impact." *Marketwire Blog* 2 Nov. 2011. Web. 26 Sept. 2012. <http://www.marketwire.com/press-release/canadian-study-reveals-lack-team-collaboration-despite-its-positive-business-impact-lse-inf-1581172.htm>.

Marques, Michelle. "Single Sourcing with FrameMaker." *TECHWR_L* 13 Mar. 2007. Web. 1 June 2010. <http://www.techwr-l.com/node/66>.

Marron, Kevin. "Close Encounters of the Faceless Kind." *Globe and Mail* 9 Feb. 2005: C1, C3. Print.

———. "Employers Are Checking You Out Online." *Globe and Mail* 25 Jan. 2006: C1. Print.

Martin, James A. "A Road Map to Graphics Web Sites." *Macworld* Oct. 1995: 135–36. Print.

Martin, Jeanette S., and Lillian H. Chaney. "Determination of Content for a Collegiate Course in Intercultural Business Communication by Three Delphi Panels." *Journal of Business Communication* 29.3 (1992): 267–83. Print.

Martucci Lamarre, Helene. *Career Focus: A Personal Job Search Guide.* Upper Saddle River, NJ: Pearson Prentice-Hall, 2006. Print.

Matson, Eric. "(Search) Engines." *Fast Company* Oct./Nov. 1997: 249–52. Print.

———. "The Seven Sins of Deadly Meetings." *Fast Company* Oct./Nov. 1997: 27–31. Print.

Max, Robert R. "Wording It Correctly." *Training and Development Journal* Mar. 1985: 5–6. Print.

McCarthy, James E. "Incinerating Trash: A Hot Issue, Getting Hotter." *Congressional Research Service Review* Apr. 1986: 19–21. Print.

McDonald, Kim A. "Covering Physics." *A Field Guide for Science Writers.* Ed. Deborah Blum and Mary Knudson. New York: Oxford, 1997. 188–95. Print.

McGuire, Gene. "Shared Minds: A Model of Collaboration." *Technical Communication* 39.3 (1992): 467–68. Print.

Meyer, Benjamin D. "The ABCs of New-Look Publications." *Technical Communication* 33.1 (1986): 13–20. Print.

Microsoft Word User's Guide: Word Processing Program for the Macintosh, Version 5.0. Redmond, WA: Microsoft Corporation, 1992. Print.

Mirel, Barbara, Susan Feinberg, and Leif Allmendinger. "Designing Manuals for Active Learning Styles." *Technical Communication* 38.1 (1991): 75–87. Print.

Mittelstaedt, Martin. "Tory Didn't See Warning, Walkerton Probe Told." *Globe and Mail* 28 June 2001. Web. 28 June 2001.

———. "Tory 'Team' Made Cuts, Walkerton Probe Told." *Globe and Mail* 27 June 2001. Web. 28 June 2001.

Modern Language Association. *MLA Handbook for Writers of Research Papers.* 7th ed. New York: Modern Language Association, 2009. Print.

Mokhiber, Russell. "Crime in the Suites." *Greenpeace* May 1989: 14–16. Print.

Monastersky, R. "Courting Reliable Science." *Science News* 153.16 (1998): 249–51. Print.

———. "Do Clouds Provide a Greenhouse Thermostat?" *Science News* 142 (1992): 69. Print.

Morgan, Meg. "Patterns of Composing: Connections between Classroom and Workplace Collaborations." *Technical Communication* 38.4 (1991): 540–42. Print.

Morgenson, Gretchen. "Would Uncle Sam Lie to You?" *Worth* Nov. 1994: 53+. Print.

Morsch, Laura. "Job Search Lies Can Be Risky." *MSN Careers* 24 Sept. 2007. Web. 1 June 2010.

Moses, Barbara. "The Challenge: How to Satisfy the New Worker's Agenda." *Globe and Mail* 10 Nov. 1998: B15. Print.

———. "What Makes Workers Tick: Trends for 2006." *Globe and Mail* 27 Jan. 2006: C1, C2. Print.

Munger, David. Unpublished review of *Technical Writing*, 8th ed., by John M. Lannon. Print.

Munter, Mary. "Meeting Technology: From Low-Tech to High-Tech." *Business Communication Quarterly* 61.2 (1998): 80–87. Print.

Nakache, Patricia. "Is It Time to Start Bragging about Yourself?" *Fortune* 27 Oct. 1997: 287–88. Print.

Nazareno, Analisa. "The Secret to a Perfect Handshake." *Globe and Mail* 25 Sept. 2004: B13. Print.

Needleman, Sarah E. "Facebook Status: Earning Money." *Globe and Mail* 1 Oct. 2009: B15. Print.

———. "Widen Social Circle for Job Hunt." *Globe and Mail* 9 Jan. 2009: C7. Print.

Nickels-Shirk, Henrietta. "'Hyper' Rhetoric: Reflections on Teaching Hypertext." *Technical Writing Teacher* 18.3 (1991): 189–200. Print.

Norco Performance Bikes. *Norco.com* n.d. Web. 20 Jan. 2010.

"Notes." *Technology Review* July 1993: 72. Print.

Nydell, Margaret K. *Understanding Arabs: A Guide for Westerners.* New York: Logan, 1987. Print.

Nystedt, Dan. "Blackberry Tops 6 Million Users, but Analysts Raise Warning Flag." *Computerworld Canada* 1 Oct. 2006. Web. 1 June 2010.

O'Connor, the Honourable Dennis R. *Report of the Walkerton Inquiry: The Events of May 2000 and Related Issues (Part One: A Summary and the Full Report).* 14 Jan. 2002. Web. 1 June 2010. <http://www.walkertoninquiry.com/report1/index.html>.

Ornatowski, Cezar M. "Between Efficiency and Politics: Rhetoric and Ethics in Technical Writing." *Technical Communication Quarterly* 1.1 (1992): 91–103. Print.

Ostrander, Elaine L. "Usability Evaluations: Rationale, Methods, and Guidelines." *Intercom* [Newsletter of the Society for Technical Communication] June 1999: 18–21. Print.

Pagens, Wayne. Personal interview. 25 Oct. 2009.

Pearce, Glenn, Iris Johnson, and Randolph Barker. "Enhancing the Student Listening Skills and Environment." *Business Communication Quarterly* 580.4 (Dec. 1995): 28–33. Print.

Peltier, Jon. "Gantt Charts in Microsoft Excel." *Tech Trax* n.d. Web. 7 Dec. 2009. <http://pubs.logicalexpressions.com/Pub0009/LPMArticle.asp?ID=343>.

Pender, Kathleen. "Dear Computer, I Need a Job." *Worth* Mar. 1995: 120–21. Print.

Perelman, Lewis, J. "How Hypermation Leaps the Learning Curve." *Forbes ASAP* 25 Oct. 1993: 78+. Print.

Perera, Rohitha. "6 Online Collaboration Tools that Take Collaboration to the Next Level." *Tripwire Magazine* January 5, 2011. Web. 18 Sept. 2013. <http://www.tripwiremagazine.com/2011/01/5-online-collaboration-tools-that-take-collaboration-to-the-next-level.html>.

Perloff, Richard M. *The Dynamics of Persuasion.* Hillsdale, NJ: Erlbaum, 1993. Print.

Peters, Tom. *The Brand You 50: Fifty Ways to Transform Yourself from an "Employee" into a Brand That Shouts Distinction, Commitment, and Passion!* New York: Knopf, 1999. Phenix Management International. Print.

Peterson, Ivars. "Web Searches Fall Short." *Science News* 153.18 (1998): 286. Print.

Petroski, Henry. *Invention by Design.* Cambridge, MA: Harvard UP, 1996. Print.

Pilieci, Vito. "Players Wanted for Canada's Thriving Video-Game Industry." *Financial Post* 25 Mar. 2009. Web. 10 Feb. 2010. <http://www.financialpost.com/careers/story.html?id=1426917>.

Pinelli, Thomas E., et al. "A Survey of Typography, Graphic Design, and Physical Media in Technical Reports." *Technical Communication* 33.2 (1986): 75–80. Print.

Plumb, Carolyn, and Jan H. Spyridakis. "Survey Research in Technical Communication: Designing and Administering Questionnaires." *Technical Communication* 39.4 (1992): 625–38. Print.

Porter, James E. "Truth in Technical Advertising: A Case Study." *IEEE Transactions on Professional Communication* 30.3 (1987): 182–89. Print.

Powell, Corey. "Science in Court." *Scientific American* Oct. 1997: 32+. Print.

PRESSFeed. "RSS Feeds—A Tutorial." *PRESSFeed* n.d. Web. 15 Nov. 2009. <http://www.press-feed.com/howitworks/rss_tutorial.php#howarefeedsdifferent>.

Price, Jonathan, and Henry Korman. *How to Communicate Technical Information*. New York: Benjamin/Cummings, 1993. Print.

Pugliano, Fiore. Unpublished review of *Technical Writing*, 5th ed., by John Lannon.

Quimby, G. Edward. "Make Text and Graphics Work Together." *Intercom* [Newsletter of the Society for Technical Communication] Jan. 1996: 34. Print.

Ray, Randy. "Web Expands Recruiting Role." *Globe and Mail* 10 Nov. 2000: E7. Print.

Reagan, Brad. "The Digital Ice Age." *Popular Mechanics* 1 Oct. 2009. Web. 1 June 2010. <http://www.popularmechanics.com/technology/gadgets/news/4201645>.

Redish, Janice C., and David A. Schell. "Writing and Testing Instructions for Usability." *Technical Writing: Theory and Practice*. Ed. Bertie E. Fearing and W. Keats Sparrow. New York: Modern Language Association, 1987: 61–71. Print.

Redish, Janice C., et al. "Making Information Accessible to Readers." *Writing in Nonacademic Settings*. Ed. Lee Odell and Dixie Goswami. New York: Guilford, 1985. Print.

Rensberger, Boyce. "Covering Science for Newspapers." *A Field Guide for Science Writers*. Ed. Deborah Blum and Mary Knudson. New York: Oxford, 1997. 7–16. Print.

Richard, K. Peter, Commissioner. "The Westray Story: A Predictable Path to Disaster." Executive summary. 11 May 2004. Web. 1 June 2010. <http://www.gov.ns.ca/enla/pubs/westray/execsumm.asp>.

Richtel, Matt. "Call, Surf, Text, Play, Watch Tilt!" *Globe and Mail* 26 Aug. 2010: L5. Print.

Rideout, Andrew. "Deploying Social Media Tools." *Backbone Magazine* Jul./Aug. 2008: 40–43. Print.

Riney, Larry A. *Technical Writing for Industry*. Englewood Cliffs, NJ: Prentice-Hall, 1989. Print.

Robert Half International. "Finding Your Transferable Skills." *CareerBuilder.com* 14 May 2008. Web. 2 June 2008. <http://www.careerbuilder.com/Custom/MSN/CareerAdvice/ViewArticle.aspx?articleid=145>.

Rockel, Nick. "To Bolivia and Beyond: How to Collaborate Abroad." *Globe and Mail* 22 Feb. 2012: B7. Print.

Rockley, Ann. "The Impact of Single Sourcing and Technology." *Communication* 48 (2001): 189–93. Print.

Rokeach, Milton. *The Nature of Human Values*. New York: Free Press, 1973. Print.

Rosman, Katherine. "Finding Drug Ties at a Medical Mag." *Brill's Content* Mar. 2000: 100. Print.

Ross, Philip E. "Lies, Damned Lies, and Medical Statistics." *Forbes* 14 Aug. 1995: 130–35. Print.

Ross, Raymond S. *Understanding Persuasion*. 3rd ed. Englewood Cliffs, NJ: Prentice-Hall, 1990. Print.

Rottenberg, Annette T. *Elements of Argument*. 3rd ed. New York: St. Martin's, 1991. Print.

Roundpeg. "New Roundpeg Elearning Platform/Adobe Connect". *Roundpeg Blog* 6 July 2012. Web. 16 Sept. 2012. <http://blog.roundpeg.com/blog/new-roundpeg-elearning-platform-adobe-connect>.

Rowland, D. *Japanese Business Etiquette: A Practical Guide to Success with the Japanese*. New York: Warner, 1985. Print.

Rubens, Philip M. "Reinventing the Wheel? Ethics for Technical Communicators." *Journal of Technical Writing and Communication* 11.4 (1981): 329–39. Print.

Ruggiero, Vincent R. *The Art of Thinking.* 3rd ed. New York: Harper, 1991. Print.

Ruhs, Michael A. "Usability Testing: A Definition Analyzed." *Boston Broadside* [Newsletter of the Society for Technical Communication] May/June 1992: 8+. Print.

Ruiz, Rebecca. "How the Internet Is Changing Health Care." *Forbes.com* (Online version) 30 Jul. 2009.Web. 30 Jan. 2013. <http://www.forbes.com/2009/08/30/health-wellness-internet-lifestyle-health-online-facebook.html>.

Samuelson, Robert. "Merchants of Mediocrity." *Newsweek* 1 Aug. 1994: 44. Print.

Sandiford, Kevin J. "Rise of The Post-PC Age." *A Web For Everyone.* 30 Apr. 2012. Web. 9 Dec. 2013.

Schwartz, Marilyn, et al. *Guidelines for Bias-Free Writing.* Bloomington, IN: Indiana UP, 1995. Print.

Scoffield, Heather, and Virginia Galt. "'Glimmers of Hope' as Some Find Work, Others Work for Themselves." *Globe and Mail* 9 May 2009: B1, B17. Print.

Scott, James C., and Diana J. Green. "British Perspectives on Organizing Bad-News Letters: Organizational Patterns Used by Major U.K. Companies." *Bulletin of the Association for Business Communication* 55.1 (1992): 17–19. Print.

Seglin, Jeffrey L. "Would You Lie to Save Your Company?" *Inc.* July 1998: 53+. Print.

Serjeant, Jill. "Even Tech Geeks Need to Unplug." *Globe and Mail* 22 Apr. 2008: L5. Print.

Shellenbarger, Sue. "Work–Life Balance? It's Working." *Globe and Mail* 19 Oct. 2007: C3. Print.

Shenouda, Judith Ellison. "A Look at Technical Communication through FAQs." In *Writer-On_Line.com* Feb. 2005. Web. 5 Sept. 2012. <http://www.writer-on-line.com/content/view/218/66/~Articles/Technical-Writing/A-Look-at-Technical-Communication-through-FAQs.html>.

Sherblom, John C., Claire F. Sullivan, and Elizabeth C. Sherblom, "The What, the Whom, and the Hows of Survey Research." *Bulletin of the Association for Business Communication* 56.4 (1993): 58–64. Print.

Sherif, Muzapher, et al. *Attitude and Attitude Change: The Social Judgment-Involvement Approach.* Philadelphia: Saunders, 1965. Print.

Shinkle, Kirk. "Running an Office by Wiki and E-mail." *US News and World Report* 10 Mar. 2008: 49.

Silcoff, Sean. "Fun Culture Doesn't Translate to Bottom Line." *Globe and Mail* 27 Feb. 2013: B6. Print.

Skeen, Dan. "Meeting Rooms in Cyberspace." *Globe and Mail* 22 Nov. 2007: B11. Print.

Snyder, Joel. "Finding It on Your Own." *Internet World* June 1995: 89–90. Print.

Society for Technical Communication. *Single-Sourcing Deconstructed* n.d. Web. 15 May 2007. <http://www.stcsig.org/ss/articles102003/SingleSourcingDeconstructed.htm>.

Solomon, Howard. "What's in a Name?" *IT World Canada* 30 Apr. 2007. Web. 7 May 2007.

Sookman, Claire. "Use 3 Ways To Communicate More Effectively With Your Team." *LinkedIn: Claire Sookman.* n.d. Web. 3 Dec. 2013.

Spencer, Sue-Ann. "Use Self-Help to Improve Document Usability." *Technical Communication* 43.1 (1996): 73–77. Print.

Spencer Lee, Katherine. "Can't Find a Job? This Might Be Why." *Computerworld Canada* 27 Oct. 2006. Web. 26 Mar. 2007.

Spruell, Geraldine. "Teaching People Who Already Learned How to Write, to Write." *Training and Development Journal* Oct. 1986: 32–35. Print.

Spyridakis, Jan H., and Michael J. Wenger. "Writing for Human Performance: Relating Reading Research to Document Design." *Technical Communication* 39.2 (1992): 202–15. Print.

Stanton, Mike. "Fiber Optics." *Occupational Outlook Quarterly* Winter 1984: 27–30. Print.

Statistics Canada. *Labour Force Survey.* [Reported by Ontario Ministry of Finance]. 2008. Web. 1 June 2010. <http://www.fin.gov.on.ca/english/budget/fallstatement/2008/08fs-ecotables.html>.

———. *Population by Sex and Age Group.* CANSIM, Table 051-0001. 30 Nov. 2009. Web. 19 Apr. 2010. <http://www40.statcan.ca/l01/cst01/demo10a-eng.htm>.

Staudenmaier, S.J. "Engineering with a Human Face." *Technology Review* July 1991: 66–67. Print.

Stone, Peter H. "Forecast Cloudy: The Limits of Global Warming Models." *Technology Review* Feb./Mar. 1992: 32–40. Print.

Stonecipher, Harry. *Editorial and Persuasive Writing.* New York: Hastings, 1979. Print.

Sullivan, Dr. John. "10 Compelling Numbers That Reveal the Power of Employee Referrals." *ERE.net* 7 May 2012. Web. 7 Feb. 2013. <http://www.ere.net/2012/05/07/10-compelling-numbers-that-reveal-the-power-of-employee-referrals>.

———. "Try Using Employee Referrals for Filling Internal Openings." *ERE.net* 14 Jan. 2013. Web. 7 Feb. 2013. <http://www.ere.net/2013/01/14/try-using-employee-referrals-for-filling-internal-openings>.

Sullivan, Matt. "7 Adobe Tools To Make Your Documents More Social." *Using FrameMaker.* 21 Sept. 2010. Web. 17 Sept. 2012. <http://framemaker.mattrsullivan.com/2010/09/21/social-documentation>.

Swallow, Erica. "15 Excellent Corporate Blogs to Learn From." *Mashable Business* 13 Aug. 2010. Web. 14 Oct. 2012.<http://mashable.com/2010/08/13/great-corporate-blogs>.

Swansburg, John, and Jeremy Singer-Vine. "Orphaned Tweets." *Slate* 8 June 2009. Web. 6 Aug. 2009. <http://www.slate.com/id2219995/?GTI=38001>.

Tahmincioglu, Eve. "How to Dig Out from the Information Avalanche." *CareerBuilder.com.* 20 May 2008. Web. 27 May 2008. <http://www.msnbc.msn.com/id/23636252>.

Taylor, Paul. "Health Care, under the Influence." *Globe and Mail* 26 Apr. 2008, natl. ed.: A10. Print.

Templeton, Brad. "10 Big Myths about Copyright Explained." Listerv posting. 29 Nov. 1994. Web. 6 May 1995.

"Testing Your Documents." *Plain English Network.* 16 Apr. 2001. Web. 4 May 2001.

Text.it. "Text Messaging Traffic Set to Double by 2010—Survey." *Text.it Media Centre.* 12 Dec. 2006. Web. 10 Jan. 2007.

Thornton, Steve. "Twitter vs. Facebook." *Twitip.* 13 Jan. 2009. Web. 8 Sept. 2009. <http://www.twitip.co/twitter-versus-facebook>.

Thrush, Emily A. "Bridging the Gap: Technical Communication in an Intercultural and Multicultural Society." *Technical Communication Quarterly* 2.3 (1993): 271–83. Print.

Timson, Judith. "Remedial Help for the Networking Challenged." *Globe and Mail* 26 Apr. 2006: C3. Print.

Tullar, William, Paula Kaiser, and Pierre L. Balthazard. "Group Work and Electronic Meeting System: From Boardroom to Classroom." *Business Communication Quarterly* 61.4 (1998): 53–65. Print.

Turner, John R. "Misconduct Scandal Shakes German Science." *Professional Ethics Report* Vol. X.3 [American Association for the Advancement of Science] Summer 1997: 2+. Print.

———. "Online Use Raises New Ethical Issues." *Intercom* [Newsletter of the Society for Technical Communication] Sept. 1995: 5+. Print.

Tyner, Ross. Personal interview. 19 Oct. 2009.

Unger, Stephen H. *Controlling Technology: Ethics and the Responsible Engineer.* New York: Holt, 1982. Print.

University of Missouri Extension. "Clear Writing: Ten Principles of Clear Statement." Oct. 1993. Web. 1 June 2010. <http://extension.missouri.edu/publications/DisplayPub.aspx?P=CM201>.

Unwalla, Mike. "Business: Beyond Plain English." 12 Nov. 2004. Web. 10 Dec. 2007.

U.S. Department of Commerce. *Statistical Abstract of the United States.* Washington, DC: GPO, 1994, 1997. Print.

U.S. Department of Labor. *Tomorrow's Jobs.* Washington, DC: GPO, 1995. Print.

U.S. General Services Administration. *Your Rights to Federal Records.* Washington, DC: GPO, 1995. Print.

Van Pelt, William. Unpublished review of *Technical Writing*, 3rd ed., by John M. Lannon. Print.

Varner, Iris I., and Carson H. Varner. "Legal Issues in Business Communications." *Journal of the American Association for Business Communication* 46.3 (1983): 31–40. Print.

Vaughan, David K. "Abstracts and Summaries: Some Clarifying Distinctions." *Technical Writing Teacher* 18.2 (1991): 132–41. Print.

Velotta, Christopher. "How to Design and Implement a Questionnaire." *Technical Communication* 38.3 (1991): 387–92. Print.

Victor, David A. *International Business Communication.* New York: Harper, 1992. Print.

"Virtual Teamwork on the Rise." *Globe and Mail* 12 Sept. 2008: C2. Print.

Walsh, Michael. "New Survey Reveals Extent, Impact of Information Overload on Workers." *LexisNexis.* Oct. 2010. Web. 3 Dec. 2013.

Walter, Charles, and Thomas F. Marsteller. "Liability for the Dissemination of Defective Information." *IEEE Transactions on Professional Communication* 30.3 (1987): 164–67. Print.

Wardin, Carla. "Using Screen Capture for Single Sourcing." *STC Single Sourcing SIG* June 2003. 13 June 2004. Print.

Weinstein, Edith K. Unpublished review of *Technical Writing*, 5th ed., by John M. Lannon. Print.

Weymouth, L.C. "Establishing Quality Standards and Trade Regulations for Technical Writing in World Trade." *Technical Communication* 37.2 (1990): 143–47. Print.

Wheeler, John. Telephone interview. 23 Feb. 2010.

White, Erin. "Employers Using 'Day-in-the-Life' Role-Playing on Job Seekers." *Globe and Mail* 25 Jan. 2006: C8. Print.

White, Jan. *Editing by Design.* 2nd ed. New York: Bowker, 1982. Print.

———. *Great Pages.* El Segundo, CA: Serif, 1990. Print.

———. *Visual Design for the Electronic Age.* New York: Watson-Guptill, 1988. Print.

Wicclair, Mark R., and David K. Farkas. "Ethical Reasoning in Technical Communication: A Practical Framework." *Technical Communication* 31.2 (1984): 15–19. Print.

Wickens, Christopher D. *Engineering Psychology and Human Performance.* 2nd ed. New York: Harper, 1992. Print.

Wight, Eleanor. "How Creativity Turns Facts into Usable Information." *Technical Communication* 32.1 (1985): 9–12. Print.

Williams, Robert I. "Playing with Format, Style, and Reader Assumptions." *Technical Communication* 30.3 (1983): 11–13. Print.

Wojahn, Patricia G. "Computer-Mediated Communication: The Great Equalizer between Men and Women?" *Technical Communication* 41.4 (1994): 747–51. Print.

Wriston, Walter. *The Twilight of Sovereignty.* New York: Scribner, 1992. Print.

Yoakum, James. "Special Report: The Advantages of Using LinkedIn." *mba4success.* 5 Jan. 2010. Web. 5 Jan. 2010. <http://www.mba4success.com>.

Yoos, George. "A Revision of the Concept of Ethical Appeal." *Philosophy and Rhetoric* 12.4 (1979): 41–58. Print.

Young, Patrick. "Writing Articles for Science Journals." *A Field Guide for Science Writers.* Ed. Deborah Blum and Mary Knudson. New York: Oxford, 1997. 110–16. Print.

INDEX

Note: Entries for tables are followed by "*t*" while entries for figures are followed by "*f*."

teamwork confused with groupthink, 93–94

unethical communication, 89–93

whistle-blowing, 102

in writing, 88–105

ethics challenge, 10, 10f, 11–12

Ethics Resource Center (ERC), 93–94

e-tipi, 66

etymology, 243–244

evaluating, in efficient writing, 42

evaluation reports, 437, 438t

evidence

assumptions made from, 138–139

certainty in, levels of, 137–138

evaluating, 136–137

interpretations of, 139

interpreting, 137–139

personal biases and, 139

sufficient, 136–137

verifiable, 137

examples

in definitions, 247–248

in inductive reasoning, 140

in instructions, 281–282

executive summary, 323

expanded definitions, 243

expansion methods, 243–248, 243f

analysis of parts, 245

comparison, 246–247

contrast, 246–247

etymology, 243–244

examples, 247–248

history and background, 244

negation, 244

operating principle, 245

required materials and conditions, 247

visuals, 246

experiments, as primary source, 117f, 124, 126

exploded diagrams, 192, 210, 211f

extemporaneous style, 521

extended definition, 183

external proposal, 313

external wikis, 550

F

Facebook, 6, 113, 458, 478, 552, 553, 554

face-to-face collaboration

in conducting meetings, 67–69

electronically mediated (virtual) collaboration vs., 67

fact sheets, 268

fair dealing

defined, 132

of electronic information, 132–133

of printed information, 131–132

fallible computer model, 145–146

familial/environmental approach, 88

faulty causal reasoning, 142–143

faulty generalizations, 141

faulty statistical reasoning, 143–146

feasibility reports, 349, 357–367, 358–366f, 432, 432t, 433f

federated searches, 112–113

feedback, misunderstandings and, 25

feeling expressers, 55

field trip reports, 429, 430t

figures in document supplements, list of, 402, 402f

first draft, 42, 46, 47–48f

flaming, 455

flow charts, 15f, 191, 205, 206f, 227f

Flowork International, 481

focus groups, 307

follow-up to interviews, 511–512

fonts, 231–232

footers, 228

formal analytical report, 374–393, 375–393f. see also analytical reports

formal outline, 174–176

formats

audience/purpose profile, 34, 35t

block format, 449, 449f

correspondence, 407

design graphics, 303–305f, 306

efficient documents and, 4

email, 451–454

full-block format, 449, 450f

headings system, 304–305, 305f

letters, 449–450, 449–450f

manuals, 303–307

memos, 450–451

page layout, 301–305f, 305–306

portfolios, 497

section identification, 303–304, 303–304f

semi-block format, 449, 449f

semi-formal, 407, 408t

short reports, 407, 408t

simplified format, 449, 450f

symbols, 306

form reports, 445–446

FrameMaker, 64

fraudulent advertising, laws against, 98

free-writing, 39

FreeYellow, 497

front matter, 300, 402

full-block format, 449, 450f

functional résumé, 488, 490f

functional sequence, 258

G

game engine, 9

gaming development, 8–9

Gantt charts, 19–20, 192, 208–209, 208–209f

GanttProject, 20

gatekeepers, 55

gender differences, group conflict and, 57

generalizations, faulty, 141

general-to-specific sequence, 181f, 182–184

Genome, 458

global audience, 12–13

global context challenge, 10, 10f, 12–13

glossary

in document supplements, 403, 404f

in manuals, 300

goals

appealing to, 79–80

contradictory, 90–91

Google, 112, 113, 115

Google Books, 112

Google Voice, 6

government agency postings, 117

government documents, access tools for, 116–117

grammar errors, 43

graphic illustrations, 209–215, 210f. see also diagrams

maps, 192

overview of, 192

photographs, 192, 213f, 214f

graphics in proposal, 345, 345t

graphs, 197–204

bar graphs, 191, 197–201

of "big picture," 190f

line graphs, 191, 201–204

overview of, 191

grid lines, 201

groups. see also virtual teams

active listening in, 55–56, 69

conflict within, 56–58

operating in, 54–59

roles in, 54–55

groupthink, 54, 93–94

H

hard copy sources

vs. electronic sources, 110, 110t

types of, 111

hard evidence, 137

harmonizers, 55

Harvard Case Study model, 444

hasty generalization, 141

headers, 228

headings

for access and orientation, 233–236

analytical reports, 369, 370f

grammatically consistent, 234

informative, 233–234

lay out by rank, 235–236, 235f, 236f

letter, 450

manuals, 304–305, 305f

specific and comprehensive, 234

visually consistent, 234

high-information words, 256

highlighting for emphasis, 232–233

highly technical document, 28–29

hiring criteria, 505

horizontal bar graphs, 198, 199f

hot lists, 116

Correction Symbols

Symbol	Meaning	Symbol	Meaning
ab	abbreviation	. . . /	ellipses
agr p	faulty pronoun/referent agreement	! /	exclamation point
agr sv	faulty subject/verb agreement	- /	hyphen
amb	ambiguity	*ital*	italics
appr	inappropriate diction	( ) /	parentheses
bias	biased tone	. /	period
ca	faulty pronoun case	? /	question mark
cap	capitalization	" / "	quotation marks
chop	choppy sentences	; /	semicolon
cl	clutter word	*qual*	needless qualifier
coh	paragraph coherence	*red*	redundancy
cont	contraction	*rep*	needless repetition
coord	faulty coordination	*ref*	faulty or vague pronoun reference
cs	comma splice	*ro*	run-on sentence
dgl	dangling modifier	*seq*	sequence of development in a paragraph
euph	euphemism		
exact	inexact word	*sexist*	sexist usage
frag	sentence fragment	*shift*	sentence shift
gen	generalization	*St mod*	stacked modifiers
jarg	needless jargon	*str*	paragraph structure
len	paragraph length	*sub*	faulty subordination
lev	level of technicality	*th op*	"th" sentence openers
mng	meaning unclear	*trans*	transition
mod	misplaced modifier	*trite*	triteness
noun ad	noun addiction	*Ts*	topic sentence
om	omitted word	*un*	paragraph unity
over	overstatement	*V*	voice
par	faulty parallelism	*var*	sentence variety
pct	punctuation	*W*	wordiness
ap/	apostrophe	*wo*	word order
[ ] /	brackets	*ww*	wrong word
: /	colon	#	numbers
, /	comma	¶	begin new paragraph
– — /	dashes		

Proofreader's Marks

/	the concluding stroke after each insertion and also used to separate two or more marks*
ℓ	delete; take it out
⌢	close up within line
-1 /# or (close up)	( ) or -1 /# close up between lines
ℰ	delete and close up
∧	insert here ⌐something
#	insert space
	insert one line space + 1 /#
(eq#)	space evenly between words
(stet)	let it stand
tr	transpose
(uc) or (cap)	set in capitals (CAPITALS)
(lc)	set in LOWERCASE (lowercase)
(sm cap)	set in small CAPITALS (SMALL CAPITALS)
ital	set in italic (italic)
(bf)	set in boldface (**boldface**)
(sp)	spell out (abbrev.)
¶ run in	Begin new paragraph. Do not begin new paragraph
break	begin new line
⊙	period
⌃	comma

⊙	colon
⌃;	semicolon
⌵	apostrophe
⌵⌵ / ⌵⌵	quotation marks
(/)	parentheses
[/]	brackets
$\frac{1}{M}$	em $\frac{1}{M}$ standard $\frac{1}{M}$ dash
$\frac{1}{N}$	en dash (2005 $\frac{1}{N}$ 2010)
=	hyphen
∨	superscript (πr^2)
∧	subscript (H_2O)
=	align horizontally
‖ ‖	align vertically
⌐ ⌐	move left
⌐	move right
⌐⌐	centre horizontally
⊢⊣	centre vertically
(wf)	wrong font
∂	inverted letter
×	broken letter
bm	bottom margin
tm	top margin
Hpg	hard page break
Spg	soft page break

*If you want to make the same change more than once in a line, indicate the change and follow it by the relevant number of slashes, such as

∧ //   Dr. Thom a pediatric surgeon completed her rounds at noon.

If more than three slashes are necessary, instead circle the number following the change:

(lc) /⑤   Complete Exercises A, B, C, D, and E.